PUBLICATIONS OF THE NEWTON INSTITUTE

Duality and Supersymmetric Theories

T0296387

Publications of the Newton Institute

Edited by H.K. Moffatt

Director, Isaac Newton Institute for Mathematical Sciences

The Isaac Newton Institute for Mathematical Sciences of the University of Cambridge exists to stimulate research in all branches of the mathematical sciences, including pure mathematics, statistics, applied mathematics, theoretical physics, theoretical computer science, mathematical biology and economics. The four six-month long research programmes it runs each year bring together leading mathematical scientists from all over the world to exchange ideas through seminars, teaching and informal interaction.

DUALITY AND
SUPERSYMMETRIC THEORIES

edited by

David I. Olive

University of Wales Swansea

and

Peter C. West

King's Collge, London

CAMBRIDGE UNIVERSITY PRESS
Cambridge, New York, Melbourne, Madrid, Cape Town, Singapore, São Paulo, Delhi

Cambridge University Press
The Edinburgh Building, Cambridge CB2 8RU, UK

Published in the United States of America by Cambridge University Press, New York

www.cambridge.org
Information on this title: www.cambridge.org/9780521103084

First published 1999
This digitally printed version 2009

A catalogue record for this publication is available from the British Library

ISBN 978-0-521-64158-6 hardback
ISBN 978-0-521-10308-4 paperback

CONTENTS

CONTRIBUTORS

Constantin Bachas, Laboratoire de Physique Theorique, Ecole Normale Superieure, 75231 Paris Cedex, France.

Tohru Eguchi, Department of Physics, Faculty of Science, University of Tokyo, Tokyo, Japan 113.

Mary K. Gaillard, Physics Department, University of California, and Theoretical Physics Group, Lawrence Berkeley National Laboratory, Berkeley, CA 94720, USA.

Jerome P. Gauntlett, Department of Physics, Queen Mary & Westfield College, Mile End Road, London E1 4NS, England.

G.W. Gibbons, Department of Applied Mathematics and Theoretical Physics, University of Cambridge, Silver St, Cambridge CB3 9EW, England.

N.S. Manton, Department of Applied Mathematics and Theoretical Physics, University of Cambridge, Silver St, Cambridge CB3 9EW, England.

David I. Olive, Department of Physics, University of Wales Swansea, Swansea SA2 8PP, Wales.

Ashoke Sen, Mehta Research Institute of Mathematics and Mathematical Physics, Chhatnag Road, Jhoosi, Allahabad 211019, India.

P.C. West, Department of Mathematics, King's College, University of London, Strand, London WC2R 2LS, England.

Bruno Zumino, Physics Department, University of California, and Theoretical Physics Group, Lawrence Berkeley National Laboratory, Berkeley, CA 94720, USA.

ACKNOWLEDGEMENTS

All the articles contained in this volume are the products of lectures delivered during a six-month programme which took place at the Newton Institute with David Olive, Peter West and Pierre van Baal as scientific organisers. All the lectures but for Sen's were part of an Easter School, known as a Summer School in the terminology of the European Commission which provided financial support for the participants (ERBFMMA-CT96-0146). We should like to thank the Newton Institute, N.M. Rothschild and Sons, EPSRC and the European Commission for financial support during the programme. DIO and PCW also acknowledge support from the EC TMR grant FMRX-CT96-0012.

We should like to thank Rachel George and Arthur Greenspoon for help with preparation of some of the manuscripts and Cambridge University Press for helpful and understanding cooperation. Finally we should like to thank all our colleagues, lecturers and participants for making the the whole experience instructive and enjoyable.

Particle Physics and Fundamental Theory: Introduction and Guide to *Duality and Supersymmetric Theories*

David Olive and Peter West

Theoretical particle physics is one of those areas of science that deals with the most fundamental questions concerning the nature of the universe in which we live and the rules that govern its behaviour. As such it is heir to a long tradition of scientific thought.

At the present time, the study is fuelled by a wealth of information supplied by experimental apparatus which is exceptional both in terms of its large scale and its extreme sophistication in a technological sense. This includes large particle accelerators such as LEP at CERN, an underground ring of 27 kilometres circumference, giant subterranean detectors and telescopes both on earth and in space.

Yet, at the same time, it is found that the resulting new laws of physics involve concepts from the most modern frontiers of pure mathematics. Thus the subject spans an enormous compass between the highest of high-tech and the most abstruse of abstractions. On the way it includes cutting-edge, high-performance computation. This delicate balance of the subject between two extremes produces a creative tension that has produced continuous and rapid progress.

The general consensus is that the outcome of the experimental data confirms what is known as the 'standard model' of particle physics. Except for gravity, this describes all the observed interactions, that is, forces between the elementary particles, and reveals a surprisingly simple unified mathematical pattern which can be expressed neatly in terms of Lie groups and their representations. The ideas embodied in the standard model differ significantly from those that held sway between 1930 and about 1965 (the four-Fermi and pseudoscalar meson theories) and relate much more to an earlier theory, Maxwell's theory of electromagnetism. This is the major physical theory of the nineteenth century and it is its success which underpins so much of the technology of the twentieth century: electric power, electronics, radio wave transmission and so on.

The standard model is actually a nonlinear extension of Maxwell's theory to include radioactive and nuclear forces. This is achieved by the 'gauge

1

principle' a statement which combines space-time structure, group theory and quantum mechanics, the key physical discovery of the opening decades of the present century. So far, the creation of the standard model has earned the award of six separate Nobel prizes, two theoretical and four experimental, involving fourteen Laureates.

The standard model also includes the idea that the nuclear constituents, the neutron and proton, are themselves composed of more elementary particles called quarks. Thus, roughly speaking, the standard model is a synthesis of Maxwell's equations and the quark model, intermediated by group theory. It gives the most detailed and precise description of nature as observed in particle accelerators but says nothing about the force of gravity. It is ironic that gravity was the first fundamental force to be understood, by Isaac Newton, three hundred years ago.

There is a deep reason for this embarrassing antipathy between gravity and the standard model. This is that the standard model, like Maxwell's theory, respects the principle of special relativity while Newtonian gravity itself does not. This discrepancy was first realised and remedied by Einstein, with his theory of general relativity, but the price exacted requires the curvature of space-time due to the presence of matter. This effect complicates all the equations of motion and, at first sight, seems to spoil the consistency with quantum theory, which is even more important than gravity on small scales.

Thus fundamental physical theory faces a dilemma: the standard model perfectly describes phenomena observed in particle accelerators and their detectors while gravity becomes important on human and astronomical scales. Yet it is difficult to reconcile the two theories so as to encompass all regimes.

This difficulty is quite longstanding and predates the standard model. The first glimmerings of a resolution appeared thirty years ago with the advent of string theory. Since 1967, the development of this theory and its cousin, superstring theory, has been alternately steady and fitful but always relentless. The importance of the role of supersymmetry whose transformations interchange bosons and fermions has likewise grown steadily. We shall explain more of this presently.

In 1994, surprising new support was found for an idea mooted seventeen years earlier concerning what is called 'electromagnetic duality' as applied to theories of a type similar to the standard model. Subsequent developments have transformed our understanding not only of this type of theory but of superstring theory and how it might fully reconcile the geometry of space-time with quantum theory. As a consequence of this and other breakthroughs in understanding, fundamental physics has seen a prolonged state of excitement at the promise of a truly unified theory which succeeds in explaining all the minutiae of particle physics as well the other observable structure of our universe in a way that is satisfying and elegant in a mathematical sense.

This book presents many of these developments through a series of coordinated articles based on lectures delivered at a pedagogical conference which was part of a longer, half-year programme on the subject, held at the Newton Institute in Cambridge.

A very encouraging feature of the new insights is how many disparate and even apparently contradictory lines of thought have been synthesised so that a once fragmented subject is gaining a gratifying coherence. Additional support arises from the fact that gains in physical understanding have come from the application of the latest and most advanced mathematical tools. In recompense, the physical ideas have gained support from the startling and successful repercussions they have had in pure mathematics in terms of conceptual breakthroughs.

In order to better introduce and interrelate the various topics covered in this book we shall devote the following sections of this introduction to describing some of the diverse lines of thought and their historical development. The reader should be warned in advance that we have deliberately omitted citing attributions except on rare occasions when the development is already recognisably labelled by the name or names of the authors concerned. Our aim is to concentrate on the ebb and flow of ideas that we are trying to describe. Likewise the reader will see that the rather ambitious aim of covering such a large subject in such a small space inevitably means that some developments are telescoped or over-simplified or even omitted altogether. This is particularly true in the following section on the standard model which covers developments over a seventy-year period. For all this we apologise.

Development of the Standard Model

String theory and the quark model shared a common origin but developed in different and independent ways before finally reuniting. Both started with attempts to interpret the spectrum of hadronic states including 'resonances' as revealed by the particle accelerator experiments of the 1960s. In the 1930s, Heisenberg's 'isotopic spin' had successfully paired the neutron and proton into a single 'nucleon' $SU(2)$ doublet. Nucleons exerted forces on each other by exchanging pi-mesons which form an isotopic triplet.

A result of the postwar generation of larger particle accelerators was the discovery of 'strangeness', a new quantum number whose inclusion led in 1961 to the enlargement of the symmetry group from $SU(2)$ to $SU(3)_{\text{Flavour}}$. The latter was originally dubbed the 'eightfold way' in order to emphasise the fact that most of the observed states fell into families described by an eight-dimensional representation (the adjoint). Yet the triplet, which is the most basic $SU(3)$ representation of all, played no ostensible role until the suggestion in 1964 that it be attributed to three fictional particles called quarks (up, down and strange). Baryons (hadrons with half-integral spin) were composed of three quarks and mesons (hadrons with integral spin) of a quark and antiquark pair.

This quark model picture provided a very successful book-keeping device for the particle quantum numbers, including angular momentum, in terms of a non-relativistic 'nuclear physics' picture of hadron structure. So it contained an important element of truth but was nevertheless too 'ad hoc' to carry conviction. The treatment of baryon states required the quarks to obey bizarre statistical properties called parastatistics. Eventually this was interpeted in terms of a new '$SU(3)_{\text{Colour}}$' symmetry whose gauging produced what was called the 'chromodynamic force' (QCD), thought to be responsible for the strong forces. This acted directly on the quarks which could each be painted in three possible colours, and only indirectly on the proton and neutron which were colourless or white. Nevertheless it was this indirect effect that held atomic nuclei together. With the discovery that there was a fourth quark carrying 'charm', the significance of $SU(3)_{\text{Flavour}}$ was relegated to mere accident while that of $SU(3)_{\text{Colour}}$ flourished. This was underlined by the more recent discovery of two more quarks, 'top' and 'bottom', completing what is thought to be the final tally, six in all.

The quarks were defictionalised once they were clearly and unambiguously 'seen' in motion inside protons at the SLAC deep inelastic scattering experiments of the early 1970s. Yet they could not be separated out individually. This was known as the problem of 'quark confinement' and was partially explained by the discovery that non-Abelian gauge theories, such as quantum chromodynamics (QCD), were 'asymptotically free'. This meant that the forces on quarks were weak at small distances, and strong at long distances, just as was seen, and in contradistinction to all examples of forces previously known.

The processes involving radioactive decays or weak forces have been intensively studied since the turn of the century, with theoretical models metamorphosing through the four-Fermi model of the 1930s and the V–A theory of the 1960s, before culminating in a gauge theory model (including a neutral current) that was confirmed experimentally in the early 1980s. The gauge group governing a unified treatment of the weak and electromagnetic forces was simply that formed of unitary 2×2 matrices with complex entries, called $U(2)$. Also essential to this success was the role played by Higgs fields, scalar fields with the unconventional property of failing to vanish in the vacuum. Although still not observed directly, these donated the observed masses to the $W^{\pm}$ and Z^{0}, the heavy gauge particles that intermediated the weak interactions. Theoretical consistency was also a driving force behind the progress to the gauge model. It was the only version that satisfied the criterion of renormalisability and it was the most attractive aesthetically.

It was then natural to combine this electroweak theory with the QCD or strong-force gauge theory. The result is the standard model and it encapsulates a wealth of experimental data in terms of the group $U(1) \times SU(2) \times SU(3)_{\text{Colour}}$, divided by a certain finite subgroup, together with particular choices of representation for the quarks and leptons (electrons etc). Using 5×5 matrices it can be seen that this gauge group can be shoehorned into what is called the grand unified group $SU(5)$, the smallest simple group containing the standard model group. We have mentioned many of the historical twists and turns that have led to this model in order to illustrate the dialogue between experiment and theory and the simplicity of the outcome in mathematical terms.

One intriguing feature of the standard model was the important role of a new guiding principle called 'the gauge principle'. This states that the equations of motion are covariant with respect to group rotations of the matter fields even if these rotations vary from point to point of space-time. The geometric and aesthetic nature of this principle is reminiscent of that used by Einstein when he unified the force of gravity with the geometry of space-time and the parallel suggests that further unification could be possible.

This is what string theory eventually achieved. Nevertheless string theory arose out of quite a different line of thought suggested by another property of the observed mesonic resonance spectrum, known as 'duality'. We shall call it 'resonance duality' so as to avoid confusion with electromagnetic duality which is completely different. The inclusion of this latter property turned out to be a crucial step. In 1967 Veneziano crystallised this idea with a concrete mathematical expression for the scattering matrix for four mesons. This soon led to the 'dual resonance model' describing the scattering of any number of mesons. Nevertheless it was possible to link this to the quark model by virtue of diagrams, called duality diagrams. These suggested in a pictorial way that the mesons could be thought of as elongated objects terminated by a quark at one end and an antiquark at the other end. It was a short step to say that the 'tension' in the string connecting the quark to the antiquark was responsible for their confinement.

This interpretation fitted in well with the concrete formulae but when the string picture was taken sufficiently seriously that its relativistic quantisation was considered carefully, there was a surprise. This split the subsequent development of string theory into two different directions with quite different physical interpretations.

String Theory in its First Incarnation

There was no conventional quantum field theory associated with the dual resonance model. Instead there were concrete formulae for the particle scattering amplitudes and these possessed a self-reproducing property that was required by general principles. At certain values of the momenta of the participating particles the amplitude became singular, behaving as a pole. This ought to correspond to the physical possibility of propagation of an intermediate particle. This meant that the residue of the pole should factorise into two subsidiary scattering amplitudes of the same type as was already proposed in the theory. Indeed this form of self-reproduction was verified explicitly even though it was not part of the input, thereby suggesting that the amplitudes, originally so ad hoc, possessed some deeper space-time significance.

In fact there were additional poles corresponding to heavier particles than those considered initially and the factorisation of their residues, if consistent, would lead to new scattering amplitudes for the heavier particles. A key question became the determination of the spectrum of these new states and this was achieved by introducing an internal set of quantum harmonic oscillators. Particle states corresponded to states in the resultant Fock space. The frequencies of the oscillators were given by all the harmonics of a fundamental frequency just as in a violin string. This connection with a string was later made tighter and more precise. The spins of the particles arose from the Lorentz indices attached to the oscillators and the highest available spins displayed a linear relation between spin and mass squared (called a 'linear Regge trajectory'), in accord with the experimental data that originally motivated the construction.

An unfortunate consequence of this inevitable feature was the occurrence of ghost states, that is negative norm states created by the time components of the oscillators. In fact photons, and more generally non-Abelian gauge particles, suffer likewise, and the remedy is provided by certain Ward identities following from the gauge principle. They ensure that the undesirable negative contributions consistently cancel with other positive contributions coming from longitudinal photons, leaving only the contributions of the transverse photons.

It was discovered that similar identities can hold in string theory providing certain additional conditions are satisfied. The proof of this depended on an understanding of the algebraic structure of the building blocks of the dual resonance model, that is the vertex operators recording the emission of a particle, and the Virasoro generators needed for the propagators and the Ward identities. These were built out of the harmonic oscillators. Both this algebraic structure and this sort of construction have become important in their own right in a branch of mathematics dealing with the representation theory of infinite-dimensional Lie algebras. What is important now are the conditions necessary for the consistent interpretation of the theory:

(1) that the leading Regge trajectory has unit intercept: equivalently the string states include massless particles of unit spin, like photons or gauge particles;

(2) space-time has 25 space dimensions and one time. Thus there are precisely 24 transverse polarisation states available to each photon.

The first condition was exciting because it suggested that the theory was only consistent if it was considered as a unified theory with massless gauge particles. This insight first arose before the standard model was finally formulated.

The second condition hinted at tantalising connections with other occurrences of the number 24, such as in the Dedekind eta-function or the Leech lattice. These hints are now fully vindicated and quite well understood, having led to breakthroughs in pure mathematics.

The idea that the correct physical role of string theory was as a truly unified theory, rather than its original role as a theory of just strong interactions, was supported by another, earlier result. The amplitudes considered so far were to be interpreted as tree amplitudes since they lacked the multiparticle branch points required by the unitarity of the scattering matrix. These were to be obtained by adding loop amplitudes obtained by consistently sewing together the tree amplitudes. The single loop amplitude so obtained was found to possess a new singularity capable of a physical interpretation only if both conditions above were satisfied. The consequence was a new spectrum of states describing 'closed string' states which were in addition to the previous 'open' string states. Included were massless states describing spin-2 particles with many of the attributes of the graviton. Since this is the particle required by any quantisation of Einstein's version of the gravitational field, the promise of a unification was extended to include gravity and hence all known forces. Indeed, as was subsequently shown, string theory at low energies produced the same graviton couplings as Einstein's theory.

The picture of a string propagating in space-time was put onto a definitive footing in 1972 by the consistent quantisation of a geometrical action for the string. This action was simply the area of the worldsheet swept out by the string as it moved in space-time, and therefore possessed a large invariance group expressing its independence of the choice of coordinate system on the worldsheet. The quantisation only worked if the two conditions above were satisfied and then guaranteed both Lorentz invariance and positive-definiteness of the quantum mechanical Hilbert space. The harmonic oscillators were a simple consequence of the quantisation procedure, being associated with the vibrations of the string.

This work crowned the bosonic string theory as a radically new type of relativistic quantum theory that was beautifully clearcut and consistent as a space-time theory. Because it described physical objects which were string-like, that is extended in space, it was intrinsically different from previous quantum field theories which had always described point-like structures. Yet two physical defects remained: the presence of a scalar tachyonic state and the absence of fermions.

The first steps in remedying these defects were already well in hand. Two new variations on the bosonic string theory had added corresponding fermionic oscillators to the bosonic oscillators and yielded extensions of the previous algebraic structures responsible for the ghost elimination. One of the two, due to Ramond, described fermionic string states emitting bosonic states while the other, due to Neveu and Schwarz, described only bosonic states. During the early 1970s it was seen how to weld these two constructions together in a consistent way when the space-time dimension was 10 rather than 26 (still with the massless gauge particles). The algebraic structure would now be recognised to be that of a superconformal field theory on the worldsheet.

Finally, in 1977, Gliozzi, Scherk and Olive realised that it was possible and consistent to impose a chiral projection that eliminated the tachyon. The resultant theory then displayed an apparent supersymmetry in space-time since there was a matching between fermions and bosons of the same mass. This theory, which became known as the 'type I superstring', produced some remarkable conventional field theories as special cases that we shall mention later.

We mentioned that string theory had suffered a schism into two factions. We have explained how one was driven by the need for theoretical consistency away from the strong interactions into the direction of unification.

But there remained the problem of quark confinement that the string could explain so neatly. Maybe the string could arise in a non-Abelian gauge theory as a soliton? This is what Nielsen and Olesen demonstrated, provided a suitable Higgs field was added. This is a scalar field that fails to vanish in the vacuum, and as a consequence can produce mass without spoiling any of the good consequences of gauge symmetry, such as renormalisability.

This string-like soliton was built of magnetic flux lines and so could only end on magnetic charges which would of course be confined, essentially by the Meissner effect. This suggested some sense in which quarks might be magnetic monopoles. The following year, 1974, 't Hooft and Polyakov constructed magnetic monopoles as solitons in a similar Yang–Mills Higgs system. Eventually the idea was entertained that an exchange of the roles of electricity and magnetism might be possible, thereby resolving the dilemma that quarks were unlikely to be magnetically charged. Ever since, this sort of idea of electromagnetic duality has permeated attempts to understand confinement. The most popular method has discretised the equations of QCD by putting them on a lattice in space-time, and treating the resultant equations numerically by means of ever more powerful computers.

Supersymmetry

Supersymmetry has already been mentioned and deserves its own treatment. It is governed by an algebra which is an extension of the familiar Poincaré algebra that expresses space-time symmetry and it is obtained by adjoining anticommuting spinor supercharges. This endows the algebra with a natural Z_2 grading, odd for fermionic generators and even for bosonic, and has the consequence that the overall symmetry rotates fermions into bosons and vice versa. The new algebra can be imagined to be a sort of square root of the old algebra, just as the Dirac operator is the square root of the Klein–Gordon operator. Such a graded algebra is a special case of what is now called a superalgebra and a large class of these has been classified, rather in analogy to the classification of simple Lie algebras.

There were two routes to this discovery of supersymmetry, one via the fermionic string theories just mentioned. The gauge conditions were related to the superconformal algebra on the two-dimensional worldsheet of the string. This contained the super-Poincaré algebra on the worldsheet as a finite-dimensional subalgebra. It was realised that this was really the symmetry algebra of a field theory defined on the worldsheet. Wess and Zumino then generalised this construction to a space-time of four dimensions by finding a field theory, describing two spin-0 particles and one spin-$\frac{1}{2}$ particle, whose invariance was governed by a supersymmetry algebra.

Quite independently, an equally valid but different line of thought was developed. In an attempt to explain the occurrence of the observed internal (flavour) symmetries, physicists searched for a larger group that combined them with the Poincaré group in a nontrivial way. However, it was proved that any theory possessing such a group must have a trivial scattering matrix. Golfand and Likhtman found that fermionic charges circumvented this no-go theorem and thereby also found the supersymmetry algebra in four-dimensional space-time. This line of argument came full circle with a more general no-go theorem. The most general super-Poincaré algebra in four dimensions was found to possess N spinor supercharges, and accordingly called type N extended supersymmetry. The appearance of additional, Lorentz-invariant conserved charges in addition to those of the internal symmetry group was also possible. These became known as central charges and their presence was consistent with the reality and Jacobi-identity properties of the Z_2 graded superalgebra. These structures are further explained and exploited in Chapters 5 and 8 of this volume.

The interest in supersymmetric theories only began in earnest with the work of Wess and Zumino in 1974. The part of the supersymmetry algebra that contains the supercharges resembles a Clifford algebra and the unique irreducible representation of the latter was exploited to construct particle supermultiplets using an extension of Wigner's method originally applied to the Poincaré algebra. This provided a list of all possible four-dimensional supersymmetric theories; the next step was the construction of their supersymmetric actions describing non-trivial interactions between the particle fields. Some supersymmetric theories in four dimensions were found rather quickly, for example, that with both $N = 1$ supersymmetry and gauge symmetry. Supersymmetry paired the spin-1 gauge field to a massless Majorana field with spin-$\frac{1}{2}$, transforming in the adjoint representation of the gauge group. Other early discoveries were two massless theories with $N = 2$ supersymmetry, one with four spin-0 and two spin-$\frac{1}{2}$ fields and the other with one field of spin-1, two of spin-$\frac{1}{2}$ and two of spin-0.

Just as it was found advantageous to maintain manifest Lorentz invariance as much as possible in ordinary relativistic quantum field theory , the same could be expected with supersymmetry. Space-time can be regarded as the coset space which is the quotient of the Poincaré group divided by the Lorentz group. Analogously it was found profitable to introduce a 'superspace' which is the quotient of the super-Poincaré group divided by the Lorentz group. As well as the usual space-time coordinates this possesses additional Grassmann (anticommuting) coordinates. Using 'superfields' depending on all these coordinates, it was possible to find a concise formulation of the theories already mentioned in which their supersymmetry was manifest.

All the supersymmetric theories so far mentioned have been invariant under supersymmetry transformation laws whose parameters are constant in space-time. These are said to possess 'rigid supersymmetry'. One can also have theories for which the parameter is a function of space-time just like gauge and general coordinate transformations. Such theories are said to possess a 'local supersymmetry'. Since two infinitesimal supersymmetry transformations combine to form a local translation in space-time, such theories must possess general coordinate invariance and hence a spin-2 particle or graviton. Thus they are supersymmetric extensions of Einstein's gravitational theory and, consequently, called 'supergravity theories'. These theories were constructed by a Noether method and did not result from a straightforward application of Einstein's ideas to superspace. Their particle content included a spin-2 graviton and N gravitini, that is, spin-$\frac{3}{2}$ fermions, as well as possible lower spin particles in the same supermultiplet when N exceeded 1. The $N = 2$ theory contained a spin-1 particle in addition to the graviton and two gravitini that mediated an antigravity force capable of cancelling gravitational attraction. This cancellation effect was strikingly similar to that for BPS solitons as discussed in Chapters 2, 5 and 6. See also Chapter 11.

The original formulations of the supergravity theories were constructed from a set of fields that carried a representation of supersymmetry only if the classical equations of motion were implemented. Eventually a formulation of supergravity that overcame this shortcoming was found by including the presence of additional 'auxiliary fields' that did not propagate any physical degrees of freedom. This new formulation could be systematically quantized and its most general coupling to a supermultiplet describing matter ('super-matter') was found using a generalisation of Einstein's tensor calculus.

In contrast to the situation in Einstein's theory of gravity, the metric did not play the leading role in the superspace formulations of supergravity. In these, the important geometric objects were the supervielbeins and superconnections whose curvatures and torsions were subject to special constraints. Initially a set of constraints was found which led to the equations of motion of supergravity theory 'on-shell', but later an improved set was found that was 'off-shell' and involved the auxiliary fields mentioned above. The latter constraints could be solved in terms of an unconstrained vector superfield, but it lacked a clear geometrical interpretation until it was realised that such a superfield could be viewed as one of the bosonic coordinates in a complexified superspace.

Despite the fact that string theory provided one of the clues leading to supersymmetry, the two subjects evolved rather independently of each other until the realisation in 1977, explained earlier, that the GSO chirally-projected version of the fermionic string theory possessed supersymmetry in space-time of ten dimensions. In particular, the low-energy limit of the open string sector yielded supersymmetric gauge theory governed by 16 real chiral supercharges. (Part of the magic of ten-dimensional space-time is the possibility of real chiral spinors). This theory possessed just two fields, the Yang–Mills gauge field and a real, chiral Dirac field (in the adjoint representation).

This could be viewed in conventional space-time of four dimensions by omitting dependence on the excess six coordinates and the result exhibited $N = 4$ supersymmetry, the maximum degree possible if no spin exceeded 1. This was the first example of a more general process called dimensional reduction which obtained supersymmetric theories in a lower-dimensional space-time from those in a higher dimension by taking the excess dimensions to describe a particular compact manifold. The amount of supersymmetry surviving in the resulting theory depended on the choice of manifold, with the simplest example, namely the torus, preserving all the supersymmetry.

Soon after, a systematic study of supersymmetry algebras and their representations in space-times of any dimensions revealed that supergravity theories, that is supersymmetric theories in which the graviton carried the highest possible spin, namely 2, could only exist if the dimension of space-time was eleven or less. In eleven dimensions there were then 32 real supercharges. This extreme supergravity theory was constructed by the Noether method and found to be remarkably simple, containing just a 3-form gauge potential in addition to the graviton and gravitino. With one exception, all the other supergravity theories with 32 supercharges could be found from it by dimensional reduction, that is ignoring dependence on the excess coordinates. Other supergravities, with fewer supercharges, could be found by truncation. An intriguing feature of the extended supergravity theories in four dimensions was that any scalar fields in the same supermultiplet as the graviton formed the coordinates of a non-compact Riemannian symmetric space of the form G/H where the group G formed symmetries of the equations of motion partly made using electromagnetic duality transformations on the spin-1 fields present. The maximal supergravity theory in four dimensions had $N = 8$ and coset symmetry $G/H = E_7/SU(8)$. (Chapter 3 discusses the general theory of duality transformations that is relevant for this).

In ten-dimensional space-time there are precisely three supersymmetry algebras with representations containing spins no larger than 2, called type I, IIA and IIB. There are 16 and 32 real supercharges in types I and II respectively. In types I and IIB all supercharges have the same chirality while in type IIA there are 16 of each chirality. The type I theory could be obtained by a dimensional reduction and subsequent truncation of the eleven-dimensional theory; however its coupling to the $N = 1$ Yang-Mills theory had to be found by applying the Noether method. The IIA supergravity theory was found by dimensional reduction of the eleven-dimensional theory on a circle whose circumference was related to the expectation value of the scalar field in the IIA theory. The IIB supergravity theory could not be deduced from the eleven-dimensional supergravity and was found using Noether and superspace techniques. It possessed two scalar fields belonging to the symmetric space $SL(2,\mathbf{R})/U(1)$. These supergravity theories are discussed in Chapter 8.

It was to be expected that the quantisation of supersymmetric field theories in four dimensions would require fewer renormalisation constants than conventional theories but the number turned out to be even smaller than naive expectation. Now the power of the superspace method became apparent, leading to super-Feynman rules and manifestly supersymmetric diagram contributions. One far-reaching result was the absence of quantum corrections to an effective potential in perturbation theory. This meant, for example, that vacuum degeneracies that often existed classically could not be removed perturbatively. Hence, in modern language, the quantum theory possessed additional 'moduli' specified by the vacuum expectation values of the degenerate scalar fields. Another implication was that supersymmetry could not be broken perturbatively and that quantum corrections preserved the magnitude of any small masses despite the presence of large mass scales. This provided a solution to what was called the 'hierarchy problem', which arose because the scale at which grand unification occurred was so much larger than the observed mass scales.

Given these results for $N = 1$ supersymmetry, even more spectacular results could be anticipated for theories with extended supersymmetry, with the highest expectation attached to the most supersymmetric rigid theory of all, the $N = 4$ Yang–Mills. Explicit calculation confirmed that the beta-function for the running of the gauge coupling indeed vanished up to 3-loop order, whilst its vanishing to all orders followed from a number of arguments. Hence the theory was ultraviolet-finite and, in fact, the first example of a conformally invariant quantum field theory in a space-time of four dimensions. It was also found that there existed a large class of conformally invariant $N = 2$ supersymmetric theories. Indeed, it was found that in the $N = 2$ gauge theory coupled to $N = 2$ matter, the beta-function received only a 1-loop contribution. For a large class of theories this vanished and so these theories were found to be again conformally invariant. In any case this 1-loop contribution also controlled the axial anomaly and so was related to the Atiyah–Singer index theorem discovered in pure mathematics. The significance of these results was more fully appreciated in the context of electromagnetic duality, as explained in Chapter 5.

On the other hand supergravity turned out to be a disappointment in this respect as the various versions all possessed non-vanishing counterterms and so were unlikely to be finite. These results forced attention back to the superstring theories as the best hope of finiteness in unified theories including gravity. The role of a supergravity theory is as a low-energy, effective action theory, rather as chiral pion models provide low-energy effective actions for QCD.

Owing to all these results, as well as the ones described in the main contents, supersymmetry has become ever more important in the theoretical model building of unified theories. But there is a dilemma as no superpartners are observed with the same mass. Rigid supersymmetry can only be broken at the classical level or via non-perturbative effects. The former mechanism still leads to mass patterns incompatible with the data unless the rigid $N = 1$ supersymmetry is made local by the addition of $N = 1$ supergravity which provides terms breaking supersymmetry with mass dimension less than four. With these 'soft' terms realistic supersymmetric versions of the standard model have been found.

The Resurgence of String Theory

The initial development of string theory, as explained above, covered the approximate ten-year period 1967–1977, after which it became relatively dormant for seven years. Part of the reason was that the wider world of theoretical physics was not ready to embrace a theory that was both so ambitious in terms of unification and so radical in the sense that it supplanted conventional quantum field theory with a different conceptual structure that was accompanied by a new brand of mathematics.

Attention shifted to other ideas, including soliton physics and, of course, supersymmetry, whose early heyday likewise covered a rough ten-year period, 1974-1984. By the end of this time most of the basic formalism, all of the models and many of their properties had been developed. Further, it had been understood that the promise of a finite quantum theory of gravity could not be fulfilled without a return to more radical ideas of string theory.

Nevertheless, the 'dormant' period of string theory saw important consolidation on a number of fronts, mainly through the establishment of connections with other areas of mathematics and physics.

The space-time supersymmetry of the type I superstring was made explicit by the discovery by Green and Schwarz who found a new expression for its action which made this manifest. It also became clear how this was connected to the triality symmetry of the group $SO(8)$ which was the Wigner little group of massless states in the space-time of ten dimensions inhabited by the superstrings. This, in turn, indicates how superstring theory is related to known exceptional structures in mathematics. While the type I closed superstring had as its low-energy limit the type I supergravity discussed above, there were found to be two superstring theories, called IIA and IIB whose low-energy limits were those of IIA and IIB supergravities. It was possible to construct formulations of all these string theories that were second quantised in the light-cone gauge.

Another development was the transfer of the BRST formalism (with its resemblance to Lie group cohomology) from gauge theory, which had been successfully quantised by this means, to string theory. As a result, the Feynman path integral formulation and the method of constructing amplitudes by sewing three vertices, found in the early days of string theory, could, at last, be properly applied to the quantum string. Moreover, multiloop amplitudes could be constructed with internal propagation confined to only those physical string states that were consistent with relativity and positivity. Previously this had only been achieved for the single-loop amplitude. Explicit expressions were found for the multiloop scattering of bosonic strings and it was then established that multiloop amplitudes in superstring theory were finite in space-time of ten dimensions. This approach was extended to include non-trivial background fields such as the graviton and dilaton. These fields defined a σ-model which turned out to be finite by virtue of the string equations of motion.

Another, and perhaps surprising, use of the BRST formalism was the construction of second quantized, gauge-covariant string field theories whose action described the propagation of the infinite number of particles of the string theory in a particularly simple form, at least at the level of the free theory.

The novelty of the mathematics of early string theory has already been mentioned. But since those days a new branch of pure mathematics had slowly been developing of its own volition. This concerned certain Lie algebras of infinite dimension, their classification and their representations. A crucial discovery was the observation that the building blocks of certain important irreducible representations were precisely the vertex operators responsible for the emission of tachyons in bosonic string theory with momenta given by the roots of the algebra. The algebras concerned were known as simply-laced affine Kac–Moody algebras but it turned out that the associated Virasoro algebra also played its part. So the algebraic structure of bosonic string theory, worked out by theoretical physicists unaided, had finally acquired the accolade of mathematical respectability. Actually, string theory possessed an even richer algebraic structure than this and these deeper structures found their place in later mathematical developments mainly due to Borcherds.

The interplay just mentioned led to a vastly improved understanding of the irreducible representations of the Virasoro algebra. This insight could be applied to the theory of second-order phase transitions of two-dimensional materials. This is because, just at the critical temperature, the behaviour of the material becomes subject to control by conformal symmetry, and hence the Virasoro algebra. As a consequence some of the group-theoretic numbers had a physical interpretation, for example, as critical exponents, and could be checked experimentally. This area of research activity became known as conformal field theory. An important class of these that were well-understood were called minimal models.

A link back to the construction of new string theories was facilitated by the realisation that the internal consistency of the string theory required only that the BRST charge of the conformal field theory should square to zero (like an exterior derivative) and that the partition function obey a property called modular invariance. The former property was equivalent to the requirement that the fields of the string, including its BRST ghosts, carried a representation of the conformal algebra with vanishing central term. Thus the construction of many string theories could be achieved simply by adding conformal multiplets until this property was satisfied.

We have already mentioned the procedure of dimensional reduction as applied to supersymmetric theories. When extended to string theories several unexpected new features arise. It had already been realised that a string could wrap around a torus in the excess dimensions, thereby giving rise to extra physical states in the space-time of lower dimension. Later it was realised that when the torus had extra discrete symmetries, then the number of massless non-Abelian gauge bosons in the resulting theory exceeded those expected simply from the dimension of the isometry group of the torus. This phenomenon provided a solitonic interpretation of the vertex operator construction already mentioned. Another surprise was a discrete symmetry interchanging Kaluza–Klein states with winding number states. For example when the torus was simply a circle the symmetry transformation inverted its radius. This sort of symmetry, and its generalisations, became known as T-duality.

As part of the general study, mentioned earlier, of quantum field theories in a space-time of generic dimension it was found that when the overall dimension equalled 2 more than a multiple of 4, the space-time translation symmetry that was manifest classically could be violated by quantum effects when there were chiral fields present. These anomalies became known as gravitational anomalies. Type IIA supergravity was safe, not being chiral, but the vulnerability of type IIB and I supergravities, which were chiral, had to be checked explicitly. Type IIB was also found to be safe whilst Green and Schwarz found that the type I supergravity coupled to super-Yang–Mills was anomaly-free only if the gauge group was $SO(32)$ or $E_8 \otimes E_8$. These two Lie groups each enjoyed the feature of possessing a weight lattice that was even and self-dual (properties crucial for the mathematical contruction of theta-functions invariant with respect to the modular group). The ranks of such groups had to be an integer multiple of 8 and these were the only two of rank 16 (which is the difference between the critical dimensions of the bosonic string and the superstring, $26 - 10$).

This result posed a dilemma since it was already known that the type I open superstring could not carry the gauge group $E_8 \otimes E_8$. This triggered a search for a new string theory with an effective low-energy theory given by type I supergravity coupled to $E_8 \otimes E_8$ Yang–Mills. The result was the discovery of a new closed superstring theory called the 'heterotic string'. The soubriquet heterotic referred to the surprising new feature that the left- and right-movers around the circle that constituted the closed string were very different. Indeed, the left-movers were those of the bosonic string while the right-movers those of the type I superstring with the deficit in dimension restored by the rank of the gauge group (which could equally be $SO(32)$). It was these discoveries and one other, still to be described, that re-ignited world-wide interest in string theory.

The realisation that one could treat left- and right-movers differently led to the formulation of string theories in an effective space-time of four dimensions. Such theories were consistent if the left- and right-momenta belonged to even self-dual indefinite lattices and the theory was modular-invariant. The large number of such lattices led to a corresponding large number of four-dimensional string theories.

A striking bonus of the $E_8 \otimes E_8$ heterotic superstring was that its gauge group contained those of the grand unified theories as subgroups and hence was of phenomenological interest. One promising construction concerned the dimensional reduction of string theories and their related supergravities between space-times of ten and four dimensions by means of Calabi–Yau manifolds of six dimensions. These are Kähler manifolds of reduced holonomy whose topological invariants translated into quantities of interest to particle phenomenology. They had the feature that the theory in four dimensions inherited just the degree of supersymmetry, $N = 1$, needed to solve the hierarchy problem, and no more. It was conjectured that this procedure was equivalent to using a particular conformal field theory with $N = 2$ supersymmetry which in turn could be represented by a two-dimensional $N = 2$ supersymmetric quantum field theory. Explicit symmetries of these theories translated into mysterious pairings of Calabi–Yau manifolds, known as mirror symmetries, which provoked the discovery of many new Calabi–Yau manifolds.

Solitons, Topology and the Index Theorem

Solitons were first observed a century and a half ago and, at the beginning of the present century identified with non-dissipative solutions to certain nonlinear differential equations governing water waves in one-dimensional channels. The subject as a whole is now studied in many areas of science, including hydrodynamics, solid state physics, engineering, biology and geology.

Skyrme was probably the first to recognise their possible relevance to particle physics, in 1958, particularly when the nonlinear equation possessed Lorentz invariance in space-time of two dimensions. This is because they then described concentrations of energy which could be stationary or could be boosted to any velocity of magnitude less than that of light. Thus they could provide a classical description of a massive particle with a finite but extended structure. This interpretation was supported by the construction of solutions describing scattering processes undergone by the solitons in which their identity was preserved.

This behaviour was rather remarkable and signalled a degree of hidden structure, later identified in many ways: with extra conservation laws involving an extension of the Poincaré algebra not possible in four-dimensional space-time; zero-curvature conditions; affine Kac–Moody algebras and structures related to quantum groups. The paradigm was provided by the sine-Gordon equation and there the soliton carried a quantum number of topological type that guaranteed its stability with respect to disintegration.

As explained in Chapters 2, 5 and 6, it was found in 1974 that certain spontaneously broken gauge theories in space-time of four dimensions possessed soliton solutions whose stability was again assured by a topological quantum number. This time it had the physical interpretation of being a magnetic charge so that the soliton was a magnetic monopole. There was no sign of the special structures mentioned above, but, instead, supersymmetry played a key role, particularly in the quantum theory, as the chapters cited explain.

A soliton describes a structure localised in space. In 1975 it was found fruitful to consider structures localised in Euclidean space-time. These became known as instantons and clearly have a less direct physical interpretation. Like solitons they gave rise to non-perturbative phenomena that are invisible in perturbation theory because the effect was not analytic in the coupling constant.

What made instantons especially interesting was the behaviour of Dirac fermion fields in their background. This effect was found completely independently by physicists and mathematicians and the insights gained have proven to be complementary.

The original instantons occur as solutions to non-Abelian gauge theories in a space of four Euclidean dimensions. There the Dirac operator in the background gauge field is a Hermitian operator that anticommutes with a spinor matrix called gamma five. This is the matrix which is formed as the product of the four gamma matrices that enter the Dirac equation. It is real and its square equals the unit matrix.

As a consequence, the eigenfunctions of the Dirac operator corresponding to eigenvalues of opposite sign are paired by the action of this matrix. But this match fails when the eigenvalue vanishes, that is for the case of physical interest since the corresponding eigenfunctions are then solutions to the Dirac equation in the given background. The degree of this mismatch is called by mathematicians the index and they showed that it could be expressed directly in terms of the instanton number which is the topological quantum number associated with the background field. In fact, a particularly elegant proof of this theorem was provided by by ideas of supersymmetric quantum mechanics on a manifold. This line of reasoning became important in some of the purest areas of mathematics.

It is this result that was related to an independent discovery by physicists studying the V–A version of the weak interactions. Quantum effects violated a classical symmetry of the theory called chiral symmetry and involving the same matrix gamma five. This effect was detected experimentally more than thirty years ago by measuring the lifetime of the neutral pi-meson.

There have been many elaborations of these basic ideas and their connections, and they have important implications for the solitons, when they are quantised. They turn out to carry additional quantum numbers inherited from solutions to the Dirac equation in their background.

The theory of solitons is intriguing and becomes even more intriguing the more they are subjected to the principles of quantum mechanics, as they must if they are to be properly interpreted as particles. Many of the developments described in this book are precisely due to such effects.

Conclusion

This brings our story into the early 1990s and time for us to draw it to a close. It was our intention to have set the stage for the many important subsequent developments presented in detail in the lectures included in this volume.

In Chapter 2, Manton describes the solitonic solutions of non-Abelian gauge theories that carry magnetic charge. Rather general electromagnetic duality transformations relating electric and magnetic fields respectively are described by Gaillard and Zumino in Chapter 3. In Chapter 4, Zumino treats many of the special geometric phenomena associated with σ-models in two dimensions, whether supersymmetric or not. This is particularly relevant to the theory of actions for the string. In supersymmetric non-Abelian gauge theories electromagnetic duality transformations can relate electrically charged particles, occurring as quanta, to magnetically charged particles occurring as solitons, as explained in Chapter 2. All this is explained at greater length by Olive and Gauntlett in Chapters 5 and 6. The development of these ideas to derive the complete low-energy effective action of the $N = 2$ supersymmetric Yang–Mills theory is explained by Eguchi in Chapter 7. In Chapter 8, West describes how the structure of the supersymmetry algebra controls the structure of the supergravity theories, in particular those in ten and eleven dimensions. This also leads naturally to a treatment of the dynamics of p-branes. Some of these p-branes arise as solitonic solutions to supergravity equations as explained by Gibbons in Chapter 9. The extension of duality to string theories is explained by Sen in Chapter 10. This also involves the interesting role of black holes and their entropy in this context. Finally, in Chapter 11, Bachas treats the properties of Dirichlet or D-branes and their role in the string version of electromagnetic duality.

We have presented the preceding 'potted history' of our subject in the hope also that it will set the recent developments in the wider context that is appropriate to them. We think that the ebb and flow of ideas we have tried to describe in this history is intrinsically interesting and suspect that it is not well-known to the younger researchers in the area, particularly in respect of all the interconnections that we have tried to emphasise.

Of course, because of its historical nature, our account is inevitably subjective and unintentionally reflects the experiences of the tellers. We have tried to be objective, but it is unavoidable that we have missed out some very important strands of the story and perhaps misrepresented some developments. If this is so, we tender our apologies, and hope to do better should we have an occasion to extend this discussion to a fuller account. Please note that we have only mentioned few people by name and then only when their name is already attached to a particular development. Thus we have omitted mention of some very distinguished figures indeed and likewise have given no references, though many of these can be found in the main text of the book.

Histories of scientific developments are of interest in their own right since they illustrate the scientific process at work. One can, with the benefit of hindsight, see missed opportunities and the transient effects of fashion. The history of string theory furnishes an interesting example. Its initial development during the period 1967–77 has been described above, together with its subsequent period of relative dormancy. Progress was intense and, in retrospect, the achievement is seen to be enormous. Yet at the time it failed to gain the interest of a wider audience, partly because it was so ambitious, with its outrageous claims of unification, partly because of the level of unfamiliar mathematical structure. As a result, the achievements of earlier times were lost to the younger generation of researchers during the period of dormancy. A similar fate befell the study of supersymmetry, whose initial development was also shunned by the community at large, only to be temporarily eclipsed by the resurrection of string theory. Now the situation has changed, and both theories have merged into part of a larger theoretical whole supported by an enormous weight of cumulative evidence of an aesthetic nature. It is this convergence of so many disparate lines of thought that has carried the day.

It is striking how this process of convergence has reconciled schools of thought in the development of particle theory that were once so opposed. Quantum field theory reached its apogee with the success of quantum electrodynamics even though there were doubters such as Paul Dirac. When this success appeared not to extend to the pseudoscalar meson theory of the strong interactions, two opposing schools of thought came into being. One preached that one should only trust the scattering matrix whilst the other that one should cling to the Lagrangian formulation of quantum field theory and focus more attention on its possible symmetries. The first approach led to the dual resonance model and hence string theory and the second to non-Abelian gauge theories, the standard model and supersymmetry. Now we see that in their different ways both approaches were correct and fruitful. Indeed both played their parts in the development of a theory that is even more satisfying than its parts. However, the recent progress has also served to underline the fact that a final and truly satisfactory formulation of superstring theory including the new understanding of the role of space-time has yet to be found. The dilemma that also remains is a challenge; namely the discovery of experimental evidence to support the new ideas. In fact, there are reasons to hope that the next generation of colliders will finally confirm the validity of some role for supersymmetry.

Solitons

N.S. Manton

Abstract

Some older and more recent results on solitons, and especially BPS monopoles, are reviewed.

1 Introduction

In the 1970s it was discovered by 't Hooft and Polyakov that a unified Yang-Mills-Higgs theory with gauge group $SU(2)$, or another simple group, can have magnetic monopole solutions without singularities [1]. Generally, the simplest monopole solutions are spherically symmetric, but despite this, they can only be obtained by numerically solving coupled nonlinear second-order ordinary differential equations. At about the same time, Julia and Zee discovered dyons in Yang-Mills-Higgs theory, which are solutions with both magnetic and electric charge [2].

Exact analytic solutions exist only in a special case, where the Higgs self-coupling is zero, i.e. where the Higgs field is massless. Prasad and Sommerfield found this exact magnetic monopole solution, somewhat fortuitously, and also the corresponding dyons, for gauge group $SU(2)$ [3]. The existence of these solutions was clarified by the work of Bogomol'nyi, who considered a variety of solitons in field theories with special couplings [4]. Such solitons are now known as BPS solitons, and we shall review them here, focussing mainly on monopoles.

Bogomol'nyi found the following soliton properties in these special field theories:

(1) The soliton solutions satisfy a first order partial differential equation;

(2) There is a lower bound on the energy proportional to the number of solitons, and the bound is attained if the first order equation is satisfied;

(3) The solitons are stable.

Subsequent work established further that:

(4) Multisoliton solutions of the same first order equation exist;

(5) There is a geometrically interesting moduli space (parameter space) for the multisoliton solutions. If one soliton has k moduli, then the N-soliton moduli space has dimension Nk;

(6) These special field theories are the bosonic truncations of supersymmetric theories;

(7) The quantized supersymmetric field theory has states describing the quantum mechanics of N solitons. The lowest energy bound states, now known as BPS states, usually lie in special, short multiplets of the supersymmetry algebra.

2   Kinks

The simplest example of a Bogomol'nyi soliton is the kink in ϕ^4 theory in one spatial dimension [4,5]. One does not need to choose special values of the couplings in this case. The field $\phi(x,t)$ has derivatives denoted by

$$\dot\phi \equiv \frac{\partial\phi}{\partial t} \quad \phi' \equiv \frac{\partial\phi}{\partial x}. \tag{2.1}$$

The Lagrangian is

$$L = \int_{-\infty}^{\infty} \left\{ \frac{1}{2}\dot\phi^2 - \frac{1}{2}\phi'^2 - \frac{1}{2}\lambda^2(\phi^2 - a^2)^2 \right\} dx \tag{2.2}$$

so static fields have energy

$$E = \int_{-\infty}^{\infty} \left\{ \frac{1}{2}\phi'^2 + \frac{1}{2}\lambda^2(\phi^2 - a^2)^2 \right\} dx. \tag{2.3}$$

There are two degenerate vacua, $\phi = a$ or $\phi = -a$.

Bogomol'nyi did not solve the static field equation, which is

$$\frac{d^2\phi}{dx^2} = 2\lambda^2(\phi^2 - a^2)\phi, \tag{2.4}$$

but found the kink solution by arguing as follows. The energy can be reexpressed as

$$E = \int_{-\infty}^{\infty} \left\{ \frac{1}{2}[\phi' + \lambda(\phi^2 - a^2)]^2 - \lambda(\phi^2 - a^2)\phi' \right\} dx \tag{2.5}$$

$$= \int_{-\infty}^{\infty} \frac{1}{2}[\phi' + \lambda(\phi^2 - a^2)]^2 dx + \left. \left(\lambda a^2\phi - \frac{1}{3}\lambda\phi^3 \right) \right|_{\phi(-\infty)}^{\phi(\infty)}. \tag{2.6}$$

The kink connects the two vacua, so $\phi(\infty) = a$ and $\phi(-\infty) = -a$. Therefore the energy of the kink is

$$E = \int_{-\infty}^{\infty} \frac{1}{2}[\phi' + \lambda(\phi^2 - a^2)]^2 dx + \frac{4}{3}\lambda a^3, \tag{2.7}$$

giving the inequality $E \geq \frac{4}{3}\lambda a^3$ for fields with these boundary conditions. The inequality becomes an equality if

$$\phi' = -\lambda(\phi^2 - a^2). \tag{2.8}$$

This is the first order Bogomol'nyi equation, which in this case is an ordinary differential equation. Solving it directly gives $\phi(x) = a \tanh \lambda a(x - x_0)$, which is the kink solution, with energy $=$ mass $= \frac{4}{3}\lambda a^3$. The position of the kink is x_0. One can get an antikink by reflection $\phi(x) = -a \tanh \lambda a(x - x_1)$. This satisfies $\phi' = \lambda(\phi^2 - a^2)$ and has the same energy as a kink.

The kink can be translated and boosted to an arbitrary subluminal speed. However there is no possibility to change the field everywhere in a finite time to $-a$ or a. There is therefore a topological obstruction to changing the kink to the vacuum, so the kink is absolutely stable. It appears that kinks exist in the quantized field theory, and are topologically stable there too.

The kink is parametrised by a collective coordinate x_0, so there is a moduli space $\mathbf{R}$ of kink solutions. One may approximate the kink motion by a motion in the moduli space. This ignores possible field oscillations around the kink, but captures the basic feature, that a kink behaves like a smooth, localised particle. One may investigate the motion of several well-separated kinks and antikinks as a motion of weakly interacting point particles. It is also possible to model the motion of kinks when they are close together, and to study the way a kink and antikink annihilate into radiation. In ref. [6], there is a discussion of the symmetric process

$$K\bar{K}K \rightarrow K + \text{radiation}, \tag{2.9}$$

where K denotes a kink and $\bar{K}$ an antikink. The annihilation sets up a large amplitude oscillation of the unique discrete mode of oscillation of the remaining kink, at frequency $2\sqrt{3}a\lambda$, which slowly radiates away because of the nonlinearities.

3   BPS monopoles

Monopole solutions in Yang-Mills-Higgs theory are possible if, via the Higgs mechanism, the gauge group is broken to a subgroup with a direct $U(1)$ factor. This factor is identified with the electromagnetic gauge group $U(1)_{\text{em}}$. Such a breaking scheme always occurs if the Higgs field transforms under the adjoint representation of the original gauge group. Asymptotically, one can project the Yang-Mills field tensor onto the $U(1)_{\text{em}}$ factor to get the Maxwell field tensor. The spatial part gives the magnetic field. For a monopole, one finds, asymptotically, that $\mathbf{B}_{\text{em}} = \frac{g}{4\pi r^3}\mathbf{x}$, where g is identified as the magnetic charge.

The unified $SU(2)$ theory has a gauge potential $A_\mu = A_\mu^a t^a \in L(SU(2))$ and Higgs field $\phi = \phi^a t^a \in L(SU(2))$, where $L(G)$ denotes the Lie algebra of G. The generators of $L(SU(2))$ are $t^a = \frac{1}{2} i \tau^a$, satisfying

$$[t^a, t^b] = -\epsilon_{abc} t^c, \quad \mathrm{Tr}(t^a t^b) = -\frac{1}{2} \delta_{ab}. \tag{3.1}$$

The covariant derivative of the Higgs field is $D_\mu \phi = \partial_\mu \phi + e[A_\mu, \phi]$. The Yang-Mills field tensor is $F_{\mu\nu} = \partial_\mu A_\nu - \partial_\nu A_\mu + e[A_\mu, A_\nu]$ and this satisfies the Bianchi identity $D_{[\lambda} F_{\mu\nu]} = 0$. The $SU(2)$ Yang-Mills magnetic field is

$$B_i = \frac{1}{2} \epsilon_{ijk} F_{jk} \tag{3.2}$$

which satisfies, by the Bianchi identity, $D_i B_i = 0$. The Lagrangian is

$$L = \int \left\{ -\frac{1}{4} |F_{\mu\nu} F^{\mu\nu}| + \frac{1}{2} |D_\mu \phi D^\mu \phi| - \frac{1}{4} \lambda (|\phi|^2 - v^2)^2 \right\} d^3 x. \tag{3.3}$$

Here, λ is the Higgs coupling, v the Higgs vacuum expectation value, and $|\ \ |$ denotes $-2\mathrm{Tr}\{\ \ \}$. The Higgs field breaks $SU(2)$ to $U(1)_{\mathrm{em}}$ if $v \neq 0$.

The BPS limit is $\lambda \to 0$ but one still requires $|\phi| = v$ in the vacuum, and also asymptotically for any finite energy field. The energy of static fields in this case is

$$E = \int \left\{ \frac{1}{2} |B_i B_i| + \frac{1}{2} |D_i \phi D_i \phi| \right\} d^3 x. \tag{3.4}$$

Bogomol'nyi showed that this energy can be rewritten as

$$E = \int \left\{ \frac{1}{2} |B_i - D_i \phi|^2 + |B_i D_i \phi| \right\} d^3 x \tag{3.5}$$

$$= \int \frac{1}{2} |B_i - D_i \phi|^2 d^3 x + \int_{S_\infty^2} |B_i \phi| dS^i, \tag{3.6}$$

where S_∞^2 is a sphere of infinite radius. The last integral is obtained using Gauss' law

$$\int_{S_\infty^2} |B_i \phi| dS^i = \int \partial_i |B_i \phi| d^3 x = \int |(D_i B_i) \phi| d^3 x + \int |B_i D_i \phi| d^3 x \tag{3.7}$$

and the Bianchi identity $D_i B_i = 0$.

$\frac{1}{v} |\mathbf{B}\phi|$ is the projection of the Yang-Mills magnetic field onto the Higgs direction, which is also the projection onto the unbroken $U(1)_{\mathrm{em}}$, so this quantity is interpreted as $\mathbf{B}_{\mathrm{em}}$. Its integral over any large sphere is the total magnetic charge g, so

$$E = \int \frac{1}{2} |B_i - D_i \phi|^2 d^3 x + vg. \tag{3.8}$$

Hence for a given positive magnetic charge g we have $E \geq vg$, which is the Bogomol'nyi energy bound. There is equality if

$$B_i = D_i\phi, \tag{3.9}$$

which is the Bogomol'nyi equation for this theory.

The Bogomol'nyi equation implies the static field equations. Starting from $B_i = D_i\phi$ one deduces:

$$
\begin{align}
\text{(i)} \quad D_i D_i \phi &= D_i B_i = 0 \tag{3.10} \\
\text{(ii)} \quad \epsilon_{ijk} D_j B_k &= \epsilon_{ijk} D_j D_k \phi \tag{3.11} \\
&= \frac{1}{2}\epsilon_{ijk}[D_j, D_k]\phi \tag{3.12} \\
&= \frac{1}{2}\epsilon_{ijk}[F_{jk}, \phi] = [B_i, \phi] = [D_i\phi, \phi]. \tag{3.13}
\end{align}
$$

There is the following topological structure. The gauge and Higgs fields are smooth on S^2_∞, and $|\phi_\infty|^2 = v^2$. Thus

$$\phi_\infty : S^2_\infty \rightarrow L(SU(2)) \tag{3.14}$$

defines a map from S^2_∞ to a sphere in $L(SU(2))$. This map has a degree N, an integer. This is related to the magnetic charge. Since $D_i\phi \rightarrow 0$ at infinity, the asymptotic gauge and Higgs fields are related, and one can show that

$$g = \frac{4\pi}{e}N. \tag{3.15}$$

The Bogomol'nyi bound becomes $E \geq \frac{4\pi v}{e}N$ with equality if $B_i = D_i\phi$. N is called the monopole number.

The Bogomol'nyi equation can be solved using the following spherically symmetric ansatz (which also works away from the BPS limit) [1]

$$\phi^a = vh(ver)\frac{x^a}{r} \tag{3.16}$$

$$A_i^a = -\epsilon^{aij}\frac{x^j}{er^2}(1 - k(ver)). \tag{3.17}$$

This involves two functions h and k of the scaled radial variable $\xi = ver$. Rotations of these fields are equivalent to gauge transformations.

From the Bogomol'nyi equation we get the radial equations

$$\frac{dh}{d\xi} = -\frac{1}{\xi^2}(k^2 - 1) \tag{3.18}$$

$$\frac{dk}{d\xi} = -kh. \tag{3.19}$$

These have the solution [3,4] $h = \coth \xi - \frac{1}{\xi}$, $k = \frac{\xi}{\sinh \xi}$, which satisfy $h \to 0$ and $k \to 1$ as $\xi \to 0$, $h \to 1$ and $k \to 0$ as $\xi \to \infty$.

This solution is an $N = 1$ monopole, with magnetic charge $g = \frac{4\pi}{e}$, since ϕ_∞ is the identity map. It has mass $M_{mono} = vg$, and the fields have no singularity at the origin. There is another solution in the theory (obeying the Bogomol'nyi equation with sign reversed). It is an antimonopole with the opposite magnetic charge but the same mass.

Semiclassically, the masses of the other particles of the quantized field theory are $M_{photon} = M_{Higgs} = 0$, and $M_{W^\pm} = ve$, where $W^\pm$ are the massive gauge bosons with electric charges $q = \pm e$. This agrees with the expectation from electromagnetic duality [7] that all masses should be given by an expression of the form $M = f(|q+ig|)$. In fact, semiclassically, $M = v|q+ig|$ in this theory, a result that becomes exact in the $N = 4$ supersymmetric extension of the field theory [8].

4   Dyons

In the BPS limit, one can also find exact dyon solutions [2,3]. These are stationary solutions of the time-dependent field equations. They satisfy $D_0 \phi = 0$, but there is a non-zero Yang-Mills electric field E_i and non-zero A_0. For fields with non-zero E_i the energy is

$$E = \int \left\{ \frac{1}{2}|E_i|^2 + \frac{1}{2}|B_i|^2 + \frac{1}{2}|D_i\phi|^2 \right\} d^3x, \tag{4.1}$$

which may be rewritten (for arbitrary constant μ) as

$$E = \int \left\{ \frac{1}{2}|E_i - \sin\mu D_i\phi|^2 + \frac{1}{2}|B_i - \cos\mu D_i\phi|^2 \right\} d^3x \tag{4.2}$$

$$+ v\sin\mu \underbrace{\int_{S^2_\infty} \mathbf{E}_{em} \cdot \mathbf{dS}}_{q} + v\cos\mu \underbrace{\int_{S^2_\infty} \mathbf{B}_{em} \cdot \mathbf{dS}}_{g}. \tag{4.3}$$

It follows that $E \geq v(q\sin\mu + g\cos\mu)$ with equality if:

$$\begin{cases} E_i = \sin\mu \ D_i\phi \\ \\ B_i = \cos\mu \ D_i\phi \end{cases} \tag{4.4}$$

which implies $\frac{q}{g} = \tan\mu$. Thus

$$\sin\mu = \frac{q}{(q^2 + g^2)^{\frac{1}{2}}}, \quad \cos\mu = \frac{g}{(q^2 + g^2)^{\frac{1}{2}}}, \tag{4.5}$$

so the energy (mass) is

$$E = v(q^2 + g^2)^{\frac{1}{2}} = v|q + ig|. \tag{4.6}$$

The first of eqs. (4.4) is solved by setting $A_0 = \sin \mu \, \phi$, for then $E_i = \partial_i A_0 + e[A_i, A_0] = \sin \mu \, D_i \phi$. The second equation is solved by taking A_i and ϕ to be the usual BPS monopole fields, but radially rescaled. The solution one obtains is a dyon, and it is smaller than the monopole. Its magnetic charge is $g = \frac{4\pi}{e}$, as before, but the electric charge q is arbitrary classically.

Quantization fixes $q = ne\hbar$. In the $N = 4$ SUSY version of theory, the mass formula $M_{\text{dyon}} = v|q + ig|$ remains exact.

5 Static Monopole-Antimonopole Solution

There is a non-trivial static solution of the full Yang-Mills-Higgs equations in the BPS limit

$$D_i D_i \phi = 0 \ , \ \ \epsilon_{ijk} D_j B_k = [D_i \phi, \phi]. \tag{5.1}$$

It has net monopole number zero, and can be interpreted as a monopole-antimonopole pair in unstable equilibrium. It is not a solution of the Bogomol'nyi equation. Taubes proved the existence of such a solution some years ago [9]. He realised that when a well-separated monopole and antimonopole are superposed, there is an arbitrary relative phase θ. If $\theta = 0$ or $\theta = 2\pi$, then the solitons annihilate when they approach. However, if $\theta = \pi$, then as they approach, the monopole doesn't know how to twist relative to the antimonopole, and they reach an unstable equilibrium. The precise properties of this solution are known, following a numerical study by Rüber [10]. (*I am grateful to W. Nahm for informing me of this.*) The solution is axisymmetric, and the monopole and antimonopole are not particularly close, nor much distorted. There is no net magnetic charge, but there is a magnetic dipole moment. The solution has energy approximately 1.7 times the mass of one monopole.

6 Multimonopoles

We return now to the Bogomol'nyi equation

$$B_i = D_i \phi. \tag{6.1}$$

This equation has solutions with N monopoles and total magnetic charge $g = \frac{4\pi N}{e}$ for any positive integer N, although analytic formulae for the fields are not obtainable, in general. The asymptotic magnitude of the Higgs field is

$$|\phi| = v - \frac{N}{er} + \cdots \tag{6.2}$$

and it follows from the Bogomol'nyi equation that the asymptotic magnetic field is $\mathbf{B}_{\text{em}} = \nabla|\phi| = \frac{N}{er^3}\mathbf{x}$.

The existence of N-monopole solutions was first proved by Jaffe and Taubes [11]. They considered N well-separated monopoles and constructed good approximate solutions to the Bogomol'nyi equation by using the BPS one monopole solution for the core regions, and a linear superposition of the asymptotic magnetic and Higgs fields (which are essentially Coulomb fields) elsewhere. Then, an iterative procedure was used to converge to a nearby exact solution. Weinberg proved, using an index theorem, that the manifold of solutions is $4N$-dimensional [12]. The interpretation is that each of the N monopoles has three moduli specifying its location, and a further modulus giving a phase, although this description breaks down when the monopoles are close together.

The existence proof does not give explicit solutions at all. However, via twistor methods, related to those used to study the self-dual Yang-Mills equation in four dimensions, Ward constructed an axisymmetric 2-monopole solution, representing two coalesced monopoles [13]. This solution was generalised to an axisymmetric N-monopole solution [14]. In these solutions, some quantities are known explicitly, but the fields are still hard to compute. The Higgs field on the axis of symmetry (the z-axis) has a rather simple form. For example, for $N = 3$

$$|\phi| = \coth z - \frac{1}{z} - \frac{1}{z + i\pi} - \frac{1}{z - i\pi}, \tag{6.3}$$

and for $N = 2$

$$|\phi| = \tanh z - \frac{1}{z + \frac{1}{2}i\pi} - \frac{1}{z - \frac{1}{2}i\pi}. \tag{6.4}$$

It is as if there are equally spaced point sources on the imaginary z-axis, and this holds for all N.

Nahm found a new approach to solving the three-dimensional Bogomol'nyi equation [15]. This exploited an analogy with the ADHM construction of self-dual Yang-Mills fields. Nahm's method transforms the Bogomol'nyi equation into a set of nonlinear ODE's. The 1-monopole and 2-monopole solutions have very simple Nahm data. Generally, the Nahm equations are no easier to solve than the Bogomol'nyi equation. However, by imposing symmetries, the number of Nahm equations decreases, and in a number of cases they can be solved exactly using Weierstrass elliptic functions. In this way, Hitchin, Manton and Murray [16], and then Houghton and Sutcliffe [17], have recently proved the existence of various multimonopoles invariant under subgroups of the rotation group SO(3). The monopole fields, in particular the Higgs field, can be reconstructed from the Nahm data by solving an auxiliary linear differential equation. In practice, this is done numerically. The energy density can be calculated using the formula $\nabla^2(|\phi|^2)$ for any solution of the Bogomol'nyi equation. Particular solutions that have been found this way are a cubic 4-monopole, a dodecahedral 7-monopole, and a family of 3-monopole solutions which interpolate between well-separated monopoles and three monopoles coalesced into a tetrahedral configuration. One surprising feature of some of these solutions is that there are more zeros of the Higgs field than the minimum possible [17]. An N-monopole must have N zeros counted with mutiplicity, but, for example, the tetrahedral 3-monopole has four positive zeros (on the vertices of a tetrahedron) and one negative zero (at the centre).

For a review of these recent developments, and many pictures of the multimonopole configurations, see ref. [18].

7 N-monopole Moduli space

The space of N-monopole solutions of the Bogomol'nyi equation $B_i = D_i\phi$, modulo gauge transformations, is a manifold M_N. The precise form of M_N depends on boundary conditions and the class of gauge transformations $g(\mathbf{x})$. A good choice is

$$\phi \to vt^3 \quad \text{as} \quad \mathbf{x} \to (0, 0, \infty) \tag{7.1}$$

$$g \to 1 \quad \text{as} \quad \mathbf{x} \to (0, 0, \infty). \tag{7.2}$$

Then M_N is smooth, connected and $\dim M_N = 4N$. There remains an action of constant gauge transformations $g = \exp i\alpha t^3$ on M_N. If this gauge transformation is time-dependent, i.e. $\alpha(t) = \omega t$, then the solution acquires an electric charge, and hence is a multidyon.

The overall geometry of M_N is quite complicated [19]. One simplification is that M_N has an exact factor $\mathbf{R}^3 \times S^1$ ($\mathbf{R}^3$ centre of mass position, S^1 overall phase). The moduli parametrise static solutions, but they can change with time and if they do so one obtains monopoles or dyons in motion. Having time-varying moduli does not generally give an exact solution of the time-dependent field equations, but in certain circumstances a good approximation is obtained. We can always project a general motion onto M_N, and at low energy the motion is effectively constrained to M_N. It makes sense to ask what the natural motion on M_N is. Since all fields parametrised by M_N have the same constant potential energy $4\pi v N/e$, only the kinetic energy matters. The kinetic energy defines a metric on M_N, and the free motion on M_N is therefore geodesic motion [20]. Stuart has recently rigorously proved that this geodesic motion is a good approximation to the full field evolution, for suitable initial conditions, at least for a finite time [21].

The metric is not easy to calculate, but in principle, is calculated as follows. The field theory defines a Lagrangian $L = T - V$ on M_N, where V is constant, and the kinetic energy is

$$T = \int \left\{ \frac{1}{2}|E_i|^2 + \frac{1}{2}|D_0\phi|^2 \right\} d^3x. \tag{7.3}$$

Note that this is quadratic in time-derivatives because $E_i = \dot{A}_i - D_i A_0$ and $D_0\phi = \dot{\phi}+e[A_0, \phi]$ are linear in time-derivatives. Let solutions of $B_i = D_i\phi$ be parametrised by moduli X^α ($1 \leq \alpha \leq 4N$), and let $X^\alpha(t)$ be time-dependent. We evaluate T from $A_i(t), \phi(t)$. Here the gauge is not fixed, and we must impose Gauss' law $D_i E_i = [\phi, D_0\phi]$ to fix A_0, and hence obtain the physical kinetic energy. (Things are simpler in background gauge. Then $A_0 = 0$ and $E_i = \dot{A}_i$, $D_0\phi = \dot{\phi}$.) In this way one obtains a quadratic kinetic energy expression

$$T = \frac{1}{2}g_{\alpha\beta}(\mathbf{X})\dot{X}^\alpha \dot{X}^\beta. \tag{7.4}$$

$g_{\alpha\beta}$ is the metric on M_N. It is curved, in general, although M_1 is flat. It is always hyper-Kähler [22,19].

The full metric on M_2 is known [19]. M_2 is 8-dimensional and has the form

$$M_2 = \mathbf{R}^3 \times \frac{S^1 \times M_2^0}{\mathbf{Z}_2} \tag{7.5}$$

where $\mathbf{R}^3 \times S^1$ is flat and M_2^0 is a non-trivial 4-dimensional manifold, now known as the Atiyah-Hitchin manifold. Atiyah and Hitchin calculated its metric indirectly using

(i) its hyper-Kähler property

(ii) rotational symmetry

(iii) its smoothness

which together determine the metric.

The detailed metric on M_N is not known for $N \geq 3$, but the metric on various geodesic submanifolds associated with symmetric configurations of monopoles is known [23]. The metric for well-separated monopoles has also been calculated, using physical arguments, assuming that monopoles/dyons behave as point particles [24]. The result has now been confirmed mathematically, using the Nahm data [25]. The asymptotic metric is a generalisation of the Taub-NUT metric. It has singularities when the monopoles are close, which simply implies that it is incorrect here. However M_N might be characterizable as the unique, smooth hyper-Kähler manifold which is asymptotic to this generalised Taub-NUT space.

Using the metrics, and their geodesics, one can study the scattering of monopoles. Two monopoles in a head-on collision scatter at 90 degrees, with the relative phase determining the scattering plane. Some genuinely three-dimensional scattering processes involving three or more monopoles are now also understood. The monopole matter is completely rearranged in these processes. One cannot follow the 'trajectory' of individual monopoles; it isn't even defined when monopoles overlap.

The metric on M_N defines the Hamiltonian for the quantum mechanics of multimonopoles. Something is known about the quantized bound states and scattering states of two monopoles, without assuming supersymmetry [26]. The bound states of monopoles in the supersymmetric theory are discussed elsewhere [27]; see also the article by Sen in this volume.

Acknowledgements

The draft of this paper was prepared from lecture notes taken by Annamaria Kiss, and typed by Rachel George. I am grateful to both for their assistance.

References

[1] G. 't Hooft, Nucl. Phys. B 79 (1974) 276; A.M. Polyakov, JETP Lett. 20 (1974) 194.

[2] B. Julia and A. Zee, Phys. Rev. D 11 (1975) 2727.

[3] M.K. Prasad and C.M. Sommerfield, Phys. Rev. Lett. 35 (1975) 760.

[4] E.B. Bogomol'nyi, Sov. J. Nucl. Phys. 24 (1976) 449.

[5] R. Rajaraman, *Solitons and Instantons*, North Holland, Amsterdam (1982).

[6] N.S. Manton and H. Merabet, Nonlinearity 10 (1997) 3.

[7] C. Montonen and D. Olive, Phys. Lett. B 72 (1977) 117.

[8] A. D'Adda, R. Horsley and P. di Vecchia, Phys. Lett. B 76 (1978) 298; E. Witten and D. Olive, Phys. Lett. B 78 (1978) 97; H. Osborn, Phys. Lett. B 83 (1979) 321.

[9] C.H. Taubes, Commun. Math. Phys. 86 (1982) 257; 86 (1982) 299.

[10] B. Rüber, Bonn diploma thesis IR-85-27 (1985) unpublished.

[11] A. Jaffe and C. Taubes, *Vortices and Monopoles*, Birkhäuser, Boston (1980).

[12] E. Weinberg, Phys. Rev. D 20 (1979) 936.

[13] R.S. Ward, Commun. Math. Phys. 79 (1981) 317.

[14] P. Forgács, Z. Horváth and L. Palla, Nucl. Phys. B 192 (1981) 141; M.K. Prasad and P. Rossi, Phys. Rev. Lett. 46 (1981) 806.

[15] W. Nahm, Bonn report HE-82-30 (1982) unpublished; in *Monopoles in Quantum Field Theory*, eds. N.S. Craigie et al., World Scientific, Singapore (1982).

[16] N.J. Hitchin, N.S. Manton and M.K. Murray, Nonlinearity 8 (1995) 661.

[17] C.J. Houghton and P.M. Sutcliffe, Commun. Math. Phys. 180 (1996) 343; Nonlinearity 9 (1996) 385; Nucl. Phys. B 464 (1996) 59.

[18] P.M. Sutcliffe, Int. J. Mod. Phys. A 12 (1997) 4663.

[19] M.F. Atiyah and N.J. Hitchin, *The Geometry and Dynamics of Magnetic Monopoles*, Princeton University Press (1988).

[20] N.S. Manton, Phys. Lett. B 110 (1982) 54.

[21] D. Stuart, Commun. Math. Phys. 166 (1994) 149.

[22] N.J. Hitchin, A. Karlhede, U. Lindström and M. Roček, Commun. Math. Phys. 108 (1987) 535.

[23] R. Bielawski, Nonlinearity 9 (1996) 1463; H.W. Braden and P.M. Sutcliffe, Phys. Lett. B 391 (1997) 366.

[24] G.W. Gibbons and N.S. Manton, Phys. Lett. B 356 (1995) 32.

[25] R. Bielawski, Bonn preprint MPI 97-107 (1997).

[26] B.J. Schroers, Nucl. Phys. B 367 (1991) 177.

[27] A. Sen, Phys. Lett. B 329 (1994) 217; G. Segal and A. Selby, Commun. Math. Phys. 177 (1996) 775.

Nonlinear Electromagnetic Self-Duality and Legendre Transformations*

Mary K. Gaillard and Bruno Zumino

Abstract

We discuss continuous duality transformations and the properties of classical theories with invariant interactions between electromagnetic fields and matter. The case of scalar fields is treated in some detail. Special discrete elements of the continuous group are shown to be related to the Legendre transformation with respect to the field strengths.

1 Duality rotations in four dimensions

The invariance of Maxwell's equations under 'duality rotations' has been known for a long time. In relativistic notation these are rotations of the electromagnetic field strength $F_{\mu\nu}$ into its dual, which is defined by

$$\tilde{F}_{\mu\nu} = \frac{1}{2}\epsilon_{\mu\nu\lambda\sigma}F^{\lambda\sigma}, \quad \tilde{\tilde{F}}_{\mu\nu} = -F_{\mu\nu}. \tag{1.1}$$

This invariance can be extended to electromagnetic fields in interaction with the gravitational field, which does not transform under duality. It is present in ungauged extended supergravity theories, in which case it generalizes to a nonabelian group [1]. In [2, 3] we studied the most general situation in which classical duality invariance of this type can occur. More recently [4] the duality invariance of the Born–Infeld theory, suitably coupled to the dilaton and axion [5], has been studied in considerable detail. In the present note we will show that most of the results of [4, 5] follow quite easily from our earlier general discussion. We shall also present some new results.

We begin by recalling and completing some basic results of [2, 3, 6]. Consider a Lagrangian which is a function of n real field strengths $F_{\mu\nu}^a$ and of some other fields χ^i and their derivatives $\chi_\mu^i = \partial_\mu\chi^i$:

$$L = L\left(F^a, \chi^i, \chi_\mu^i\right). \tag{1.2}$$

*This work was supported in part by the Director, Office of Energy Research, Office of High Energy and Nuclear Physics, Division of High Energy Physics of the US Department of Energy under Contract DE–AC03–76SF00098 and in part by the National Science Foundation under grant PHY–95–14797.

33

Since

$$F^a_{\mu\nu} = \partial_\mu A^a_\nu - \partial_\nu A^a_\mu, \tag{1.3}$$

we have the Bianchi identities

$$\partial^\mu \tilde{F}^a_{\mu\nu} = 0. \tag{1.4}$$

On the other hand, if we define

$$\tilde{G}^a_{\mu\nu} = \frac{1}{2}\epsilon_{\mu\nu\lambda\sigma}G^{a\lambda\sigma} \equiv 2\frac{\partial L}{\partial F^{\mu\nu}_a}, \tag{1.5}$$

we have the equations of motion

$$\partial^\mu \tilde{G}^a_{\mu\nu} = 0. \tag{1.6}$$

We consider an infinitesimal transformation of the form

$$\delta\begin{pmatrix} G \\ F \end{pmatrix} = \begin{pmatrix} A & B \\ C & D \end{pmatrix}\begin{pmatrix} G \\ F \end{pmatrix}, \tag{1.7}$$

$$\delta\chi^i = \xi^i(\chi), \tag{1.8}$$

where A, B, C, D are real $n \times n$ constant infinitesimal matrices and $\xi^i(\chi)$ functions of the fields χ^i (but not of their derivatives), and ask under what circumstances the system of the equations of motion (1.4) and (1.6), as well as the equation of motion for the fields χ^i are invariant. The analysis of [2] shows that this is true if the matrices satisfy

$$A^T = -D, \quad B^T = B, \quad C^T = C, \tag{1.9}$$

(where the superscript T denotes the transposed matrix) and in addition the Lagrangian changes under (1.7) and (1.8) as

$$\delta L = \frac{1}{4}\left(FB\tilde{F} + GC\tilde{G}\right). \tag{1.10}$$

The relations (1.9) show that (1.7) is an infinitesimal transformation of the real noncompact symplectic group $Sp(2n, \mathbf{R})$ which has $U(n)$ as maximal compact subgroup. The finite form is

$$\begin{pmatrix} G' \\ F' \end{pmatrix} = \begin{pmatrix} a & b \\ c & d \end{pmatrix}\begin{pmatrix} G \\ F \end{pmatrix}, \tag{1.11}$$

where the $n \times n$ real submatrices satisfy

$$c^T a = a^T c, \quad b^T d = d^T b, \quad d^T a - b^T c = 1. \tag{1.12}$$

For the $U(n)$ subgroup, one has in addition

$$A = D, \quad B = -C, \tag{1.13}$$

or, in finite form,

$$a = d, \quad b = -c. \tag{1.14}$$

Notice that the Lagrangian is not invariant. In [2] we showed, however, that the derivative of the Lagrangian with respect to an invariant parameter *is* invariant. The invariant parameter could be a coupling constant or an external background field, such as the gravitational field, which does not change under duality rotations. It follows that the energy-momentum tensor, which can be obtained as the variational derivative of the Lagrangian with respect to the gravitational field, is invariant under duality rotations. No explicit check of its invariance, as was done in [4, 5, 7, 8], is necessary. Using (1.7) and (1.9) it is easy to verify that

$$\delta \left(L - \frac{1}{4} F \tilde{G} \right) = \delta L - \frac{1}{4} \left(F B \tilde{F} + G C \tilde{G} \right), \tag{1.15}$$

so (1.10) is *equivalent* to the invariance of $L - \frac{1}{4} F \tilde{G}$.

The symplectic transformation (1.11) can be written in a complex basis as

$$\begin{pmatrix} F' + iG' \\ F' - iG' \end{pmatrix} = \begin{pmatrix} \phi_0 & \phi_1^* \\ \phi_1 & \phi_0^* \end{pmatrix} \begin{pmatrix} F + iG \\ F - iG \end{pmatrix}, \tag{1.16}$$

where $*$ means complex conjugation and the submatrices satisfy

$$\phi_0^T \phi_1 = \phi_1^T \phi_0, \quad \phi_0^\dagger \phi_0 - \phi_1^\dagger \phi_1 = 1. \tag{1.17}$$

The relation between the real and the complex basis is

$$\begin{aligned} 2a &= \phi_0 + \phi_0^* - \phi_1 - \phi_1^*, & 2ib &= \phi_0 - \phi_0^* - \phi_1 + \phi_1^*, \\ -2ic &= \phi_0 - \phi_0^* + \phi_1 - \phi_1^*, & 2d &= \phi_0 + \phi_0^* + \phi_1 + \phi_1^*. \end{aligned} \tag{1.18}$$

In [2, 3] we also described scalar fields valued in the quotient space $Sp(2n, \mathbf{R})/U(n)$. The quotient space can be parameterized by a complex symmetric $n \times n$ matrix $K = K^T$ whose real part has positive eigenvalues, or equivalently by a complex symmetric matrix $Z = Z^T$ such that $Z^\dagger Z$ has eigenvalues smaller than 1. They are related by

$$K = \frac{1 - Z^*}{1 + Z^*}, \quad Z = \frac{1 - K^*}{1 + K^*}. \tag{1.19}$$

These formulae are the generalization of the well-known map between the Lobachevskiĭ unit disk and the Poincaré upper half-plane: Z corresponds to the single complex variable parameterizing the unit disk, iK to the one parameterizing the upper half plane.

Under $Sp(2n, \mathbf{R})$

$$K \to K' = (-ib + aK)(d + icK)^{-1}, \quad Z \to Z' = (\phi_1 + \phi_0^* Z)(\phi_0 + \phi_1^* Z)^{-1}, \tag{1.20}$$

or, infinitesimally,

$$\delta K = -iB + AK - KD - iKCK, \quad \delta Z = V + T^*Z - ZT - iZV^*Z, \quad (1.21)$$

where

$$T = -T^\dagger, \quad V = V^T. \quad (1.22)$$

The invariant nonlinear kinetic term for the scalar fields can be obtained from the Kähler metric [9]

$$\text{Tr}\left(dK^* \frac{1}{K + K^*} dK \frac{1}{K + K^*}\right) = \text{Tr}\left(dZ \frac{1}{1 - Z^*Z} dZ^* \frac{1}{1 - ZZ^*}\right) \quad (1.23)$$

which follows from the Kähler potential

$$\text{Tr}\ln(1 - ZZ^*) \quad \text{or} \quad \text{Tr}\ln(K + K^*), \quad (1.24)$$

which are equivalent up to a Kähler transformation. It is not hard to show that the metric (1.23) is positive definite. In this section the normalization of the fields $F_{\mu\nu}^a$ has been chosen to be canonical when iK is set equal to the unit matrix, i.e., when the self-duality group reduces to the $U(n)$ subgroup; the full $Sp(2n, \mathbf{R})$ self-duality can be realized when the matrix K is a function of scalar fields. Throughout this paper we assume a flat background space-time metric; the generalization to a nonvanishing gravitational field is straightforward, [2]–[5].

2   Born–Infeld theory

As a particularly simple example we consider the case when there is only one tensor $F_{\mu\nu}$ and no additional fields. Our equations become

$$\tilde{G} = 2\frac{\partial L}{\partial F}, \quad (2.1)$$

$$\delta F = \lambda G, \quad \delta G = -\lambda F \quad (2.2)$$

and

$$\delta L = \frac{1}{4}\lambda\left(G\tilde{G} - F\tilde{F}\right). \quad (2.3)$$

We have restricted the duality transformation to the compact subgroup $U(1) \cong SO(2)$, as appropriate when no additional fields are present. So $A = D = 0$, $C = -B = \lambda$.

Since L is a function of F alone, we can also write

$$\delta L = \delta F \frac{\partial L}{\partial F} = \lambda G \frac{1}{2}\tilde{G}. \quad (2.4)$$

Comparing (2.3) and (2.4), which must agree, we find

$$G\tilde{G} + F\tilde{F} = 0. \tag{2.5}$$

Together with (2.1), this is a partial differential equation for $L(F)$, which is the condition for the theory to be duality invariant. If we introduce the complex field

$$M = F - iG, \tag{2.6}$$

(2.5) can also be written as

$$M\widetilde{M}^* = 0. \tag{2.7}$$

Clearly, Maxwell's theory in vacuum satisfies (2.5), or (2.7), as expected. A more interesting example is the Born–Infeld theory [7], given by the Lagrangian

$$L = \frac{1}{g^2}\left(-\Delta^{\frac{1}{2}} + 1\right), \tag{2.8}$$

where

$$\Delta = -\det\left(\eta_{\mu\nu} + gF_{\mu\nu}\right) = 1 + \frac{1}{2}g^2F^2 - g^4\left(\frac{1}{4}F\tilde{F}\right)^2. \tag{2.9}$$

For small values of the coupling constant g (or for weak fields) L approaches the Maxwell Lagrangian. We shall use the abbreviation

$$\beta = \frac{1}{4}F\tilde{F}. \tag{2.10}$$

Then

$$\frac{\partial\Delta}{\partial F} = g^2F - \beta g^4\tilde{F}, \tag{2.11}$$

$$\tilde{G} = 2\frac{\partial L}{\partial F} = -\Delta^{-\frac{1}{2}}\left(F - \beta g^2\tilde{F}\right), \tag{2.12}$$

and

$$G = \Delta^{-\frac{1}{2}}\left(\tilde{F} + \beta g^2F\right). \tag{2.13}$$

Using (2.12) and (2.13), it is very easy to check that $G\tilde{G} = -F\tilde{F}$: the Born–Infeld theory is duality invariant. It is also not too difficult to check that $\partial L/\partial g^2$ is actually *invariant* under (2.2) and the same applies to $L - \frac{1}{4}F\tilde{G}$ (which in this case turns out to be equal to $-g^2\partial L/\partial g^2$). These invariances are expected from our general theory.

It is natural to ask oneself whether the Born–Infeld theory is the most general physically acceptable solution of (2.5). This was investigated in [4] where a negative result was reached: more general Lagrangians satisfy (2.5), the arbitrariness depending on a function of one variable. We discuss this in detail in Section 6.

3   Schrödinger's formulation of Born's theory

Schrödinger [8] noticed that, for the Born–Infeld theory (2.8), F and G satisfy not only (2.5) [or (2.7)], but also the more restrictive relation

$$M\left(M\widetilde{M}\right) - \widetilde{M}M^2 = \frac{g^2}{8}\widetilde{M}^*\left(M\widetilde{M}\right)^2. \tag{3.1}$$

We have verified this by an explicit, although lengthy, calculation using (2.6), (2.12), (2.13) and (2.9). Schrödinger did not give the details of the calculation, presenting instead convincing arguments based on particular choices of reference systems. One can write (3.1) as

$$\frac{\partial \mathcal{L}}{\partial M} = g^2 \widetilde{M}^*, \tag{3.2}$$

where

$$\mathcal{L} = 4\frac{M^2}{\left(M\widetilde{M}\right)}, \tag{3.3}$$

and Schrödinger proposed $\mathcal{L}$ as the Lagrangian of the theory, instead of (2.8). Of course, $\mathcal{L}$ is a Lagrangian in a different sense than L, which is a field Lagrangian in the usual sense. Multiplying (3.1) by M and saturating the unwritten indices $\mu\nu$, the left hand side vanishes, so that (2.7) follows. Using (3.1) it is easy to see that $\mathcal{L}$ is pure imaginary: $\mathcal{L} = -\mathcal{L}^*$. Schrödinger also pointed out that, if we introduce a map

$$\frac{1}{g^2}\frac{\partial \mathcal{L}}{\partial M} = f(M), \tag{3.4}$$

so that (3.1) or (3.2) can be written as

$$f(M) = \widetilde{M}^*, \tag{3.5}$$

the square of the map is the identity map

$$f\left(f(M)\right) = M. \tag{3.6}$$

This, together with the properties

$$f(\widetilde{M}) = -\tilde{f}(M), \quad f(M^*) = f(M)^*, \tag{3.7}$$

ensures the consistency of (3.1). Schrödinger used the Lagrangian (3.3) to construct a conserved, symmetric energy-momentum tensor. We have checked that, when suitably normalized, his energy-momentum tensor agrees with that of Born and Infeld up to an additive term proportional to $\eta_{\mu\nu}$.

Schrödinger's formulation is very clever and elegant and it has the advantage of being *manifestly* covariant under the duality rotation $M \to Me^{i\lambda}$

which is the finite form of (2.2). It is also likely that, as he seems to imply, his formulation is fully equivalent to the Born–Infeld theory (2.8), which would mean that the more restrictive equation (3.1) eliminates the remaining ambiguity in the solutions of (2.7). This virtue could actually be a weakness if one is looking for more general duality invariant theories.

4   General solution of the self-duality equation

The self-duality equation (2.5) can be solved in general as follows. Assuming Lorentz invariance in four-dimensional space-time, the Lagrangian must be a function of the two invariants

$$\alpha = \frac{1}{4}F^2, \quad \beta = \frac{1}{4}F\tilde{F}, \quad L = L(\alpha, \beta). \tag{4.1}$$

Now

$$\tilde{G} = 2\frac{\partial L}{\partial F} = L_\alpha F + L_\beta \tilde{F}, \quad G = -L_\alpha \tilde{F} + L_\beta F, \tag{4.2}$$

where we have used the standard notation $L_\alpha = \partial L/\partial \alpha$, $L_\beta = \partial L/\partial \beta$. Substituting these expressions in (2.5) we obtain

$$\left[(L_\beta)^2 - (L_\alpha)^2 + 1 \right] \beta + 2 L_\alpha L_\beta \alpha = 0. \tag{4.3}$$

This partial differential equation for L can be simplified by the change of variables

$$x = \alpha, \quad y = \left(\alpha^2 + \beta^2 \right)^{\frac{1}{2}}, \tag{4.4}$$

which gives

$$(L_x)^2 - (L_y)^2 = 1. \tag{4.5}$$

Alternatively one can use the variables

$$p = \frac{1}{2}(x + y), \quad q = \frac{1}{2}(x - y), \tag{4.6}$$

to obtain the form

$$L_p L_q = 1. \tag{4.7}$$

The equation (4.5), or (4.7), has been studied extensively in mathematics and there are several methods to obtain its general solution [10]. (It is interesting that the same equation occurs in a study of 5-dimensional Born–Infeld theory [11].) In our case we must also impose the physical boundary condition that the Lagrangian should approximate the Maxwell Lagrangian

$$L_M = -\alpha = -x = -p - q \tag{4.8}$$

when the field strength F is small.

According to one of the methods given in Courant–Hilbert, the general solution of (4.7) is given by

$$L = \frac{2p}{v'(s)} + v(s), \tag{4.9}$$

$$q = \frac{p}{[v'(s)]^2} + s, \tag{4.10}$$

where the arbitrary function $v(s)$ is determined by the initial values:

$$L(p = 0, q) = v(q), \tag{4.11}$$

$$L_p(p = 0, q) = \frac{1}{v'(q)}. \tag{4.12}$$

One must solve for $s(p, q)$ from (4.10) and substitute into (4.9). To verify [11] that these equations solve (4.7), differentiate (4.9) and (4.10):

$$dL = \frac{2dp}{v'} + \left(v' - \frac{2p}{[v']^2}v''\right)ds, \tag{4.13}$$

$$dq = \frac{dp}{v'^2} + \left(1 - \frac{2p}{[v']^3}v''\right)ds, \tag{4.14}$$

and eliminate ds between (4.13) and (4.14) to obtain

$$dL = \frac{1}{v'}dp + v'dq, \tag{4.15}$$

i.e.,

$$L_p = \frac{1}{v'}, \quad L_q = v', \quad L_pL_q = 1. \tag{4.16}$$

The condition that L should approach the Maxwell Lagrangian for small field strengths implies that

$$v(s) = L(p = 0, s) \cong -s \tag{4.17}$$

for small s.

It is trivial to check the above procedure for the Maxwell Lagrangian, and we shall not do it here. The Born–Infeld Lagrangian (with $g = 1$ for simplicity) is given by

$$L_{BI} = -\Delta^{\frac{1}{2}} + 1, \tag{4.18}$$

$$\Delta = (1 + 2p)(1 + 2q), \tag{4.19}$$

in terms of the variables p and q. Setting $p = 0$ we see that this corresponds to

$$v(s) = -(1 + 2s)^{\frac{1}{2}} + 1, \tag{4.20}$$

$$v'(s) = -(1 + 2s)^{-\frac{1}{2}}. \tag{4.21}$$

Then (4.10) gives

$$q = p(1 + 2s) + s, \qquad (4.22)$$

which is solved by

$$s = \frac{q - p}{1 + 2p}, \qquad 1 + 2s = \frac{1 + 2q}{1 + 2p}. \qquad (4.23)$$

Using (4.9), we reconstruct the Lagrangian

$$L_{BI} = -2p \left(\frac{1 + 2q}{1 + 2p}\right)^{\frac{1}{2}} - \left(\frac{1 + 2q}{1 + 2p}\right)^{\frac{1}{2}} + 1 = -[(1 + 2p)(1 + 2q)]^{\frac{1}{2}} + 1. \qquad (4.24)$$

Unfortunately, in spite of this elegant method for finding solutions of the self-duality equation, it seems very difficult to find new explicit solutions given in terms of simple functions. The reason is that, even for a simple function $v(s)$, solving the equation (4.10) for s gives complicated functions $s(p, q)$.

5   Axion, dilaton and $SL(2, \mathbf{R})$

It is well known that, if there are additional scalar fields which transform nonlinearly, the compact group duality invariance can be enhanced to a duality invariance under a larger noncompact group (see, e.g., [2] and references therein). In the case of the Born–Infeld theory, just as for Maxwell's theory, one complex scalar field suffices to enhance the $U(1) \cong SO(2)$ invariance to the $SU(1, 1) \cong SL(2, \mathbf{R})$ noncompact duality invariance. This is pointed out in [5], but it also follows from the considerations of our paper [2]. In the example under consideration, K is a single complex field, not an $n \times n$ matrix. In order to agree with today's more standard notation we shall use

$$S = iK = S_1 + iS_2 = a + ie^{-\phi}, \qquad S_2 > 0, \qquad (5.1)$$

where ϕ is the dilaton and a is the axion. For $SL(2, \mathbf{R}) \cong Sp(2, \mathbf{R})$, the matrices A, B, C, D are real numbers and $A = -D$, B and C are independent. Then the infinitesimal $SL(2, \mathbf{R})$ transformation is

$$\delta S = B + 2AS - CS^2, \qquad (5.2)$$

and the finite transformation is

$$S' = \frac{aS + b}{cS + d}, \qquad ad - bc = 1. \qquad (5.3)$$

For the $SO(2) \cong U(1)$ subgroup, $A = 0$, $C = -B = \lambda$,

$$\delta S = -\lambda - \lambda S^2. \qquad (5.4)$$

The scalar kinetic term, proportional to

$$\frac{\partial_\mu S^* \partial^\mu S}{(S - S^*)^2},$$ (5.5)

is invariant under the nonlinear transformation (5.2) which, in terms of S_1, S_2, takes the form

$$\delta S_1 = B + 2AS_1 - C\left(S_1^2 - S_2^2\right), \quad \delta S_2 = 2AS_2 - 2CS_1S_2.$$ (5.6)

Since the scalar kinetic term is separately invariant, we consider from now on the part $\hat{L}(S, F)$ of the Lagrangian that does not depend on the derivatives of S.

The full noncompact duality transformation on $F_{\mu\nu}$ is now

$$\delta G = AG + BF, \quad \delta F = CG + DF, \quad D = -A,$$ (5.7)

and we are seeking a Lagrangian $\hat{L}(S, F)$ which satisfies

$$\delta \hat{L} = \frac{1}{4}\left(FB\tilde{F} + GC\tilde{G}\right),$$ (5.8)

where

$$\delta \hat{L} = \delta F \frac{\partial \hat{L}}{\partial F} + \delta S_1 \frac{\partial \hat{L}}{\partial S_1} + \delta S_2 \frac{\partial \hat{L}}{\partial S_2},$$ (5.9)

and now

$$\tilde{G} = 2\frac{\partial \hat{L}}{\partial F}.$$ (5.10)

Equating (5.8) and (5.9) we see that $\hat{L}$ must satisfy

$$\frac{1}{4}\left(CG\tilde{G} - BF\tilde{F}\right) - \frac{1}{2}AF\tilde{G} + \delta S_1 \frac{\partial \hat{L}}{\partial S_1} + \delta S_2 \frac{\partial \hat{L}}{\partial S_2} = 0.$$ (5.11)

This equation can be solved as follows. Assume that $L(\mathcal{F})$ satisfies (2.1) and (2.5), i.e.

$$\mathcal{G}\tilde{\mathcal{G}} + \mathcal{F}\tilde{\mathcal{F}} = 0,$$ (5.12)

where

$$\tilde{\mathcal{G}} = 2\frac{\partial L}{\partial \mathcal{F}}.$$ (5.13)

For instance, the Born–Infeld Lagrangian $L(\mathcal{F})$ does this. Then

$$\hat{L}(S, F) = L(S_2^{\frac{1}{2}}F) + \frac{1}{4}S_1F\tilde{F}$$ (5.14)

satisfies (5.11). Indeed

$$\frac{\partial \hat{L}(S, F)}{\partial F} = \frac{\partial L}{\partial \mathcal{F}}S_2^{\frac{1}{2}} + \frac{1}{2}S_1\tilde{F}.$$ (5.15)

So

$$\tilde{G} = \tilde{\mathcal{G}} S_2^{\frac{1}{2}} + S_1 \tilde{F}, \tag{5.16}$$

$$G = \mathcal{G} S_2^{\frac{1}{2}} + S_1 F, \tag{5.17}$$

where we have defined

$$\mathcal{F} = S_2^{\frac{1}{2}} F, \tag{5.18}$$

and $\tilde{\mathcal{G}}$ is given by (5.13). Now

$$G\tilde{G} = \mathcal{G}\tilde{\mathcal{G}} S_2 + S_1^2 F \tilde{F} + 2 S_1 \mathcal{F} \tilde{\mathcal{G}}. \tag{5.19}$$

Using (5.12) in this equation we find

$$G\tilde{G} = \left(S_1^2 - S_2^2 \right) F \tilde{F} + 2 S_1 \mathcal{F} \tilde{\mathcal{G}}. \tag{5.20}$$

We also have

$$F\tilde{G} = \mathcal{F}\tilde{\mathcal{G}} + S_1 F \tilde{F}. \tag{5.21}$$

On the other hand, since

$$\frac{\partial L}{\partial S_2^{\frac{1}{2}}} = \frac{\partial L}{\partial \mathcal{F}} F = \frac{1}{2} \tilde{\mathcal{G}} F, \tag{5.22}$$

we obtain

$$\frac{\partial \hat{L}}{\partial S_2} = \frac{\partial L}{\partial S_2^{\frac{1}{2}}} \frac{1}{2} S_2^{-\frac{1}{2}} = \frac{1}{4} \tilde{\mathcal{G}} S_2^{-\frac{1}{2}} F = \frac{1}{4} \tilde{\mathcal{G}} \mathcal{F} S_2^{-1}. \tag{5.23}$$

In addition

$$\frac{\partial \hat{L}}{\partial S_1} = \frac{1}{4} F \tilde{F}. \tag{5.24}$$

Using (5.20), (5.21), (5.23) and (5.24), together with (5.6), we see that (5.11) is satisfied. It is easy to check that the scale invariant combinations $\mathcal{F}$ and $\mathcal{G}$, given by (5.18) and (5.13), have the very simple transformation law

$$\delta \mathcal{F} = S_2 C \mathcal{G}, \quad \delta \mathcal{G} = -S_2 C \mathcal{F}, \tag{5.25}$$

i.e., they transform according to the $U(1) \cong SO(2)$ compact subgroup just as F and G in (2.2), but with the parameter λ replaced by $S_2 C$. If $L(\mathcal{F})$ is the Born–Infeld Lagrangian, the theory with scalar fields given by $\hat{L}$ in (5.14) can also be reformulated à la Schrödinger. From (5.17) and (5.18) solve for $\mathcal{F}$ and $\mathcal{G}$ in terms of F, G, S_1 and S_2. Then $\mathcal{M} = \mathcal{F} - i\mathcal{G}$ must satisfy the same equation (3.1) that M does when no scalar fields are present.

6   Duality as a Legendre transformation

We have observed that, even in the general case of $Sp(2n, \mathbf{R})$, although the Lagrangian is not invariant, the combination [see (1.15)]

$$\hat{L} - \frac{1}{4} F \tilde{G} \tag{6.1}$$

is invariant. Here we restrict ourselves to the case of $SL(2, \mathbf{R})$, one tensor $F_{\mu\nu}$ and one complex scalar field $S = S_1 + i S_2$. As in Section 5, we use the notation $\hat{L}$ to denote the part of the Lagrangian that depends on the scalar fields, as well as on $F_{\mu\nu}$, but not on scalar derivatives. Then

$$\hat{L}(S_1, S_2, F) - \frac{1}{4} F \tilde{G} = \hat{L}(S_1', S_2', F') - \frac{1}{4} F' \tilde{G}', \tag{6.2}$$

where

$$\begin{pmatrix} G' \\ F' \end{pmatrix} = \begin{pmatrix} a & b \\ c & d \end{pmatrix} \begin{pmatrix} G \\ F \end{pmatrix}, \quad S' = \frac{aS + b}{cS + d}, \quad ad - bd = 1, \tag{6.3}$$

$$\tilde{G} = 2 \frac{\partial \hat{L}}{\partial F}. \tag{6.4}$$

There are several interesting special cases of this invariance statement. The first corresponds to $a = d = 1$, $c = 0$, b arbitrary, which gives

$$G' = G + bF, \quad F' = F, \quad S_1' = S_1 + b, \quad S_2' = S_2. \tag{6.5}$$

The second corresponds to $b = c = 0$, $d = 1/a$, a arbitrary, which gives

$$G' = aG, \quad F' = \frac{1}{a} F, \quad S' = a^2 S, \quad S_1' = a^2 S_1, \quad S_2' = a^2 S_2. \tag{6.6}$$

The third corresponds to $a = d = 0$, $b = -1/c$, c arbitrary, which gives

$$G' = -\frac{1}{c} F, \quad F' = cG, \quad S' = -\frac{1}{c^2 S}, \quad S_1' = -\frac{S_1}{c^2 |S|^2}, \quad S_2' = \frac{S_2}{c^2 |S|^2}. \tag{6.7}$$

Using (6.5) in (6.2) we find

$$\hat{L}(S_1, S_2, F) - \frac{1}{4} F \tilde{G} = \hat{L}(S_1 + b, S_2, F) - \frac{1}{4} F \left(\tilde{G} + b \tilde{F} \right). \tag{6.8}$$

Taking $b = -S_1$, we obtain

$$\hat{L}(S_1, S_2, F) = \hat{L}(0, S_2, F) + \frac{1}{4} S_1 F \tilde{F}, \tag{6.9}$$

which gives the dependence of $\hat{L}$ on S_1, in agreement with (5.14). This choice for the constant b is allowed because this part of the Lagrangian, which does not include the kinetic term for the scalar fields, does not contain derivatives of the scalar fields. Using (6.6) in (6.2) we find

$$\hat{L}(S_1, S_2, F) - \frac{1}{4}F\tilde{G} = \hat{L}\left(a^2 S_1, a^2 S_2, \frac{1}{a}F\right) - \frac{1}{4}F\tilde{G}, \qquad (6.10)$$

i.e.,

$$\hat{L}(S_1, S_2, F) = \hat{L}\left(a^2 S_1, a^2 S_2, \frac{1}{a}F\right). \qquad (6.11)$$

Setting $S_2 = 0$ in this equation, we see that $\hat{L}(S_1, 0, F)$ is a function of $S_1^{\frac{1}{2}}F$, in agreement with the more precise statement (6.9). Setting instead $S_1 = 0$, we find that $\hat{L}(0, S_2, F)$ is a function of $S_2^{\frac{1}{2}}F$, in agreement with (5.14).

Using (6.7) in (6.2) we find

$$\hat{L}(S_1, S_2, F) - \frac{1}{4}F\tilde{G} = \hat{L}\left(-\frac{S_1}{c^2|S|^2}, \frac{S_2}{c^2|S|^2}, cG\right) + \frac{1}{4}G\tilde{F}, \qquad (6.12)$$

i.e.,

$$\hat{L}\left(-\frac{S_1}{c^2|S|^2}, \frac{S_2}{c^2|S|^2}, cG\right) = \hat{L}(S_1, S_2, F) - \frac{1}{2}F\tilde{G}, \qquad (6.13)$$

or

$$\hat{L}\left(-\frac{1}{c^2 S}, cG\right) = \hat{L}(S, F) - \frac{1}{2}F\tilde{G}. \qquad (6.14)$$

We have shown that the Ansatz (5.14) of Section 5 is a natural consequence of the invariance of $\hat{L} - \frac{1}{4}F\tilde{G}$. Equation (6.14) with (6.4) can be interpreted as a Legendre transformation. Given a Lagrangian $\hat{L}(S, F)$, *define* the dual Lagrangian $\hat{L}_D(S, F_D)$, a function of the dual field F_D, by

$$\hat{L}_D(S, F_D) + \hat{L}(S, F) = \frac{1}{2}FF_D, \qquad (6.15)$$

$$F_D = 2\frac{\partial \hat{L}}{\partial F}, \quad F = 2\frac{\partial \hat{L}_D}{\partial F_D}. \qquad (6.16)$$

With these definitions, the dual of the dual of a function equals the original function.[1] In general, the dual Lagrangian is a very different function from the original Lagrangian. For a self-dual theory, if we set

$$F_D = \tilde{G}, \quad \tilde{F}_D = -G, \qquad (6.17)$$

we see from (6.14) that

$$-\hat{L}\left(-\frac{1}{c^2 S}, cG\right) = \hat{L}_D(S, \tilde{G}), \qquad (6.18)$$

---

[1] The unconventional factor $1/2$ on the right hand side of (6.15) is introduced to avoid overcounting when summing over the indices of the antisymmetric tensors F and F_D.

which must be independent of c, since G is.

The above argument can be inverted. Let the Legendre transformation (6.15) produce a dual Lagrangian given by (6.18) with $c = 1$, or

$$\hat{L}_D(S, F_D) = -\hat{L}\left(-\frac{1}{S}, -\tilde{F}_D\right) = -\hat{L}\left(-\frac{1}{S}, G\right). \tag{6.19}$$

It then follows that $\hat{L} - \frac{1}{4}F\tilde{G}$ is invariant under (6.7) with $c = 1$, i.e.,

$$G' = -F, \quad F' = G, \quad S' = -\frac{1}{S}. \tag{6.20}$$

If we now assume that it is also invariant under (6.5) with arbitrary b, it follows that it is invariant under the entire group $SL(2, \mathbf{R})$. Indeed, if we call t_b the transformation (6.5) and s the transformation (6.20), the product $t_b s t_{b'} s t_{b''}$ gives the most general transformation of $SL(2, \mathbf{R})$.

If we normalize the scalar field differently, taking e.g., instead of S,

$$\tau = cS, \quad L'(\tau, F) = \hat{L}(S, F), \tag{6.21}$$

$$L'_D(\tau, F_D') + L'(\tau, F) = \frac{1}{2c}FF_D', \tag{6.22}$$

and write the Legendre transformation as

$$2\frac{\partial L'(\tau, F)}{\partial F} = \frac{1}{c}F_D', \quad 2\frac{\partial L'_D(\tau, F_D')}{\partial F_D'} = \frac{1}{c}F, \tag{6.23}$$

we see that

$$F_D' = cF_D = c\tilde{G}, \tag{6.24}$$

and

$$
\begin{aligned}
L'_D(\tau, F_D') &= \hat{L}_D(S, F_D) = -\hat{L}\left(-\frac{1}{c^2 S}, -c\tilde{F}_D\right) \\
&= -L'\left(-\frac{1}{cS}, -c\tilde{F}_D\right) = -L'\left(-\frac{1}{\tau}, -\tilde{F}_D'\right),
\end{aligned} \tag{6.25}
$$

for a self-dual theory.

A standard normalization [12, 13] is $c = 4\pi$, in which case the expectation value of the field τ is

$$\langle\tau\rangle = \frac{\theta}{2\pi} + i\frac{4\pi}{g^2}. \tag{6.26}$$

In the presence of magnetically charged particles and dyons (both electrically and magnetically charged) the invariance of the charge lattice restricts [14] the $SL(2, \mathbf{R})$ group to the $SL(2, \mathbf{Z})$ subgroup generated by

$$\tau \to -\frac{1}{\tau}, \quad \tau \to \tau + 1. \tag{6.27}$$

At the quantum level the Legendre transformation corresponds to the integration over the field F in the functional integral, after adding to the Lagrangian $\hat{L}$ a term $-\frac{1}{2}FF_D$.

7 Concluding remarks

Nonlinear electromagnetic Lagrangians, like the Born–Infeld Lagrangian, can be supersymmetrized [15, 16] by means of the four-dimensional $N = 1$ superfield formalism, and this can be done even in the presence of supergravity. When the Lagrangian is self-dual, it is natural to ask whether its supersymmetric extension possesses a self-duality property that can be formulated in a supersymmetric way. We were not able to do this in the nonlinear case. When the Lagrangian is quadratic in the fields $F^a_{\mu\nu}$, the problem has been solved in [17], where the combined requirements of supersymmetry and self-duality were used to constrain the form of the weak coupling $(S_2 \to \infty)$ limit of the effective Lagrangian from string theory, in which one neglects the nonabelian nature of the gauge fields.

The $SL(2, \mathbf{Z})$ subgroup of $SL(2, \mathbf{R})$ that is generated by the elements $4\pi S \to -1/4\pi S$ and $S \to S + 1/4\pi$ relates different string theories [18] to one another.

The generalization of [2] to two-dimensional theories [19] has been used to derive the Kähler potential for moduli and matter fields in effective field theories from superstrings. In this case the scalars are valued on a coset space $\mathcal{K}/\mathcal{H}$, $\mathcal{K} \in SO(n, n)$, $\mathcal{H} \in SO(n) \times SO(n)$. The kinetic energy is invariant under $\mathcal{K}$, and the full classical theory is invariant under a subgroup of $\mathcal{K}$. String loop corrections reduce the invariance to a discrete subgroup that contains the $SL(2, \mathbf{Z})$ group generated by $T \to 1/T$, $T \to T - i$, where $\mathrm{Re}\,T$ is the squared radius of compactification in string units.

Acknowledgements

We are grateful for the hospitality provided by the Isaac Newton Institute where this work was initiated. We thank Gary Gibbons, David Jackson, Bogdan Morariu, David Olive, Harold Steinacker, Kelly Stelle and Peter West for inspiring conversations. This work was supported in part by the Director, Office of Energy Research, Office of High Energy and Nuclear Physics, Division of High Energy Physics of the US Department of Energy under Contract DE–AC03–76SF00098 and in part by the National Science Foundation under grant PHY–95–14797.

References

[1] S. Ferrara, J. Scherk and B. Zumino, *Nucl. Phys.* **B121:** 393 (1977); E. Cremmer and B. Julia, *Nucl. Phys.* **B159:** 141 (1979).

[2] M.K. Gaillard and B. Zumino, *Nucl. Phys.* **B193:** 221 (1981).

[3] B. Zumino, *Quantum Structure of Space and Time*, eds. M.J. Duff and C.J. Isham (Cambridge University Press) p. 363 (1982).

[4] G.W. Gibbons and D.A. Rasheed, *Nucl. Phys.* **B454**: 185 (1995).

[5] G.W. Gibbons and D.A. Rasheed, *Phys. Lett.* **B365**: 46 (1996).

[6] M.K. Gaillard and B. Zumino, Berkeley preprint LBNL-40370, UCB-PTH-97/29, hep-th/9705226 (1997), to be published in the memorial volume for D.V. Volkov.

[7] M. Born and L. Infeld, *Proc. Roy. Soc.* (London) **A144**: 425 (1934).

[8] E. Schrödinger, *Proc. Roy. Soc.* (London) **A150**: 465 (1935).

[9] P. Binétruy and M.K. Gaillard, *Phys. Rev.* **D32**: 931 (1985).

[10] R. Courant and D. Hilbert, *Methods of Mathematical Physics* **Vol. II**, Interscience (1962), p. 93 and Chapters I and II *passim*.

[11] M. Perry and J.H. Schwarz, *Nucl. Phys.* **B489**: 47 (1997).

[12] E. Witten, *Phys. Lett.* **B86**: 283 (1979).

[13] N. Seiberg and E. Witten, *Nucl. Phys.* **B426**: 19 (1994).

[14] See, *e.g.* D.I. Olive, *Nucl. Phys.* B (Proc. Suppl.) **46**: 1 (1996).

[15] S. Deser and R. Puzalowski, *J. Phys.* **A13**: 2501 (1980).

[16] S. Cecotti and S. Ferrara, *Phys. Lett.* **B187**: 335 (1987).

[17] P. Binétruy and M.K. Gaillard, *Phys. Lett.* **B365**: 87 (1996).

[18] J.H. Schwarz and A. Sen, *Phys. Lett.* **B312**: 105 (1993) and *Nucl. Phys.* **B411**: 35 (1994); M. Duff, *Nucl. Phys.* **B442**: 47 (1995); E. Witten, *Nucl. Phys.* **B443**: 85 (1995).

[19] S. Cecotti, S. Ferrara and L. Girardello, *Nucl. Phys.* **B308**: 436 (1988).

Supersymmetric σ-Models in 2 Dimensions

Bruno Zumino

1 The bosonic σ-model

I have been asked to give a brief introduction to supersymmetric σ-models in two space-time dimensions.

Let us recall first the properties of the bosonic σ-models. Let $x^\mu, \mu = 0, 1$ be the two-dimensional coordinates. The fields

$$\phi^i(x) \quad i = 1, \dots, N \tag{1.1}$$

are valued in a Riemannian manifold of metric $G_{ij}(\phi)$, called the target space. The action is

$$I = -\frac{1}{2} \int d^2x G_{ij}(\phi) \partial_\mu \phi^i(x) \partial^\mu \phi^j(x). \tag{1.2}$$

A simple example can be obtained by starting from the free theory

$$I = -\int d^2x \frac{1}{2} \partial_\mu \phi^a \partial^\mu \phi^a \quad a = 1, \dots, N+1 \tag{1.3}$$

and imposing the constraint

$$\sum \phi^a \phi^a = 1 \tag{1.4}$$

Solving for ϕ^{N+1} we obtain an action of the form (1.2) where

$$G_{ij}(\phi) = \delta_{ij} + \frac{\phi^i \phi^j}{1 - \Sigma \phi^k \phi^k}, \quad i, j, k = 1, \dots, N \tag{1.5}$$

is the metric on the sphere S^N.

Another well-known example is the chiral model given by the action

$$I = -\frac{1}{2} \int d^2x \operatorname{Tr} \partial_\mu U(x) \partial^\mu U^{-1}(x), \tag{1.6}$$

where U is a matrix representation of a (usually compact) Lie group and Tr denotes the trace. This also can be written in the form (1.2) by introducing parameters ϕ^i on the group manifold.

A change of coordinates on the Riemannian manifold $\phi^i \to \phi'^i, G_{ij}(\phi) \to G'_{ij}(\phi')$, where

$$\phi^i = \phi^i(\phi'), \tag{1.7}$$

$$G'_{ij}(\phi') = \frac{\partial \phi^k}{\partial \phi'^i} \frac{\partial \phi^l}{\partial \phi'^j} G_{kl}(\phi), \qquad (1.8)$$

leaves the action (1.2) invariant. This reparameterization invariance gives the action a geometric meaning. Isometries of the metric of the Riemannian manifold correspond to the internal symmetries of the model. Let us recall that an isometry is given in infinitesimal form by

$$\delta \phi^i = \xi^i(\phi), \qquad (1.9)$$
$$D_j \xi_k + D_k \xi_j = 0, \qquad (1.10)$$
$$D_j \xi_k = \partial_j \xi_k - \Gamma^i{}_{jk}(\phi)\xi_i, \qquad (1.11)$$

where $\Gamma^i{}_{jk}$ is the Levi-Civita connection

$$\Gamma^i{}_{jk} = \frac{1}{2} G^{il}(\partial_j G_{kl} + \partial_k G_{jl} - \partial_l G_{jk}). \qquad (1.12)$$

(1.10) is Killing's equation. It should be obvious which are the isometries for the above examples of the spherical model and of the chiral model.

It is often convenient to introduce light-cone coordinates in two dimensions

$$x^{\pm} = \frac{1}{\sqrt{2}}(x^0 \pm x^1). \qquad (1.13)$$

When written in light cone coordinates the action (1.2) is invariant under the transformation $x^{\pm} \to x'^{+}(x^{+}), x^{-} \to x'^{-}(x^{-})$ (in the Euclidean setting the light-cone coordinates become two complex conjugate coordinates z and $\bar{z}$ and the action is invariant under holomorphic transformations of these coordinates). A special case of this transformation is the scale transformation $x^{\mu} \to x'^{\mu} = \lambda x^{\mu}$ where λ is a constant.

One can add to the action (1.2) the interaction

$$I_{WZW} = -\frac{1}{2} \int d^2 x \varepsilon^{\mu\nu} B_{ij}(\phi) \partial_\mu \phi^i \partial_\nu \phi^j, \qquad (1.14)$$

where

$$\varepsilon_{01} = -\varepsilon_{10} = 1, \quad \varepsilon_{00} = \varepsilon_{11} = 0, \qquad (1.15)$$
$$B_{ij}(\phi) = -B_{ji}(\phi). \qquad (1.16)$$

This interaction is usually called the Wess–Zumino–Witten (WZW) term [1], [2]. The entire action is invariant under reparameterization of the field manifold, which can be written in infinitesimal form as

$$\delta B_{ij}(\phi) = D_i V_j(\phi) - D_j V_i(\phi)$$
$$= \partial_i V_j(\phi) - \partial_j V_i(\phi) \qquad (1.17)$$
$$\delta \phi^i = -V^i(\phi) \qquad (1.18)$$
$$\delta G_{ij}(\phi) = D_i V_j(\phi) + D_j V_i(\phi). \qquad (1.19)$$

The antisymmetric quantity B_{ij} may not transform like a tensor when we go from one coordinate patch to another on the field manifold, but may change by a gauge transformation (1.17). If the manifold has a nontrivial topology the WZW term may become multivalued and need to be multiplied by a quantized coefficient so that the exponentiated action in the functional path integral is single valued.

Let us write the total action in light-cone coordinates

$$I = \int dx^+ dx^- (G_{ij} - B_{ij})\partial_+\phi^i \partial_-\phi^j. \tag{1.20}$$

It is easy to see that the corresponding equations of motion are

$$\partial_+\partial_-\phi^i + (\Gamma^i{}_{jk} + H^i{}_{jk})\partial_+\phi^j \partial_-\phi^k = 0, \tag{1.21}$$

$$H_{ijk}(\phi) = \frac{1}{2}(\partial_i B_{jk} + \partial_j B_{ki} + \partial_k B_{ij}), \tag{1.22}$$

i.e.

$$\hat{D}_+\partial_-\phi^i = 0 \quad \text{or} \quad \hat{D}_-\partial_+\phi^i = 0, \tag{1.23}$$

where $\hat{D}_\pm$ are defined with the connection

$$\hat{\Gamma}^i_{\pm\ jk} = \Gamma^i{}_{jk} \pm H^i{}_{jk}, \tag{1.24}$$

which is not symmetric in j and k. The antisymmetric part of the connection is the torsion *tensor* (which in this case is totally antisymmetric in all three indices, when the upper index is lowered). One can say that the WZW term introduces torsion on the field manifold, the torsion being given by the curl of B_{ij} as in (1.22).

2 Supersymmetry in two dimensions

In Minkowski two-dimensional space-time, Majorana-Weyl spinors transform as

$$\psi^i_+ \quad \rightarrow \quad e^{\frac{1}{2}l}\psi^i_+, \tag{2.1}$$

$$\psi^i_- \quad \rightarrow \quad e^{-\frac{1}{2}l}\psi^i_- \tag{2.2}$$

under Lorentz transformations of parameter l. Here $+$ and $-$ denote the two different chiralities. Vectors, e.g. $v^i_\pm = \partial_\pm\phi^i$, transform as

$$v^i_+ \quad \rightarrow \quad e^l v^i_+, \tag{2.3}$$

$$v^i_- \quad \rightarrow \quad e^{-l} v^i_-. \tag{2.4}$$

The $(1,1)$ supersymmetry algebra has two fermionic generators Q_+ and Q_- which satisfy

$$Q_+^2 = P_+ = -i\partial_+, \quad Q_-^2 = P_- = -i\partial_-, \quad Q_+Q_- + Q_-Q_+ = 0. \qquad (2.5)$$

Other brackets vanish, e.g. $[Q_\pm, P_\pm] = 0$, etc. (In general, the notation (a, b) means that there are a femionic generators of positive and b of negative chirality.) A representation of the algebra in terms of fields can be obtained by introducing a superspace of bosonic coordinates x^+, x^- and fermionic (Grassmannian) coordinates θ_+, θ_-. A superfield can be expanded

$$\Phi(x^+, x^-, \theta_+, \theta_-) = \phi(x) + i\theta_-\psi_+(x) + i\theta_+\psi_-(x) + i\theta_+\theta_-F(x). \qquad (2.6)$$

where the coefficient (component) fields have obvious statistics and Lorentz properties. $F(x)$ is a nonpropagating auxiliary field, whose equations of motion are algebraic, i.e. contain no derivatives of $F(x)$. On superfields the supersymmetry generators are represented as

$$Q_+ = i\frac{\partial}{\partial\theta_-} - \theta_-\partial_+, \quad Q_- = i\frac{\partial}{\partial\theta_+} - \theta_+\partial_-. \qquad (2.7)$$

The "supercovariant" derivatives

$$\mathcal{D}_+ = i\frac{\partial}{\partial\theta_-} + \theta_-\partial_+, \quad \mathcal{D}_- = i\frac{\partial}{\partial\theta_+} + \theta_+\partial_- \qquad (2.8)$$

anticommute with both Q's and satisfy

$$\mathcal{D}_+^2 = i\partial_+, \quad \mathcal{D}_-^2 = i\partial_-, \quad \mathcal{D}_+\mathcal{D}_- + \mathcal{D}_-\mathcal{D}_+ = 0. \qquad (2.9)$$

The action can be written as a superspace integral

$$I = \int dx^+ dx^- d\theta_- d\theta_+ G_{ij}(\Phi)\mathcal{D}_+\Phi^i\mathcal{D}_-\Phi^j, \qquad (2.10)$$

where the Grassmannian integrals can be defined, according to Berezin, as

$$\int d\theta_+ \equiv \frac{\partial}{\partial\theta_+}, \quad \int d\theta_- \equiv \frac{\partial}{\partial\theta_-}. \qquad (2.11)$$

In order to find the ordinary space-time Lagrangian one must perform the integration over the Grassmannian variables and eliminate the auxiliary fields by using their equations of motion.

A convenient way to evaluate the action (2.10) in component form is to observe that the θ integrations, in the form of θ derivatives, can be replaced by supercovariant derivatives, because the additional x derivatives one introduces in this way integrate to zero upon x integration. So one can compute the Lagrangian as

$$L = [\mathcal{D}_+\mathcal{D}_-(G_{ij}\mathcal{D}_+\phi^i\mathcal{D}_-\phi^j)]_{\theta=0}. \qquad (2.12)$$

In doing the computation one must use the algebra (2.9) and the relations

$$[\Phi^i]_{\theta=0} = \phi^i, \quad [\mathcal{D}_\pm \Phi^i]_{\theta=0} = -\psi^i_\pm, \quad [\mathcal{D}_+\mathcal{D}_-\Phi^i]_{\theta=0} = -iF^i. \tag{2.13}$$

This method is especially convenient when the superfield satisfies supercovariant constraints, which is not the case here, but will be the case below (see Section 4, Eq. (4.10)).

The result for the Lagrangian (2.12) is

$$\begin{aligned} L &= G_{ij}\partial_+\phi^i\partial_-\phi^j + iG_{ij}(\psi^i_+ D_-\psi^j_+ + \psi^i_- \mathcal{D}_+\psi^j_-) + \\ &\quad +2iF_k\Gamma^k_{ij}\psi^i_+\psi^j_- + G_{ij}F^iF^j + \partial_k\partial_l G_{ij}\psi^k_+\psi^l_-\psi^i_+\psi^j_-, \end{aligned} \tag{2.14}$$

where Γ is the Levi-Civita connection and the covariant derivatives on the spinors are defined by

$$\mathcal{D}_-\psi^j_+ = \partial_-\psi^j_+ + \Gamma^j_{kl}\partial_-\phi^k\psi^l_+, \quad \mathcal{D}_+\psi^j_- = \partial_+\psi^j_- + \Gamma^j_{kl}\partial_+\phi^k\psi^l_-. \tag{2.15}$$

Using standard formulas of Riemannian geometry the last three terms in (2.14) can be rearranged to give

$$\begin{aligned} L &= G_{ij}\partial_+\phi^i\partial_-\phi^j + iG_{ij}(\psi^i_+ D_-\psi^j_+ + \psi^i_- D_+\psi^j_-) \\ &\quad +\frac{1}{2}R_{ijkl}\psi^i_+\psi^j_+\psi^k_-\psi^l_- \\ &\quad +G_{ij}(F^i + i\Gamma^i_{kl}\psi^k_+\psi^l_-)(F^j + i\Gamma^j_{mn}\psi^m_+\psi^n_-). \end{aligned} \tag{2.16}$$

In either form (2.14) or (2.16) we see that the equations of motion for the auxiliary fields are

$$F^i + i\Gamma^i_{kl}\psi^k_+\psi^l_- = 0. \tag{2.17}$$

Using these equations the Lagrangian reduces to the first three terms in (2.16), which have an obvious geometric meaning, and the action is given by [3]

$$\begin{aligned} I &= \int dx^+ dx^- [G_{ij}(\partial_+\phi^i\partial_-\phi^j + i\psi^i_+ D_-\psi^j_+ + i\psi^i_- D_+\psi^j_-) \\ &\quad +\frac{1}{2}R_{ijkl}\psi^i_+\psi^j_+\psi^k_-\psi^l_-]. \end{aligned} \tag{2.18}$$

This model is invariant under supersymmetry transformations which, in superspace, are given by

$$\delta\Phi^i = -i(\varepsilon_- Q_+ + \varepsilon_+ Q_-)\Phi^i, \tag{2.19}$$

where ε_- and ε_+ are Grassmannian infinitesimal parameters. The action (2.10) was constructed with supercovariant derivatives so that this would be true. In terms of component fields the transformations become

$$\begin{aligned} \delta\phi^i &= i\varepsilon_-\psi^i_+ + i\varepsilon_+\psi^i_-, \tag{2.20} \\ \delta\psi^i_+ &= -\varepsilon_-\partial_+\phi^i - \varepsilon_+ F^i, \quad \delta\psi^i_- = -\varepsilon_+\partial_-\phi^i + \varepsilon_- F^i, \tag{2.21} \\ \delta F^i &= -i\varepsilon_-\partial_+\psi_- + i\varepsilon_+\partial_-\psi_+. \tag{2.22} \end{aligned}$$

The algebra of these transformations closes: the commutator of two transformations δ and δ' of parameters $\varepsilon_\pm$ and $\varepsilon'_\pm$ is a two-dimensional translation of parameter $2i\varepsilon'_-\varepsilon_-, 2i\varepsilon'_+\varepsilon_+$. For instance,

$$(\delta'\delta - \delta\delta')\phi^i = 2i\varepsilon'_-\varepsilon_-\partial_+\phi^i + 2i\varepsilon'_+\varepsilon_+\partial_-\phi^i. \qquad (2.23)$$

One can shorten (2.20)–(2.21) by replacing the auxiliary fields by their value from the equations of motion, i.e. taking

$$\delta\psi^i_\pm = -\varepsilon_\mp\partial_\pm\phi^i \pm \varepsilon_\pm i\Gamma^i{}_{kl}\psi^k_+\psi^l_-. \qquad (2.24)$$

The action is still invariant under (2.20) with (2.24). However now the algebra closes only on the mass shell, i.e. by use of the spinor fields equations of motion.

For certain geometries of the target space, the $(1,1)$ supersymmetry can be enlarged to a $(2,2)$ supersymmetry (when the Riemannian manifold is a complex Kähler manifold) or even to a $(4,4)$ supersymmetry (when it is a hyper-Kähler manifold). This is discussed in the next two sections.

3   Complex manifolds

In this section I discuss briefly those properties of complex manifolds which will be needed later [4]. An almost complex structure on a real $2n$-dimensional differentiable orientable manifold is a tensor field $J^i_j(\phi)$ such that

$$J^i{}_k J^k{}_j = -\delta^i{}_j. \qquad (3.1)$$

A manifold endowed with an almost complex structure is called an almost complex manifold. One defines the Nijenhuis torsion of J to be the tensor field

$$N^i{}_{jk} = 2(J^h_j\partial_h J^i{}_k - J^h_k\partial_h J^i{}_j - J^i{}_h\partial_j J^h{}_k + J^i{}_h\partial_k J^h{}_j) = -N^i{}_{kj}, \qquad (3.2)$$

where ∂_h is the ordinary partial derivative. It is remarkable that N is actually a tensor. When N vanishes, J is called a complex structure and the manifold a complex manifold. In this case it is possible to define in local patches n complex coordinates ϕ^α and their complex conjugates $\overline{\phi}^\alpha$ such that $J\phi = -i\phi$, $J\overline{\phi} = i\overline{\phi}$. The transition functions between different coordinate patches are holomorphic functions of the coordinates ϕ (anti-holomorphic functions of $\overline{\phi}$).

If the $2n$-dimensional manifold is a Riemannian manifold one can require the complex structure to satisfy

$$G_{ij}J^i{}_k J^j{}_l = G_{kl}, \qquad (3.3)$$

(invariance of the metric) and

$$D_k J^i{}_j \equiv \partial_k J^i{}_j + \Gamma^i{}_{lk} J^l{}_j - J^i{}_l \Gamma^l{}_{jk} = 0, \qquad (3.4)$$

(the tensor J is covariantly constant) where Γ is the Levi-Civita connection. With these conditions the complex manifold is called a Kähler manifold. Notice that the vanishing of the Nijenhuis torsion (3.2) follows from (3.4) and the symmetry of Γ, i.e. the absence of torsion on the Riemannian manifold.

The Riemann tensor can be defined by considering the action of the commutator of two covariant derivatives on tensors. Using this definition and (3.4), one can easily show that the complex structure satisfies

$$R_{ijkl} J^k{}_m J^l{}_n = R_{ijmn} = R_{klmn} J^k{}_i J^l{}_j. \qquad (3.5)$$

In terms of the complex coordinates (3.3) implies that the metric satisfies

$$\begin{aligned}
G_{\alpha\beta}(\phi, \bar{\phi}) &= G_{\bar{\alpha}\bar{\beta}}(\phi, \bar{\phi}) = 0, \\
G_{\alpha\bar{\beta}}(\phi, \bar{\phi}) &= G_{\bar{\beta}\alpha}(\phi, \bar{\phi}),
\end{aligned} \qquad (3.6)$$

while from (3.4) one can derive that the two-form

$$\Omega = -2i G_{\alpha\bar{\beta}} d\phi^\alpha \wedge d\overline{\phi^\beta} \qquad (3.7)$$

is closed, i.e.

$$\partial_\alpha G_{\beta, \bar{\gamma}} = \partial_\beta G_{\alpha\bar{\gamma}}, \quad \partial_{\bar{\alpha}} G_{\beta, \bar{\gamma}} = \partial_{\bar{\gamma}} G_{\beta\bar{\alpha}}, \qquad (3.8)$$

where $\partial_\alpha = \partial/\partial\phi^\alpha$, $\partial_{\bar{\alpha}} = \partial/\partial\overline{\phi^\alpha}$. These equations imply that one can find locally a real function K such that

$$G_{\alpha\bar{\beta}}(\phi, \bar{\phi}) = \partial_\alpha \partial_{\bar{\beta}} K(\phi, \bar{\phi}). \qquad (3.9)$$

Ω is called the Kähler form, and K is called the Kähler potential. K does not transform like a scalar from one patch to another and is defined only up to so-called Kähler transformations

$$K(\phi, \bar{\phi}) \rightarrow K(\phi, \bar{\phi}) + f(\phi) + \overline{f(\phi)} \qquad (3.10)$$

which leave the metric invariant. The Kähler potentials in two patches will in general be connected by a suitable Kähler transformation.

If a Kähler manifold admits two complex structures J^1 and J^2 which anticommute

$$(J^1)^i{}_k (J^2)^k{}_j + (J^2)^i{}_k (J^1)^k{}_j = 0, \qquad (3.11)$$

then the product $J^3 = J^1 J^2$ satisfies all conditions (3.1), (3.3) and (3.4) and is also a complex structure. For instance

$$(J^3)^2 = J^1 J^2 J^1 J^2 = -J^1 J^1 J^2 J^2 = -1. \qquad (3.12)$$

Furthermore

$$J^2 J^3 = J^2 J^1 J^2 = -J^1 J^2 J^2 = J^1 = -J^3 J^2$$
$$J^3 J^1 = J^1 J^2 J^1 = -J^2 J^1 J^1 = J^2 = -J^1 J^3. \tag{3.13}$$

So the tensors J^1, J^2 and J^3 satisfy (by matrix multiplication) the hyper-complex algebra of the quaternion units. The manifold is called a hyper-Kähler manifold; its real dimension is necessarily a multiple of four.

4   The Kähler and hyper-Kähler case

If the scalar fields of a supersymmetric σ-model are valued in a Kähler manifold, the Lagrangian of (2.18) is invariant under the transformation

$$\phi \to \phi' = \phi, \quad \psi_+ \to \psi'_+ = J\psi_+, \quad \psi_- \to \psi'_- = -J\psi_-, \tag{4.1}$$

where we have omitted the target space indices. This is easy to see using the equations (3.3) to (3.5). As a consequence one can define a second supersymmetry transformation

$$\delta'\phi = i\varepsilon'_- \psi'_+ + i\varepsilon'_+ \psi'_-, \quad \delta'\psi'_\pm = -\partial_\pm \phi \varepsilon'_\mp \mp F'\varepsilon'_\pm, \tag{4.2}$$

where the value of the new auxiliary field is taken to be

$$F'^k = -i\Gamma^k_{lm} \psi'^l_+ \psi'^m_-. \tag{4.3}$$

With some algebra one can check that the two supersymmetry transformations (2.20) with (2.24) and (4.2) commute (on the mass shell of the auxiliary fields). Thus the $(1,1)$ supersymmetry is enlarged to a $(2,2)$ supersymmetry [5].

If one does not want to use the equation of motion of the auxiliary fields one can complete the transformation (4.1) by

$$F^k \to F'^k = F^k + iM^k_{lm}\psi^l_+ \psi^m_-, \tag{4.4}$$

where

$$M^k_{lm}[J] = (\partial_l J^k_n - \partial_n J^k_l)J^n_m, \tag{4.5}$$

and (4.2) by

$$\delta'F' = -i\varepsilon'_- \partial_+ \psi'_- + i\varepsilon'_+ \partial_- \psi'_+. \tag{4.6}$$

Using (4.1) and (4.4), the two equations of motion (2.17) and (4.3) go into each other.

Notice that the inverse of the relation (4.4) between F and F' has the same form

$$F^k = F'^k + iM^k_{lm}\psi'^l_+ \psi'^m_-. \tag{4.7}$$

Also, comparing with (3.2), we see that

$$M^k{}_{lm} - M^k{}_{ml} = \frac{1}{2}N^k{}_{lm} \tag{4.8}$$

which vanishes for a complex structure. Therefore $M^k{}_{lm}$ is symmetric in l, m.

The $(2, 2)$ model can be formulated more symmetrically by introducing from the beginning all four supersymmetry generators. This was indeed the first example of a geometric formulation of nonlinear σ-models [6]. In a superspace of coordinates x^+, x^-, θ_+, θ_-, $\overline{\theta_+}$, $\overline{\theta_-}$, the generators and the supercovariant derivatives satisfy

$$[Q_\pm, Q_\pm]_+ = [Q_\pm, Q_\mp]_+ = 0 \quad [Q_\pm, \overline{Q_\pm}]_+ = 2P_\pm.$$

$$Q_\pm = i\frac{\partial}{\partial\overline{\theta_\mp}} - \theta_\mp\partial_\pm, \quad \overline{Q_\pm} = i\frac{\partial}{\partial\theta_\mp} - \overline{\theta_\mp}\partial_\pm,$$

$$\mathcal{D}_\pm = i\frac{\partial}{\partial\overline{\theta_\mp}} + \theta_\mp\partial_\pm, \quad \overline{\mathcal{D}_\pm} = i\frac{\partial}{\partial\theta_\mp} + \overline{\theta_\mp}\partial_\pm. \tag{4.9}$$

A "chiral" superfield (necessarily complex) satisfies

$$\Phi(x^+, x^-, \theta_+, \theta_-, \overline{\theta_+}, \overline{\theta_-}), \quad \mathcal{D}_\pm\bar{\Phi} = 0, \quad \overline{\mathcal{D}_\pm}\Phi = 0. \tag{4.10}$$

These constraints can be solved by observing that the four combinations $y^\pm = x^\pm + i\theta_\mp\overline{\theta_\mp}$ and $\overline{\theta_\mp}$ are annihilated by $\overline{\mathcal{D}_\pm}$, so that we can take

$$\Phi(y^+, y^-, \overline{\theta_+}, \overline{\theta_-}) = \phi(y^+, y^-) + i\overline{\theta_+}\psi_-(y^+, y^-) + i\overline{\theta_-}\psi_+(y^+, y^-)$$
$$+ i\overline{\theta_+}\,\overline{\theta_-}F(y^+, y^-). \tag{4.11}$$

The superspace action takes the extremely simple form

$$I = \int dx^+ dx^- d\theta_+ d\theta_- d\overline{\theta_+} d\overline{\theta_-} K(\Phi^\alpha, \overline{\Phi^\alpha})$$
$$= \int dx^+ dx^- G_{\alpha\bar{\beta}}(\partial_+\phi^\alpha\partial_-\overline{\phi^\beta} + \partial_-\phi^\alpha\partial_+\overline{\phi^\beta}) + \cdots,$$
$$G_{\alpha\bar{\beta}} = \partial_\alpha\partial_{\bar{\beta}}K, \tag{4.12}$$

where the dots denote additional terms involving also fermions (see [6]). It is easy to see that the action is invariant under the Kähler transformation (3.10). It should be noticed that (4.9) to (4.12) parallel very closely the formulas for the $N = 1$ (four Majorana components) supersymmetry in four space-time dimensions. Indeed the present model can be obtained by dimensional reduction from the four-dimensional model, simply by assuming that all fields are independent of the two space coordinates x^2 and x^3. This is especially clear if the four-dimensional theory is formulated using the van der Waerden two-component spinor notation.

If the manifold admits more than one complex structure, one can define more supersymmetries. For the hyper-Kähler case

$$\begin{aligned}
\delta^a \phi &= i\varepsilon_-^a \psi_+^a + i\varepsilon_+^a \psi_-^a, \\
\delta^a \psi_\pm^a &= \mp\varepsilon_\pm^a F^a - \varepsilon_\mp^a \partial_\pm \phi, \\
\delta^a F^a &= -i\varepsilon_-^a \partial_+ \psi_-^a + i\varepsilon_+^a \partial_- \psi_+^a,
\end{aligned} \tag{4.13}$$

where $a = 1, 2, 3$. Here

$$\begin{aligned}
\psi_\pm^a &= \pm J^a \psi_\pm, \\
F^a &= F + iM[J^a]\psi_+ \psi_-,
\end{aligned} \tag{4.14}$$

the omitted target space index structure being obvious. The three new auxiliary fields are related by

$$F^{1k} + F^{2k} = i(\partial_l J_m^{3k} - \partial_m J_l^{3k})\psi_+^{2m}\psi_-^{1l} \tag{4.15}$$

plus the two equations obtained by rotating cyclically the indices $1, 2, 3$. The three supersymmetries δ^a commute with each other and with the original δ of (2.20) to (2.22). Thus the algebra of the four super-generators is now

$$[Q^a, \overline{Q}^b]_+ = 2\delta^{ab} \, \mathcal{P}, \tag{4.16}$$

where $a = 0, 1, 2, 3$ and Q^0 is the generator of the original supersymmetry δ. The $(1, 1)$ supersymmetry has been enhanced to a $(4, 4)$ supersymmetry [5].

Just as a natural way to understand the Kähler $N = 2$ supersymmetry model in two dimensions is to relate it to the $N = 1$ model in four dimensions [6], the $N = 4$ model in two dimensions can be obtained by dimensional reduction from the $N = 1$ in six, or the $N = 2$ in four. The $N = 1$ in six dimensions has been studied in [7] [8], where a more detailed description of hyper-Kähler manifolds was given than we have done here, and the six dimensional Lagrangian (on the auxiliary shell, no other is known) was written out explicitly. Various dimensional reductions can be obtained from it.

5 Chiral supersymmetries

In two dimensions one can formulate various supersymmetric models which have only supersymmetry generators of given chirality [10]. For instance, using a superspace of coordinates x^+, x^- and θ_- and using the superfield

$$\Phi(x^+, x^-, \theta_-) = \phi(x^+, x^-) + i\theta_- \psi_+(x^+, x^-), \tag{5.1}$$

one can write the action for the chiral model $(1, 0)$ with torsion

$$\begin{aligned}
I &= \int dx^+ dx^- d\theta_- (G_{ij}(\Phi) - B_{ij}(\Phi))\mathcal{D}_+ \Phi^i \partial_- \Phi^j \\
&= \int dx^+ dx^- [(G_{ij}(\phi) - B_{ij}(\phi))\partial_+ \phi^i \partial_- \phi^j + iG_{ij}(\phi)\psi_+^i \hat{D}_- \psi_+^j], \\
\hat{D}_- \psi_+^j &= \partial_- \psi_+^j + \hat{\Gamma}^j_{-kl}\partial_- \phi^k \psi_+^l,
\end{aligned} \tag{5.2}$$

where the connection is $\hat{\Gamma}^j_{-\,kl} = \Gamma^j{}_{kl} - H^j{}_{kl}$ as in (1.24).

In absence of torsion, if the manifold admits a complex structure and is Kähler, the $(1,0)$ model can be elevated to a $(2,0)$ model by a procedure analagous to that we employed above to go from $(1,1)$ to $(2,2)$. In the presence of torsion the complex structure must satisfy

$$\hat{D}_i J^j{}_k(\phi) = 0, \tag{5.3}$$

where $\hat{D}_i$ is the covariant derivative with torsion. It is easy to see that in complex coordinates this implies $\partial_\beta G_{\alpha\bar\gamma} - \partial_\alpha G_{\beta\bar\gamma} = -2H_{\alpha\beta\bar\gamma}$, $H_{\alpha\beta\gamma} = 0$, so that the manifold is not Kähler (which would require $H_{\alpha\beta\bar\gamma} = 0$). With a little algebra one can show that the metric can be written locally as $G_{\alpha\bar\beta} = \partial_\alpha \bar{X}_{\bar\beta} + \partial_{\bar\beta} X_\alpha$. The $(2,0)$ model with torsion is specified by the superspace action [11]

$$I = \frac{i}{2} \int dx^+ dx^- d\theta_- d\overline{\theta_-} (X_\alpha \partial_- \Phi^\alpha - \bar{X}_{\bar\beta} \partial_- \bar\Phi^{\bar\beta}) \tag{5.4}$$

(for the Kähler case, i.e. no torsion, $X_\alpha \propto \partial_\alpha K(\Phi, \bar\Phi)$), and the constraints

$$\begin{aligned}
\mathcal{D}_+ \bar\Phi^{\bar\beta} &= \overline{\mathcal{D}_+} \Phi^\alpha = 0, \\
\mathcal{D}_+ &= i\frac{\partial}{\partial\overline{\theta_-}} + \theta_- \partial_+, \quad \overline{\mathcal{D}_+} = i\frac{\partial}{\partial\theta_-} + \overline{\theta_-}\partial_+, \\
\{\mathcal{D}_+, \overline{\mathcal{D}_+}\} &= 2i\partial_+.
\end{aligned} \tag{5.5}$$

The $(1,0)$ model can be coupled to a fermionic multiplet of negative chirality [10], by using the spinor superfield

$$\Lambda_-(x^+, x^-, \theta_-) = \lambda_-(x^+, x^-) + \theta_- F(x^+, x^-). \tag{5.6}$$

One can make the model more interesting by assuming that this superfield belongs to a representation V of a gauge group, $\Lambda^A_-(A = 1, \ldots, P)$, with gauge field $A_i{}^A{}_B(\Phi)$, and introducing a metric $G_{AB}(\Phi)$. Then we can add to the action given in (5.2) the action

$$I_V = -i \int dx^+ dx^- d\theta_- G_{AB}(\Phi) \Lambda^A_- (\mathcal{D}_+ + A_+)^B_C \Lambda^C_-, \quad (A_+)^B_C = A_i{}^B{}_C \mathcal{D}_+ \Phi^i. \tag{5.7}$$

In component form the total action is

$$\begin{aligned}
I = \int dx^+ dx^- [&(G_{ij} - B_{ij})\partial_+\phi^i \partial_-\phi^j + iG_{ij}\psi^i_+ \hat{D}_-\psi^j_+ \\
&+ iG_{AB}\lambda^A_- \nabla_+ \lambda^B_- + \frac{1}{2}F_{ijAB}\psi^i_+ \psi^j_+ \lambda^A_- \lambda^B_-],
\end{aligned} \tag{5.8}$$

where the field strength and gauge covariant derivative are

$$F_{ijAB} = (\partial_i \hat{A}_j - \partial_j \hat{A}_i + [\hat{A}_i, \hat{A}_j]_-)_{AB} \tag{5.9}$$

$$\hat{A}_i{}^A{}_B = A_i{}^A{}_B + \frac{1}{2}G^{AC}G_{BC,i}, \quad \nabla_+{}^B{}_C = \partial_+ \delta^B_C + A_i{}^B{}_C \partial_+ \phi^i. \tag{5.10}$$

In general G_{ij}, B_{ij} and $A_i{}^A_B$ are independent external fields. In the particular case in which the gauge group is the structure group of the field manifold (tangent space group) the indices $A, B, \ldots$ becomes the tangent space indices $a, b, \ldots, A_i{}^a_b \equiv \omega_i{}^a_b$ and $F_{ij}{}^a_b \equiv R_{ij}{}^a_b$, so the present model reduces to the model $(1,1)$ with torsion described earlier, but now in tangent space notation.

There exists also an interesting modification of the $(2,2)$ model described earlier, where the superfield satisfies the chirality constraints (4.10). It is the $(2,2)$ twisted model [14] where the superfield X satisfies the twisted chirality constraints

$$\mathcal{D}_+X = \overline{\mathcal{D}_-}X = 0 \quad \overline{\mathcal{D}_+}\bar{X} = \mathcal{D}_-\bar{X} = 0. \tag{5.11}$$

A general classification of $(2,2)$ models with torsion is given in [9], where an off shell formulation can be found.

6 Concluding remarks

The qualitative lesson to be learned from the above discussion is that the number of supersymmetries is intimately related to the geometric structure of the target space manifold: more geometric structure corresponds to more supersymmetries.

The list of publications in the bibliography of the present article is very incomplete. A variety of other supersymmetric σ-models can be constructed (see, *e.g.*, [12], [13]). An excellent review up to 1986 is that of S. Mukhi [15] which contains numerous references. He also discusses the cancellations of ultraviolet divergences due to supersymmetry and some of the relevance to string theory. After 1986 the literature on the subject has exploded and no comprehensive review is known to me. I would like to mention only two relatively more recent references, [16] and [17], on the ultraviolet divergences respectively of the $(4,0)$ model with torsion and of the $(2,2)$ model without torsion. These papers also contain numerous references.

Acknowledgements

I am very grateful to Bogdan Morariu and Laura Scott for considerable help. This work was supported in part by the Director, Office of Energy Research, Office of High Energy and Nuclear Physics, Division of High Energy Physics of the US Department of Energy under DE–AC03–76SF00098 and in part by the National Science Foundation under grant PHY–95–14797.

References

[1] Wess, J. and Zumino, B., *Phys. Lett.* **37B**, 95 (1971).

[2] Witten, E., *Comm. Math. Phys.* **92**, 455 (1984).

[3] Freedman, D.Z., and Townsend, P.K., *Nucl. Phys.* **B177**, 433 (1981).

[4] Kobayashi, S., and Nomizu, K., *Foundations of Differential Geometry, Vol. II, Interscience* (1969).

[5] Alvarez-Gaumé, L., and Freedman, D.Z., *Comm. Math. Phys.* **80**, 443 (1981).

[6] Zumino, B., *Phys. Lett.* **87B**, 203 (1979).

[7] Rozansky, L., and Witten, E., IASSN-HEP-96/128, hep-th/9612216. *Selecta Math (N.S.)*, **3**(3), 401–458, (1997).

[8] Figueroa-O'Farrill, J.M., Köhl, C., and Spence, B., *Nucl. Phys.* **B503** [PM], 614 (1997).

[9] Sevrin, A., and Troost, J., *Nucl. Phys.* **B492** [PM], 623 (1997); VUB/TENA/96/07, hep-th/9610103.

[10] Hull, C.M., and Witten, E., *Phys. Lett.* **160B**, 398 (1985).

[11] Brooks, R., Muhammad, F., and Gates, S.J., *Nucl. Phys.* **B268**, 599 (1986).

[12] Curtwright, T., and Zachos, C.K., *Phys. Rev. Lett.* **53**, 1799 (1984).

[13] Hull, C.M., *Nucl. Phys.* **B267**, 266 (1986).

[14] Gates, S.J., Hull, C.M., and Rocek, M., *Nucl. Phys.* **B248**, 157 (1984).

[15] Mukhi, S., in Proceedings of the Panchgani Winter 1986 Indian School in Theoretical Physics, V. Singh and S. Wadia, eds., World Scientific, pp. 111-144 (1987).

[16] Howe, P.S., and Papadopoulos, G., *Nucl. Phys.* **B381**, 360 (1992).

[17] Jack, I., Jones, D.R.T., and Panvel, J., *Int. J. Mod. Phys.* **A8**, 2591 (1993).

Introduction to Duality

David I. Olive

1 Introduction

Electromagnetic duality is an idea with a long pedigree that addresses a number of old questions in theoretical physics, for example:

- Why does space-time possess four dimensions?

- Why is electric charge quantised?

- What is the origin of mass?

- What is the internal structure of the elementary particles?

- How are quarks confined?

etc.

During the last forty years our understanding of quantum field theory, whilst remaining incomplete, has advanced dramatically on a number of fronts that, at first sight, appeared unrelated:

- Unified gauge theories with Higgs

- Supersymmetry

- Instanton theory

- The theory of solitons

- The idea of integrable quantum field theories as deformations of conformally invariant QFTs, ...

and so on. As I shall explain, the old idea of electromagnetic duality can be considerably enhanced in the light of these developments. The bonus is a compelling framework of ideas within which these apparently disparate developments become much more unified. Despite these advantages, and some initial enthusiasm and progress, the ideas were largely resisted for a number of years. Probably the main reason was that the duality transformation interchanges weak and strong coupling regimes of a quantum field theory, thereby exchanging a regime in which calculations can be performed easily by perturbative methods with one in which calculation is difficult and traditionally unreliable. Nevertheless the duality conjecture implies that these regimes are similar and it is this that seemed so unreasonable.

Although the idea remains unproven, the importation of powerful new mathematical techniques into quantum field theory, such as the Atiyah–Singer index theorem, hyper-Kähler geometry etc, have stimulated the discovery of strong supporting evidence for the idea of duality. Now that the idea is accepted it is ironic that one of its features that appears most attractive and indeed is most exploited is this very insight provided concerning the strong coupling regime.

The new vantage point on elementary particle theory then provides a springboard for extensions of the basic ideas to more ambitious theories such as supergravity and the various versions of string theory. This will be covered by many of the other lecturers.

Let me begin by elaborating three of the themes that will ultimately be woven together.

Theme I – Quantum excitations and solitons Because the principles of special relativity and quantum mechanics are so soundly based in the physics of the present century, any theory of particle physics must combine the two in a seamless way. The most promising approach involves relativistic quantum field theory, even though after more than sixty years of development it has to be admitted that it remains not totally satisfactory. Traditionally, particles have appeared as the quantum excitations of the fields entering the equations of motion. However, there is another, conceptually distinct notion of how a particle can appear in a field theory. This is as a soliton, originally a concept in hydrodynamic theory. This is a classical solution to the field equations possible when they exhibit a particular sort of intense nonlinearity. When stationary, its energy concentrates in a small region of space and does not dissipate as time evolves. It can be boosted to move at any speed slower than light. Its stability is often associated with the fact that it carries some knottedness property that cannot be unravelled. All this enables it to be interpreted as a particle with structure.

Is it possible that the distinction between these two apparently different notions of particle phenomena dissolves when the theory is fully quantised? Skyrme [1961] was probably the first to ask this question and he found that in a certain model, in a space-time of two dimensions, where the question was tractable, that the distinction did indeed disappear. He and Coleman [1975] showed that the soliton solution of the sine-Gordon equation describing a single scalar field could be considered equivalently as the quantum fluctuation of what is known as the massive Thirring field. Indeed the two actions, sine-Gordon and massive Thirring, are quantum equivalent. This was established by a version of what is now known as the vertex operator construction (Mandelstam [1975]).

Theme II – Unified Theories and Conformal Symmetry   With the increasing flow of data from high energy particle accelerators, so has grown the urge to construct unified theories that codify the observed patterns of particle behaviour in a simple mathematical structure. Happily, the most successful theories have been based on principles of an aesthetic and geometrical nature. Examples are the Yang–Mills gauge theories generalising Maxwell's equations in a non-linear way, and Einstein's theory of general relativity, relating gravitation to the Riemannian geometry of space-time. These theories have two distinctive features, nonlinearity and conformal symmetry. Unifying them are the string theories in their various guises and these also display a conformal symmetry, this time on the world sheet. Conformal transformations in space-time are those that preserve angles but not necessarily distances. Hence a conformally invariant field theory will exclude particle masses. This is perfectly acceptable for the photon, the quantum of the Maxwell field, but not for the more general Yang–Mills equations, as the intermediate bosons W^+ and W^- are known to be massive. Therefore a mechanism must be found that destroys the conformal symmetry without corrupting too much of the aesthetic attraction of the theory.

Two possibilities are known, one now more than thirty years old, the Higgs mechanism whereby it is the vacuum, rather than the equations of motion, that fail to respect all the gauge symmetry. As a consequence certain scalar fields, called Higgs fields, may fail to vanish there, and so produce masses for those gauge particles associated with generators of the gauge group not annihilating the vacuum. A second type of mechanism, due to Zamololodchikov [1989], came to light ten years ago. This insight came from the study of second-order phase transitions in two dimensions in terms of models that possessed certain integrability properties. It was realised that the integrability was a relic of the conformal symmetry that held at the critical temperature. This can happen because the integrable theory is a special kind of deformation of the conformal theory. The sinh-Gordon theory illustrates this phenomenon. Having broken the sinh term into its two constituent exponentials we can multiply one of these pieces by a real positive coefficient which can be absorbed by a redefinition of the scalar field and the mass parameter. However, when this coefficient vanishes, the equation changes radically, collapsing to the Liouville equation, so named because Liouville was the first to solve it, exploiting its conformal symmetry. Reversing this process, we can say that the sinh-Gordon equation is an integrable deformation of the Liouville equation. Making the coupling in sinh-Gordon imaginary produces the sine-Gordon theory already mentioned.

Does this procedure which works so well in two dimensions have an analogue in a space-time of four dimensions? Can it then be applied to unified gauge theories? If so, what is its relation to the Higgs mechanism?

Theme III – Duality symmetry of Maxwell theory The basic idea is surely very old; simply that there is a tantalizing similarity between the electric and magnetic fields $\underline{E}$ and $\underline{B}$. This resemblance was made more precise once Maxwell's equations governing their behaviour were established. In vacuo these equations display several symmetries of physical importance, Poincaré rather than Galilean space-time symmetry, and beyond that, conformal symmetry (with respect to transformations preserving angles and not just lengths). Even more sensitive to the precise nature of the Minkowski space-time metric is the electromagnetic duality rotation symmetry of these equations

$$\underline{E} + i\underline{B} \to e^{i\phi}(\underline{E} + i\underline{B}). \tag{1.1}$$

It is the extension of this fascinating symmetry that is the guiding theme in what follows. Notice that we can form two real, quadratic expressions invariant with respect to (1.1);

$$\frac{1}{2}|\underline{E} + i\underline{B}|^2 = \frac{1}{2}(E^2 + B^2), \tag{1.2a}$$

$$\frac{1}{2i}(\underline{E} + i\underline{B})^* \wedge (\underline{E} + i\underline{B}) = \underline{E} \wedge \underline{B}. \tag{1.2b}$$

These are both physically observable quantities, respectively recognised as the energy and momentum densities of the electromagnetic field.

It is natural to try to extend the duality rotation symmetry (1.1) to include matter carrying charges and, beyond that, to the non-abelian gauge theories of the type already mentioned as being thought to unify the fundamental particle interactions.

The first step is easy: point particles carrying electric and magnetic charges q and g can satisfy Newton's equations of motion with the driving force specified by a natural generalisation of the Lorentz force law. The complete system of equations is indeed duality rotation invariant if (1.1) is augmented by

$$q + ig \to e^{i\phi}(q + ig). \tag{1.3}$$

The price to be paid for this is the necessity of magnetic charge g never observed in isolation. A second difficulty is that the symmetry only works classically and not when quantum mechanics is introduced. The reason is that electrically charged particles must then be described by complex wave functions whose equations of motion require the introduction of the gauge potentials ϕ and $\underline{A}$ forming a four-vector A_μ.

The possibility of a gauge four-potential A_μ such that

$$F_{\mu\nu} = \partial_\mu A_\nu - \partial_\nu A_\mu. \tag{1.4}$$

was already implied by Maxwell's equations for the field strengths $F_{\mu\nu}$. These latter are observable physical quantities and so exist globally in any sort of region of space-time, whatever its topology. The gauge potentials are not so directly physical and so need not exist globally. From a strict mathematical point of view they can only be constructed locally, that is, in regions of space-time that are topologically trivial in the sense that they can be contracted continuously to a point.

This mathematical nicety will fit in with another feature of (1.4), that the there are ambiguities in the potentials giving rise to a given field strength. Thus for any reasonable function $\chi(x)$ on space-time, two gauge potentials A_μ and A'_μ give rise to the same field strength if they are related by

$$A'_\mu = A_\mu + \partial_\mu \chi. \tag{1.5}$$

So perhaps these 'gauge transformations' should be used to relate gauge potentials valid in different patches of space-time in a region where they overlap.

After the discovery of quantum mechanics, in which electrically charged particles are described by complex wave functions $\psi(x)$, it was realised by Weyl and by Fock that gauge covariance of the overall equations of motion could only be achieved if wave functions corresponding to different choices of gauge potential were related by

$$\psi'(x) = e^{iq\chi(x)/\hbar}\psi(x). \tag{1.6}$$

where q is the electric charge carried by the particle with the wave function in question and the gauge function χ is the same as in (1.5).

Now consider a spherical surface S_2 fixed in three-dimensional space surrounding a centre which is an excluded point. This is an example of a space-time region with non-trivial topology since it cannot be contracted continuously to a point without a tear whilst avoiding the origin. This sphere can be covered by two patches which we can think of as northern and southern hemispheres overlapping on the equator. These hemispheres are topologically trivial as they can be retracted to their respective poles. If the gauge potential A_μ were to exist globally on S_2, the flux of magnetic field out of the sphere would vanish by Stokes' theorem since $\underline{B} = \nabla \wedge \underline{A}$ and

$$\int_{S_2} \underline{B}.d\underline{S} = \int_{S_2} \nabla \wedge \underline{A} \cdot d\underline{S} = 0$$

This means that the net magnetic charge inside the sphere would vanish. On the other hand, if we allow two different gauge potentials, A_μ and A'_μ in the northern and southern hemispheres respectively, Stokes' theorem can only be applied to the hemispheres and yields for the magnetic flux the following integral around the equator, S_1,

$$\int_{S_2} \underline{B} \cdot d\underline{S} = \int_{S_1} \left(\underline{A}(x) - \underline{A}'(x) \right) \cdot d\underline{x}$$

By (1.5) this is the line integral of the gradient of the gauge function χ around the equator. This integrates to give

$$\chi(\phi = 2\pi) - \chi(\phi = 0)$$

where ϕ is the longitude (azimuthal angle). On the other hand, for the wave functions ψ and ψ' to be single-valued in their respective hemispheres, equation (1.6) implies that this jump in χ around the equator must equal an integral multiple of $2\pi\hbar/q$. But the flux evaluated equals the magnetic charge g inside the sphere. The final conclusion is not that this vanishes, but rather that it satisfies the condition

$$qg = 2\pi\hbar\mathbb{Z}$$

This is the celebrated Dirac [1931] quantisation condition, derived according to an argument of Wu and Yang [1975].

This condition was extended to greater generality by Zwanziger [1968] and Schwinger [1969] and restricted the charges on any pair of isolated dyons to satisfy

$$q_1 g_2 - q_2 g_1 = 2\pi n\hbar, \qquad n = 0, \pm 1, \pm 2, \pm 3 \dots \ . \tag{1.7}$$

Notice that this condition respects the duality rotation (1.3) applied to dyons 1 and 2 simultaneously. Dyon is the epithet coined by Schwinger [1969] for a particle carrying both electric and magnetic charges.

In 1979 Witten realised that it is possible to solve these quantisation conditions. The most general family of solutions is given by

$$q + ig = q_0(m\tau + n), \qquad m, n \in \mathbb{Z} \tag{1.8}$$

where the complex parameter τ takes the form

$$\tau \equiv \frac{\theta}{2\pi} + \frac{2\pi i n_0 \hbar}{q_0^2}. \tag{1.9}$$

Notice that apart from an integer n_0, the imaginary part of τ is the inverse of the fine structure constant and hence fundamentally positive. The real part is given by a new angular parameter θ that Witten [1979] recognised as being what is called the vacuum angle already familiar in certain specific theories.

Equation (1.8) means that the allowed dyon charges constitute points of a discrete lattice, to be called the 'charge lattice'. Notice that this lattice structure 'spontaneously breaks' the duality rotation symmetry (1.3). Associated with this is the attractive consequence that electric and magnetic charges are quantised. To make this clearer it is helpful to look at the charge lattice (1.8) in the $q + ig$ plane depicted below:-

```
x   .   x   .   x   .   x   .   x   .   x   .   x   .   x
x   .   x   x   .   x   x   .   x   x   .   x   x   .   x
x   .   x   .   x   .   x   .   x   .   x   .   x   .   x
x   x   x   x   x   x   x   x   x   x   x   x   x   x   x
.   .   .   .   .   .   x   o   x   .   .   .   .   .   .
x   x   x   x   x   x   x   x   x   x   x   x   x   x   x
x   .   x   .   x   .   x   .   x   .   x   .   x   .   x
```

The origin, denoted by zero, represents particles carrying no charge, such as the photon. The points of the lattice with coordinate m equal to zero lie on the horizontal line through the origin and correspond to states carrying only electric charge q and no magnetic charge. They form a one-dimensional lattice and this means that for magnetically neutral states electric charge is quantised, always occurring as an integer multiple of q_0, the charge of the state corresponding to the $\times$ to the right of the origin. This can be thought of as the electron, with the positron occurring to the left of the origin. Of course the quantisation of electric charge has long been one of the most visible features of the spectrum of observed elementary particles. Dirac [1931] was very impressed that this fact could be explained by the existence of a single magnetic charge somewhere in the universe, not otherwise observed. Since alternative explanations turn out to be elusive, the theoretical study of magnetic monopoles has provided a source of fascination ever since this insight.

The reader may wonder why there is a distinction in the diagram between the points of the charge lattice outside the origin, some being denoted by dots and some by crosses. The crosses correspond to primitive vectors of the lattice, namely vectors that can be joined to the origin by a straight line avoiding any other points of the lattice. Equivalently, the integer coordinates m, n in (1.8) are coprime. The physical significance of this will become clearer later when specific models are considered.

Although the lattice has spontaneously broken the duality rotation symmetry (1.3) of the quantisation condition (1.7) which it solves, it is still conceivable that the symmetry survives in a formula for the particle masses just as it applies to the energy density (1.2a). Could particle mass be a function of $|q + ig|$?

2   Unified Gauge Theories

Themes I and III of the previous chapter date back to the previous century whereas the ideas of unified gauge theories are relatively new. Because they allow the electromagnetic gauge group $U(1)$ to be unified with the other gauge groups they offer the possibility of an alternative explanation of the quantisation of electric charge that avoids introducing magnetic charge. To see this in more detail consider the Lagrangian density describing a non-abelian gauge theory, with a Higgs field in the adjoint representation,

$$\mathcal{L} = -\frac{1}{4}\text{Tr}\left(\mathbf{F}_{\mu\nu}\mathbf{F}^{\mu\nu}\right) + \frac{1}{2}\text{Tr}\left(\mathcal{D}_{\mu}\Phi\mathcal{D}^{\mu}\Phi\right) - V(\Phi), \tag{2.1}$$

where the covariant derivative of a field in the representation D of the gauge group G (assumed simple and compact) satisfies the standard properties

$$\mathcal{D}_{\mu} \equiv \partial_{\mu} + ieD(\mathbf{W}_{\mu}), \tag{2.2a}$$

$$[\mathcal{D}_{\mu}, \mathcal{D}_{\nu}] = ieD(\mathbf{F}_{\mu\nu}) = ieD\left(\partial_{\mu}\mathbf{W}_{\nu} - \partial_{\nu}\mathbf{W}_{\mu} + ie[\mathbf{W}_{\mu}, \mathbf{W}_{\nu}]\right). \tag{2.2b}$$

Just as $\mathbf{W}_{\mu}$ and $\mathbf{F}_{\mu\nu}$, written in boldface, denote Lie algebra valued gauge potentials and field strengths respectively, so Φ denotes a scalar Higgs field which is also Lie algebra valued. Hence we can write $\Phi = \phi \cdot \mathbf{T}$ where the $\mathbf{T}$ denote the generators of the Lie algebra of G, and ϕ the adjoint Higgs field. Because ϕ is a Higgs field it is supposed not to vanish in the vacuum. Consequently G is broken to the subgroup whose adjoint action on

$$Q = e\hbar\Phi/a = e\hbar\phi \cdot \mathbf{T}/a \tag{2.3}$$

preserves (2.3). Here, a denotes the magnitude of the Higgs field in vacuo, $a^2 = \phi^2$, and is a constant as the covariant derivative of the Higgs field $\mathcal{D}_{\mu}\Phi$ vanishes there. This subgroup, which is associated with long-range gauge fields, has the structure

$$(U(1)_Q \times K)/Z \tag{2.4}$$

where Z is a discrete diagonal subgroup that need not concern us. In particular, if G is $SU(2)$, the Higgs field is an isovector of length a in vacuo. The only rotations leaving this axis invariant are precisely those about the axis itself as given by (2.4) since in this case K is trivial. For more detailed reviews see Goddard and Olive [1978], Olive [1982].

As given by (2.3) Q can be interpreted as the electric charge operator as it generates an invariant $U(1)$ subgroup of the exact gauge group (2.4) just as the Maxwell $U(1)$ should. Furthermore it is correctly normalised. The eigenvalues q of the matrix $D(Q)$ then specify the electric charges of the particles created by a field in the representation D. These eigenvalues are indeed independent of the space-time point at which the Higgs field is evaluated since the matrix varies by gauge conjugation in vacuo.

According to the Higgs mechanism, (Englert and Brout [1964], Higgs [1966], Kibble [1967]), the formula for the mass of a gauge particle with electric charge q in this situation is simply

$$\text{MASS(Gauge particle)} = a|q|, \tag{2.5}$$

whatever the gauge group G and whatever the direction the adjoint Higgs field chooses in the vacuum.

Putting these results together we see that when G is $SU(2)$ the three gauge particles have masses

$$\text{MASS}(W^{\pm}) = a|e\hbar|, \qquad \text{MASS}(\gamma) = 0, \qquad (2.6)$$

where γ denotes the photon.

Notice that in this case the electric charge (2.3) is proportional to an angular momentum matrix and hence has quantised eigenvalues. Thus unification has indeed led to the quantisation of electric charge without overt magnetic monopoles.

This $SU(2)$ model is known as the Georgi–Glashow model but was superseded by the Salam–Weinberg model as a description of electroweak gauge interactions because of its lack of a weak neutral current. Despite this we shall continue studying it because of the interesting properties that emerge and the possibilities of different sorts of generalisation.

Amongst these properties is the possibility of knotted solutions of a type first realised by 't Hooft [1974] and Polyakov [1974]. The occurrence of these can be inferred by considering the energy E corresponding to the above Lagrangian (2.1) for a coupled adjoint Higgs-gauge system:

$$2E = \int d^3x \left(\text{Tr} \left(\mathbf{E}_i^2 + \mathbf{B}_i^2 \right) + (\mathcal{D}_i \Phi)^2 + (\mathcal{D}_0 \Phi)^2 + 2V(\Phi) \right).$$

Since the suffix i is summed over the three space axes, it is expressed as a sum of positive terms. In order for any classical field configuration to have finite energy, all the terms must vanish faster than the inverse of the radius of a very large two-sphere in three-space, when evaluated on that sphere. In particular the space components of the covariant derivative of the Higgs field must vanish there, so

$$\mathcal{D}_i \hat{\underline{\phi}} \equiv \nabla_i \hat{\underline{\phi}} + e\underline{W}_i \wedge \hat{\underline{\phi}} = 0$$

where the hat indicates that the Higgs field has been divided by a so as to be a unit vector in the vacuum. We are now considering the gauge group to be $SU(2)$ and use appropriate vector notation. Taking the vector product of this with the Higgs field yields the following behaviour for the gauge potential in terms of the Higgs field

$$\underline{W}_i = A_i \hat{\underline{\phi}} - \frac{1}{e} \hat{\underline{\phi}} \wedge \nabla_i \hat{\underline{\phi}},$$

where the $SU(2)$ invariant potential A_i is undetermined and will eventually drop out of the argument. Notice that the undetermined term is parallel to the Higgs field while the behaviour perpendicular to it is fully determined. Since the commutator of two covariant derivatives gives the field strength we have

$$0 = [\mathcal{D}_i, \mathcal{D}_j]\hat{\underline{\phi}} = e\underline{F}_{ij} \wedge \hat{\underline{\phi}}.$$

Hence, at larger distances, the magnetic field is parallel to the Higgs field in $SU(2)$ group space. In fact explicit calculation yields

$$\underline{F}_{ij} = \hat{\underline{\phi}}\left(\nabla_i A_j - \nabla_j A_i - \frac{1}{e}\hat{\underline{\phi}} \cdot (\nabla_i \hat{\underline{\phi}} \wedge \nabla_j \hat{\underline{\phi}})\right) = F_{ij}\hat{\underline{\phi}}.$$

The Maxwell magnetic charge of the field configuration of finite energy can be found my calculating the magnetic flux out of the large sphere surrounding it. Since the above expression for the magnetic field holds on this sphere, it yields

$$g = \frac{1}{2}\int dS_i \epsilon_{ijk} F_{jk} = \frac{-1}{2e}\int dS_i \epsilon_{ijk} \hat{\underline{\phi}} \cdot (\nabla_j \hat{\underline{\phi}} \wedge \nabla_k \hat{\underline{\phi}})$$

Notice that the undetermined potential A_μ has integrated out by Stokes' theorem leaving an integrand that has a very nice geometrical interpretation. It is the Jacobian of the map provided by the Higgs field mapping from the two-sphere at large distances to the two-sphere that is the Higgs vacuum in the $SU(2)$ theory ($\hat{\underline{\phi}}^2 = 1$). Hence the integral is the winding number m, say, of this map between two two-spheres times the area, 4π, of the unit sphere. Thus

$$g = -\frac{4\pi}{e}m = -\frac{4\pi\hbar}{q(W)}m. \tag{2.7}$$

Thus the winding number m counts the number of units of magnetic charge and so can be identified precisely with the coordinate in equation (1.8). If fields carrying half-integral $SU(2)$ spin are introduced the smallest possible electric charge in the theory, q_0, equals $q(W)/2$. Otherwise it is $q(W)$, the charge carried by the W^+ gauge particle. Hence, taking the imaginary part of (1.8) and comparing with (2.7), we see that the Dirac–Schwinger–Zwanziger condition is satisfied, with n_0 equal to 1 or 2 in the respective situations.

There are actually solutions to the equations of motion for all integer values of m and the winding number 1 solution (corresponding to the identification map between the spheres) will describe a magnetic monopole soliton and -1 its antiparticle, the antimonopole, also a soliton.

We learn that the idea that the quantisation of electric charge could be explained by unification without the occurrence of magnetic charge is illusory. Dirac's original argument has been vindicated.

Given a stationary solution, it should be possible to insert it in the energy integral following from (2.1) and hence evaluate its mass. In fact, by neatly completing a square in the energy expression above, Bogomol'nyi [1976] found the lower bound

$$\text{MASS(Magnetic monopole)} \geq a|g|. \tag{2.8}$$

Because of the remarkable similarity to the Higgs formula (2.5) it is natural to enquire whether this 'Bogomol'nyi bound' could be saturated. Indeed, the energy integral shows that this can be done if two conditions are satisfied. Firstly, the self-interaction of the Higgs field specified by $V(\Phi)$ in (2.1) should vanish identically. This is known as the Prasad–Sommerfield limit [1975]. Secondly, the following first-order field equations, known as the Bogomol'nyi equations [1976], should hold:

$$\mathcal{D}_0 \Phi = 0, \qquad \mathbf{E}_i = 0, \qquad i = 1, 2, 3, \tag{2.9a}$$
$$\mathcal{D}_i \Phi = \pm \mathbf{B}_i, \qquad i = 1, 2, 3. \tag{2.9b}$$

It is easy to check that any solution to (2.9) automatically solves the full second-order equations of motion in the Prasad-Sommerfield limit. The innocuous facade that the Bogomol'nyi equations present is deceptive. It will emerge that they exhibit a remarkably rich mathematical structure which will have profound physical implications in what follows.

As a first step note that, if a fifth, space-like dimension x^5 is introduced in which

$$\mathbf{W}^5 = \Phi, \tag{2.10}$$

then equations (2.9b) can be regarded as the self-dual Yang–Mills equations in four Euclidean dimensions (1,2,3 and 5). Here we have exploited the fact that the Higgs field Φ lies in the same adjoint representation of the gauge group as the gauge potentials $\mathbf{W}^\mu$. Normally the solutions to such equations describe instantons but here the solutions are independent of both x^5 and time and will have a different character, describing what are called BPS monopoles. To justify this we have to determine the nature of the moduli spaces $\mathcal{M}(m)$ of solutions to the Bogomol'nyi equations (2.9b) labelled by the different values of the winding number m which also specifies the magnetic charge.

Suppose we seek small deviations about a solution that still solve the equations. These will satisfy linear differential equations in the background of the original solution. After discounting small gauge transforms of the original equation as irrelevant this linear equation can be recast as a Dirac equation for adjoint spinors in the Bogomol'nyi background. Its solutions form a vector space whose dimension is finite and given by a version of the Atiyah–Singer index theorem as being

$$\text{Dim}\,\mathcal{M}(m) = 4|m|. \tag{2.11}$$

Furthermore $\mathcal{M}(m)$ is a manifold. Thus it has no singular points but is not compact, for good physical reasons that we shall see. These results are due to E. Weinberg [1979].

The physical consequences are remarkable. Equation (2.11) can be restated as saying that a solution with magnetic quantum number m has $4m$ bosonic zero modes. In other words there is a $4m$-dimensional space of continuous and physically significant alterations that can be made without changing the energy. When $m = 1$ three of the four possibilities correspond to moving the monopole soliton sideways. Thus three of the coordinates on the moduli space $\mathcal{M}(1)$ are the spatial coordinates of the monopole. The fourth is more subtle: it describes a coordinate conjugate to the electric charge, just as the space coordinate is conjugate to momentum. Momentum is excited when the space coordinate moves in time. Likewise electric charge can be excited when its conjugate coordinate moves in time. Thus when we consider motions on the moduli space of BPS solutions we can endow monopoles with momentum and with electric charge (so that they become dyons). However, this is jumping ahead and we must just accept this charge coordinate for the time being.

Since $\text{Dim}\,\mathcal{M}(m) = m\text{Dim}\,\mathcal{M}(1)$, we conclude that $\mathcal{M}(m)$ describes m magnetic monopoles, each with three space coordinates and its own charge coordinate.

So each point of $\mathcal{M}(m)$ describes a configuration of m like monopoles. Since the solution is independent of time, this configuration is stationary. By Newton's equations of motion it follows that the monopoles, rather surprisingly, exert no forces on each other when they are at relative rest. In other words, intermonopole forces vanish at relative rest and so must be velocity dependent. Notice that this differs from the situation with sine-Gordon solitons. There the analogue of the Bogomol'nyi first-order differential equation is satisfied by the soliton and by the antisoliton individually but not by any multisoliton configuration as, in this case, solitons do exert static forces on each other.

The Bogomol'nyi equations do not describe mixed configurations of monopoles and antimonopoles, which must be described by solutions of the full second-order equations and so be time dependent. Thus monopole–antimonopole pairs must exert static forces on each other.

Finally notice that there is no zero mode corresponding to the angular orientation of any monopole. This means that they cannot acquire intrinsic angular momentum by classical motion.

We can now summarise the particle spectrum of the $SU(2)$ gauge theory as far as it is understood so far. Four particles arise as quantum excitations of the field in (2.1), the photon, the two heavy gauge particles $W^{\pm}$ and the Higgs which is neutral and massless, because its mass term vanishes in the Prasad-Sommerfield limit. The remaining spectrum consists of the solitons with winding number ± 1. These can be monopoles as already discussed or, possibly, dyons (Julia and Zee [1975]). A Bogomol'nyi style argument

[Coleman *et al.* 1977] shows that their mass is given by

$$\text{MASS}(q,g) = a\sqrt{q^2 + g^2} = a|q + ig|. \tag{2.12}$$

In fact all the particles we have mentioned satisfy this single universal mass formula which unifies Higgs and Bogomol'nyi mechanisms, quantum excitations and classical topological solitons. Furthermore it does respect the duality rotation symmetry (1.3), in the simplest possible way. It is quite remarkable that this theory should display such properties and it is natural to seek a deeper understanding of it.

It is interesting that the vacuum expectation value parameter is explicitly responsible for all forms of mass, as it alone breaks the conformal symmetry of the theory. In other words, this theory seems to provide a version in four space-time dimensions of Zamolodchikov's mechanism [1989] for the deformation of a conformally invariant field theory (cf. Theme II above). Maybe it is indeed close to being integrable.

3 Electromagnetic Duality Conjectures

We have just seen that an $SU(2)$ gauge theory with triplet Higgs field has remarkable and unexpected properties. Some of the particle states, those with magnetic charge, arise as solitons and so have an extended classical structure, quite unlike the quantum excitations of the original fields which are magnetically neutral and point-like. Yet, despite this, the mass formula displays an unusual symmetry with respect to rotations in the complex charge space, $q + ig$, between these states. In view of Skyrme's result that sine-Gordon solitons can be viewed as being created by massive Thirring fields in a reformulation of the theory, it is tempting to investigate the possibility of a second formulation of the gauge theory with quantum fields ascribed to the monopole solitons rather than the original gauge particles.

Because the monopole solitons carry magnetic charge, they emanate a Coulomb magnetic field which is presumably described by a gauge theory whose potential couples to the magnetic charge. This means that there is a 'magnetic' gauge group with coupling strength inversely related to the original 'electric coupling' by

$$q_0 \rightarrow g_0 = \frac{4\pi\hbar}{q_0}, \tag{3.1}$$

because of the relation (2.7), respecting the Dirac–Schwinger–Zwanziger quantisation condition (1.4). Notice the characteristic feature that this transformation exchanges strong and weak coupling regimes.

Thinking along these lines, two specific conjectures were proposed in 1977. First, considering a more general context, with a less restrictive exact symmetry group H than (2.4), Goddard, Nuyts and Olive [1977] established a generalised version of the Dirac quantisation condition and used it to propose conjectures for the precise global structure of the magnetic, or dual gauge group $H^\vee$. However it is only the situation leading to (2.4) that yields a specific mass formula. Montonen and Olive [1977] therefore considered the setup as described above, restricted to $G = SU(2)$, even though the mass formula described holds for any gauge group G with adjoint Higgs.

The possible quantum states of the theory correspond to points of the lattice considered in the first chapter. Ignoring the dyons, the single-particle states correspond to five points of this lattice. The photon and Higgs particle correspond to the origin, the heavy gauge particles $W^\pm$ to the points $(\pm q_0, 0)$. Thus the particles created by the fundamental fields in the original 'electric' formulation of the action all lie on the real, electric axis. Supposing they have no electric charge, the magnetic monopole solitons $M^\pm$ lie on the imaginary axis at $(0, \pm g_0)$ while any dyon solitons would lie on the horizontal lines through these points, but we shall temporarily ignore them.

Now, if we follow the transformation (3.1) by a rotation through a right angle in the $q + ig$ plane, the four points just described away from the origin are rearranged. This suggests that the 'dual' or magnetic formulation in which the $M^\pm$ monopoles are created by fields present in the action will likewise be a spontaneously broken gauge theory, but with coupling constants changed according to (3.1). In this alternative formulation it is the $W^\pm$ states that will occur as solitons.

This is the electromagnetic duality conjecture of Montonen and myself as originally formulated [1977]. In principle, it could be proven by finding a generalisation of Skyrme's vertex operator construction, but, this still seems beyond reach despite the intervening advances in mathematical knowledge. Notice that the sine-Gordon quantum field theory was described by two quite dissimilar actions, whereas in the present case the two hypothetical actions have a very similar structure but refer to electric and magnetic formulations with inversely related coupling strengths.

Physical predictions should coincide whichever action is chosen as the starting point. The conjecture will immediately pass at least two tests of this kind, showing that it is not obviously inconsistent. First, the mass formula (2.12) satisfies this criterion, precisely because of the universal property that has already been emphasised, and indeed motivated the idea of two dual formulations.

The fact that the forces between an M^+ pair are velocity dependent thereby vanishing at relative rest, as discussed in the previous chapter, had just been realised by Manton [1977]. It was therefore a valid test of the conjecture to establish whether this force vanishes in the magnetic formulation. This is equivalent to checking the absence of the static $W^+ - W^+$ force in the electric formulation and this is simply a question of Feynman rules in the

Born approximation. Then two diagrams contribute, photon exchange and Higgs exchange. Photon exchange produces the expected q^2/r^2 Coulombic repulsion which is precisely counterbalanced by a q^2/r^2 Coulombic attraction due to the Higgs exchange. This is possible because the adoption of the Prasad–Sommerfield limit forces the Higgs to be massless. To check this in more detail we have to extract the couplings from the original Lagrangian density (2.1).

The first step is to choose a gauge in which the Higgs field is aligned in a standard direction, the 3-axis in $SU(2)$ space, say:

$$\phi_i(x) = \delta_{i3}(a + \sigma(x)). \tag{3.2}$$

This is possible because we are considering fluctuations about the vacuum rather than a monopole background. The field $\sigma(x)$ is the neutral Higgs left after its charged partners have been absorbed by $W^\pm$.

Inserting (3.2) into the term in (2.1) describing the Higgs kinetic energy we obtain

$$\frac{(\mathcal{D}_\mu \phi)^2}{2} = \frac{(\partial_\mu \sigma)^2}{2} + \frac{e^2 a^2}{2} W_\mu^+ W_\mu^- + e^2 a \sigma W_\mu^+ W_\mu^- + \cdots \tag{3.3}$$

As well as confirming that σ is massless and that $W^\pm$ has a mass $M = ea\hbar$, this expression shows that σ has a Yukawa coupling to $W^\pm$ of strength $e^2 a = eM/\hbar$. Thus the Feynman rule for the corresponding vertex is $\epsilon_1 . \epsilon_2 Me/\hbar$, where ϵ_1 and ϵ_2 are the polarisation vectors of the two W^+ lines incident at the vertex. This is to be compared with the corresponding rule for the photon coupling $\epsilon_1 \cdot \epsilon_2 (p_1 + p_2)_\mu e/2\hbar$.

Now it is easy to calculate the contributions to $W^+ - W^+$ scattering due to Higgs and photon exchange respectively. These are

$$\epsilon_1 \cdot \epsilon_2 \epsilon_3 . \epsilon_4 \frac{M^2 e^2}{\hbar^2 k^2}$$

and

$$-\epsilon_1 \cdot \epsilon_2 \epsilon_3 \cdot \epsilon_4 \frac{e^2 (p_1 + p_2).(p_3 + p_4)}{4\hbar^2 k^2}$$

where k^μ is the momentum transfer, that is, the momentum carried by the virtual particles exchanged. In the static limit the momentum transfer is very small and the momenta p_1, p_2, p_3 and p_4 are equal. Then the two contributions above are equal and opposite and hence cancel, yielding the claimed result. On the other hand, if $W^+ - W^-$ scattering were considered p_1, p_2, $-p_3$ and $-p_4$ would be equal in the static limit, leading to the equality of the two contributions and hence the doubling of the expected Coulomb attraction.

The electromagnetic duality conjecture immediately provokes the following questions:

(1) How can the magnetic monopole solitons possess the unit spin characteristic of heavy gauge particles?

(2) Surely the effect of quantum corrections will vitiate the universal mass formula?

(3) How would the proper inclusion of dyon states affect the picture?

We have already noted that the monopoles are spinless on the classical level, as the moduli space of Bogomol'nyi solutions has no degree of freedom to accommodate their rotation. Hence any spin must originate in the quantisation procedure applied to the theory.

In fact it was soon realised that there were several options in quantising the theory. It is possible to add fermionic and other fields to the original theory which play no role in the solutions discussed so far, since setting them to zero is consistent with the new equations of motion. Furthermore, this can be done in a way that enhances the symmetry of the theory so that it achieves supersymmetry (D'Adda, Horsley and Di Vecchia [1978]). The quantum fluctuations of the added fields do have an important effect. They can cancel the existing fluctuations so as to ensure that the mass formula is quantum exact and they can create spin. Before understanding how supersymmetry achieves these feats we shall now explain in more detail what supersymmetry is and the relevant features of its algebraic structure.

4 Algebraic Structure of Supersymmetry

Many of the key features of supersymmetry are very well illustrated by the supersymmetric harmonic oscillator, probably the simplest toy model for it. Consider two annihilation-creation oscillators a and b which are bosonic and fermionic respectively:

$$[a, a^\dagger] = 1 = \{b, b^\dagger\}, \qquad b^2 = 0 \qquad (4.1)$$

With these, form the supercharges

$$Q = a^\dagger b, \qquad Q^\dagger = b^\dagger a \qquad (4.2)$$

whose anticommutator is the Hamiltonian H

$$\{Q, Q^\dagger\} = H = a^\dagger a + b^\dagger b = \left(a^\dagger a + \frac{1}{2}\right) + \left(b^\dagger b - \frac{1}{2}\right) \qquad (4.3a)$$

The two bracketed terms are the Hamiltonians for the individual bosonic and fermionic oscillators respectively with their zero point energies. Notice that these are equal and opposite and so cancel from the total Hamiltonian. Furthermore

$$Q^2 = Q^{\dagger 2} = 0 = [Q, H] \qquad (4.3b)$$

Equations (4.3) constitute the supersymmetry algebra satisfied by the super-charges and Hamiltonian. The Heisenberg equation of motion for a reads:

$$\frac{da}{dt} = i[a, H] = ia$$

Hence, repeating, a indeed satisfies the harmonic oscillator equation with unit frequency, as does b by a similar calculation.

One of the important general features of a supersymmetry algebra is that

$$H \geq 0 \qquad (4.4)$$

as can be seen from (4.3a). In fact, if $|\Omega\rangle$ is the ground state, that is the state for which H has its smallest possible eigenvalue, zero, then it is the state annihilated by both a and b, the oscillator vacuum. Thus the vanishing of the zero-point energy is an intrinsic feature of a supersymmetric system and this is the reason why quantum corrections tend to cancel to zero. Linear fluctuations in a more complex supersymmetric system can be decomposed into independent supersymmetric oscillators, each of whose zero point energy vanishes.

It is easy to calculate all the energy eigenstates of the supersymmetric oscillator and an interesting pattern emerges. For energy eigenvalue n, $(n \geq 1)$ there are precisely two eigenstates, $a^{\dagger n}|\Omega\rangle$ and $a^{\dagger n-1}b^{\dagger}|\Omega\rangle$. Thus the dimension of the energy eigenspace depends on whether the energy eigenvalue is zero or not, being one when it is zero and two otherwise. Corresponding to this we talk of 'short' and 'long' representations of supersymmetry.

Now let us see how these features extend to supersymmetry algebras in Minkowski space-time, that is, with signature (1,3). There the (Dirac) gamma matrices have some special properties that we want to exploit. The Dirac equation reads

$$(i\gamma \cdot \partial + m)\psi = 0,$$

provided the γ^{μ} satisfy the Clifford algebra

$$\{\gamma^{\mu}, \gamma^{\nu}\} = 2g^{\mu\nu} = 2\,\text{diag}\,(1, -1, -1, -1).$$

Anticommuting with these is

$$\gamma^5 = \gamma^0\gamma^1\gamma^2\gamma^3, \qquad (\gamma^5)^2 = -1.$$

It will be very important that γ^5 behaves like the imaginary unit i, as does the Hodge star operator, implicit in the duality transformation (1.1). The Clifford algebra has only one irreducible representation and that has dimension 4. It is possible to choose a 'Majorana' basis in which the matrices are pure imaginary. In it $i\gamma^\mu$ are real and given by

$$\begin{pmatrix} 0 & 0 & 0 & 1 \\ 0 & 0 & -1 & 0 \\ 0 & 1 & 0 & 0 \\ -1 & 0 & 0 & 0 \end{pmatrix}, \begin{pmatrix} 0 & 0 & 0 & 1 \\ 0 & 0 & 1 & 0 \\ 0 & 1 & 0 & 0 \\ 1 & 0 & 0 & 0 \end{pmatrix}, \begin{pmatrix} 0 & 0 & 1 & 0 \\ 0 & 0 & 0 & -1 \\ 1 & 0 & 0 & 0 \\ 0 & -1 & 0 & 0 \end{pmatrix}, \begin{pmatrix} 1 & 0 & 0 & 0 \\ 0 & 1 & 0 & 0 \\ 0 & 0 & -1 & 0 \\ 0 & 0 & 0 & -1 \end{pmatrix}.$$

This means that it is perfectly consistent for the solution ψ to the Dirac equation to be real and this will be exploited. Notice that the last three matrices are symmetric while the first, $i\gamma^0$, is real antisymmetric, as is

$$\gamma^5 = \begin{pmatrix} 0 & 0 & -1 & 0 \\ 0 & 0 & 0 & -1 \\ 1 & 0 & 0 & 0 \\ 0 & 1 & 0 & 0 \end{pmatrix}$$

Now we can introduce the supercharges Q^i_α carrying a Dirac spinor index α running from 1 to 4 and an internal symmetry index i running from 1 to N. Because of the previous comment these can conveniently be taken to be Hermitian. Because of their spinor nature they satisfy the anticommutation relations of type N supersymmetry:

$$\{Q^i_\alpha, Q^j_\beta\} = 2(\gamma^\mu P_\mu \gamma^0)_{\alpha\beta}\delta^{ij} \qquad (4.5a)$$

Note that everything in this equation is real and that the right hand side of the equation displays symmetry under interchange of $\alpha \leftrightarrow \beta$ and $i \leftrightarrow j$ separately, even though the left hand side only requires the joint symmetry. There are further commutation relations expressing translation invariance,

$$[P^\mu, Q^i_\alpha] = 0, \qquad (4.5b)$$
$$[P^\mu, P^\nu] = 0, \qquad (4.5c)$$

as well as the statement of the Lorentz transformation properties. We can solve (4.5a) for the Hamiltonian by taking the trace in the spinor indices α and β and choosing i equal to j without summing:

$$H = P^0 = \frac{1}{4}\sum_{\alpha=1}^4 (Q^i_\alpha)^2, \qquad i = 1, 2, \ldots, N. \qquad (4.6)$$

Hence $H \geq 0$ because the supercharges are Hermitian. Again, as in the toy oscillator, positivity of energy is guaranteed by the algebraic structure.

Furthermore the vacuum, the state with zero energy, is annihilated by all the supercharges and hence is totally supersymmetric. It automatically has zero momentum by (4.5a).

Equations (4.5b) and (4.5c) mean that the superalgebra must be represented on the set of states with any given momentum. Let us consider massive particle states, choosing a rest frame in which $P^\mu = (M, 0, 0, 0)$ so that (4.5a) reads

$$\{Q^i_\alpha, Q^j_\beta\} = 2M\delta_{\alpha\beta}\delta^{ij} \qquad (4.7)$$

This is recognisable as a Clifford algebra in $4N$ Euclidean dimensions [Nahm 1978]. Its unique irreducible representation therefore has dimension 2^{2N}.

This is to be contrasted with a massless particle state whose momentum can, at best, be chosen as $P^\mu = (E, E, 0, 0)$ so that (4.5a) reads

$$\{Q^i_\alpha, Q^j_\beta\} = 2E(1 - \gamma^1\gamma^0)_{\alpha\beta}\delta^{ij} \qquad (4.8)$$

Because $(1 - \gamma^1\gamma^0)$ is 4×4 matrix with two eigenvalues equal to 0 and the other two equal to 2 we see that half the supercharges must annihilate the massless state and hence effectively vanish while the other half generate a Clifford algebra in $2N$ Euclidean dimensions with a unique irreducible representation of dimension 2^N (the square root of the dimension in the massive case).

In practice, N, the number of spinor supercharges, will not exceed 4 and we can tabulate the number of dimensions required to represent the Clifford algebra on the different types of state:

	$N = 1$	$N = 2$	$N = 4$
$P^2 > 0$	4	16	256
$P^2 = 0$	2	4	16
$P_\mu = 0$	1	1	1

The massless case is said to form the 'short' representation.

To understand the spin content of the massless multiplets better, consider the helicity, the component of angular momentum in the direction of motion,

$$J_1 = \frac{p \cdot J}{|p|} = \frac{i\gamma^2\gamma^3}{2} = \frac{-i\gamma^5\gamma^1\gamma^0}{2}$$

using the definitions. But according to (4.8) the non-vanishing supercharges satisfy $\gamma^1\gamma^0 Q = -Q$. Hence

$$J_1 Q = \frac{i\gamma^5}{2}Q. \qquad (4.9)$$

As a consequence, of the $4N$ spinor supercharges, $2N$ vanish leaving N with helicity $1/2$ and N with helicity $-1/2$. These can be thought of as N fermionic annihilation-creation operators destroying and creating helicity in units of $1/2$. Acting on a ground state they form a Fock space of dimension 2^N upon which the range of helicities is

$$\Delta J_1 = \frac{N}{2} \qquad (4.10)$$

(allowing for the ground state to have a helicity value). This range equals 1/2, 1 or 2 according as $N = 1$, 2 or 4.

In contrast the corresponding result for massive states is more complicated to prove but yields

$$\Delta J_1 = N \tag{4.11}$$

which is twice as large.

Returning to the spontaneously broken gauge theory of chapters 2 and 3 we recall that, in the first instance, the two fields, the Higgs and gauge field, are massless and carry helicities ± 1 and 0. Since they transform similarly under the gauge group it is tempting to fit them into a single irreducible representation of the Clifford algebra on massless states. This requires Δ(helicity) = 1, at least, so that $N \geq 2$. Consider first $N = 2$. The representation of the supercharge algebra on massless states is four-dimensional with a helicity span of one. Letting the helicity axis be vertical this yields a supermultiplet as follows

$$
\begin{array}{ccc}
 & \times & \\
\times & & \times \\
 & \times &
\end{array}
$$

If we want to describe helicity one and no higher we must include helicity -1 and this is done with two copies of the above pattern

$$
\begin{array}{ccc}
 & \times & \\
\times & & \times \\
\times & & \otimes \\
\otimes & & \otimes \\
 & \otimes &
\end{array}
\tag{4.12}
$$

Here the helicity runs from one at the top down to minus one at the bottom. Thus, in order to construct an $N = 2$ massless supermultiplet containing the gauge and Higgs field, we have to add a second scalar field as well as the two adjoint spin half fields. We shall call this the gauge multiplet. There is a second sort of massless $N = 2$ multiplet possible which has no helicity ± 1 components, called the hypermultiplet, also made of two copies of the fundamental unit:

$$
\begin{array}{cccccc}
 & \times & & & \otimes & \\
\times & & \times & \otimes & & \otimes \\
 & \times & & & \otimes &
\end{array}
\tag{4.13}
$$

The second copy is needed to realise the internal symmetry associated with the supersymmetry, known as $\mathcal{R}$-symmetry. The fundamental anticommutation relation (4.5a) has a manifest $O(N)$ symmetry according to which the supercharges transform as an N. This $O(N)$ can be enlarged to $U(N)$ by a complexification in which the role of the imaginary unit acting on the supercharges is simply γ_5, as suggested above. In the $N = 2$ massless gauge multiplet the fermions form an $SU(2)_\mathcal{R}$ doublet and the two spinless particles form two singlets. On the other hand, in the hypermultiplet the fermions are singlets while the spinless particles form a doublet. Since the doublet is a complex (pseudoreal) representation and γ^5 does not act on spinless particles there have to be four real scalars.

Any other possible massless $N = 2$ multiplets must possess helicities exceeding one in magnitude and hence are not considered.

It turns out that $N = 3$ gives the same structure as $N = 4$ so we shall move on to the latter. Since the helicity span given by (4.9) is 2, there is only one possible massless $N = 4$ supermultiplet with helicity not exceeding one. Its helicities run from 1 down to -1:

$$
\begin{array}{ccccccc}
 & & & \times & & & \\
 & \times & \times & & \times & \times & \\
\times & \times & \times & & \times & \times & \times \\
 & \times & \times & & \times & \times & \\
 & & & \times & & &
\end{array}
\qquad (4.14)
$$

In this case the fermions form a $4_L \oplus \bar{4}_R$ of $SU(4)_\mathcal{R} \equiv SO(6)_\mathcal{R}$ and the spinless particles a 6 (which is real). Notice that this supermultiplet consists of a single irreducible representation of the Clifford algebra (4.7).

Higher values of N imply massless supermultiplets with helicities larger than one in magnitude and that is why those values are excluded. The problems worsen if massive supermultiplets are considered. This leads to a paradox that we now discuss.

5   Supersymmetry to the Rescue

We have seen how the massless states of the spontaneously broken gauge theory can be fitted into a single (gauge) supermultiplet of either the $N = 2$ or the $N = 4$ theory but it still has to be determined which of these degrees of supersymmetry is appropriate. Before addressing this there is a more urgent problem. It was seen earlier that mass was acquired from the Higgs mechanism and it is a familiar feature of this mechanism that the total number of helicity states does not change. Yet, according to the Clifford algebra analysis just discussed, the dimensionality of the representation changes radically with mass thereby, begetting a paradox. (This is a problem peculiar to $N = 2$ and $N = 4$ and not $N = 1$ supersymmetry because in $N = 1$ the massless gauge and Higgs particles have to be in different supermultiplets). The resolution of this paradox is that the supersymmetry algebra (4.5a) as it stands is incomplete, as extra terms may occur on the right hand side (Haag, Łopuszański and Sohnius [1975]). For example, if $N = 2$ we can have

$$
\{Q^i_\alpha, Q^j_\beta\} = 2(\gamma^\mu P_\mu \gamma^0)_{\alpha\beta}\delta^{ij} + 2i\epsilon^{ij}\left((Z_1 + \gamma^5 Z_2)\gamma^0\right)_{\alpha\beta}, \qquad (5.1)
$$

where ϵ^{ij} is the antisymmetric matrix with two rows and columns. The quantities Z_1 and Z_2 are real, called central charges because they commute with the supercharges and all the other generators of the algebra. Notice that the right hand side still reproduces the joint symmetry under interchange of $\alpha \leftrightarrow \beta$ and $i \leftrightarrow j$ of the left hand side. Furthermore the structure is consistent with all the generalised Jacobi identities.

The reason that the superalgebra realised on massless states, (4.8), unlike the superalgebra on massive states (4.7) had a shorter representation was that the matrix on the right hand side was singular. This possibility is now attainable for massive states with the extra terms on the right hand side of (5.1). To see this consider the algebra acting on a massive particle at rest carrying charges Z_1 and Z_2. Then

$$\{Q^i_\alpha, Q^j_\beta\} = 2M\delta_{\alpha\beta}\delta^{ij} + 2i\epsilon^{ij}\left((Z_1 + \gamma^5 Z_2)\gamma^0\right)_{\alpha\beta}. \tag{5.2}$$

As $((Z_1 + \gamma^5 Z_2)\gamma^0)^2 = (Z_1 + \gamma^5 Z_2)(Z_1 - \gamma^5 Z_2) = Z_1^2 + Z_2^2$, the right hand side of (5.2) has eigenvalues $2(M \pm \sqrt{Z_1^2 + Z_2^2})$, each with fourfold multiplicity. Because of the hermiticity of the supercharges, the left hand side of (5.2) is a positive definite matrix and hence the mass M is bounded below

$$M \geq \sqrt{Z_1^2 + Z_2^2}. \tag{5.3}$$

When this bound is saturated, that is, an equality, half the eigenvalues of the right hand side of (5.2) vanish and the other half equal $4M$. As a consequence the superalgebra has a short representation exactly as in the massless case (Witten and Olive [1978]). This means that mass can be acquired without changing the number of states, provided the bound (5.3) is saturated.

Because the supercharges are conserved so are the central charges. As the only conserved charges available are the electric and magnetic ones, Q and G, we expect the central charges to be functions of these. A simple and traditional way of checking a superalgebra is to check the algebra of supertransformations on the field of the theory and this results in

$$Z_1 = aQ, \tag{5.4a}$$

while Z_2 apparently vanishes. This method has two limitations:

(1) As it involves transformations of classical fields it is immune to possible quantum corrections.

(2) As none of the fields in the action carry magnetic charge its possible appearance is invisible to this argument.

Nevertheless we already see something interesting: the Higgs mass formula is a consequence of the supersymmetry algebra, combined with the notion that mass requires no extra states.

The calculation can be refined by expressing the supercharges as integrals over space, thereby demonstrating that

$$Z_2 = aG, \tag{5.4b}$$

at least semiclassically (Witten and Olive 1978]). Thus the saturation of the bound (5.3) is tantamount to the universal mass formulae (2.12) that originally provoked the electromagnetic duality conjecture of Section 3.

In the $N = 4$ theory the superalgebra can also be modified by the addition of extra charges on the right hand side. In this case $\epsilon^{ij} Z_p$ on the right hand side of (5.1) is replaced by $Z_{(p)}^{ij}$ where $p = 1, 2$ and the matrices $Z_{(1)}$ and $Z_{(2)}$ are real antisymmetric matrices with four rows and columns each transforming as the real 6 of $SU(4)_\mathcal{R}$. These matrices carry information about the electric and magnetic charges and again lead to the validity of the mass formula (2.12) being equivalent to the shortness of the relevant supermultiplet.

The argument that the states of a given momentum must represent the supersymmetry algebra is perfectly democratic: it applies equally to any sort of particle whether it is a quantum excitation of an original field, whether it is a soliton or whether it arises by some other mechanism (as indeed will be the case). In particular it means that monopole dyon solitons do, willy nilly, possess spin. If the maximum helicity does not exceed one in magnitude there are two possibilities in $N = 2$, the gauge multiplet (4.12) or the hypermultiplet (4.13). In $N = 4$ there is only one possibility, the gauge multiplet (4.14). As these are short representations the mass formula has to hold.

In Section 2 we mentioned that within the moduli space of Bogomol'nyi solutions describing configurations of like monopole solitons there was no bosonic zero mode that could be interpreted as an angular coordinate for any monopole and hence capable of excitation to give angular momentum. What then is the origin of the spin required by supersymmetry? The answer has to to with quantum fluctuations of zero modes of the fermion fields needed for supersymmetry. In a Bogomol'nyi background, the equations of motion for the fermion fields take the form of a massless Dirac equation coupled to the Euclidean self dual gauge configuration described in Section 2. The number of fermionic zero modes is then determined by the version of the Atiyah–Singer index theorem appropriate to the monopole background, as elucidated by Callias [1978] and E. Weinberg [1979]. It is proportional to the magnetic charge and depends on the transformation properties of the fermions under the $SU(2)$ gauge group as well as the supersymmetry algebra.

When quantised canonically, these zero modes generate a Clifford algebra whose representation space yields a space of states all with the same energy (because they are zero modes). However, the states carry a variety of quantum numbers of the type carried by the fermion fields in question, such as spin. This phenomenon first became familiar in the context of the Ramond fermionic string in the early 1970s.

So far there is only one type of fermion field, that in the gauge multiplet and so transforming under the adjoint representation of the gauge group. The effects on spin of its zero modes is to create a hypermultiplet of monopole states in the $N = 2$ theory and a gauge multiplet in the $N = 4$ theory. Thus it is only in the $N = 4$ theory that monopoles can be gauge particles and hence this is the only case in which the electromagnetic duality conjecture of Montonen and myself can be valid, as first realised by Osborn [1979]. The idea cannot be valid as it stands in the $N = 2$ theory but nevertheless a variety of interesting phenomena can occur.

It is possible to introduce extra hypermultiplets in the $N = 2$ theory that transform as doublets under the gauge $SU(2)$. If there are N_F of these, all massless in the first instance, this introduces an $U(N_F)$ flavour symmetry accidentally augmented to $SO(2N_F)$. These can be thought of as quarks. Their zero modes do not affect the monopole spin but they do contribute $SO(2N_F)$ spinor flavour quantum numbers to them. This is the phenomenon sometimes known as charge fractionalisation. The results mean that in the $N = 2$ theory it is possible to contemplate a duality symmetry which exchanges monopoles with quarks, rather than gauge particles, as in the $N = 4$ theory. This is the idea that has been developed and refined by Seiberg and Witten.

The internal $U(N)_\mathcal{R}$ symmetry plays an important role in both the $N = 2$ and $N = 4$ theories. Because the imaginary unit i acting on the fermions is realised by γ_5, the invariant subgroup $U(1)_\mathcal{R}$ is a chiral symmetry and hence vulnerable to quantum anomalies of the type related to the Atiyah–Singer index theorem. In the $N = 2$ supersymmetric gauge theory the divergence of the $U(1)_\mathcal{R}$ current is proportional to the instanton number density with a coefficient which is one-loop exact and possesses a nice group theoretic factor expressing the gauge group transformation properties of the field content of the theory,

$$\partial_\mu j_5^\mu = \frac{e^3 \hbar}{(4\pi)^2}(\tilde{h}(G) - x(H))\mathrm{Tr}\,(*\mathbf{F}_{\mu\nu}\mathbf{F}^{\mu\nu}). \tag{5.5}$$

Here $x(H)$ is the Dynkin index for the (possibly reducible) representation of the gauge group G carried by the hypermultiplet H. Since the gauge multiplet is automatically in the adjoint representation of G its Dynkin index is given by the dual Coxeter number of G, $\tilde{h}(G)$. Such quantities are familiar from work on conformal field theory (see Goddard and Olive [1986]). Recall that the definition of the Dynkin index $x(D)$ for a representation D of a group G is given in terms of an orthonormal basis of Lie algebra generators T^i by

$$\mathrm{Tr}\,(D(T^i)D(T^j)) = x(D)\psi^2\delta^{ij}.$$

The reason that the difference of these two integers (or half-integers) occurs as a factor on the right hand side of (5.5) is special to $N = 2$ supersymmetry. It is that the fermions of given helicity possess equal and opposite $U(1)_\mathcal{R}$

charges in the gauge and hypermultiplets. This is because the supercharges possess $U(1)_{\mathcal{R}}$ charge equal to twice their helicity and the gauge particles and the hypermultiplet scalars possess zero charge.

The result has several consequences. First the $U(1)_{\mathcal{R}}$ symmetry is broken by the presence of instantons to a cyclic subgroup, usually $Z(4(\tilde{h}(G) - x(H)))$, but there may be special effects causing an alteration of the order by a factor of 2. The nonvanishing of the vacuum expectation value of the Higgs (gauge multiplet scalars) will break this cyclic subgroup further. All this is important in the Seiberg-Witten analysis of $N = 2$ and its generalisations.

A second feature is that $N = 2$ supersymmetry relates this breaking of $U(1)_{\mathcal{R}}$ symmetry to the breaking of conformal symmetry. Hence the Callan–Symanzik beta function for the running of the gauge coupling constant is proportional to the same coefficient as appears on the right hand side of (5.5). This means that the necessary and sufficient condition for asymptotic freedom is that

$$x(H) \leq \tilde{h}(G)$$

with equality corresponding to exact conformal symmetry.

An obvious way of achieving the latter is to consider the hypermultiplet to be in the adjoint representation of the gauge group, just like the gauge multiplet. Then the eight states in each multiplet add to sixteen, the same number and helicity content as in the $N = 4$ gauge multiplet. Indeed this yields the $N = 4$ supersymmetric gauge theory which is thus seen to have a Callan–Symanzik beta function which vanishes identically, at least in perturbation theory (Sohnius and West [1981], Brink, Lindgren and Nilsson [1983], Mandelstam [1983]).

This ensures several features favourable to the validity of the electromagnetic duality conjecture. It means that the couplings, electric or magnetic, do not run, eliminating the quandary as to whether it is the renormalised or unrenormalised couplings which should satisfy the Dirac–Schwinger–Zwanziger quantisation condition (Rossi [1981]). The $N = 4$ theory was the first quantum field theory with exact quantum conformal symmetry to be identified in four dimensions. It seems that in some sense, yet to be fully understood, it is the quantum field theory closest to its classical version because of the vanishing of quantum effects such as the beta function. It is certainly the most symmetrical renormalisable quantum field theory in four dimensions and is the most inviting candidate for exact construction, presumably by mathematical technology not yet available.

The masses occur because of the scale introduced by the non-vanishing Higgs vacuum expectation values which supply the deformation from exact conformal symmetry in accordance with the insight of Zamolodchikov in two dimensions.

It is easy to spot other ways for the beta function to vanish. For example,

$\tilde{h}(SU(N_c)) = N_c$ while each flavour of hypermultiplet quark in the defining representation contributes $1/2$ to the Dynkin index. Hence $2N_c$ copies of these yield conformal symmetry. This has been understood in detail by Seiberg and Witten [1994b] when $N_c = 2$.

We now have two strong reasons for thinking that $N = 4$ supersymmetry is the correct arena for the exact electromagnetic duality conjecture of Montonen and myself, the fact that only then do the monopole-dyon solitons form a gauge multiplet and the conformal symmetry. The third and decisive reason was provided by Sen and concerns the dyons of higher magnetic charge, as we now discuss.

6 Electromagnetic Duality and the Modular Group

Now that it is established that the $N = 4$ supersymmetric $SU(2)$ gauge theory is the most promising candidate framework for the exact electromagnetic duality conjecture of Section 3, it is instructive to reexamine the charge lattice introduced at the end of Section 1 in this new light. We have an explicit theory whose particle spectrum is partly known. In particular it is clear that the only magnetically neutral particles are the three gauge $N = 4$ supermultiplets corresponding to the photon and the $W^{\pm}$. These have electrical charges 0 and $\pm q_0$, that is, coordinates $m = 0$ and $n = 0, \pm 1$. The monopole dyon solitons have $m = \pm 1$ with any value of n and all likewise form gauge supermultiplets according to the analysis of fermionic zero modes. Thus the single particle states with $m = 0, \pm 1$ are all known and, for these values of m correspond precisely to the primitive vectors of the charge lattice, that is, the points denoted by a cross in the figure at the end of chapter 1. This contrasts with the $N = 2$ supersymmetric $SU(2)$ gauge theory including $SU(2)$ doublet flavours. There the points $m = 0$, $n = 0$, ± 1, ± 2 all correspond to particles created by fields in the defining Lagrangian.

Another specific feature of the $N = 4$ supersymmetric gauge theory concerns the angle θ occurring in the real part of the complex variable τ, (1.9) which was defined by the ratio of periods of the charge lattice. As Witten explained [1979], this variable can be identified with the 'vacuum angle' which is an extra coupling parameter occurring in the $N = 4$ supersymmetric gauge Lagrangian. It multiplies an appropriate multiple of the instanton number density. Because this density is a total derivative it affects neither the classical equations of motion nor the energy-momentum density. But, as Witten showed by a Noether argument, it does affect the identification of electric charge in the way indicated by equation (1.8). Sometimes this parameter can be defined away by changing the phase in the elementary fermion fields, but not when there is $N = 4$ supersymmetry which lacks $U(1)_\mathcal{R}$ chiral symmetry.

Because of the absence of quantum corrections, the universal mass formula (2.12) is believed to be quantum exact in the $N = 4$ supersymmetric gauge theory. This means that mass is given by Euclidean distance from the origin in the charge lattice as far as concerns the short supermultiplets (with sixteen components) corresponding to single particle states. An attractive consequence of this mass formula is that any state satisfying the mass formula and given by a primitive vector is absolutely stable with respect to any decays conserving the two charges, electric and magnetic. This is because the straight lines defined by the original state and its two decay products form the sides of a non-degenerate triangle. Hence, by the triangle inequality, the decay products are heavier. This argument breaks down when the decay of a non-primitive vector state is considered. Its coordinate vector is automatically an integer multiple, k, say, of a primitive vector and its mass therefore k times the mass of the primitive vector state and so on the verge of instability.

Now let us recall the electromagnetic duality conjecture of Montonen and myself. According to that, a new action can be chosen with monopoles created by elementary fields. This is defined by a choice of primitive vector (and its negative). A second primitive vector is needed to define the θ angle. Thus the choice of action is defined by a pair of primitive vectors in the charge lattice, that is, a basis for the lattice. As remarked earlier, with this new action the roles of quantum excitations and solitons change, and we now expect to find a new set of single-particle states. These must still correspond to primitive vectors of the charge lattice, as that concept is basis-independent. This led Ashoke Sen [1994] to conjecture that in the $N = 4$ supersymmetric $SU(2)$ gauge theory there is a precise correspondence between all the single-particle states and all the primitive vectors of the charge lattice. This would imply a quite concrete statement as to which of the states $(m = \pm 2, n)$ with double magnetic charge could be single-particle states, namely only those in which the coordinate n is an odd integer, as depicted in the lattice diagram at the end of chapter 1.

It would therefore be a test of the electromagnetic duality conjecture to find the dyon states with double magnetic charge and verify that n is odd and this is what Sen did in 1994. Since from our previous work we know that these missing states can be neither quantum excitations of the original fields nor classical soliton solutions, a new mechanism is needed for single-particle states. Sen realised that a pair of like $m = 1$ monopoles could form quantum bound states in a very precise sense.

To do this he considered the moduli manifold $\mathcal{M}(2)$ of solutions to the Bogomol'nyi equations (2.9) with $m = 2$. This is a manifold whose physical significance implies certain nice mathematical properties that we need to describe. The coordinates on the manifold are the bosonic zero modes which are interpreted as the collective coordinates of the two-monopole system as already discussed. It is possible to set these coordinates in motion and find an approximate form of the action expressed in terms of these coordinates and their time derivatives. For slow motions it is quadratic in these velocities and the corresponding quadratic form for the kinetic energy automatically defines a Riemannian metric for the manifold (Manton [1982]). In principle this works for any $\mathcal{M}(m)$, as Manton first realised, but it is only when $m = 2$ that it has been possible to find a closed analytic expression for this metric. This construction, due to Atiyah and Hitchin [1985], involves elliptic functions and was discovered by indirect mathematical arguments. This metric exhibits the special feature that there are three anticommuting complex structures on it, and is an example of a hyper-Kähler manifold (as are all the $\mathcal{M}(m)$).

It follows that the slow motion of a configuration of monopoles is determined by the Euler–Lagrange equations of motion of this action and hence follow geodesics on the monopole moduli spaces. This is sufficient to determine the classical scattering of two monopoles at low relative velocity and yields surprisingly complex behaviour (Gibbons and Manton [1986]), including a type of incipient breathing motion perpendicular to the scattering plane, visible on a video prepared by IBM. Despite these beautiful results, there is no idea of how to describe relative motion of monopole solitons with unlike charge. The duality conjecture does predict pair annihilation, in contrast with the behaviour of conventional solitons in two dimensions.

When the fermionic zero modes are included the action describes a supersymmetric mechanics on $\mathcal{M}$, made possible because of the hyper-Kähler metric.

Sen realised that, when this is quantised so that the wave function of the theory satisfies a Schrödinger equation involving the Laplacian for the Riemannian manifold, then zero energy eigenvalues, that is, solutions to Laplace's equation, yield bound states satisfying the mass formula. Duality requires that this bound state be unique for odd n and absent for even n. This means that the 'function' would have to be a two-form (when the overall motion is factored out leaving the relative motion between the two monopoles described by a four-manifold). This is exactly what Sen [1994] found, namely two-monopole quantum bound states occurring if and only if n is odd. Further, these fell into the 16-dimensional short $N = 4$ gauge supermultiplet.

This leaves the question of establishing a similar result for primitive vectors of the charge lattice at higher values of $|m|$. Sen's argument cannot be repeated without an explicit form of the metric but it is possible to make encouraging confirmatory progress with indirect arguments using Hodge's theorem relating zero modes of the Laplacian to the Betti numbers of $\mathcal{M}(m)$ (Segal and Selby [1996]). Topological information about this manifold is available through its characterisation in terms of rational maps between two Riemann spheres.

Armed with the new insight that the spectrum of single-particle states corresponds to the primitive vectors of the charge lattice, augmented by the origin, rather than the five points previously considered, we can see that the original Montonen–Olive conjecture was too modest. Instead of possessing two equivalent choices of action, the $N = 4$ supersymmetric gauge theory apparently possesses an infinite number of them, all with an isomorphic structure, but with different values of the parameters.

Roughly speaking, the reason is that physical reality is described by the charge lattice endowed with an orientation given by the sign of the imaginary part of τ, (1.9), or, since $n_0 = 2$ in the $N = 4$ supersymmetric theory,

$$\tau = \frac{\theta}{2\pi} + \frac{4\pi i \hbar}{q_0^2}. \tag{6.1}$$

Choices of action correspond to choices of basis in the oriented lattice, that is, an ordered pair of non-collinear primitive vectors (or periods). As the charge lattice is two-dimensional, these choices are related by the action of the modular group, an infinite discrete group containing the previous transformation (3.1).

Let us choose a primitive vector in the charge lattice, represented by a complex number, q_0', say. Then we may ascribe short $N = 4$ supermultiplets of quantum fields to each of the three points $\pm q_0'$ and 0. The particles corresponding to the origin are massless and neutral whereas the particles corresponding to $\pm q_0'$ possess complex charge $\pm q_0'$ and mass $a|q_0'|$. We may form an $N = 4$ supersymmetric action with these fields. It is unique, given the coupling $|q_0'|$, apart from the vacuum angle whose specification requires a second primitive vector, $q_0' \tau'$, say, non-collinear with q_0'. The remaining single-particle states are expected to arise as monopole solitons or as quantum bound states of them as discussed above.

Since the two non-collinear primitive vectors q_0' and $q_0' \tau'$ form an alternative basis for the charge lattice, they can be expressed as integer linear combinations of the original basis, q_0 and $q_0 \tau$:

$$q_0' \tau' = a q_0 \tau + b q_0, \tag{6.2a}$$
$$q_0' = c q_0 \tau + d q_0, \tag{6.2b}$$

where

$$a, b, c, d \in \mathbf{Z}. \tag{6.2c}$$

Equally, $q_0 \tau$ and q_0 can be expressed as integer linear combinations of $q_0' \tau'$ and q_0'. This requires that the matrix of coefficients in (6.2a) and (6.2b) has determinant equal to ± 1,

$$ad - bc = \pm 1. \tag{6.2d}$$

Only the plus sign preserves the sign of the imaginary part of τ, (6.1), and hence the sign of the fine structure constant, which is the inverse of this. Such matrices

$$\begin{pmatrix} a & b \\ c & d \end{pmatrix}$$

form a group, $SL(2, \mathbf{Z})$, whose quotient by its centre, $(\pm I)$, is called the 'modular group'. This is an infinite discrete group, of course. Equation (6.2a) divided by (6.2b) yields

$$\tau' = \frac{a\tau + b}{c\tau + d}.$$

These are the transformations which form the modular group and preserve the sign of the imaginary part of τ. This gives the relation between the values of the dimensionless parameters in the two choices of action corresponding to the two choices of basis. It is customary to think of the modular group as being generated by elements T and S, where

$$T : \tau \to \tau + 1 \qquad S : \tau \to -\frac{1}{\tau}.$$

According to (6.1), T increases the vacuum angle by 2π. This is a trivial symmetry of the quantum theory. If the vacuum angle vanishes, S yields the transformation (3.1) above, as considered by Montonen and myself.

Notice that the modular group can be used to map between any pair of primitive vectors of the charge lattice. Hence the infinite number of single-particle massive states, each forming a 16-dimensional gauge supermultiplet of $N = 4$ supersymmetry and corresponding to a primitive vector, can be thought of as an infinite multiplet with respect to the modular group. Beware that the masses of these states can be indefinitely large, even for $m = 1$.

Proof of the quantum equivalence of all the actions associated with each choice of basis in the charge lattice would presumably require a generalised vertex operator transformation relating the corresponding quantum fields, and still provides a formidable challenge.

What we have described so far are the results that are believed to be exact. This faith is based on the remarkable confluence of a large number of quite disparate arguments, some of them of quite a sophisticated mathematical nature. For example, the Atiyah–Singer index theorem has played an ubitiquous role. Nevertheless, final proof presumably awaits further mathematical breakthroughs.

One of the hopes underlying the $N = 4$ supersymmetric theory is that it is the most symmetrical quantum field theory possible in flat space time with four dimensions. This means that gravity is ignored but it would be undeniably attractive to reinstate it. According to current wisdom, quantum consistency requires consideration of a superstring theory in ten-dimensional space-time. Now it is a historical fact that the $N = 4$ supersymmetric gauge theory was first discovered by dimensionally reducing a rather simple supersymmetric gauge theory in ten dimensions down to four (Gliozzi, Scherk and Olive [1977]). Furthermore, that supersymmetric gauge theory was constructed as what was then called the zero slope limit of the type I superstring. This is the superstring with just one real chiral supercharge in ten dimensions. Because of this pedigree it is not unreasonable to imagine that superstrings possess duality properties too, and indeed, this is what has been learnt, leading to a revitalisation of the subject. Indeed, it has led to quite a decisive advance in the conceptual understanding of string theory as a truly unified theory incorporating gravity in a convincing way.

The way is now open for the other speakers to describe these and the other more recent developments.

Acknowledgements

I am grateful to the Newton Institute for Mathematical Sciences in Cambridge and I wish to thank it for support. I would like to thank many of my colleagues for numerous helpful discussions.

References

Atiyah M.F. and Hitchin N.J. (1985) *Phys. Lett.* **107A** 21–25, 'Low energy scattering of non-abelian monopoles'.

Bogomol'nyi E.B. (1976) *Sov. J. Nucl. Phys.* **24** 449–454, 'The stability of classical solutions'.

Brink L., Lindgren O. and Nilsson B.E.W. (1983) *Phys. Lett.* **123B** 323–328 'The ultra-violet finiteness of the $N = 4$ Yang–Mills theory'.

Callias C. (1978) *Comm. Math. Phys.* **62** 213–234, 'Axial anomalies and index theorems on open spaces'.

Coleman S. (1975) *Phys. Rev.* **D11** 2088–2097, 'Quantum sine–Gordon equation as the massive Thirring model'.

Coleman S., Parke S., Neveu A., and Sommerfield C.M. (1977) *Phys. Rev.* **D15** (77) 544–545, 'Can one dent a dyon?'.

D'Adda A., Horsley R. and Di Vecchia P. (1978) *Phys. Lett.* **76B** 298–302 'Supersymmetric monopoles and dyons'.

Dirac P.A.M. (1931) *Proc. Roy. Soc.* **A33** 60–72, 'Quantised singularities in the electromagnetic field'.

Englert F. and Brout R. (1964) *Phys. Rev. Lett.* **13** 321–323, 'Broken symmetry and the mass of gauge vector bosons'.

Gibbons G. and Manton N. (1986) *Nucl. Phys.* **B274** 183–224, 'Classical and quantum dynamics of monopoles'.

Gliozzi F., Scherk J. and Olive D.I. (1977) *Nucl. Phys.* **B122** 253–290, 'Super-symmetry, supergravity theories and the dual spinor model'.

Goddard P., Nuyts, J. and Olive D.I. (1977) *Nucl. Phys.* **B125** 1–28, 'Gauge theories and magnetic charge'.

Goddard P. and Olive D.I. (1978) *Reports Prog. Phys.* 41 1357–1437, 'Magnetic monopoles in gauge field theories'.

Goddard P. and Olive D.I. (1986) *Int. J. Mod. Phys.* **A1** 303–414, 'Kac-Moody and Virasoro algebras in relation to quantum physics'.

Haag R., Łopuszański J.T. and Sohnius M. (1975) *Nucl. Phys.* **B88** 257–274, 'All possible generators of supersymmetry of the S-matrix'.

Higgs P. (1966) *Phys. Rev.* **145** 1156–1163, 'Spontaneous symmetry breakdown without massless bosons'.

't Hooft G. (1974) *Nucl. Phys.* **B79** 276–284, 'Magnetic monopoles in unified gauge theories'.

Julia B. and Zee A. (1975) *Phys. Rev.* **D11** 2227–2232, 'Poles with both magnetic and electric charges in non-abelian gauge theory'.

Kibble T.W.B. (1967) *Phys. Rev.* **155** 1554–1561, 'Symmetry breaking in non-abelian gauge theories'.

Mandelstam S. (1975) *Phys. Rev.* **D11** 3026–3030, 'Soliton operators for the quantized sine–Gordon equation'.

Mandelstam S. (1983) *Nucl. Phys.* **B213** 149–168, 'Light-cone superspace and the ultraviolet finiteness of the $N = 4$ model'.

Manton N. (1977) *Nucl. Phys.* **B126** (77) 525–541, 'The force between 't Hooft–Polyakov monopoles'.

Manton N. (1982) *Phys. Lett.* **110B** 54–56, 'A remark on the scattering of BPS monopoles'.

Montonen C. and Olive D.I. (1977) *Phys. Lett.* **72B** 117–120, 'Magnetic monopoles as gauge particles?'.

Nahm W. (1978) *Nucl. Phys.* **B135** 149–166, 'Supersymmetries and their repre-sentations'.

Olive D.I. (1982) 'Magnetic monopoles and electromagnetic duality conjectures'. In *Monopoles in Quantum Field theory*, edited by N.S. Craigie, P. Goddard and W. Nahm, World Scientific, 157–191.

Osborn H. (1979) *Phys. Lett.* **83B** 321–326, 'Topological charges for $N = 4$ supersymmetric gauge theories and monopoles of spin 1'.

Polyakov A.M. (1974) *JETP Lett.* **20** 194–195, 'Particle spectrum in quantum field theory'.

Prasad M.K. and Sommerfield C.M. (1975) *Phys. Rev. Lett,* **35** (75) 760–762, 'Exact classical solution for the 't Hooft monopole and the Julia-Zee dyon'.

Rossi P. (1981) *Phys. Lett.* **99B** 229–231, '$N = 4$ supersymmetric monopoles and the vanishing of the β function'.

Schwinger J. (1969) *Science* **165** 757–761, 'A magnetic model of matter'.

Segal G. and Selby A. (1996) *Comm. Math. Phys.* **177** 775–787, 'The cohomology of the space of magnetic monopoles'.

Seiberg N. and Witten E. (1994a) *Nucl. Phys.* **B426** 19–52, *Erratum* **B430** 485–486, 'Electromagnetic duality, monopole condensation, and confinement in $N = 2$ supersymmetric Yang–Mills theory'.

Seiberg N. and Witten E. (1994b) *Nucl. Phys.* **B431** 484–550, 'Monopoles, duality and chiral symmetry breaking in $N = 2$ supersymmetric QCD'.

Sen A. (1994) *Phys. Lett.* **329B** 217–221, 'Dyon–monopole bound states, self-dual harmonic forms on the multi-monopole moduli space, and $SL(2, \mathbf{Z})$ invariance in string theory'.

Skyrme T.H.R. (1961) *Proc. Roy. Soc.* **A262** 237–245, 'Particle states of a quantized meson field'.

Sohnius M.F. and West P.C. (1981) *Phys. Lett.* **100B** 245–250, 'Conformal invariance in $N = 4$ Yang–Mills theory'.

Weinberg E. (1979) *Phys. Rev.* **D20** 936–944, 'Parameter counting for multi-monopole solutions'.

Witten E. (1979) *Phys. Lett.* **86B** 283–287, 'Dyons of charge $e\theta/2\pi$'.

Witten E. and Olive D.I. (1978) *Phys. Lett.* **78B** 97–101, 'Supersymmetry algebras that include topological charges'.

Wu T.T. and Yang C.N. (1975) *Phys. Rev.* **D12** 3845–3857, 'Concept of non-integrable phase factors and global formulation of gauge fields'.

Zamolodchikov A.B. (1989) 'Integrable Field Theory from Conformal field Theory', in *Advanced Studies in Pure Mathematics* **19** 642–674.

Zwanziger D. (1968) *Phys. Rev.* **176** 1489–1495, 'Quantum field theory of particles with both electric and magnetic charges'.

Supersymmetric Monopoles and Duality

Jerome P. Gauntlett

Abstract

S-duality in supersymmetric gauge theories leads to precise predictions concerning the spectrum of BPS monopoles and dyons. In a semi-classical analysis these predictions translate into precise conjectures about the existence of harmonic forms or spinors on the moduli spaces of BPS monopole solutions. These notes review the current status of these investigations.

1 Introduction

Electromagnetic duality has emerged as a powerful tool to study strongly coupled quantum fields. In N=4 super-Yang–Mills theory and some special theories with N=2 supersymmetry, the duality is conjectured to be exact in the sense that it is valid at all energy scales. These theories provide the most natural setting for Montonen and Olive's original idea [1] since they have vanishing β-functions and hence the quantum corrections are under precise control. In many theories with N=2 and N=1 supersymmetry duality plays an important role in elucidating the infrared dynamics. In these models one can study strong coupling phenomenon such as confinement and chiral symmetry breaking in an exact context [2, 3].

The purpose of these lectures is to review some aspects of theories with exact duality or 'S-duality'. The duality group in these theories involves $SL(2, \mathbf{Z})$ which includes a $\mathbf{Z}_2$ corresponding to the interchange of electric and magnetic charges along with the interchange of strong and weak coupling. Sen was the first to realise [4] that this enhancement to $SL(2, \mathbf{Z})$ leads to highly non-trivial predictions about the BPS spectrum of magnetic monopoles and dyons in the theory. BPS states are important for testing duality because they form short representations of the supersymmetry algebra and hence we have some control over their behaviour as we vary the coupling. At weak coupling the predicted BPS spectrum can be translated into statements about the existence of harmonic forms or spinors on the moduli space of BPS monopole solutions.

We begin with bosonic $SU(2)$ BPS monopoles, reviewing some aspects of the moduli space approximation and discussing how quantised dyons appear in the semiclassical spectrum. Next we describe some features of $N=4$ super-Yang–Mills theory before studying the S-duality predictions. We analyse the $SU(2)$ case followed by the higher rank gauge groups. This is followed by a discussion of the S-duality for $N = 2$ super-Yang–Mills theory with gauge group $SU(2)$ and $N_f = 4$ hypermultiplets in the fundamental representation. We conclude by outlining some open problems in the study of exact duality.

2 $SU(2)$ BPS Monopoles and the Moduli Space Approximation

Consider the Yang–Mills–Higgs Lagrangian

$$\mathcal{L} = -\frac{1}{4}F^a_{\mu\nu}F^{a\mu\nu} - \frac{1}{2}D_\mu\Phi^a D^\mu\Phi^a , \qquad (2.1)$$

where $A_\mu = A^a_\mu T^a$ is an SU(2) connection with field strength $F_{\mu\nu} = \partial_\mu A_\nu - \partial_\nu A_\mu + e[A_\mu, A_\nu]$, and $\Phi = \Phi^a T^a$ transforms in the adjoint representation with the covariant derivative given by $D_\mu\Phi = \partial_\mu\Phi + e[A_\mu, \Phi]$. We choose the Lie algebra generators T^a to be anti-Hermitian. There is no potential term for the Higgs field and we are thus considering the 'BPS limit' [5, 6] which is relevant for the supersymmetric extension. The moduli space of Higgs vacua is obtained by imposing $\langle\Phi^a\Phi^a\rangle = v^2$ and is thus a two-sphere. If $v^2 \neq 0$ then $SU(2)$ is spontaneously broken to $U(1)$. The electric and magnetic charge with respect to the $U(1)$ specified by the Higgs field are given by

$$\begin{aligned} Q_e &= \frac{1}{v}\int dS^i (E^a_i \Phi^a) , \\ Q_m &= \frac{1}{v}\int dS^i (B^a_i \Phi^a) , \end{aligned} \qquad (2.2)$$

where $E^a_i = F_{0i}$ and $B^a_i = \frac{1}{2}\epsilon_{ijk}F^a_{jk}$ are the non-abelian electric and magnetic field strengths, respectively, and the integration is over a surface at spatial infinity.

The perturbative states consist of a massless photon, a massless neutral scalar and $W^\pm$ bosons with electric charge $Q_e = \pm e$ and mass ev. To analyse the monopole and dyon spectrum we need to construct classical static monopole solutions and then perform a semiclassical analysis. Let us begin by noting that for all finite energy configurations the Higgs field must lie in the vacuum at spatial infinity. The Higgs field of these configurations thus provides a map from the two-sphere at spatial infinity to the two-sphere of Higgs vacua. These maps are characterised by a topological winding number k and one can show that this implies that the magnetic charge is quantised:

$$Q_m = \frac{4\pi}{e}k . \qquad (2.3)$$

The minimal magnetic monopole charge is twice the Dirac unit because we could add electrically charged fields in the fundamental representation of $SU(2)$ that would carry $\frac{1}{2}$-integer electric charges in contrast to the integer charged W-bosons.

To proceed with the construction of static monopole solutions it will be convenient to work in the $A_0 = 0$ gauge. We must then impose Gauss' Law, the A_0 equation of motion, as a constraint on the physical fields:

$$D_i\dot{A}_i + e[\Phi, \dot{\Phi}] = 0. \tag{2.4}$$

In this gauge the Hamiltonian is $H = T + V$, where the kinetic and potential energies are given by

$$T = \frac{1}{2}\int d^3x(\dot{A}_i^a\dot{A}_i^a + \dot{\Phi}^a\dot{\Phi}^a), \tag{2.5}$$

$$V = \frac{1}{2}\int d^3x(B_i^aB_i^a + D_i\Phi^aD_i\Phi^a), \tag{2.6}$$

respectively. Noting that V can be rewritten [6] as

$$V = \frac{1}{2}\int d^3x[(B_i^a \mp D_i\Phi^a)(B_i^a \mp D_i\Phi^a)] \pm vQ_m, \tag{2.7}$$

we deduce that in each topological class k corresponding to magnetic charge given by (2.3) there is a Bogomol'nyi bound on the mass of any static classical monopole solution:

$$M \geq v|Q_m| = \frac{4\pi v}{e}|k|. \tag{2.8}$$

The static energy is minimised when the bound is saturated, which is equivalent to the Bogomol'nyi (or BPS) equations

$$B_i = \pm D_i\Phi. \tag{2.9}$$

The upper sign corresponds to positive k or 'monopoles' and the lower sign corresponds to negative k or 'anti-monopoles'. For the most part we will restrict our considerations to monopoles, the extension to anti-monopoles being straightforward. In the $A_0 = 0$ gauge there are no static dyon solutions; the dyons emerge as time dependent solutions as we will see.

The moduli space of gauge inequivalent solutions to the Bogomol'nyi equations will be denoted by $\mathcal{M}_k$. Let us discuss some of the geometry of this manifold. We begin by recalling that in the $A_0 = 0$ gauge the configuration space of fields is given by $\mathcal{C} = \mathcal{A}/\mathcal{G}$, where $\mathcal{A} = \{A_i(x), \Phi(x)\}$ is the space of finite energy field configurations and we have divided out by $\mathcal{G}$, the group of gauge transformations that go to the identity at spatial infinity. Tangent vectors $\{\dot{A}, \dot{\Phi}\}$ to $\mathcal{C}$ must satisfy Gauss Law (2.4). From this point of view, the kinetic energy in (2.5) is simply the metric on $\mathcal{C}$. The moduli Z^α that appear in the general solution to the Bogomol'nyi equations, $\{A(x, Z), \Phi(x, Z)\}$, are natural coordinates on $\mathcal{M}_k \subset \mathcal{C}$. Tangent vectors to $\mathcal{M}_k$ must also satisfy the linearised Bogomol'nyi equations

$$\epsilon_{ijk} D_j \dot{A}_k = D_i \dot{\Phi} + e[\dot{A}_i, \Phi] . \tag{2.10}$$

Using the coordinates Z^α we have

$$\{\dot{A}, \dot{\Phi}\} = \dot{Z}^\alpha \{\delta_\alpha A_i, \delta_\alpha \Phi\} , \tag{2.11}$$

where $\{\delta_\alpha A_i, \delta_\alpha \Phi\}$ satisfy

$$D_i \delta_\alpha A_i + e[\Phi, \delta_\alpha \Phi] = 0 , \tag{2.12}$$
$$\epsilon_{ijk} D_j \delta_\alpha A_k = D_i \delta_\alpha \Phi + e[\delta_\alpha A_i, \Phi] , \tag{2.13}$$

which are simply the equations for a physical zero mode in the fluctuations about a given monopole solution. The zero modes can be obtained by differentiating the general solution with respect to the moduli but in general one has to include a gauge transformation to ensure that it satisfies (2.13), i.e.,

$$\begin{aligned} \delta_\alpha A_i &= \partial_\alpha A_i - D_i \epsilon_\alpha , \\ \delta_\alpha \Phi &= \partial_\alpha \Phi - e[\Phi, \epsilon_\alpha] . \end{aligned} \tag{2.14}$$

The metric on $\mathcal{C}$ gives rise to a metric on $\mathcal{M}_k$ which can be written in terms of the zero modes:

$$\mathcal{G}_{\alpha\beta}(Z) = \int d^3 x [\delta_\alpha A_i^a \delta_\beta A_i^a + \delta_\alpha \Phi^a \delta_\beta \Phi^a] . \tag{2.15}$$

The moduli space $\mathcal{M}_k$ is $4k$-dimensional, which can be established, for example, by counting zero modes using an index theorem [7]. The space of field configurations $\mathcal{A}$ inherits three almost complex structures from those on $\mathbf{R}^4$ and they descend to give a hyper-Kähler structure on $\mathcal{M}_k$. Explicit formulae for the complex structures on $\mathcal{M}_k$ in terms of the zero modes can be found in [8]. More details on the geometry of $\mathcal{M}_k$ can be found in [9]. Note that the moduli space of k anti-monopoles is isometric to that of k monopoles.

The moduli space for a single BPS monopole can be determined by explicitly constructing the most general solution and we find $\mathcal{M}_1 = \mathbf{R}^3 \times S^1$. The $\mathbf{R}^3$ piece simply corresponds to the position of the monopole in space. The S^1 arises from the gauge transformation $g = e^{\chi \Phi / v}$ on any solution. Since this does not go to the identity at infinity, it is a 'large' gauge transformation, it corresponds to a physical motion. Since all fields are in the adjoint a 2π rotation in $SU(2)$ is the identity and we conclude that $0 \leq \chi < 2\pi$. We will see that this coordinate is a dyon degree of freedom.

We have noted that the dimension of $\mathcal{M}_k$ is $4k$. The physical reason for the existence of these multimonopole configurations is that in the BPS limit there is a cancellation between the vector repulsion and scalar attraction between two monopoles. Heuristically one can think of the $4k$ dimensions as corresponding to a position in $\mathbf{R}^3$ and a phase for each monopole, but the structure of $\mathcal{M}_k$ turns out to be much more subtle and interesting. For general k we can separate out a piece corresponding to the motion of the centre of mass of the multi-monopole configuration and we have $\mathcal{M}_k = \mathbf{R}^3 \times (S^1 \times \tilde{\mathcal{M}}_k^0)/\mathbf{Z}_k$. The S^1 factor is related to the total electric charge. $\tilde{\mathcal{M}}_k^0$ is $4(k-1)$ dimensional and hyper-Kähler. It admits an $SO(3)$ group of isometries which corresponds to a rotation of the multi-monopole configuration in space. Although the topology of these spaces is well understood, the metric is explicitly known only for $k = 2$ [9].

To determine the semi-classical spectrum of states with magnetic charge k we start with a classical solution $(A^{cl}(x, Z), \Phi^{cl}(x, Z))$. To have a well defined perturbation scheme with $e \ll 1$, we need to introduce a collective coordinate for each zero mode; these are the moduli Z^α. We then expand an arbitrary time dependent field as a sum of the massive modes with time dependent coefficients and allow the collective coordinates to become time-dependent (see, e.g., [10]). A low-energy ansatz for the fields is obtained by ignoring the massive modes and demanding that the only time dependence is via the collective coordinates. Thus we are led to the ansatz[1]

$$
\begin{aligned}
A_i(x, t) &= A_i^{cl}(x, Z(t)) , \\
\Phi(x, t) &= \Phi^{cl}(x, Z(t)), \\
A_0 &= \dot{Z}^\alpha \epsilon_\alpha .
\end{aligned}
\qquad (2.16)
$$

After substituting this into the action (2.1) we obtain an effective action

$$
S = \frac{1}{2} \int dt \mathcal{G}_{\alpha\beta} \dot{Z}^\alpha \dot{Z}^\beta - \frac{4\pi v}{e} k , \qquad (2.17)
$$

which is precisely that of a free particle propagating on the moduli space $\mathcal{M}_k$ with metric (2.15). This is the moduli space approximation [12]. The classical equations of motion are simply the geodesics on $\mathcal{M}_k$.

To proceed with the semiclassical analysis we need to study the quantum mechanics of (2.17). Let us show how a quantised spectrum of dyons emerges in the quantum theory. For $k = 1$ we have $\mathcal{M}_1 = \mathbf{R}^3 \times S^1$ and including various constants we have:

$$
S = \frac{1}{2} \int dt \left[(\frac{4\pi v}{e}) \dot{\mathbf{Z}}^2 + \frac{4\pi}{v e^3} \dot{\chi}^2 \right] - \frac{4\pi v}{e} . \qquad (2.18)
$$

The wavefunctions are plane waves of the form $e^{i\mathbf{P}\cdot\mathbf{Z}} e^{i n_e \chi}$ where n_e is an integer. In the moduli space approximation $Q_e = -i e \partial_\chi$ and we see that

---

[1]Note that the A_0 term is included to ensure that the motion is orthogonal to gauge transformations. One could do a gauge transformation if one wants to remain in the $A_0 = 0$ gauge (see also the discussion in [11]).

we have a tower of dyons with $Q_e = n_e e$. The mass of these states can be calculated and we get

$$\begin{aligned} M &= n_e^2 v e^3 / 8\pi + 4\pi v / e \\ &\approx v[Q_e^2 + Q_m^2]^{1/2} \, , \end{aligned} \qquad (2.19)$$

where we have used the fact that we are assuming $e \ll 1$ in our approximations. By generalising the argument that led to (2.8) it can be shown [13] that $M \geq v[Q_e^2 + Q_m^2]^{1/2}$ for all classical solutions to the equations of motion (static dyons can be obtained if we do not work in the $A_0 = 0$ gauge). We thus see that in the moduli space approximation the bound is saturated. Of course in the purely bosonic theory we are considering here this could get disturbed by quantum corrections.

For $k > 1$ we can perform a similar analysis on $\mathcal{M}_k$, looking for scattering states and bound states of the Hamiltonian in the usual fashion. This has been pursued for $k = 2$ in the bosonic theory in [14, 15]. The momentum conjugate to the coordinate on the S^1 gives the total electric charge Q_e of the configuration[2] and the bound states have masses $M = v[Q_e^2 + Q_m^2]^{1/2} + \Delta E$ where ΔE is the relative kinetic energy.

We conclude this section by considering a renormalisable term that we can add to the Lagrangian (2.1) that plays an important role in duality:

$$\delta \mathcal{L} = -\frac{\theta e^2}{32\pi^2} F_{\mu\nu}^a * F^{a\mu\nu} \, . \qquad (2.20)$$

Since it is a total derivative it doesn't affect the equations of motion. However, it is related to instanton effects and it also affects the electric charge of dyons. Recall that the dyon collective coordinate arose from doing a gauge transformation about the Φ axis. The Noether charge picks up a θ dependent contribution and one finds that $Q_e = n_e e + (e\theta/2\pi)n_m$ [16]. In the moduli space approximation this manifests itself via $Q_e = -ie\partial_\chi + (e\theta/2\pi)n_m$. At this point it is convenient to rescale the fields $\{A, \Phi\} \to \{A, \Phi\}/e$. Our combined Lagrangian then takes the simple form

$$\mathcal{L} = -\frac{1}{16\pi} \mathrm{Im}\, \tau [F_{\mu\nu}^a F^{a\mu\nu} + iF_{\mu\nu}^a * F^{a\mu\nu}] - \frac{1}{2e^2} D\Phi^2 \, , \qquad (2.21)$$

where we have introduced the complex parameter

$$\tau = \frac{\theta}{2\pi} + \frac{4\pi i}{e^2} \, . \qquad (2.22)$$

The BPS mass formula for dyons (2.19) is then given by

$$M = v|n_e + n_m \tau| \, . \qquad (2.23)$$

Due to the rescaling, here and in the following v contains a hidden factor of the coupling constant e.

---

[2] The individual electric charges of each monopole are not good quantum numbers due to the possibility of W-boson exchange which the moduli space approximation incorporates.

3 $N = 4$ Super-Yang–Mills

N=4 super-Yang-Mills theory has the maximal amount of supersymmetry with spins less than or equal to one. It has vanishing beta-function and is conjectured to exhibit S-duality, which we shall define below. We consider N=4 super-Yang–Mills with arbitrary simple gauge group G. It can be obtained as the dimensional reduction on a six-torus of N=1 super-Yang–Mills theory in ten dimensions (see, e.g., [17]). The ten-dimensional Lorentz group reduces to $SO(3,1) \times SO(6)$ and $SO(6)$, or more precisely Spin(6), becomes a global symmetry of the theory. The bosonic fields in the supermultiplet come from the ten dimensional gauge field and consist of a gauge field and six Higgs fields ϕ^I, transforming as a **6** of Spin(6), all taking values in the adjoint representation of G. There are four Weyl fermions in the adjoint transforming as a **4** of Spin(6) that come from the reduction of the Majorana-Weyl spinor in ten dimensions. Including a θ parameter, the bosonic part of the action is

$$\begin{aligned} S &= -\frac{1}{16\pi}\text{Im} \int \tau \text{Tr} \left(F^2 + iF * F \right) \\ &- \frac{1}{2e^2} \int \left[\text{Tr}\, D\phi^I D\phi^I + V(\phi^I) \right] , \end{aligned} \tag{3.1}$$

where the potential is given by

$$V(\phi^I) = \sum_{1 \leq I < J \leq 6} \text{Tr}\, [\phi^I, \phi^J]^2 , \tag{3.2}$$

and here we have taken $\text{Tr}\, T^a T^b = \delta^{ab}$.

The classical vacua of the theory are given by $V(\phi^I) = 0$ or equivalently $[\phi^I, \phi^J] = 0$ for all I, J. In this theory, there are no quantum corrections to the moduli space of vacua. For generic vacua, i.e., generic expectation values $\langle \phi^I \rangle$, the gauge symmetry is broken down to $U(1)^r$, where r is the rank of the gauge group. A given N=4 theory is specified by G, $\langle \phi^I \rangle$ and τ.

The six Higgs fields define a set of conserved electric and magnetic charges which appear as central charges in the N=4 supersymmetry algebra:

$$\begin{aligned} Q_e^I &= \frac{1}{ev} \int d\mathbf{S} \cdot \text{Tr}\, (\mathbf{E}\phi^I) , \\ Q_m^I &= \frac{1}{ev} \int d\mathbf{S} \cdot \text{Tr}\, (\mathbf{B}\phi^I) , \end{aligned} \tag{3.3}$$

For BPS saturated states, i.e., states in the short 16 dimensional representation of the supersymmetry algebra, the mass is exactly given by the formula [17]:

$$M^2 = \frac{v^2}{e^2}((Q_e^I)^2 + (Q_m^I)^2) . \tag{3.4}$$

The spin content of the short BPS multiplet is the same as the massless multiplet and has spins ≤ 1. The S-duality predictions of these short BPS multiplets will be the main focus of these notes. There are also medium sized representations consisting of 64 states with spins $\leq 3/2$, but these only arise when Q_e^I is not proportional to Q_m^I. These can only appear when the rank of the gauge group is greater than one and we shall later briefly discuss the S-duality predictions concerning them. Although the medium sized representations saturate a BPS bound we shall only refer to short representations as BPS states in the following for ease of exposition. The generic representation of the $N=4$ algebra has 256 states with spins ≤ 2 and the masses can be renormalised.

It is important to emphasise that the mass formula for BPS states (3.4) is derived from the supersymmetry algebra and hence it is valid in the quantum theory [18, 17], in contrast to the bosonic case. Thus the mass of BPS states is exactly given by their electric and magnetic quantum numbers. This is an important property of BPS states which enables us to use them to test S-duality. It will also be useful to note that half of the supersymmetry generators are realised as zero on a BPS multiplet. This is sometimes rephrased as saying that BPS states preserve (or break) half of the supersymmetry.

In a generic vacuum $\langle \phi^I \rangle$ at weak coupling we deduce that there are massive W-boson BPS multiplets. To determine the dyon spectrum we need to quantise the BPS monopole solutions in a semiclassical context. For simplicity we will restrict our attention in the following to a single direction in the moduli space of vacua:

$$\langle \phi^2 \rangle = \ldots = \langle \phi^6 \rangle = 0 \,,$$
$$\phi^1 \equiv \Phi \,, \qquad \langle \mathrm{Tr}\, \Phi^2 \rangle = v^2, \tag{3.5}$$

which clearly satisfies $V(\phi^I) = 0$. Classical BPS monopole solutions with zero electric charge are then obtained by solving the Bogomol'nyi equations we considered before

$$B_i = D_i \Phi \,. \tag{3.6}$$

Note that for the vacua (3.5) only the first component of the electric and magnetic charges are non-zero and we will write $Q_e^1 = Q_e$, $Q_m^1 = Q_m$. More general vacua have been considered in [19] but the region we will analyse seems to lead to the richest BPS monopole physics.

4 $N = 4$, $G = SU(2)$ and S-Duality

We now restrict our attention to gauge group $G = SU(2)$. Since we are focusing on a single Higgs field (3.5) we will be able to directly use many of the results in Section 2. We will assume $v^2 \neq 0$ so that $SU(2) \rightarrow U(1)$.

BPS states with charges (n_m, n_e) satisfy the mass formula (2.23). It is an important fact that BPS states with (n_m, n_e) relatively prime integers are absolutely stable for all values of τ. This is deduced by charge conservation and the triangle inequality.

We now state the S-duality conjecture: the $SL(2, \mathbf{Z})$ transformations

$$\tau \rightarrow \frac{a\tau + b}{c\tau + d},$$

$$(n_m, n_e) \rightarrow (n_m, n_e) \begin{pmatrix} a & b \\ c & d \end{pmatrix}^{-1}, \qquad (4.1)$$

where $a, b, c, d \in \mathbf{Z}$, $ad - bc = 1$, give the same theory [4]. The $SL(2, \mathbf{Z})$ group is generated by $T : \tau \rightarrow \tau + 1$, which is equivalent to the transformation $\theta \rightarrow \theta + 2\pi$ that can be deduced in perturbation theory (after a relabeling of states), and $S : \tau \rightarrow -1/\tau$, which for $\theta = 0$ is equivalent to strong-weak coupling and electric-magnetic duality originally considered in [1, 17].

Since the BPS mass formula (2.23) can be derived from the supersymmetry algebra and hence it holds in the quantum theory, it must be invariant under S-duality. To see the invariance one should note that $v \rightarrow v' = v|c\tau+d|$ under a $SL(2, \mathbf{Z})$ transformation because we rescaled the Higgs field by a factor of the coupling constant e.

We now argue that there are more sophisticated tests of S-duality. We begin by noting that the perturbative spectrum can be determined at weak coupling and consists of a neutral massless photon multiplet $(0, 0)$ and massive $W^{\pm}$-boson BPS multiplets with charge $(0, \pm1)$. S-duality maps the W-boson multiplets to BPS states (k, l), with k and l relatively prime integers, typically at strong coupling. But since these are precisely the absolutely stable BPS states they cannot decay as we vary τ and we deduce that they must also exist at weak coupling where we can search for them using semiclassical techniques. We will argue that they can be translated into the existence of certain geometric structures on the moduli space $\mathcal{M}_k$. Note that the photon BPS multiplet lies on its own $SL(2, \mathbf{Z})$ orbit. The lattice of BPS states predicted by S-duality is drawn in Figure 1.

If we assume that the entire spectrum of BPS states does not vary as we change the coupling then we can deduce that the above BPS states are the only BPS states in the theory. Any extra states would necessarily be BPS states at threshold, i.e., at threshold to decay into other BPS states. For example, the mass of a potential BPS state $(2, 2)$ is only marginally stable into the decay of two $(1, 1)$ states. If there were such states at threshold then we could use S-duality to map them to purely electrically charged states $(0, n)$ with $n \neq \pm1$. Using our assumption that the spectrum of BPS states doesn't change as we vary the coupling, we conclude that these states should exist

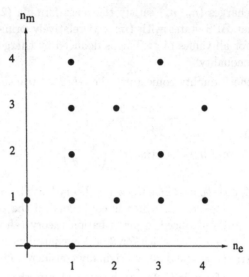

Figure 1. The lattice of BPS states in N=4 super-Yang–Mills theory with gauge group $SU(2)$.

at weak coupling but this contradicts what we see in perturbation theory[3]. We believe that the additional assumption is weak due to the very strong constraints that $N{=}4$ supersymmetry imposes on the quantum theory. What is now known about the BPS spectrum, and will be reviewed below, supports this assumption[4].

Let us now translate the prediction of the spectrum of BPS states with relatively prime charges (k, l) into statements about the moduli space of monopoles. The semiclassical analysis begins with the moduli space $\mathcal{M}_k$ of BPS monopole solutions. We have noted that the $4k$ coordinates on $\mathcal{M}_k$ can be interpreted as collective coordinates that must be introduced for $4k$ bosonic zero modes. In the $N{=}4$ context we also have fermionic zero modes. These arise from solving the Dirac equation for the fermion fields in the presence of a given monopole solution. There are four Weyl or two Dirac spinors in the adjoint of $SU(2)$ and an index theorem [21] tells us that there are $4k$ fermionic c-number zero modes that require the introduction of $4k$ complex Grassmann odd fermionic 'collective coordinates' ψ^α. This means the low-energy ansatz for the fermions will include terms of the schematic form

$$\lambda(x,t) \sim \psi(t)\lambda^{cl}(x, Z(t)) \tag{4.2}$$

---

[3] Assuming that we can trust perturbation theory to provide us with an accurate picture of the purely electrically charged spectrum at weak coupling.

[4] See [20] for an alternative way of determining the BPS spectrum that also bears on this issue.

where $\lambda^{cl}(x, Z)$ is a c-number fermion zero mode for the monopole solution specified by the moduli Z. We noted above that BPS states preserve half of the supersymmetry. This manifests itself in the fact that half of the super-symmetry generators leave the classical BPS monopole solution invariant. It can be shown that the bosonic and fermionic zero modes form a multiplet of the unbroken supersymmetries. This is essential in obtaining a supersymmetric low-energy ansatz for the fields. The ansatz for the low-energy fields is technically quite involved and has been carried out in [8, 22]. The result of substituting the ansatz into the spacetime Lagrangian leads to the following supersymmetric quantum mechanics:

$$S = \frac{1}{2} \int dt (\mathcal{G}_{\alpha\beta}[\dot{Z}^\alpha \dot{Z}^\beta + i\bar{\psi}^\alpha \gamma^0 D_t \psi^\beta]$$
$$+ \frac{1}{6} R_{\alpha\beta\gamma\delta} \bar{\psi}^\alpha \psi^\gamma \bar{\psi}^\beta \psi^\delta) - \frac{4\pi v}{e^2} k , \qquad (4.3)$$

where we have traded the complex ψ^α for a real two-component Majorana spinor ψ^α_i and the covariant derivative of these fermions is obtained using the pullback of the Christoffel connection: $D_t \psi^\alpha = \dot{\psi}^\alpha + \Gamma^\alpha_{\beta\gamma} \dot{Z}^\beta \psi^\gamma$. For a general metric the supersymmetric quantum mechanics has N=1 supersymmetry specified by a real two-component spinor ϵ. In the case that the target is hyper-Kähler there are an additional three supersymmetries with parameters $\epsilon^{(m)}$ [23]. Since the monopole moduli spaces are hyper-Kähler there are eight real supersymmetry parameters which precisely correspond to the half of the spacetime supersymmetry that is preserved by BPS states.

The quantisation of this model is discussed in [24]. The states are in one to one correspondence with differential forms on $\mathcal{M}_k$. There are four real two-component supercharges. Replacing one of these with a complex one-component charge Q we can write the Hamiltonian as $H = \{Q, Q^\dagger\} + 4\pi n_m v/e^2$ where we have included the topological term. The supersymmetry charge Q is realised as the exterior derivative acting on forms, $Q = d$, and $Q^\dagger$ as its adjoint $Q^\dagger = d^\dagger = *d*$ with $*$ being the Hodge star acting on forms. As a consequence, the Hamiltonian is the Laplacian acting on differential forms:

$$H = dd^\dagger + d^\dagger d + \frac{4\pi v}{e^2} n_m. \qquad (4.4)$$

For $n_m = 1$ we have $\mathcal{M}_1 = \mathbf{R}^3 \times S^1$. A basis of forms is given by $\{1, dZ^\alpha, \dots, dZ^1 \wedge dZ^2 \wedge dZ^3 \wedge dZ^4\}$ which gives 16 states corresponding to a BPS multiplet. To be more precise we need to check that the spins of these states are the same as those of the BPS multiplet. For $n_m = 1$ all of the fermionic zero modes can be constructed explicitly as Goldstinos by acting with the broken supersymmetry generators. One can check the angular momentum content and one finds that the spin content is that of a BPS multiplet [17]. The wave functions multiplying these forms are just as in the bosonic case, $e^{i\mathbf{P}\cdot\mathbf{Z}}e^{in_e\chi}$, corresponding to dyons with $Q_e = n_e e + e\theta/2\pi$. The Laplacian on $\mathbf{R}^3 \times S^1$ is trivial and by following the same arguments as in the bosonic case, we deduce that the mass of these states is given by (2.23). Putting this together we deduce that for $n_m = 1$ there is a tower of BPS dyon states $(n_m, n_e) = (1, n_e)$ exactly as predicted by duality.

Now we turn to $n_m = k > 1$. In this case $\mathcal{M}_k = \mathbf{R}^3 \times (S^1 \times \tilde{\mathcal{M}}_k^0)/\mathbf{Z}_k$. If we first ignore the $\mathbf{Z}_k$ identification, then the states are tensor products of forms on $\mathbf{R}^3 \times S^1$ with forms on $\tilde{\mathcal{M}}_k^0$, respectively, $|s\rangle = |\omega\rangle_{n_e} \otimes |\alpha\rangle$. The analysis for the states $|\omega\rangle_{n_e}$ is similar to the $n_m = 1$ case: there are 16 differential forms that are again associated with Goldstinos and these make up the spin content of a BPS multiplet. The wave functions give rise to quantised electric charge with $Q_e = n_e e + e\theta n_m/2\pi$. The energy of the states $|s\rangle$ can be determined and we find

$$H|s\rangle = (\frac{P^2}{2M} + M)|\omega_{n_e}\rangle \otimes |\alpha\rangle + |\omega\rangle_{n_e} \otimes |\Delta\alpha\rangle \qquad (4.5)$$

with M given by the BPS mass formula (2.23). Thus to get a BPS state with charges (n_m, n_e) we need a normalisable (i.e., L^2) harmonic[5] form α on $\tilde{\mathcal{M}}_k^0$. The action of $\mathbf{Z}_k$ on the S^1 is a cyclic shift which leads to the action $|\omega\rangle_{n_e} \to e^{2\pi i n_e/k}|\omega\rangle_{n_e}$. Hence for the state $|s\rangle$ to be well defined on $\mathcal{M}_k$ we need the form $|\alpha\rangle$ to transform as $|\alpha\rangle \to e^{-2\pi i n_e/k}|\alpha\rangle$.

Recalling that duality predicts that for each pair of relatively prime integers (k, l) there is a unique BPS state, we conclude that $\tilde{\mathcal{M}}_k^0$ must have a unique normalisable harmonic form which picks up a phase $e^{-2\pi i l/k}$ under the $\mathbf{Z}_k$ action, for every relatively prime pair of integers (k, l). The uniqueness implies that the form must either be self-dual or anti-self dual, since otherwise acting with the Hodge star $*$ would generate another harmonic form with the above properties. This conjecture was formulated by Sen who also found the harmonic form for $k = 2$. [4]. For $k > 2$ substantial evidence was provided in [25] (see also [26]).

5 $N = 4$, **Rank** $(G) > 1$

We now turn to $N=4$ theories with simple gauge groups G with rank $r > 1$ with maximal symmetry breaking to $U(1)^r$. For simplicity of notation we will often discuss the case $G = SU(3) \to U(1)^2$. The Lie algebra of G has a maximal abelian subalgebra H with r generators H_i. We can define raising

---

[5]Note that if $\Delta\alpha = \epsilon\alpha$ with $\epsilon \neq 0$ then by acting with the supersymmetry charges one can show that it always comes in multiplets of 16. Combining this with the 16 states $|\omega\rangle$ gives rise to a 256 multiplet of $N=4$ supersymmetry. These states are relevant for studying the scattering of BPS states.

and lowering operators $E_{\pm\alpha}$ that satisfy

$$[H_i, E_\alpha] = \alpha_i E_\alpha,$$
$$[E_\alpha, E_{-\alpha}] = \sum_{i=1}^{r} \alpha^i H_i \ . \tag{5.1}$$

(a linear combination of these generators give the T^a satisfying $\mathrm{Tr}\, T^a T^b = \delta_{ab}$ that we used before). $\boldsymbol{\alpha}$ is an r-component root vector. A basis of simple roots, $\boldsymbol{\beta}^{(a)}$ ($a = 1, \cdots, r$), may be chosen such that any root is a linear combination of $\boldsymbol{\beta}^{(a)}$ with integral coefficients all of the same sign. Positive roots are those with positive coefficients.

We continue to work with a single Higgs field Φ by restricting our attention to the special part of moduli space (3.5). We may choose the Cartan subalgebra such that our vacuum is specified by $\langle\Phi\rangle = v\mathbf{h}\cdot\mathbf{H}$ with $v^2 = \langle\mathrm{Tr}\Phi^2\rangle$. If $\boldsymbol{\alpha}\cdot\mathbf{h} = 0$ for some root $\boldsymbol{\alpha}$ then the unbroken gauge group is nonabelian. Otherwise, maximal symmetry breaking occurs, and $\langle\Phi\rangle$ picks out a unique set of simple roots which satisfy the condition $\mathbf{h}\cdot\boldsymbol{\beta}^{(a)} > 0$ [27].

Since the fields are in the adjoint representation, the electric quantum numbers of states live on the r-dimensional root lattice spanned by the simple roots $\boldsymbol{\beta}^{(a)}$,

$$\mathbf{q} = \sum n_a^e \boldsymbol{\beta}^{(a)} \ , \tag{5.2}$$

where the n_a^e are integers. The electric charge (for $\theta = 0$) is then given by

$$Q_e = e\mathbf{h}\cdot\mathbf{q} \ . \tag{5.3}$$

At weak coupling we deduce that for each root $\boldsymbol{\alpha}$ there is a BPS W-boson with $\mathbf{q} = \boldsymbol{\alpha}$. For $SU(3)$ we have W-bosons with $\mathbf{n}^e = \pm(1,0), \pm(0,1)$ and $\pm(1,1)$ corresponding to the two simple roots $\boldsymbol{\beta}^{(1)}, \boldsymbol{\beta}^{(2)}$ and the non-simple positive root $\boldsymbol{\gamma} = \boldsymbol{\beta}^{(1)} + \boldsymbol{\beta}^{(2)}$, respectively. From (3.4) we deduce that the W-bosons corresponding to simple roots are stable, while those corresponding to the non-simple roots are only neutrally stable. In $SU(3)$ we have that $M_\gamma = M_{\boldsymbol{\beta}^{(1)}} + M_{\boldsymbol{\beta}^{(2)}}$.

Magnetic quantum numbers arise from topologically nontrivial field configurations. For any finite energy solution the Higgs field must approach the vacuum: let the asymptotic value along the positive z-axis be $\Phi_0 = v\mathbf{h}\cdot\mathbf{H}$ (the value in any other direction can only differ from this by a gauge transformation). Asymptotically we also have

$$B_i = \frac{r_i}{4\pi r^3}G(\Omega) \ , \tag{5.4}$$

where G is covariantly constant, and takes the value G_0 along the positive z-axis. The Cartan subalgebra may be chosen so that $G_0 = \mathbf{g}\cdot\mathbf{H}$. For a smooth solution this quantity must satisfy a topological quantization condition [28, 29]

$$e^{iG_0} = I \ . \tag{5.5}$$

The solution to this equation is

$$\mathbf{g} = 4\pi \sum n_a^m \boldsymbol{\beta}^{(a)*} \ , \tag{5.6}$$

where the n_a^m are integers and the $\boldsymbol{\beta}^{(a)*}$ are the simple coroots, defined as

$$\boldsymbol{\beta}^{(a)*} = \frac{\boldsymbol{\beta}^{(a)}}{\boldsymbol{\beta}^{(a)2}} \ . \tag{5.7}$$

The magnetic quantum numbers thus live on the coroot lattice spanned by the $\boldsymbol{\beta}^{(a)*}$. For maximal symmetry breaking, all of the n_a^m are conserved topological charges, labeling the homotopy class of the Higgs field configuration. For solutions of the Bogomol'nyi equations (3.6) all of the integers in n_a^m have the same sign. The topological charge $\mathbf{g}$ determines the magnetic charge by the formula

$$Q_m = \frac{1}{e}\mathbf{g} \cdot \mathbf{h} \ . \tag{5.8}$$

A general dyon state may be labeled either by the electric and magnetic charge r-vectors $\mathbf{q}$, $\mathbf{g}$ or by the integer valued r-vectors $\mathbf{n}^e$ and $\mathbf{n}^m$. For a BPS state the mass is given by the BPS mass formula (3.4) which, using (5.3) and (5.8), can be recast in the form

$$M = v|(\mathbf{h} \cdot \boldsymbol{\beta}^{(a)})n_a^e + \tau(\mathbf{h} \cdot \boldsymbol{\beta}^{(a)*})n_a^m| \ , \tag{5.9}$$

where we have reinstated θ.

We now have enough definitions to define the action of S-duality. It is the natural generalisation of the $SU(2)$ case (4.1): the $SL(2,\mathbf{Z})$ duality on a general dyon state is given by

$$\tau \ \rightarrow \ \frac{a\tau + b}{c\tau + d} \ ,$$

$$(\mathbf{n}^m, \mathbf{n}^e) \ \rightarrow \ (\mathbf{n}^m, \mathbf{n}^e)\begin{pmatrix} a & b \\ c & d \end{pmatrix}^{-1} \ , \tag{5.10}$$

and when we act with the S-generator $S : \tau \rightarrow -1/\tau$, we must replace the group G with its dual group G^* [28]. For simply-laced groups i.e., all the roots have the same length (the ADE groups), the $N=4$ supersymmetric Lagrangian with gauge group G is the same as that of G^* since all fields are in the adjoint representation. For non-simply-laced groups this is not true since for example $SO(2N + 1)^* = Sp(N)$. We shall restrict our considerations to simply-laced gauge groups in the following, the duality groups and the predictions of the dyon spectrum for the non-simply-laced groups are discussed in [30, 31].

Just as in the $SU(2)$ case S-duality maps the perturbative W-boson states into an infinite number of dyon BPS states. For the $SU(3)$ case we generate the following $SL(2, \mathbf{Z})$ orbits:

$$
\begin{aligned}
(\mathbf{n}^m, \mathbf{n}^e) = \quad & (k(1,0), l(1,0)), \\
& (k(0,1), l(0,1)), \\
& (k(1,1), l(1,1)) ,
\end{aligned}
\tag{5.11}
$$

for relatively prime integers k and l. Like the $SU(2)$ case, we have again typically been mapped to strong coupling. For the first two classes of states we note from the BPS mass formula (5.9) that they are absolutely stable and hence we conclude that they also exist at weak coupling. The whole orbit of states coming from the $(1,1)$ W-boson are only marginally stable. Consequently we have to again employ the additional assumption that in the $N{=}4$ theory the spectrum of marginal states does not change as we vary the coupling. In this case we should see these states at weak coupling also.

It is worth noting here that by starting with the perturbative spectrum of W-bosons, whose electric and magnetic charge vectors Q_e^I and Q_m^I are trivially parallel, S-duality only makes predictions about the short BPS representations of the $N{=}4$ supersymmetry algebra. If any medium sized representations of the $N{=}4$ algebra existed they would necessarily have non-parallel charge vectors and lie on separate $SL(2, \mathbf{Z})$ orbits. By realising the field theories on the world-volume of D-3-branes in string theory, it has recently been suggested that such orbits do exist [32]. It would be interesting to investigate this further.

To test the S-duality predictions (5.11) we begin by reviewing some aspects of BPS monopole solutions. Using an index theorem, Weinberg has argued [27] that the moduli space of monopoles of charge $\mathbf{n}^m$ has dimension

$$
d = 4 \sum_a n_a^m .
\tag{5.12}
$$

A number of explicit monopole solutions can be constructed by embedding $SU(2)$ monopoles as follows [33]. Let ϕ^s, A_i^s be an $SU(2)$ monopole solution with charge k and Higgs expectation value λ. If we let $\boldsymbol{\alpha}$ be any root satisfying $\boldsymbol{\alpha} \cdot \mathbf{h} > 0$ then we can define an $SU(2)$ subgroup with generators

$$
\begin{aligned}
t^1 &= (2\alpha^2)^{-1/2}(E_{\boldsymbol{\alpha}} + E_{-\boldsymbol{\alpha}}) \\
t^2 &= -i(2\alpha^2)^{-1/2}(E_{\boldsymbol{\alpha}} - E_{-\boldsymbol{\alpha}}) \\
t^3 &= (\alpha^2)^{-1}\boldsymbol{\alpha} \cdot \mathbf{H} .
\end{aligned}
\tag{5.13}
$$

A monopole with magnetic charge

$$\mathbf{g} = 4\pi k \boldsymbol{\alpha}^* \tag{5.14}$$

is then given by

$$
\begin{aligned}
\Phi &= \sum_s \phi^s t^s + v(\mathbf{h} - \frac{\mathbf{h}\cdot\boldsymbol{\alpha}}{\alpha^2}\boldsymbol{\alpha})\cdot\mathbf{H} \\
A_i &= \sum_s A_i^s t^s \\
\lambda &= v\mathbf{h}\cdot\boldsymbol{\alpha} .
\end{aligned}
\tag{5.15}
$$

Since the moduli space of $SU(2)$ monopoles with charge k has dimension $4k$ these solutions provide a $4k$-dimensional submanifold of monopole solutions with charge (5.14). Note that by embedding an $SU(2)$ monopole with charge one we obtain spherically symmetric monopole solutions.

Weinberg has shown [27] that there is a distinguished set of r 'fundamental monopoles' with $\mathbf{g} = 4\pi\boldsymbol{\beta}^{(a)*}$, i.e., they have magnetic charge vectors $\mathbf{n}^m$ consisting of a one in the ath position and zeroes elsewhere. The reason for calling them fundamental is twofold. First, they have no 'internal' degrees of freedom: all of these solutions can be constructed by embedding an $SU(2)$ monopole of unit charge using the corresponding simple root and consequently they have only four zero modes: three translation zero modes and a $U(1)$ phase zero mode corresponding to dyonic excitations of the same $U(1)$ as where the magnetic charge lies[6]. Secondly, the index theorem (5.12) is consistent with thinking of a general monopole with charge $\mathbf{n}^m$ as a multimonopole configuration consisting of n_a^m fundamental monopoles of type a.

Note that in the special case of magnetic monopoles with charge vector $\mathbf{g} = 4\pi k \boldsymbol{\beta}^{(a)*}$, i.e., consisting of k fundamental monopoles of the same type, the dimension of moduli space is $4k$. Thus we deduce that these solutions can all be obtained by embedding $SU(2)$ monopoles of charge k, using the embedding based on the same simple root.

Let us now return to the BPS states predicted by S-duality. We need to study the semiclassical quantisation for a given magnetic charge $\mathbf{n}^m$. Just as in the $SU(2)$ case the bosonic zero and fermionic zero modes are paired by the unbroken supersymmetry and a low-energy ansatz again leads to the $N=4$ supersymmetric quantum mechanics (4.3) on the moduli space of solutions $\mathcal{M}_{\mathbf{n}^m}$. First consider monopoles with $\mathbf{n}^m = (k,0)$ or $\mathbf{n}^m = (0,k)$, i.e., k fundamental monopoles of the same type. The moduli space of these monopoles is the $SU(2)$ moduli space $\mathcal{M}_k$. The dyonic states with charges

---

[6]One can check that the embedded $SU(2)$ solutions are invariant under gauge transformations of the other $U(1)$'s.

$(k(1,0), l(1,0))$ and $(k(0,1), l(0,1))$ predicted by duality are equivalent to the harmonic forms on $\mathcal{M}_k$ required by S-duality in the $SU(2)$ theory. The results of [4, 25] thus constitute tests of duality for higher rank gauge groups [34].

The new predictions for $SU(3)$ monopoles arise in the sectors with both magnetic quantum numbers non-zero. In particular, the $(k(1,1), l(1,1))$ dyon states should arise as bound states of k $(1,0)$ and k $(0,1)$ monopoles. Note from the BPS mass formula that these states are only neutrally stable and consequently they should emerge as bound states at threshold. At present these states have been shown to exist only for $k = 1$. Let us make some comments on this case.

It was shown in [34, 35, 36] that the moduli space for $\mathbf{n}^m = (1,1)$ is given by

$$\mathcal{M}_{(1,1)} = \mathbf{R}^3 \times \frac{\mathbf{R} \times \mathcal{M}_{\text{TN}}}{\mathbf{Z}} \tag{5.16}$$

where $\mathcal{M}_{\text{TN}}$ is four-dimensional Taub-NUT space. The $\mathbf{R}^3$ factor corresponds to the centre of mass of the $(1,0)$ and $(0,1)$ monopole configuration. Taub-NUT space is a hyper-Kähler manifold as required for the quantum mechanics (4.3) to have $N=4$ supersymmetry. Taub-NUT space has $U(2)$ isometry, of which an $SU(2)_L$ subgroup corresponds to the action of rotating the multi-monopole configuration in space. That this is $SU(2)$ and not $SO(3)$ can be demonstrated by studying the zero modes about the spherically symmetric $(1,1)$ solution that can be obtained via the $SU(2)$ embedding using the root γ [34]. Note that the fixed point set of the $SU(2)_L$ action is a single point in Taub-NUT space (the 'nut') and this corresponds to the spherically symmetric solutions. The extra $U(1)_R$ isometry in $U(2)$ combined with the factor $\mathbf{R}$ in (5.16) and the identification under the integers $\mathbf{Z}$ lead to dyon states with electric charge $\mathbf{n}^e$. One might have expected an S^1 factor rather than $\mathbf{R}$ for the total electric charge in the $(1,1)$ direction (i.e., parallel to the magnetic charge), but this is not quite correct due to the fact that in general the masses of the two fundamental monopoles $(1,0)$ and $(0,1)$ are not equal [35].

The basis of 16 forms on $\mathbf{R}^3 \times \mathbf{R}$ leads to a BPS supermultiplet of 16 states in the $N=4$ supersymmetric quantum mechanics. In order to get the dyon BPS states predicted by duality (5.11) with magnetic charge $(1,1)$ there must exist a unique normalizable harmonic (anti)-self-dual two-form on Taub-NUT space that is invariant under the $U(1)_R$ isometry to ensure that the electric charge is $l(1,1)$. Such a harmonic form exists [34, 35].

The $SU(N)$ BPS spectra contain the embedded BPS spectra of the $SU(N')$ case for all $N' < N$. This can be illustrated by considering $G = SU(4) \to U(1)^3$. In this case S-duality predicts BPS dyon states with magnetic charge vectors $\mathbf{n}^m = k(1,0,0)$ $k(0,1,0)$, $k(0,0,1)$, (and l units of parallel electric charge), which are equivalent to the $SU(2)$ predictions, $\mathbf{n}^m = k(1,1,0)$,

$k(0,1,1)$ which are equivalent to the $SU(3)$ predictions, and $k(1,1,1)$ which are the new $SU(4)$ predictions. Apart from the cases we have discussed there is only one more class of moduli spaces whose metrics are explicitly known: when there are no more than a single fundamental monopole of each type (e.g, including the case $(1,1,1)$ for the $SU(4)$ example) [37, 38, 39]. After factoring out the centre of mass, the moduli spaces are natural generalisations of Taub-NUT space and the harmonic forms predicted by duality have been shown to exist [40].

6 $N = 2$, $G = SU(2)$, $N_f = 4$

Pure $N = 2$ super Yang–Mills theory can be obtained from the dimensional reduction on a two-torus of $N = 1$ supersymmetric Yang–Mills theory in six dimensions [41]. There is a single vector supermultiplet consisting of an $N = 1$ vector multiplet and an $N = 1$ chiral superfield Φ both transforming in the adjoint representation of the gauge group. In component form the bosonic part of the Lagrangian is given by (3.1) with just two real Higgs fields which are the real and imaginary parts of ϕ, the bosonic component of Φ. The potential is given by

$$V(\phi) = \frac{1}{g^2} \mathrm{Tr}\, [\phi, \phi^\dagger]^2. \qquad (6.1)$$

If we choose a gauge in which $\phi = a\sigma^3$, the classical moduli space of vacua is the complex u plane with $u = 2a^2 = \mathrm{Tr}\,\phi^2$. The gauge symmetry is spontaneously broken from $SU(2)$ to $U(1)$ on the u-plane ($u \neq 0$) and hence it is known as the 'Coulomb branch'.

Theories with $N = 2$ supersymmetry can also contain hypermultiplets. Each hypermultiplet consists of two $N = 1$ chiral superfields Q and $\tilde{Q}$ transforming in conjugate representations of the gauge group. We consider N_f hypermultiplets in the fundamental representation of $SU(2)$, Q^I, $\tilde{Q}_I$, $I = 1, 2, \ldots, N_f$. The terms in the Lagrangian involving the hypermultiplets consist of canonical kinetic energy terms as well as a coupling term given in $N = 1$ superfield language by the superpotential

$$W = \sqrt{2}\sum_I \tilde{Q}_I \Phi Q^I. \qquad (6.2)$$

We will set the possible mass terms for the hypermultiplets to be zero.

For general gauge group there is an $SU(N_f) \times U(1)$ flavor symmetry which acts on the fields Q^I, $\tilde{Q}_I$ which transform as N_f, $\bar{N}_f$, respectively. However for $SU(2)$ the fundamental representation is pseudoreal rather than complex and as a result Q^I and $\tilde{Q}_I$ lie in equivalent representations. This leads to an $SO(2N_f)$ flavor symmetry which can be made evident through a change of basis: $(Q^I, \tilde{Q}_I) \rightarrow Q'^i$, $i = 1, \ldots 2N_f$. As we shall see, there exist monopoles and dyons transforming as spinors of $SO(2N_f)$ and so more precisely the flavor symmetry is $\text{Spin}(2N_f)$.

We will now focus on the $N_f = 4$ theory, which has vanishing β-function and is conjectured to exhibit exact S-duality [3]. The classical vacua of this theory are the same as in the quantum theory: in addition to the Coulomb branch there can also be 'Higgs branches' when the hypermultiplets get non-zero expectation values and $SU(2)$ is completely broken. It is on the Coulomb branch where duality leads to non-trivial predictions about the spectrum of dyons.

The perturbative spectrum on the Coulomb branch arising from the vector multiplets consists of a massless photon multiplet with $(n_m, n_e) = (0,0)_1$ and massive W-boson multiplets with charge $\pm(0,2)_1$, where the subscript indicates the states are singlets of $\text{Spin}(8)$ and we have changed our normalisation of the electric charge for convenience. The hypermultiplets give rise to states $\pm(0,1)_{8_v}$ that transform in the 8_v representation of $\text{Spin}(8)$.

¿From the $N = 2$ supersymmetry algebra we can deduce the BPS bound

$$M \geq \sqrt{2}|Z| = \sqrt{2}|a(n_e + \tau n_m)|, \qquad (6.3)$$

where, because of our normalisation of the electric charges, we now have $\tau = \theta/\pi + i\frac{8\pi}{g^2}$ [18, 2]. The irreducible representations of the $N = 2$ supersymmetry algebra for BPS states contain four states. All the perturbative states are BPS states: the hypermultiplet is made up of two irreducible representations each with spin content $S_z = \{0, 0, \pm 1/2\}$, where S_z is the component of spin along the z-axis, while the W-boson multiplet has $S_z = \{0, 1/2, 1/2, 1\}$ and its CPT conjugate.

The precise S-duality group for this theory is conjectured to be the semi-direct product of $\text{Spin}(8) \ltimes SL(2, \|Z)$ [3]. The mod 2 reduction of $SL(2, Z)$ is homomorphic to the permutation group of three objects, S_3, which is also the group of outer automorphisms of the flavour symmetry $\text{Spin}(8)$. Thus $SL(2, \|Z)$ acts on $\text{Spin}(8)$ via this homomorphism. In detail the S-duality action is given by

$$\tau \rightarrow \frac{p\tau + q}{r\tau + s} \qquad (6.4)$$

$$a \rightarrow |r\tau + s|a \qquad (6.5)$$

$$(n_m, n_e)_r \rightarrow [(n_m, n_e)\begin{pmatrix} p & q \\ r & s \end{pmatrix}^{-1}]_{r'}. \qquad (6.6)$$

$$(6.7)$$

and the representation r' is determined by triality. The vector (v), spinor (s) and conjugate spinor (c) representations r are transformed via the

$SL(2, \|Z) \rightarrow S_3$ homomorphism. Explicitly, the mod 2 reduction of the $SL(2, \|Z)$ matrix gives the following permutations:

$$\begin{pmatrix} 0 & 1 \\ 1 & 0 \end{pmatrix} \rightarrow \begin{cases} v & \rightarrow & s \\ s & \rightarrow & v \\ c & \rightarrow & c \end{cases} \tag{6.8}$$

$$\begin{pmatrix} 1 & 0 \\ 1 & 1 \end{pmatrix} \rightarrow \begin{cases} v & \rightarrow & c \\ s & \rightarrow & s \\ c & \rightarrow & v \end{cases} \quad etc. \tag{6.9}$$

$$\tag{6.10}$$

Beginning with the hypermultiplet of states $(0, \pm1)_{\mathbf{8_v}}$, the $SL(2, \|Z)$ action generates an orbit of states $\pm(k, l)$ with k and l relatively prime, with the Spin(8) representation determined by the following mod 2 grading of charges:

$$(0, 1) \rightarrow \mathbf{8_v}; \qquad (1, 0) \rightarrow \mathbf{8_s}; \qquad (1, 1) \rightarrow \mathbf{8_c}. \tag{6.11}$$

As these BPS states are absolutely stable, duality predicts that this orbit of hypermultiplets should be observable at weak coupling.

The W-boson multiplets with charges $(0, \pm2)_1$ generate an orbit of vector multiplets with charges $(2k, 2l)_1$. All of these BPS states are only neutrally stable to decay into two states with charge (k, l). Assuming that the spectrum of marginal states does not change as we vary the coupling, these states should emerge at weak coupling as bound states at threshold. The BPS states predicted by S-duality now fill out a 'decorated lattice' which is illustrated in Figure 2.

The semiclassical search for the predicted BPS states mirrors that of the $N=4$ case. We start with a magnetic monopole with magnetic charge k. As we have noted, the bosonic zero modes give rise to $4k$ collective coordinates which are coordinates on the moduli space $\mathcal{M}_k$. There are $2k$ complex fermionic zero modes arising from the adjoint fermions, which require $4k$ real Grassmann odd collective coordinates ψ^α. In addition there are k real zero modes for each fundamental Weyl fermion [21], not related to any bosonic zero modes by supersymmetry.

Since the fermionic zero modes exist for each monopole solution, i.e., each point on $\mathcal{M}_k$, the proper geometric description is a fibre bundle over $\mathcal{M}_k$. For a single Weyl fermion the 'index bundle' of the Dirac operators in the fundamental representation, Ind_k, is actually an $O(k)$ bundle over the moduli space $\mathcal{M}_k$ [42]. This bundle has a natural connection given by

$$A_\alpha^{AB}(Z) = \int d^3x \lambda^A(x, Z)^\dagger \frac{\partial}{\partial Z^\alpha} \lambda^B(x, Z), \tag{6.12}$$

where $\lambda^A(x, Z^\alpha)$, $A = 1, \ldots, k$, are the c-number zero modes around a monopole solution specified by the coordinates Z^α on $\mathcal{M}_k$. It can be shown that

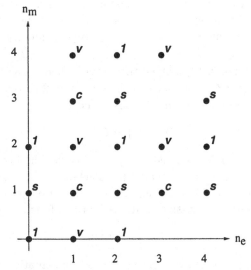

Figure 2. The decorated lattice of BPS states in N=2 super-Yang–Mills theory with gauge group $SU(2)$ and $N_f = 4$ hypermultiplets in the fundamental representation.

the curvature of this connection is of type $(1, 1)$ with respect to each of the three complex structures on $\mathcal{M}_k$, and hence the curvature is anti-self-dual [43]. Note that the analogue of the connection (6.12) for the index bundle of the adjoint fermions is simply the Levi-Civita connection on $\mathcal{M}_k$. For the case of $N_f = 4$ hypermultiplets we have 8 Weyl fermions in the fundamental representation and this gives rise to $8k$ fermion zero modes which leads to 8 copies of the $O(k)$ bundle. Correspondingly we must introduce Grassmann-odd collective coordinates ρ^{iA}, $i = 1, \ldots 8$.

By constructing a suitable low-energy ansatz for the fields and substituting into the Lagrangian and integrating over space, we obtain the following supersymmetric quantum mechanics [44, 45, 46]:

$$S = \frac{1}{2} \int dt \left(\mathcal{G}_{\alpha\beta} \left[\dot{Z}^\alpha \dot{Z}^\beta + i\psi^\alpha D_t \psi^\beta \right] + i\rho^{iA} \mathcal{D}_t \rho^{iA} + \frac{1}{2} F^{AB}_{\alpha\beta} \psi^\alpha \psi^\beta \rho^{iA} \rho^{iB} \right).$$
(6.13)

where $\mathcal{D}_t \rho^{iA} = \dot{\rho}^{iA} + A^{AB}_\alpha \dot{Z}^\alpha \rho^{iB}$ and $F = dA$. The bosonic monopole solution breaks half of the supersymmetry. Given that the number of real components in the supersymmetries of the $N=2$ theory is 8, we expect to have a quantum mechanics with four real parameters. The action (6.13) automatically has a supersymmetry with one real parameter, which is sometimes called $N = 1/2$ supersymmetry [47] (although there is a mismatch between the number of bosons and fermions this action still admits supersymmetries that are non-linearly realized). Since the moduli space is hype-rKähler and the field strength of the gauge connection is $(1,1)$ with respect to each of the complex structures, the action has four real supersymmetry parameters as required.

The quantisation of (6.13) without the matter fermions ρ and its connection to the spectrum of BPS monopoles in the pure $N = 2$ theory was described in [8, 48]. The anticommutation relations of the ψ^α are given by

$$\{\psi^\alpha, \psi^\beta\} = \mathcal{G}^{\alpha\beta} \tag{6.14}$$

and hence the ψ^α can be realised as gamma matrices acting on spinors on the moduli space (one can also realise the algebra on holomorphic forms on the moduli space, but this is an equivalent description on a hyper-Kähler manifold). The anticommutation relations for the ρ fermions are given by

$$\{\rho^{iA}, \rho^{jB}\} = \delta^{ij}\delta^{AB}. \tag{6.15}$$

The monopole and dyon states must provide a representation of this Clifford algebra. The Hamiltonian can be written as $H = \{Q, Q^\dagger\} + \sqrt{2}|a\tau k|$, where Q is one of the supercharges and we have included the topological term. The supersymmetry charge Q is realised as Dirac operator on $\mathcal{M}_k$ coupled to the $O(k)$ connection on the index bundle.

For $k = 1$, $\mathcal{M}_1 = \|R^3 \times S^1$. The quantisation of the ψ^α fermions give rise to four-component spinors on $\|R^3 \times S^1$ corresponding to four different states in the field theory with spin 0 and 1/2, making up a short BPS multiplet [17]. If we combine these states with similar states that come from quantizing the anti-monopoles we obtain a complete hypermultiplet of $N = 2$ supersymmetry. Quantization of the bosonic coordinates on $R^3 \times S^1$ leads to a spectrum of BPS dyons $\pm(1, n_e)$ as usual.

Let us now analyze the effect of the ρ fermions. For $k = 1$ we have the anticommutation relations

$$\{\rho^i, \rho^j\} = \delta^{ij}. \tag{6.16}$$

The representation of this Clifford algebra consists of the 16-dimensional spinor representation of Spin(8). This representation is reducible and splits into two irreducible representations, both of dimension 8, under projection by the chirality operator in Spin(8) which we denote by $(-1)^H$.

However, this is not quite the end of the story. Even for $k = 1$ there is still a non-trivial bundle structure. The $O(1)$ connection on the index bundle is non-trivial over the S^1 factor and leads to

$$\mathrm{Ind}_1 = \|R^3 \times \mathrm{M\ddot{o}b}, \tag{6.17}$$

where Möb is the Möbius bundle over S^1. Physically this arises because the gauge transformation which generates a 2π rotation about the S^1 factor acts as the non-trivial element of the center of $SU(2)$ [16, 3]. Since the $N_f = 4$ matter fermions transform in the fundamental representation of $SU(2)$, there must be a correlation between the $U(1)$ charge (as measured by rotation about the S^1 factor) and the $SO(8)$ chirality (as measured by $(-1)^H$). Specifically, it is expressed by the constraint

$$e^{i\pi Q} = (-1)^H, \qquad (6.18)$$

where Q is the charge operator. Thus we see that states with $(-1)^H = 1$ have even electric charge while states with $(-1)^H = -1$ carry odd electric charge. Putting this altogether, we obtain a tower of hypermultiplets $\pm(1, 2n_e)$ in the $\mathbf{8_s}$ representation and another tower $\pm(1, 2n_e + 1)$ in the $\mathbf{8_c}$ representation in agreement with Spin(8) $\ltimes SL(2, \|Z)$ duality [3].

Let us now briefly sketch the main ideas for $k = 2$, referring the reader to [44, 45] for further details. The quantization of the ψ^α implies that the states are tensor products of spinors on $\|R^3 \times S^1$ with spinors on $\tilde{M}_2^0$. The discussion of the spinors on $\|R^3 \times S^1$ is essentially the same as for the single monopole case: the dyon states $(2, n_e)$ are in a short BPS multiplet with spin 0 and 1/2. Thus, to find new BPS states in the spectrum we must look for zero energy states on $\tilde{M}_2^0$.

Now we consider the matter fermions. The index bundle Ind_2 is a real two-dimensional vector bundle over M_2 with structure group $O(2)$ which is described in detail in [42]. There is an obstruction to obtaining an orientable bundle on the non-simply connected manifold $(S^1 \times \tilde{M}_2^0)/\|Z_2$. One obtains an orientable bundle $\widetilde{\mathrm{Ind}}_2$ by pulling Ind_2 back to $S^1 \times \tilde{M}_2^0$. Let's work with the $U(1)$ bundle $\widetilde{\mathrm{Ind}}_2$. Replacing the real Grassmann parameters ρ^{iA} with complex parameters ρ^i, the anticommutation relations become those of annihilation and creation operators:

$$\{\rho^i, \rho^{j*}\} = \delta^{ij}. \qquad (6.19)$$

The states of the supersymmetric quantum mechanics are now spinors on M_2, $|\Psi\rangle$, on which the algebra (6.19) is realized. Starting with a state $|\Psi\rangle$ satisfying $\rho^i|\Psi\rangle = 0$ we can build up the ρ Fock space by acting with the ρ^{i*} in the usual manner. The supersymmetry charge acting on these states takes the form

$$Q = \not{D} - i(N_\rho - 4)A, \qquad (6.20)$$

where A is the $U(1)$ gauge connection, $N_\rho = \rho^{i*}\rho^i$, and the factor 4 arises as a normal ordering constant fixed by a discrete charge conjugation symmetry. From (6.20) we see that there is a correlation between the number of ρ^{i*}'s excited, N_ρ, and the $U(1)$ charge carried by the corresponding spinor on M_2. Since the ρ^i carry the $\mathbf{8_v}$ representation of Spin(8), there is also a correlation between the Spin(8) representation carried by the state and N_ρ

q	Spin(8) rep
±0	$35 + 35$
±1	56
±2	28
±3	$8_{\mathbf{v}}$
±4	1

Table 1. Correlation between $U(1)$ charge of spinors on moduli space and Spin(8) flavour symmetry representations of states.

and consequently the $U(1)$ charge q. We display the explicit correlation in Table 1.

BPS states correspond to zero modes of the Dirac operator on $\tilde{M}_2^0$ coupled to the $U(1)$ connection which have finite L^2 norm. Recall that the duality conjectures predicts a tower of states $\pm(2, 2n_e + 1)$ forming a hypermultiplet transforming as a $8_{\mathbf{v}}$ of $SO(8)$ and another tower of states $(2, 2n_e)$ transforming as $SO(8)$ singlets and filling out a vector multiplet.

First consider the hypermultiplets. Recall that the spin content of a hypermultiplet is $S_z = (0, 0, 0, 0, \pm 1/2, \pm 1/2)$ where S_z is the component of spin along the z-axis. We have noted that the spinor on $\|R^3 \times S^1$ corresponds to four states in a short BPS multiplet with $S_z = (0, 0, \pm 1/2)$. If these spinors are combined with one zero mode of the Dirac operator on $\tilde{M}_2^0$ with zero angular momentum we will obtain a BPS hypermultiplet after we also include the corresponding states that come from quantizing the zero modes around the charge 2 anti-monopole. If in addition the zero mode of the Dirac operator has $U(1)$ charge $q = \pm 3$ then according to Table 1 the hypermultiplet will transform in the $8_{\mathbf{v}}$ representation.

Now consider the vector multiplets. The spin content of the vector multiplet is $S_z = (0, 0, \pm 1/2, \pm 1/2, \pm 1)$. To obtain this spin content we need to combine the spinor on $\|R^3 \times S^1$ which has spin $S_z = (0, 0, \pm 1/2)$ with two zero modes of the Dirac operator on $\tilde{M}_2^0$, one with $S_z = 1/2$, the other with $S_z = -1/2$. To form a singlet representation of $SO(8)$ the spinors on $\tilde{M}_2^0$ must be zero modes of the Dirac operator with charge $q = \pm 4$.

The existence of the conjectured zero modes of the Dirac operator were established in [44, 45] using index theory. There are a number of subtleties involving the distinction between the oriented bundle $\widetilde{\mathrm{Ind}}_2$ versus Ind_2, the $\|Z_2$ discrete identification on the moduli space and the proper identification of the angular momentum and electric charges. We refer the reader to [44, 45] for more details.

7   Some Open Problems

As we have discussed, a number of highly non-trivial checks of exact S-duality can be carried out by studying the geometry of monopole moduli spaces. Let us conclude by discussing some open issues.

In the $N=4$ theory with maximal symmetry breaking, the BPS states predicted by duality correspond to normalisable harmonic forms on BPS monopole moduli spaces. For gauge group $SU(2)$ the work of [25] provided substantial evidence for the appropriate harmonic forms. For higher rank gauge groups, the harmonic forms on the monopole moduli spaces for the cases when no more than a single fundamental monopole of each type is present have been verified [34, 35, 40]. It remains to verify that the rest of the moduli spaces have the appropriate harmonic forms as predicted by S-duality. Perhaps the approach of [25] can be generalised to the case of higher rank gauge groups.

New issues arise in the $N=4$ case when a non-abelian gauge group remains unbroken. In this case the existence of massless W-bosons might seem to require dual massless monopoles which cannot be studied as conventional semiclassical solitons. There are also massive monopoles that can be studied. If they carry non-abelian magnetic charge there are subtleties to do with the moduli space approximation due to the non-normalisability of zero modes corresponding to global gauge rotations (see e.g., [49, 50]). The moduli spaces of monopoles that have net abelian magnetic charge can in some cases be determined as limits of moduli spaces in which the symmetry is maximally broken. Curiously, it is claimed that the harmonic forms found by [40] become non-normalisable in this limit [50].

Another direction to explore for the $N = 4$ theory is the presence of medium sized BPS multiplets when the rank of the gauge group is greater than one. Recently it has been shown in string theory that certain three-pronged strings can end on D-3-branes and that they correspond to medium sized BPS states in the low-energy effective $N=4$ gauge theory on the D-3-brane [32]. It would be interesting to find evidence for these states directly in $N = 4$ super-Yang–Mills theory.

For the $N = 2$ $SU(2)$ theory with $N_f = 4$ hypermultiplets in the fundamental representation, the Spin(8) $\ltimes$ $SL(2, \mathbf{Z})$ duality predictions have been verified for $k = 2$ using index theory [44, 45]. It would be of interest to explicitly construct the harmonic spinors to clarify the angular momentum structure. In [44] the predictions for harmonic spinors on $\mathcal{M}_k$ for $k > 2$ were discussed. It would be interesting if an analysis similar to [25] is possible for higher monopole charge.

For higher rank theories with $N=2$ supersymmetry and vanishing β-function less is known about exact duality and the spectrum of BPS dyons. A straightforward attempt to find the duality group acting on the lattice of electric and magnetic charges was attempted in [51] for $G=SU(3)$ (see also [52]) but the results were inconclusive.

All of these tests we have been discussing concern the spectrum of BPS states. If the exact duality conjectures are true then they should also apply to non-BPS states, for example the scattering of BPS states. Since such processes are not protected by supersymmetry it remains a challenging problem to find evidence for S-duality in this sector.

More generally, we would like to know the underlying reasons for duality in field theory. String theory duality would seem to provide one answer since we can embed these gauge theories in various string theory settings. Of course this still leaves the more involved issue of elucidating the deeper principles that underly string theory duality.

References

[1] C. Montonen and D. Olive, *Phys. Lett.* **72B** (1977) 117.

[2] N. Seiberg and E. Witten, *Nucl. Phys.* **B426** (1994) 19.

[3] N. Seiberg and E. Witten, *Nucl. Phys.* **B431** (1994) 484.

[4] A. Sen, *Phys. Lett.* **B329** (1994) 217.

[5] M.K. Prasad and C.M. Sommerfeld, *Phys. Rev. Lett.* **35** (1975) 760.

[6] E.B. Bogomol'nyi, *Sov. J. Nucl. Phys.* **24** (1976) 449.

[7] E. Weinberg, *Phys. Rev.* **20** (1979) 936.

[8] J.P. Gauntlett, *Nucl. Phys.* **B411** (1994) 443.

[9] M.F. Atiyah and N.J. Hitchin, *The Geometry and Dynamics of Magnetic Monopoles*, Princeton Univ. Press, Princeton, NJ, 1988.

[10] R. Rajaraman, *Solitons and Instantons*, North-Holland 1982.

[11] E.J. Weinberg, hep-th/9610065.

[12] N.S. Manton, *Phys. Lett.* **110B** (1982) 54.

[13] S. Coleman, S. Parke, A. Neveu and C.M. Sommerfield, *Phys. Rev.* **D15** (1977) 544.

[14] G.W. Gibbons and N. S. Manton, *Nucl. Phys.* **274** (1986) 183.

[15] B. Schroers, *Nucl. Phys.* **367** (1991) 177.

[16] E. Witten, *Phys. Lett.* **86B** (1979) 283.

[17] H. Osborn, *Phys. Lett.* **83B** (1979) 321.

[18] E. Witten and D. Olive, *Phys. Lett.* **78B** (1978) 97.

[19] C. Fraser and T. Hollowood, *Phys. Lett.* **B402** (1997) 106, hep-th/9704011.

[20] F. Ferrari, *Nucl. Phys.* **B501** (1997) 53, hep-th/9702166.

[21] C. Callias, *Comm. Math. Phys.* **62** (1978) 213.

[22] J. D. Blum, *Phys. Lett.* **B333** (1994) 92.

[23] L. Alvarez-Gaumé and D. Freedman, *Comm. Math. Phys.* **80** (1981) 443.

[24] E. Witten, *Nucl. Phys.* **B202** (1982) 253.

[25] G. Segal and A. Selby, *Comm. Math. Phys.* **177** (1996) 775.

[26] M. Porrati, *Phys. Lett.* **B377** (1996) 67.

[27] E. J. Weinberg, *Nucl. Phys.* **B167** (1980) 500.

[28] P. Goddard, J. Nuyts and D. Olive, *Nucl. Phys.* **B125** (1977) 1.

[29] F. Englert and P. Windey, *Phys. Rev.* **D14** (1976) 2728.

[30] L. Girardello, A. Giveon, M. Porrati and A. Zaffaroni, *Phys. Lett.* **B334** (1994) 331; *Nucl. Phys.* **B448** (1995) 127.

[31] N. Dorey, C. Fraser, T. Hollowood and M.A. Kneipp, *Phys. Lett.* **383B** (1996) 422.

[32] O. Bergman, *Nucl. Phys.* **B525** (1998) 104; hep-th/9712211.

[33] F. A. Bais, *Phys. Rev.* **D18** (1978) 1206. ·

[34] J.P. Gauntlett and D.A. Lowe, *Nucl. Phys.* **B472** (1996) 194.

[35] K. Lee, E. Weinberg and P. Yi, *Phys. Lett.* **B376** (1996) 97.

[36] S.A. Connell, *The dynamics of the SU(3) charge (1,1) magnetic monopoles*, University of South Australia preprint. ftp://maths.adelaide.edu.au/pure/mmurray/oneone.tex

[37] K. Lee, E. Weinberg and P. Yi, *Phys. Rev.* **D54** (1996) 1633.

[38] M.K.Murray, *J.Geom.Phys.* **23** (1997) 31, hep-th/9605054.

[39] G. Chalmers, hep-th/9605182.

[40] G.W. Gibbons, *Phys. Lett.* **B382** (1996) 53.

[41] A. d'Adda, R. Horsley and P. di Vecchia, *Phys. Lett.* **76B** (1978) 298.

[42] N.S. Manton and B.J. Schroers, *Ann. Phys.* **225** (1993) 290.

[43] N. Hitchin as quoted in [42].

[44] J.P. Gauntlett and J. Harvey *Nucl. Phys.* **B463** (1996) 287.

[45] S. Sethi, M. Stern and E. Zaslow, *Nucl. Phys.* **B457** (1995) 484.

[46] M. Cederwall, G. Ferretti, B.E.W. Nilsson, P. Salomonson,
      Mod. Phys. Lett. **A11** (1996) 367.

[47] L. Alvarez-Gaumé, *J. Phys.* **A16** (1983) 4177.

[48] J. P. Gauntlett, *Nucl. Phys.* **B400** (1993) 103.

[49] N. Dorey, C. Fraser, T. Hollowood and M.A. Kneipp, hep-th/9512116.

[50] K. Lee, E. Weinberg and P. Yi, *Phys. Rev.* **D54** (1996) 6354.

[51] M. Cederwall and M. Holm, hep-th/9603134.

[52] O. Aharony and S. Yankielowicz, *Nucl. Phys.* **B473** (1996) 93.

Seiberg–Witten Theory and S-Duality

Tohru Eguchi

Abstract

We review the exact solutions of $N = 2$ supersymmetric Yang–Mills theories. We analyze the mechanism for solitons converting into elementary particles in the moduli space of the Coulomb phase and also classify $N = 2$ superconformal field theories.

1 Introduction

In these lectures I am going to give an elementary introduction to Seiberg–Witten theory, i.e. the exact solutions of $N = 2$ supersymmetric Yang–Mills theories. The Seiberg–Witten solution [1, 2] was the first exact solution in strongly-coupled quantum gauge theories in 4 dimensions. The solution was first constructed for $SU(2)$ Yang–Mills theory with $0 \leq N_f \leq 4$ hypermultiplets in vector representations. It was subsequently generalized to the case of classical gauge groups $SU(N), SO(N), Sp(N)$ [3, 4, 5, 7, 8, 9, 10] and provides a wealth of information on the strong coupling dynamics of gauge theories. There are excellent pedagogical reviews of Seiberg–Witten theory [11, 12, 13].

Let us first recall that there exist two types of supermultiplets in $N = 2$ supersymmetric field theories: vector and hypermultiplets. An $N = 2$ vector multiplet consists of a gauge field A_μ^a, two Weyl spinors λ^a, ψ^a and a complex scalar field ϕ^a. Here the suffix 'a' runs over the generators of the gauge group G, $a = 1, \ldots, \dim G$. An $N = 2$ vector multiplet decomposes into a sum of an $N = 1$ vector multiplet made of (A_μ^a, λ^a) and a chiral multiplet made of (ϕ^a, ψ^a).

On the other hand an $N = 2$ hypermultiplet consists of a pair of $N = 1$ chiral fields (q^i, ψ_q^i) and $(\tilde{q}^j, \tilde{\psi}_q^j)$ which belong to a representation R of G, $i = 1, 2 \ldots, \dim(R)$, and its conjugate R^*, $j = 1, 2, \ldots, \dim(R^*)$, respectively. In the following we restrict ourselves to the case of $G = SU(2)$ with R in the vector representation.

The $N = 1$ vector and chiral superfields in an $N = 2$ vector multiplet are denoted by W_α^a and A^a, respectively. The $N = 1$ chiral fields of an $N = 2$ hypermultiplet are denoted by Q^i and $\tilde{Q}^j$. Under a $U(1)_R$ rotation they transform as

$$U(1)_R : \quad \begin{cases} \Phi(\theta) \rightarrow e^{2i\alpha}\Phi(e^{-i\alpha}\theta), & W_\alpha(\theta) \rightarrow e^{i\alpha}W_\alpha(e^{-i\alpha}\theta) \\ Q \rightarrow Q(e^{-i\alpha}\theta), & \tilde{Q} \rightarrow \tilde{Q}(e^{-i\alpha}\theta). \end{cases} \tag{1.1}$$

On the other hand, under the third component of $SU(2)_R$, fields transform as

$$U(1)_J : \quad \begin{cases} \Phi(\theta) \to \Phi(e^{-i\alpha}\theta), & W_\alpha(\theta) \to e^{i\alpha}W_\alpha(e^{-i\alpha}\theta) \\ Q \to e^{i\alpha}Q(e^{-i\alpha}\theta), & \tilde{Q} \to e^{i\alpha}\tilde{Q}(e^{-i\alpha}\theta). \end{cases} \tag{1.2}$$

In the following we first consider the case of $SU(2)$ gauge theory without matter hypermultiplets, $N_f = 0$. If we denote the scalar field in the vector multiplet by a 2×2 matrix, $\phi = \sum_a \phi^a \sigma_a/2$, the potential energy of the scalar fields is expressed as

$$V(\phi) = \frac{1}{g^2}\mathrm{Tr}\left[\phi, \phi^\dagger\right]^2. \tag{1.3}$$

Thus the potential has a flat direction

$$\langle\phi^3\rangle = a \neq 0, \quad \langle\phi^1\rangle = \langle\phi^2\rangle = 0. \tag{1.4}$$

When the scalar field develops a vacuum value $a \neq 0$, the gauge symmetry breaks down to $U(1)$ and the theory is in the Coulomb phase. The vacuum value a can take an arbitrary value and the theory possesses an infinitely degenerate vacuum. The vacuum value becomes the moduli of the $N = 2$ gauge theory. It turns out that the moduli space of $N = 2$ theory has the structure of a special Kähler geometry with its Kähler potential written in terms of some holomorphic function.

On the other hand if there exists matter in the theory, scalar fields in the matter hypermultiplets could develop non-zero vacuum values $\langle q^i\rangle$ which in general completely break gauge invariance. A theory with vacua $\langle q^i\rangle \neq 0$ is said to be in the Higgs phase. Due to Yukawa-like couplings between scalars of vector and hypermultiplets, Coulomb and Higgs phases mutually exclude each other. The moduli space of the Higgs phase has the structure of a hyper-Kähler manifold. Hyper-Kähler manifolds are not deformed by quantum corrections and do not possess interesting dynamics in the present model. We will study the structure of the Coulomb phase later.

Let us introduce a gauge invariant 'order parameter' $u = \mathrm{Tr}\,\phi^2$. Then the complex u-plane becomes the moduli space. Classically $u = a^2/2$; however, this relation is modified at the quantum level. As is well-known, the nature of the quantum theory strongly depends on the sign of the beta function. In $N = 2$ supersymmetric theories a perturbative beta function is given only by the one-loop contribution. A well-known formula for the one-loop beta function is given by

$$\beta = \frac{-bg^3}{16\pi^2}, \quad b = \frac{11}{3}c_2(G) - \sum_{R_S}\frac{1}{6}T(R_S) - \sum_{R_F}\frac{2}{3}T(R_F) \tag{1.5}$$

$$T(R)\delta^{ab} = \mathrm{Tr}\,R^a R^b, \quad c_2(G)\delta^{ab} = f_{acd}f_{bcd}, \tag{1.6}$$

where $c_2(G)$ is the value of the second-order Casimir invariant for the group G, R^a are representation matrices and $T(R)$ is the so-called Dynkin index for the representation R. In (1.5) $S(F)$ stands for the contributions of the scalar (spinor) fields (the coefficient is 1/6 (2/3) for a real scalar (spinor) field). In the case of $SU(N)$, $c_2(SU(N)) = T(\text{adjoint rep.}) = N$, and $T(\text{vector rep.}) = 1/2$.

An $N = 2$ vector multiplet consists of a vector, two real spinors and one complex scalar field all in the adjoint representation. Thus in the case of $SU(N_c)$ gauge group

$$b = \left(\frac{11}{3} - \frac{1}{6} \times 2 - \frac{2}{3} \times 2\right) N_c = 2N_c. \tag{1.7}$$

Similarly an $N = 2$ hypermultiplet with two complex scalars and two real spinors in the vector or adjoint representation of $SU(N_c)$ yield

$$\text{vector representation}: \quad b = -\left(\frac{1}{6} \times 4 + \frac{2}{3} \times 2\right) \times \frac{1}{2} = -1, \tag{1.8}$$

$$\text{adjoint representation}: \quad b = -\left(\frac{1}{6} \times 4 + \frac{2}{3} \times 2\right) \times N_c = -2N_c. \tag{1.9}$$

Therefore the beta function for an $SU(N_c)$ gauge theory coupled to N_f vector matter or one adjoint matter is given by

$$\begin{cases} \beta = (N_f - 2N_c)g^3/16\pi^2 : & N_f \text{ vector matter}, \\ \beta = 0 : & \text{one adjoint matter}. \end{cases} \tag{1.10}$$

Thus the theory is asymptotically free in the range

$$0 \leq N_f < 2N_c. \tag{1.11}$$

The theory with $N_f = 2N_c$ vector matter or one adjoint matter has a vanishing beta function. Thus these are ultraviolet finite theories. It is known that the theory with adjoint matter has in fact an $N = 4$ supersymmetry.

2 $SU(2)$ matter-free gauge theory

Let us first look at the dynamics in the Coulomb branch of $N = 2$ $SU(2)$ pure gauge theory. When the scalar field develops a vacuum value $\phi^3 \neq 0$, the gauge fields A_μ^1, A_μ^2 (together with their fermionic partners) become massive while the $U(1)$ sector of the theory $(A_\mu^3, \phi^3, \lambda^3, \psi^3)$ remains massless. Due to asymptotic freedom regions of large vacuum value $u = \text{Tr}\,\phi^2$ correspond to the weak coupling regime. In the limit of $u \to \infty$, A_μ^1, A_μ^2 become very heavy and almost decouple from the theory. Thus it seems reasonable to rewrite the theory in this region in terms of the remaining massless degrees of freedom.

Let us drop the suffix 3 of the $U(1)$ fields $(A^3_\mu, \phi^3, \lambda^3, \psi^3)$ and write their $N = 1$ gauge and chiral superfield as W_α and A, respectively. Then the most general effective action for the $N = 2$, $U(1)$ gauge theory is written as

$$S = \frac{1}{4\pi} \mathrm{Im} \left(\int d\theta^4 \bar{A} \frac{\partial F}{\partial A} + \frac{1}{2} \int d\theta^2 \frac{\partial^2 F}{\partial A^2} W_\alpha W^\alpha \right). \qquad (2.1)$$

Here $F(A)$ is the prepotential of the theory and is a holomorphic function of the chiral field A. In (2.1) the kinetic term of the chiral field A and the coefficient of the gauge kinetic term are both described in terms of a single function F. This constraint is due to $N = 2$ supersymmetry. The Kähler potential of the chiral field is given by

$$K = \frac{1}{4\pi} \mathrm{Im} \left(\bar{A} \frac{\partial F}{\partial A} \right), \qquad (2.2)$$

which is the characteristic equation of the special Kähler geometry.

In the case of the standard action for the $N = 2$, $U(1)$ gauge field the prepotential is simply quadratic in A

$$F = \frac{1}{2} \tau_{cl} A^2, \qquad \tau_{cl} = \frac{\theta}{2\pi} + \frac{4\pi i}{g^2}, \qquad K = \frac{1}{4\pi} \mathrm{Im}\, \tau \cdot A\bar{A}. \qquad (2.3)$$

Here θ denotes the vacuum angle. Then the above action (2.1) reduces to the sum of free kinetic terms of $A_\mu, \lambda, \psi, \phi$ and the θ term. In the present situation we originally had an interacting $SU(2)$ gauge theory and have eliminated the charged fields to recast the theory in terms of massless neutral fields. Then all the renormalizaion effects due to massive fields are encoded in the prepotential $F(A)$. Let us introduce a field-dependent coupling constant by

$$\tau(A) \equiv \frac{\partial^2 F(A)}{\partial A^2} = \frac{\theta(A)}{2\pi} + \frac{4\pi i}{g^2(A)}. \qquad (2.4)$$

In practice, the superfield A is identified as a (the vacuum value of the scalar component) and thus $F(A) = F(a)$.

Due to the asymptotic freedom one expects that F approaches its weak coupling behavior (2.3) at large $|u|$. We can then use the value of the one-loop beta function to obtain

$$F(a) \approx \frac{ia^2}{2\pi} \log \frac{a^2}{\Lambda^2} + \sum_{k=1}^{\infty} F_k \left(\frac{\Lambda}{a} \right)^{4k} a^2. \qquad (2.5)$$

Here Λ denotes the mass scale of the theory generated by the dimensional transmutation and the sum over k is the non-perturbative instanton contribution. In fact the perturbative part gives

$$\tau(a) = \frac{\partial^2 F}{\partial a^2} \approx \frac{2i}{\pi} \log \frac{a}{\Lambda}, \qquad \frac{1}{g(a)^2} \approx \frac{1}{2\pi^2} \log \frac{a}{\Lambda}. \qquad (2.6)$$

This reproduces the beta function (1.10).

In (2.5) the k-instanton contribution is proportional to a^{-4k} and violates the $U(1)_R$ charge conservation by $-8k$. This is due to the $U(1)_R$ axial anomaly

$$\partial_\mu J_R^\mu = -\frac{2 \times 2}{16\pi^2} F_{\mu\nu}^a \, {}^* F^{a\mu\nu} \tag{2.7}$$

which yields

$$\Delta R = -8 \times \text{instanton number.} \tag{2.8}$$

The instanton expansion coefficients F_k may be evaluated directly by perturbation theory around instanton backgrounds [14, 15, 16]. However, the calculation would soon become prohibitively difficult as one went to higher orders in the expansion. The analysis of the strong coupling region $|u| \leq \Lambda^2$ needs a powerful new technique which goes beyond perturbation theory.

In order to determine the behavior of the prepotential F in the strong coupling region one can use the following information:

(1) F has the asymptotic behavior (2.5) at $|u| \to \infty$.

(2) F is a holomorphic function of a.

(3) The imaginary part of the second derivative of F is proportional to the coupling constant and hence must be positive definite.

The second derivative of a holomorphic function is holomorphic. If the function has no singularities other than the one at ∞, then the positivity (3) becomes violated because of the property of a holomorphic function (mean-value theorem). Thus F must have at least one additional singularity in the strong coupling region. It turns out, however, if there exists only one extra singularity in the strong coupling region, the behavior of the theory around the singularity is in conflict with physical expectations. Thus there must exist at least two singularities in the strong coupling regime.

What then happens at these singularities in the complex u-plane where the description by the effective Lagrangian (2.1) breaks down? The expression for the effective action is based on the assumption that the $N = 2$, $U(1)$ vector multiplet is the only massless degree of freedom in the Coulomb phase. Thus if a new massless degree of freedom appears at some point in the u-plane, describing the theory by means of the $U(1)$ multiplet is no longer adequate and the effective Lagrangian necessarily develops a singularity. Seiberg and Witten made the brilliant assumption that in the strong coupling region of the $N = 2$ gauge theory, solitons (rather than elementary particles) become massless at singular points.

We now recall that there exists a special class of solitons, i.e. BPS states in supersymmetric field theories with extended supersymmetries. BPS states saturate the Bogomolnyi bound on the soliton mass and form 'small' representations of extended supersymmetry algebra. Since the BPS and non-BPS states belong to different representations and have a different number of degrees of freedom, BPS states cannot be converted into non-BPS states by quantum corrections and vice versa. In the case of $N = 2$ supersymmetric gauge theory 't Hooft–Polyakov monopoles appear as the BPS states. At the classical level, masses of the BPS states with n_e electric and n_m magnetic charges are given by

$$M_{BPS} = \sqrt{2}\left| n_e a + n_m \frac{4\pi i}{g^2} a \right| \qquad (n_e, n_m) \in \mathbf{Z}^2; \qquad (2.9)$$

note that the vacuum value a of the Higgs field is related to its conventional value v by $a = gv$. We recall that in classical theory $F = \tau a^2/2$ and hence $4\pi i a/g^2 = \partial F/\partial a$ (when $\theta = 0$). At the quantum level the RHS of (2.9) is modified due to quantum effect. Let us define the dual scalar field as

$$a_D = \frac{\partial F}{\partial a}. \qquad (2.10)$$

Then the quantum version of (2.9) is given by

$$M_{BPS}(u) = \sqrt{2}\left| n_e a(u) + n_m a_D(u) \right| \qquad (2.11)$$

In (2.11) a and a_D are regarded as functions of u.

Let us suppose that the function $a_D(u)$ vanishes at $u = u_0$. Then at u_0, monopoles become massless and the effective action develops a singularity. In order to restore a regular description we have to incorporate the monopole degrees of freedom explicitly in the effective Lagrangian. It is known that monopoles occur in hypermultiplets in the $N = 2$ theory. Let us denote monopole fields by the chiral superfields M, $\widetilde{M}$. We also perform a duality transformation of the $U(1)$ vector multiplet so that it has a local coupling to the monopole fields. Then around the singularity $u = u_0$ we have an effective action

$$S = \frac{1}{4\pi}\left(\int d^4\theta \mathrm{Im}\left(h_D(A_D)\bar{A}_D \right) + \frac{1}{2}\int d^2\theta \mathrm{Im}\left(\tau_D W_{\alpha D} W^{\alpha D} \right) \right)$$
$$+ \int d^4\theta M^\dagger e^{V_D} M + \int d^4\theta \widetilde{M} e^{-V_D} \widetilde{M}^\dagger. \qquad (2.12)$$

Here h_D and τ_D are defined by

$$h(A) \equiv A_D, \qquad h_D(A_D) \equiv -A, \qquad (2.13)$$

$$\frac{-1}{\tau(A)} = \frac{-1}{\partial h(A)/\partial A} = \frac{\partial h_D(A_D)}{\partial A_D} = \tau_D(A_D) \equiv \frac{\theta_D}{2\pi} + \frac{4\pi i}{g_D^2}. \qquad (2.14)$$

The kinetic term of the chiral field of (2.1) is rewritten in (2.12) as

$$\text{Im} \int d^4\theta h(A)\overline{A} = \text{Im} \int d^4\theta A_D \overline{A} = \text{Im} \int d^4\theta h_D(A_D)\overline{A}_D. \qquad (2.15)$$

In the above, $V_D(W_{\alpha D})$ denotes the dual gauge superfield containing the dual photon which couples to the monopoles in a standard manner. Thus (2.12) describes a dual $N = 2$ QED where a magnetic photon interacts with the magnetically charged matter field. The term τ_D encodes the renormalization of the magnetic charge due to the vacuum polarization by monopoles.

It turns out that the positions of two strong coupling singularities in the u-plane are given by Λ^2 and $-\Lambda^2$ (after breakdown by anomaly the Z_2 subgroup $u \to -u$ of $U(1)_R$ symmetry survives which maps the two singularities into each other). Thus at $u = \Lambda^2$, a_D vanishes and the monopole becomes massless. On the other hand at $u = -\Lambda^2$, $a_D - a$ vanishes and the dyons become massless. Near the dyon point the theory is described by a Lagrangian similar to (2.12) with a dyon hypermultiplet replacing the monopole, and a mixture of a photon and its dual replacing the magnetic photon. Around these singularities we can compute the one-loop beta function and determine the behavior of the prepotential. At the monopole point, for instance, renormalization of the magnetic charge g_D is given by $a_D \partial \tau_D / \partial a_D = -8\pi i / g_D^3 \times a_D \partial g_D / \partial a_D = -2 \times 8\pi i / 16\pi^2$. This leads to $\tau_D \approx (-i/\pi) \log a_D$. We find

At the monopole point :

$$\tau_D \approx \frac{-i}{\pi} \log a_D, \quad a_D \approx \text{const.} \, (u - \Lambda^2),$$

$$a(u) \approx a_0 + \frac{i}{\pi} a_D \log a_D \qquad (2.16)$$

At the dyon point :

$$\tau_D \approx \frac{-i}{\pi} \log(a_D - a), \quad a_D - a \approx \text{const.} \, (u + \Lambda^2),$$

$$a \approx a_0 + \frac{i}{\pi}(a_D - a) \log(a_D - a) \qquad (2.17)$$

From these formula we can determine the monodromy of a, a_D as one goes around the singularities

At the monopole point :

$$a_D \to a_D, \quad a \to a - 2a_D$$

$$\begin{pmatrix} a_D \\ a \end{pmatrix} \longrightarrow \begin{pmatrix} 1 & 0 \\ -2 & 1 \end{pmatrix} \begin{pmatrix} a_D \\ a \end{pmatrix}, \quad M_{\Lambda^2} \equiv \begin{pmatrix} 1 & 0 \\ -2 & 1 \end{pmatrix}$$

$$(2.18)$$

At the dyon point :

$$a_D \rightarrow -a_D + 2a, \qquad a \rightarrow -2a_D + 3a$$

(2.19)

$$\begin{pmatrix} a_D & a \end{pmatrix} \longrightarrow \begin{pmatrix} a_D & a \end{pmatrix} \begin{pmatrix} -1 & 2 \\ -2 & 3 \end{pmatrix}, \quad M_{-\Lambda^2} \equiv \begin{pmatrix} -1 & 2 \\ -2 & 3 \end{pmatrix}$$

If we introduce the S and T matrices which generate $SL(2, \mathbf{Z})$, the $M_{\pm \Lambda^2}$ are represented as

$$S = \begin{pmatrix} 0 & 1 \\ -1 & 0 \end{pmatrix}, \quad T = \begin{pmatrix} 1 & 1 \\ 0 & 1 \end{pmatrix},$$

(2.20)

$$M_{\Lambda^2} = ST^2 S^{-1}, \quad M_{-\Lambda^2} = (TS)T^2(TS)^{-1}$$

(2.21)

Monodromy at ∞ is given by

$$a \approx \sqrt{2u} \longrightarrow -a, \quad a_D = \frac{\partial F}{\partial a} \approx \frac{2ia}{\pi} \log \frac{a}{\Lambda} + i\frac{a}{\pi} \longrightarrow -a_D + 2a$$

$$M_\infty = \begin{pmatrix} -1 & 2 \\ 0 & -1 \end{pmatrix} = PT^{-2}, \quad P \equiv \begin{pmatrix} -1 & 0 \\ 0 & -1 \end{pmatrix}$$

(2.22)

The complex u-plane has the topology of a sphere. One may verify

$$M_{\Lambda^2} M_{-\Lambda^2} = M_\infty$$

(2.23)

and see that the monodromies are consistent with each other.

Now it is possible to present an exact solution to $N = 2$ pure $SU(2)$ gauge theory. One introduces the so-called Seiberg–Witten curve $\mathcal{C}$ and differential

$$\text{curve } \mathcal{C}: \qquad y^2 = (x^2 - \Lambda^4)(x - u),$$

(2.24)

$$\text{differential}: \qquad \lambda = \frac{\sqrt{2}}{2\pi}\sqrt{\frac{x-u}{x^2 - \Lambda^4}}dx.$$

(2.25)

These curves and differentials are designed so that their period integrals produce the functions $a(u)$ and $a_D(u)$

$$a(u) = \oint_\alpha \lambda = \frac{\sqrt{2}}{\pi} \int_{-\Lambda^2}^{\Lambda^2} \sqrt{\frac{x-u}{x^2 - \Lambda^4}}dx,$$

$$a_D(u) = \oint_\beta \lambda = \frac{\sqrt{2}}{\pi} \int_{\Lambda^2}^{u} \sqrt{\frac{x-u}{x^2 - \Lambda^4}}dx$$

(2.26)

where α, β are the standard homology cycles of a torus. At generic values of u the cubic equation $(x^2 - \Lambda^4)(x - u) = 0$ has three distinct roots $x_i = u, \Lambda^2, -\Lambda^2$, $i = 1, 2, 3$. When some of these roots coincide, the curve becomes degenerate and we have a singular situation. Thus we have the monopole $(u = \Lambda^2)$ and dyon singularity $(u = -\Lambda^2)$ in the present case. In general the discriminant Δ of a cubic equation $x^3 + fx + g = 0$ is given by $\Delta = 4f^3 + 27g^2$ and the positions of the singularities in the u-plane are located by $\Delta(u) = 0$.

It is easy to recover the behavior at the three singularities (2.16), (2.17) and (2.22) from (2.26). In fact the above periods obey a differential equation with regular singular points (the Picard–Fuchs equation)

$$\frac{d^2\Omega(u)}{du^2} + \frac{\Omega(u)}{4(u^2 - \Lambda^4)} = 0, \qquad (2.27)$$

where a and a_D are two independent solutions of this equation. Equation (2.27) is nothing but the Gauss hypergeometric equation with singular points at $u = \pm\Lambda^2$ and ∞. Expansions (2.5), (2.16) and (2.17) are just the standard Taylor expansions around singular points. It is easy to see that the indices of the equation are degenerate at each singularity, which corresponds to the non-vanishing beta function and the appearance of the logarithmic solution.

A special feature of (2.27) is the absence of a first-derivative term. Then the Wronskian $a(u)\partial a_D(u)/\partial u - a_D(u)\partial a(u)/\partial u$ becomes a constant. Using $a_D = \partial F/\partial a$ we can integrate this equation to obtain [17]

$$a\frac{\partial F}{\partial a} - 2F = \text{const.} \times u. \qquad (2.28)$$

The value of const. can be fixed by comparing the behavior of the left-hand and right-hand sides at $u = \infty$. It turns out that it is proportional to the coefficient of the one-loop beta function, const. $= ib/2\pi$ (in the present case, $b = 4$). The above equation shows the breakdown of scale invariance of the prepotential due to the non-vanishing beta function. A generalization of (2.28) is known to hold in all $N = 2$ supersymmetric gauge theories including massless matter fields [18, 19].

3 $SU(2)$ gauge theory with matter

Let us next turn to the case of $SU(2)$ gauge theory coupled to matter in the vector representations. Since it turns out that the ultraviolet-finite theory $N_f = 4$ is somewhat special, we shall restrict ourselves to the case of $1 \leq N_f \leq 3$. A special feature of the $N = 2$ gauge theory coupled to matter is the existence of a Yukawa-like interaction between scalars of the vector and hypermultiplets. The superpotential consists of a Yukawa-type term and a bare mass term of matter fields

$$W = \sum_i \sqrt{2}\tilde{M}_i A M_i + \sum_i m_i \tilde{M}_i M_i. \qquad (3.1)$$

Thus in the Coulomb phase with the vacuum value $\phi^3 = a$, the matter field acquires an effective mass $m_i \pm \frac{1}{\sqrt{2}}a$.

Let us first look at the case of large masses $m_i \gg \Lambda$ and study regions in the moduli space around $u \approx m_i^2$ ($i = 1, \ldots, N_f$). These are in the weak coupling regime where an analysis based on the bare Lagrangian is reliable. Then there exists a massless particle at each $u = m_i^2$ ($i = 1, \ldots, N_f$) with the bare mass m_i being canceled by the vacuum value a. Therefore, when masses are large, there are N_f additional singular points in the complex u-plane at large $|u|$. We call these singularities the *squark* singularities. If the mass m_i goes to ∞, the corresponding singularity moves away to ∞ in the u-plane and we eventually lose one flavor, $N_f \to N_f - 1$. Thus the squark singularity describes an elementary particle which becomes massless at $u = m_i$. On the other hand if the masses m_i decrease, squark singularities move into a region of smaller values of $|u|$. In the limit of zero mass, squark singularities will be located in the strong coupling region $|u| \approx \Lambda^2$ and appear as massless solitons. We will later discuss how an elementary particle could convert itself into a soliton as the mass is varied.

Let us first look at the massless case. In the massless limit there exists a global discrete symmetry (a remnant of the $U(1)_R$ symmetry broken by anomaly) which acts on the u-plane and determines the the location of the singularities. We list below the global symmetry, electric and magnetic charges (n_m, n_e) carried by the strong-coupling singularities and their monodromies:

$$\begin{array}{ccc} & \text{global symmetry} & \text{monodromies} \end{array} \qquad (3.2)$$

$$N_f = 0 \qquad \begin{array}{c} Z_2 \\ (1,0),(1,2); \end{array} \qquad STS^{-1},(T^2 S)T(T^2 S)^{-1} \qquad (3.3)$$

$$N_f = 1 \qquad \begin{array}{c} Z_3 \\ (1,0),(1,1),(1,2); \end{array} \quad STS^{-1},(TS)T(TS)^{-1},(T^2 S)T(T^2 S)^{-1} \qquad (3.4)$$

$$N_f = 2 \qquad \begin{array}{c} Z_2 \\ (1,0),(1,1); \end{array} \qquad ST^2 S^{-1},(TS)T^2(TS)^{-1} \qquad (3.5)$$

$$N_f = 3 \qquad \begin{array}{c} \text{none} \\ (1,0),(2,1); \end{array} \qquad ST^4(S)^{-1},(ST^2 S)T(ST^2 S)^{-1} \cdot \qquad (3.6)$$

(In the case $N_f = 0$ the normalization of electric charge is changed by a factor 2 from the previous section.) Note that in the massless limit the theory with N_f flavors has a global $SO(2N_f)$ symmetry (**2** and **2*** representations of $SU(2)$ are equivalent and thus M_i and $\overline{M}_i$ form a $2N_f$-dimensional representation of $SO(2N_f)$), and the singularities fall into representations of $SO(2N_f)$. Both of the singularities in $N_f = 2$ theory have multiplicity 2 and form representations $(\mathbf{2},\mathbf{1}),(\mathbf{1},\mathbf{2})$ of $SO(4)$, while those of $N_f = 3$ have multiplicity 4 and 1 and form **4** and **1** representations of $SO(6)$. Note also that the monodromy around a singularity of multiplicity k with charge $(n_m, n_e) = (1, n)$

is given by $(T^n S)T^k (T^n S)^{-1}$ in the above table. We can also check that the strong-coupling monodromies add up to the weak-coupling one at ∞: $\prod_i M_i = M_\infty$ where $M_\infty = PT^{-(4-N_f)}$.

One of the curious features of the massless solution is the appearance of so-called jumping lines in the u-plane. A jumping line is defined by

$$\{a_D(u)/a(u) \text{ is real} \mid u \in \mathbf{C}\}. \tag{3.7}$$

It passes through strong-coupling singularities and divides the u-plane into two parts, inner and outer, with the latter being accessible to the observation at ∞ while the former is not. In the case of $N_f = 1$, for instance, the jumping line looks like a circle passing through singularities at $u = u_1$, u_2 and u_3, where $u_j = \exp((2j - 1)\pi i/3)(27/4)^{1/3}\Lambda^2/4$; see Figure 1 [20, 21].

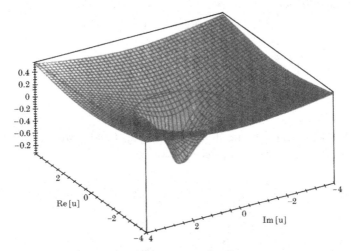

Figure 1. $\operatorname{Im} a_D(u)/a(u)$, $(N_f = 1, m = 0)$

On the jumping line the distinction between the electric and magnetic charge becomes ambiguous. In the BPS mass formula (2.11), a and a_D span a 2-dimensional lattice whose coordinates define the electric and magnetic charges. When a_D/a is real, the lattice collapses and one cannot distinguish n_e from n_m. Thus a soliton may change its identity when it crosses a jumping line. We note that since all the strong-coupling singularities are located on the jumping line, they can be reached from $u = \infty$ without crossing the line. Thus these singularities have well-defined quantum numbers.

Let us, for instance, examine the quantum numbers of three singularities in the $N_f = 1$ case. The Seiberg–Witten curve for the massless $N_f = 1$ theory is given by

$$y^2 = x^2(x - u) - \frac{\Lambda^6}{64}. \tag{3.8}$$

Its discriminant is $\Delta = 256u^3 + 27\Lambda^6$ and hence the singularities are located at $u_2 = -(27/4)^{1/3}\Lambda^2/4$, $u_1 = \exp(-2i\pi/3)u_2$ and $u_3 = \exp(2i\pi/3)u_2$ in the u-plane. We denote the zeros of the cubic polynomial $x^2(x-u) - \Lambda^6/64$ in the x-plane by x_1, x_2, x_3; see Figure 2.

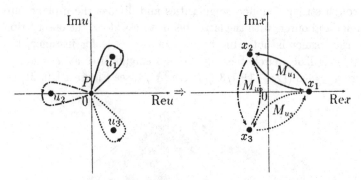

Figure 2. Singularities in the u- and x-planes

We also define three cycles in the x-plane, α, β and γ which go around a pair of points among x_1, x_2, x_3; seee Figure 3.

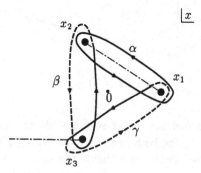

Figure 3. α, β, γ cycles

Note that the sum of these cycles vanishes: $\alpha + \beta + \gamma = 0$. We then consider contours C_i which start from $u = 0$, go around u_i and return to $u = 0$. When one travels along C_1, for instance, x_1 and x_2 are interchanged while x_3 remains invariant as in Figure 2. Then it is easy to see that by going along C_i the three cycles transform as

$$C_1 : \begin{cases} \alpha & \to & \alpha \\ \beta & \to & \beta + \alpha \\ \gamma & \to & \gamma - \alpha \end{cases}, \; C_2 : \begin{cases} \alpha & \to & \alpha - \beta \\ \beta & \to & \beta, \\ \gamma & \to & \gamma + \beta \end{cases} \; C_3 : \begin{cases} \alpha & \to & \alpha + \gamma \\ \beta & \to & \beta - \gamma. \\ \gamma & \to & \gamma \end{cases} \quad (3.9)$$

If we take α, β as the basis of the cycles, we obtain the monodromy matrices

$$M_{u_1} = \begin{pmatrix} 1 & 0 \\ 1 & 1 \end{pmatrix}, M_{u_2} = \begin{pmatrix} 1 & -1 \\ 0 & 1 \end{pmatrix}, M_{u_3} = \begin{pmatrix} 0 & -1 \\ 1 & 2 \end{pmatrix}. \quad (3.10)$$

Monodromy at ∞ is then given by

$$M_\infty = M_{u_1} M_{u_2} M_{u_3} = \begin{pmatrix} -1 & -3 \\ 0 & -1 \end{pmatrix}. \quad (3.11)$$

On the other hand, from the known weak-coupling behavior of a and a_D, namely $a \sim \sqrt{u}$, and $a_D \sim 3ia/2\pi \log u$, one can derive their monodromy property at ∞:

$$a_D \to -a_D + 3a, \qquad a \to -a. \quad (3.12)$$

Hence a is identified as the period around the cycle β, and a_D as the period around the cycle $-\alpha$. If we rewrite the monodromy matrices in the basis of a_D and a, they become

$$M_{u_1} = \begin{pmatrix} 1 & 0 \\ -1 & 1 \end{pmatrix}, M_{u_2} = \begin{pmatrix} 1 & 1 \\ 0 & 1 \end{pmatrix}, M_{u_3} = \begin{pmatrix} 0 & 1 \\ -1 & 2 \end{pmatrix}. \quad (3.13)$$

In order to define the charges of the singularities u_i properly we next consider paths extending from infinity to u_i as in Figure 4.

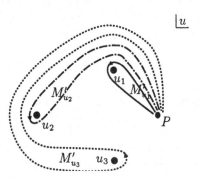

Figure 4. Charge of solitons

These paths avoid entering into the area inside the jumping line. They generate monodromies as

$$M'_{u_1} = M_{u_1} = \begin{pmatrix} 1 & 0 \\ -1 & 1 \end{pmatrix}, \quad (3.14)$$

$$M'_{u_2} = M_{u_1} M_{u_2} M_{u_1}^{-1} = \begin{pmatrix} 2 & 1 \\ -1 & 0 \end{pmatrix}, \qquad (3.15)$$

$$M'_{u_3} = (M_{u_1} M_{u_2}) M_{u_3} (M_{u_1} M_{u_2})^{-1} = \begin{pmatrix} 3 & 4 \\ -1 & -1 \end{pmatrix}. \qquad (3.16)$$

By considering eigenvectors with eigenvalue 1 we deduce the charges of massless solitons: $(1,0)$ at $u = u_1$, $(1,1)$ at $u = u_2$, $(1,2)$ at $u = u_3$, respectively. This reproduces the result of (3.4).

Let us now generalize our discussions and consider the case of massive matter fields. Seiberg–Witten curves for massive theories are constructed by requiring the correct singularity structure and decoupling limit $m_i \to \infty$. They are given by [2]

$$N_f = 1 : \quad y^2 = x^2(x-u) + \frac{1}{4}m\Lambda_1^3 x - \frac{1}{64}\Lambda_1^6, \qquad (3.17)$$

$$N_f = 2 : \quad y^2 = (x^2 - \frac{1}{64}\Lambda_2^4)(x-u) + \frac{1}{4}m_1 m_2 \Lambda_2^2 x - \frac{1}{64}(m_1^2 + m_2^2)\Lambda_2^4, \quad (3.18)$$

$$N_f = 3 : \quad y^2 = x^2(x-u) - \frac{1}{64}\Lambda_3^2(x-u)^2 - \frac{1}{64}(m_1^2 + m_2^2 + m_3^2)\Lambda_3^2(x-u)$$

$$+ \frac{1}{4}m_1 m_2 m_3 \Lambda_3 x - \frac{1}{64}(m_1^2 m_2^2 + m_2^2 m_3^2 + m_3^2 m_1^2)\Lambda_3^2. \qquad (3.19)$$

Here Λ_{N_f} stands for the Λ parameter of the N_f-flavor theory and obeys the decoupling condition

$$\lim_{m_i \to \infty} m_i \Lambda_{N_f}^{4-N_f} = \Lambda_{N_f-1}^{4-(N_f-1)}. \qquad (3.20)$$

In fact in (3.17)–(3.19) a curve of N_f-theory reduces to one of (N_f-1)-theory in the decoupling limit $m_i \to \infty$.

The discriminants of the curves are

$$N_f = 1 : \quad \Delta_1 = 27\Lambda_1^6 + 256\Lambda_1 m^3 - 288\Lambda_1^3 mu - 256m^2 u^2 + 256u^3, \quad (3.21)$$

$$N_f = 2 : \quad \Delta_2 = (\Lambda_2^2 + 8m^2 - 8u)^2(\Lambda_2^4 - 64\Lambda_2^2 m^2 + 16\Lambda_2^2 u + 64u^2), \quad (3.22)$$

$$N_f = 3 : \quad \Delta_3 = (\Lambda_3 m + 8m^2 - 8u)^3(3\Lambda_3^3 m + 24\Lambda_3^2 m^2 + 2048\Lambda_3 m^3$$

$$+84\Lambda_3^2 u - 768\Lambda_3 mu - 2048u^2). \qquad (3.23)$$

Here we have put the masses to be equal for simplicity: $m_i = m$. When matter fields have the same mass, the flavor symmetry of the theory becomes $SU(N_f)$ and the singularities fall into their representations. In fact in the $N_f = 2$ case the singularity at $u = m^2 - \Lambda_2^2/8$ has multiplicity 2 and hence forms the **2** representation of $SU(2)$. Similarly the singularity at $u = m^2 - \Lambda_3 m/8$ in $N_f = 3$ theory is a **3** representation of $SU(3)$.

The Seiberg–Witten differentials are now given by

$$N_f = 1 : \quad \lambda = \frac{-\sqrt{2}}{4\pi}\frac{ydx}{x^2}, \qquad (3.24)$$

$$N_f = 2 : \quad \lambda = \frac{-\sqrt{2}}{4\pi} \frac{y\,dx}{(x^2 - \frac{1}{64}\Lambda_2^4)}, \tag{3.25}$$

$$N_f = 3 : \quad \lambda = \frac{\sqrt{2}}{\pi\Lambda_3} dx \log\left(y + i\frac{\Lambda_3}{8}(x - u + \frac{1}{2}\sum_i m_i^2 - \frac{32}{\Lambda_3^2}x^2)\right). \tag{3.26}$$

A new aspect of the massive theory is that the differential has poles with non-zero residues proportional to the mass of matter fields. In the cases of $N_f = 1, 2$, residues are given by

$$2\pi i \mathrm{Res}\,(\lambda, x = 0) = \pm\frac{m}{\sqrt{2}}, \tag{3.27}$$

$$2\pi i \mathrm{Res}\,(\lambda, x = \pm\Lambda^2/8) = \pm\frac{1}{2}\frac{m_1 \mp m_2}{\sqrt{2}}. \tag{3.28}$$

Then the values of the periods of λ depend explicitly on the choice of contours (rather than their homology classes) and the integral picks up an additive factor when the contour is deformed across a pole. This is in accord with the BPS mass formula in the massive theory which has an additional term proportional to the mass of matter fields:

$$M_{BPS} = \sqrt{2}\left|n_e a(u) + n_m a_D(u) + \sum_i S_i \frac{m_i}{\sqrt{2}}\right|. \tag{3.29}$$

Quantum numbers S_i appear in the central charge of the $N = 2$ algebra when there exist Abelian conserved quantities. For a soliton in the singlet (spinor) representation of flavor symmetry we have $S_i = 1\,(1/2)$. A monodromy transformation in massive theory then takes the form

$$\begin{pmatrix} m/\sqrt{2} \\ n_m \\ n_e \end{pmatrix} \longrightarrow M \begin{pmatrix} m/\sqrt{2} \\ n_m \\ n_e \end{pmatrix} \tag{3.30}$$

where M is a 3×3 matrix.

In the case of the massive $N_f = 1$ theory, the sum of periods around the cycles α, β and γ is now given by $\alpha + \beta + \gamma = 2m/\sqrt{2}$ (for an integral of λ around a cycle we use the same notation as the cycle itself). When one goes around the trajectories C_i, cycles transform as:

$$C_1 : \begin{cases} \alpha & \to & \alpha \\ \beta & \to & \beta + \alpha, \end{cases} C_2 : \begin{cases} \alpha & \to & \alpha - \beta \\ \beta & \to & \beta, \end{cases} C_3 : \begin{cases} \alpha & \to & 2\frac{m}{\sqrt{2}} - \beta \\ \beta & \to & \alpha + 2\beta - 2\frac{m}{\sqrt{2}}. \end{cases} \tag{3.31}$$

It is easy to see that the monodromy matrices are now given by

$$M_{u_1} = \begin{pmatrix} 1 & 0 & 0 \\ 0 & 1 & 0 \\ 0 & 1 & 1 \end{pmatrix}, M_{u_2} = \begin{pmatrix} 1 & 0 & 0 \\ 0 & 1 & -1 \\ 0 & 0 & 1 \end{pmatrix}, M_{u_3} = \begin{pmatrix} 1 & 0 & 0 \\ 2 & 0 & -1 \\ -2 & 1 & 2 \end{pmatrix}. \tag{3.32}$$

Let us next go back to the discriminant (3.21) and follow the motion of singularities as the mass m varies. When the mass is increased (in the real positive direction), the singularity u_2 moves to the right and eventually goes away to ∞ as $u_2 \approx m^2$ while the other two singularities u_1, u_3 remain in the strong-coupling region. They turn into $N_f = 0$ singularities in the decoupling limit

$$\Delta_1 \approx (u - m^2)(u^2 - m\Lambda_1^3) = (u - m^2)(u^2 - \Lambda_0^4). \tag{3.33}$$

Thus u_2 is the squark singularity in this case (which of the three singularities becomes the squark depends on the direction along which m goes to ∞). Then we have an apparent paradox: the singularity at u_2 was originally a dyon with charges $(1,1)$ but now in the weak-coupling region it behaves like an elementary particle. How does a soliton throws away its magnetic charge when it moves out to infinity? The precise mechanism of this metamorphosis seems to be not well understood; however, it is obviously related to the existence of the jumping line. In Figure 3 we have defined the charge of the singularity u_2 by making use of a contour which avoids entering into the region inside the jumping line. On the other hand when the singularity is in the weak-coupling region close to ∞, we may define the charge of the singularity simply by a contour as in Figure 5.

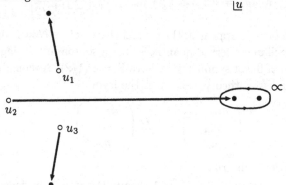

Figure 5. The charge of a squark

The monodromy matrix at ∞ is given by

$$M_\infty = M_{u_1} M_{u_2} M_{u_3} = \begin{pmatrix} 1 & 0 & 0 \\ 4 & -1 & -3 \\ 2 & 0 & -1 \end{pmatrix}. \tag{3.34}$$

The period a transforms under $u \to \exp(2\pi i)u$ as $a \to -a$ and thus corresponds to an eigenvector of (3.34) with an eigenvalue -1. It is easy to see that the eigenvector is given by $(-1, 0, 1)$ and hence $a = \beta - m/\sqrt{2}$ or $\beta = a + m/\sqrt{2}$. Thus the cycle β which gave rise to a massless dyon in the strong-coupling region now behaves like an elementary particle at infinity. A flip in the quantum number has taken place due to the existence of jumping line.

4 Four–dimensional conformal field theory

Another interesting feature of the $SU(2)$ theory with matter is the existence of superconformal points. Four-dimensional conformal field theory with $N = 2$ supersymmetry was first discovered by Argyres and Douglas in the $SU(3)$ theory [22]. In this section, however, I will follow the approach of [23] and locate the superconformal points in $SU(2)$ theories coupled to matter.

We saw in the previous section that when we increased the value of the mass, the would-be squark singularity passed in between the other singularities and moved into the weak-coupling region. Hence if we fine tune the value of the mass, the squark singularity could collide with the other singularities and generate a new phase of the $N = 2$ field theory.

We know that the locations of singularities are given by the zeros of the discriminant $\Delta(u, m)$. We may then fine tune m to a critical value m^* so that $\Delta(u, m^*)$ has a multiple root at $u = u^*$. Values of u^* and m^* are easily found by calculating the discriminant of $\Delta(u, m)$ as a function of u. At these critical points the Seiberg–Witten curves become completely degenerate:

$$N_f = 1 : y^2 \approx (x - x^*)^3, \quad m^* = \frac{3\Lambda}{4}, \quad u^* = \frac{3\Lambda^2}{4}, \quad x^* = \frac{\Lambda^2}{4}, \tag{4.1}$$

$$N_f = 2 : y^2 \approx (x - x^*)^3, \quad m^* = \frac{\Lambda}{2}, \quad u^* = \frac{3\Lambda^2}{8}, \quad x^* = \frac{\Lambda^2}{8}, \tag{4.2}$$

$$N_f = 3 : y^2 \approx (x - x^*)^3, \quad m^* = \frac{\Lambda}{8}, \quad u^* = \frac{\Lambda^2}{32}, \quad x^* = \frac{\Lambda^2}{64}. \tag{4.3}$$

Thus both α and β cycles vanish at these points. In order to see in more detail what is happening here let us look at the differential equations satisfied by the periods a and a_D. Since in the case of massive theories, monodromy is described by 3×3 matrices, the Picard–Fuchs equation becomes a third-order differential equation

$$D(u)\Omega'''(u) + P(u)\Omega''(u) + Q(u)\Omega'(u) = 0. \tag{4.4}$$

We always have a trivial solution $\Omega = $ const. which corresponds to a cycle surrounding the simple pole. The coefficient functions are given by

$$N_f = 1 :$$

$$\left\{ \begin{array}{l} P(u) = (81\Lambda^6 - 384\Lambda^3 m^3 - 2048 m^4 u + 3840 m^2 u^2 - 1536 u^3), \\ Q(u) = 8(-9\Lambda^3 m - 32 m^4 + 72 m^2 u - 24 u^2), \\ D(u) = (4m^2 - 3u)(27\Lambda_1^6 + 256\Lambda_1^3 m^3 - 288\Lambda_1^3 mu - 256 m^2 u^2 + 256 u^3). \end{array} \right. \tag{4.5}$$

$$N_f = 2:$$

$$\begin{cases} P(u) = 64(15\Lambda^4 m^2 - 96\Lambda^2 m^4 - 2\Lambda^4 u + 48\Lambda^2 m^2 u \\ \qquad\qquad -256m^4 u - 32\Lambda^2 u^2 + 448m^2 u^2 - 128u^3), \\ Q(u) = 16(-\Lambda^4 - 24\Lambda^2 m^2 - 128m^4 - 16\Lambda^2 u + 320m^2 u - 64u^2), \\ D(u) = (\Lambda^2 + 8m^2 - 8u)(\Lambda^2 - 16m^2 + 8u)(\Lambda^4 - 64\Lambda^2 m^2 + 16\Lambda^2 u + 64u^2). \end{cases} \tag{4.6}$$

$$N_f = 3:$$

$$\begin{cases} P(u) = 4(-7\Lambda^4 m^2 - 208\Lambda^3 m^3 - 4288\Lambda^2 m^4 + 73728\Lambda m^5 - 32\Lambda^3 mu \\ \quad +2560\Lambda^2 m^2 u - 40960\Lambda m^3 u + 131072m^4 u - 64\Lambda^2 u^2 + 16384\Lambda mu^2 \\ \quad -212992m^2 u^2 + 32768u^3), \\ Q(u) = 256(7\Lambda^2 m^2 - 8\Lambda m^3 + 256m^4 + 32\Lambda mu - 704m^2 u + 64u^2), \\ D(u) = (\Lambda m + 8m^2 - 8u)(\Lambda m - 16m^2 + 4u)(3\Lambda^3 m + 24\Lambda^2 m^2 \\ \qquad\qquad +2048\Lambda m^3 + 8\Lambda^2 u - 768\Lambda mu - 2048u^2). \end{cases} \tag{4.7}$$

In the $N_f = 1$ case, for instance, at the critical value $m = m^*$ we have

$$\Omega'''(u) + \frac{1}{(u - u^*)}\Omega''(u) - \frac{1}{36(u - u^*)^2}\Omega'(u) = 0. \tag{4.8}$$

Now the exponents of Ω around $u = u^*$, $\Omega(u) \approx (u - u^*)^\delta$ are non-degenerate and given by $\delta = \frac{7}{6}, \frac{5}{6}$ and 0. Thus the degeneracy among the exponents is lifted and both of the solutions obey a power law: $a(u), a_D(u) \approx (u - u^*)^{5/6}$ (the $(u - u^*)^{7/6}$ component vanishes faster at $u \approx u^*$). The logarithmic solution has now disappeared. (The explicit evaluation of periods at critical points is discussed in [24].)

We may interpret this as the vanishing of the beta function and the restoration of the scale invariance at the critical point. As we noted before, both the α and the β cycles vanish at the critical point. In general a cycle corresponds to a massless soliton with some charge (n_m, n_e) and the intersection number of cycles on the torus is proportional to the 'cross-product' of the charges $n_m n'_e - n_e n'_m$. When the cross-product does not vanish, two set of charges represents solitons which are mutually non-local to each other (there are no Lagrangians where both of the fields have a local description). Thus the general criterion for the appearance of conformal invariant theories is the coexistence of massless solitons with mutually non-local charges [22]. If we normalize the scaling dimensions of periods a, a_D to 1, the scaling dimension of the order parameter u becomes 6/5 in the $N_f = 1$ theory. Similarly we find that the exponents of the order parameter are given by 4/3 (3/2) in $N_f = 2, 3$ theories.

The detailed physical nature of these superconformal theories is as yet unknown; however, one may attempt to classify them by means of their global symmetries and critical exponents. In [25] a detailed study has been made by making use of all solutions of $N = 2$ Yang–Mills theories with classical gauge groups and fine tuning the parameters to critical points. It turns out that the critical points obtained from $SU(n+1)$, $SO(2n+1)$ and $Sp(2n)$ gauge theories have the same exponents and global symmetries and thus seem to define identical conformal field thories. On the other hand $SO(2n)$ and E_n gauge theories lead to different exponents. Thus we have some evidence for an ADE type classification of $N = 2$ conformal field theories. The critical exponents are given by

$$A_n: \quad 2\frac{e+1}{h+2}, \quad e = 1, 2, \dots, n, \qquad\qquad h = n+1 \qquad (4.9)$$

$$D_n: \quad 2\frac{e+1}{h+2}, \quad e = 1, 3, \dots, 2n-3, n-1 \qquad h = 2n-2 \quad (4.10)$$

$$E_6: \quad 2\frac{e+1}{h+2}, \quad e = 1, 4, 5, 7, 8, 11, \qquad\qquad h = 12 \qquad (4.11)$$

$$E_7: \quad 2\frac{e+1}{h+2}, \quad e = 1, 5, 7, 9, 11, 13, 17, \qquad\quad h = 18 \qquad (4.12)$$

$$E_8: \quad 2\frac{e+1}{h+2}, \quad e = 1, 7, 11, 13, 17, 19, 23, 29, \quad h = 30. \qquad (4.13)$$

In all these cases e runs over the Dynkin exponents of the group and h is the dual Coxeter number. The $SU(2)$ theory with $N_f = 1, 2$ belongs to the class A_2, A_3 above and the $SU(2)$ theory with $N_f = 3$ belongs to D_4.

5 Heterotic-type II duality

Now we would like to argue that the above ADE classification of 4-dimensional superconformal theories may be motivated by a string-theoretic derivation of $N = 2$ Yang–Mills gauge theories [26, 27]. In [27] the authors have streamlined the derivation of Seiberg–Witten curves and differentials of the Yang–Mills theory by considering Calabi–Yau manifolds with a $K3$ fibration [26, 28, 29]. For instance, we start from a Calabi–Yau hypersurface in a weighted projective space $WP^5_{1,1,2,8,12}$

$$
\begin{aligned}
W &= \frac{1}{24}(x_1^{24} + x_2^{24}) + \frac{1}{12}x_3^{12} + \frac{1}{3}x_4^3 + \frac{1}{2}x_5^2 + \psi_0(x_1 x_2 x_3 x_4 x_5) \\
&\quad + \frac{1}{6}\psi_1(x_1 x_2 x_3)^6 + \frac{1}{12}\psi_2(x_1 x_2)^{12} = 0.
\end{aligned}
\qquad (5.1)
$$

This equation is known to describe the $SU(3)$ pure gauge theory in the field theory limit [26]. By introducing new parameters $a = -\psi_0^6/\psi_1$, $b = \psi_2^{-2}$ and $c = -\psi_2/\psi_1^2$ and changing of variables $x_1/x_2 = \zeta^{\frac{1}{12}}b^{\frac{-1}{24}}$, $x_1^2 = x_0\zeta^{\frac{1}{12}}$, we can rewrite W as

$$
\begin{aligned}
W(\zeta, a, b, c) &= \frac{1}{24}\left(\zeta + \frac{b}{\zeta} + 2\right)x_0^{12} + \frac{1}{12}x_3^{12} + \frac{1}{3}x_4^3 + \frac{1}{2}x_5^2 \\
&\quad + \frac{1}{6\sqrt{c}}(x_0 x_3)^6 + \left(\frac{a}{\sqrt{c}}\right)^{\frac{1}{6}} x_0 x_3 x_4 x_5.
\end{aligned}
\qquad (5.2)
$$

For each fixed value of ζ, $W = 0$ describes a $K3$ surface.

The discriminant of the Calabi–Yau manifold (5.2) is given by

$$\Delta \approx (b-1)((1-c)^2 - bc^2)(((1-a)^2 - c)^2 - bc^2). \qquad (5.3)$$

The field theory limit is achieved by taking

$$a = -2\epsilon u^{3/2}/3\sqrt{3}, \quad b = \epsilon^2 \Lambda^6, \quad c = 1 - \epsilon(-2u^{2/3}/3\sqrt{3} + v) \qquad (5.4)$$

and letting $\epsilon \equiv (\alpha')^{3/2} \to 0$. Here u and v are gauge-invariant Casimirs of $SU(3)$. After a suitable change of variable W takes the form (we choose a patch $x_0 = 1$)

$$W = z + \frac{\Lambda^6}{z} + 2C_{A_2}(x, u, v) + s^2 + w^2 \qquad (5.5)$$

where $C_{A_2}(x, u, v) = x^3 - ux - v$.

Now one considers the period integral of the holomorphic 3-form on the Calabi–Yau manifold

$$\omega = \int \Omega = \int \frac{dz}{z} \wedge \frac{ds \wedge dx}{\left.\frac{\partial W}{\partial w}\right|_{W=0}} \qquad (5.6)$$

In the case of the A_2 singularity

$$\left.\frac{\partial W}{\partial w}\right|_{W=0} = 2\sqrt{z + \Lambda^6/z + 2C_{A_2}(x, u, v) + s^2}. \qquad (5.7)$$

The integral over s is trivial and equals 2π. The boundary of the s-integration is given by

$$z + \frac{\Lambda^6}{z} + 2C_{A_2}(x, u, v) = 0 \qquad (5.8)$$

which describes the spectral curve [30, 31]. If one shifts $z \to y - C_{A_2}$, one recovers the hyperelliptic curve

$$y^2 = C_{A_2}(x, u, v)^2 - \Lambda^6. \qquad (5.9)$$

The period is rewritten as

$$\begin{aligned}
\omega &= \pi \int \frac{dz}{z} \wedge dx = \pi \oint x \frac{d(y - C_{A_2})}{(y - C_{A_2})}, \\
&= \pi \oint x d\log(y - C_{A_2}) = \frac{\pi}{2} \oint x d\log \frac{(y - C_{A_2})}{(y + C_{A_2})}
\end{aligned} \qquad (5.10)$$

which reproduces the Seiberg–Witten differential of the $SU(3)$ gauge theory.

In the above we have considered the case when the $K3$ surface (ALE space) degenerates to have an A_2-type singularity. We may generalize this construction and consider the case when the K_3 surface degenerates into more general types of ADE singularities. In the case of E_n singularities one can no longer perform the s-integral and reduce ω to an integral over a Riemann surface. It turns out, however, that we can still analyze the critical behavior of ω and determine the exponents of the E_n gauge theories.

Let us first discuss the E_6 theory,

$$
\begin{aligned}
W &= z + \frac{\Lambda^{24}}{z} + 2C_{E_6}(x,s) + w^2, \\
C_{E_6}(x,s) &= x^4 + s^3 + u_1 x^2 s + u_4 x s + u_5 x^2 + u_7 s + u_8 x + u_{11}, \quad (5.11)
\end{aligned}
$$

where u_i are the perturbations of the E_6 singularity. The period is given by

$$
\omega = \int \frac{dz}{z} \wedge \frac{ds \wedge dx}{\sqrt{z + \frac{\Lambda^{24}}{z} + 2C_{E_6}(x,s)}}. \tag{5.12}
$$

The critical point is located at

$$
\begin{aligned}
x^* &= 0, \quad s^* = 0, \quad z^* = \Lambda^{12}, \quad u_{11}^* = -\Lambda^{12}, \\
u_1^* &= u_4^* = u_5^* = u_7^* = u_8^* = 0. \tag{5.13}
\end{aligned}
$$

Let us first determine the exponent of the parameter u_1. By perturbing away from the critical point

$$
z = \Lambda^{12} + t^{1/2}\Lambda^6 \tilde{z}, \quad x = t^{1/4}\tilde{x}, \quad s = t^{1/3}\tilde{s}, \quad u_1 = t^{1/6}\tilde{u}_1 \tag{5.14}
$$

we have

$$
\omega \approx t^{1/2 - 1/2 + 1/4 + 1/3} \int d\tilde{z} \wedge \frac{d\tilde{s} \wedge d\tilde{x}}{\sqrt{\tilde{z}^2 + \tilde{x}^4 + \tilde{s}^3 + \tilde{u}_1 \tilde{x}^2 \tilde{s}}} \approx t^{7/12}. \tag{5.15}
$$

By introducing the unit of mass μ we find

$$
t \approx \mu^{12/7}, \quad u_1 \approx \mu^{2/7}. \tag{5.16}
$$

Thus the perturbation u_1 has the exponent $2/7$. We can similarly determine the exponents of the parameters u_i $(i = 4,5,7,8,11)$. They are

$$
\frac{2(e_i + 1)}{14}, \quad i = 1,4,5,7,8,11 \tag{5.17}
$$

and hence again fit the formula $2(e_i + 1)/(h + 2)$ where the dual Coxeter number h of E_6 is 12.

We may also compute exponents for the E_7 and E_8 gauge theories. Singularities are described by the polynomials

$$
\begin{aligned}
C_{E_7}(x,s) &= x^3 + x s^3 + u_1 s^4 + u_5 s^3 + u_7 x s + u_9 s^2 + u_{11} x \\
&\quad + u_{13} s + u_{17}, \tag{5.18} \\
C_{E_8}(x,s) &= x^5 + s^3 + u_1 x^3 s + u_7 x^2 s + u_{11} x^3 + u_{13} x s + u_{17} x^2 \\
&\quad + u_{19} s + u_{23} x + u_{29}. \tag{5.19}
\end{aligned}
$$

We find that their critical exponents are given by

$$\frac{2(e_i + 1)}{20}, \quad i = 1, 5, 7, 9, 11, 13, 17 \qquad \text{for } E_7 \qquad (5.20)$$

$$\frac{2(e_i + 1)}{32}, \quad i = 1, 7, 11, 13, 17, 19, 23, 29 \qquad \text{for } E_8. \qquad (5.21)$$

These two expressions again fit the formulas (4.12) and (4.13).

It is easy to check that the above construction also reproduces the exponents of (4.9), (4.10) in the case of A_n and D_n singularities. Thus we have some considerable evidence for the ADE classification of SCFTs originating from pure $N = 2$ gauge theories. The pattern of the classification follows that of the degeneration of the $K3$ surface which features in the heterotic-type II duality based on $K3$-fibered Calabi–Yau manifolds. It will be very interesting to understand the OPE structure and the selection rule of correlation functions of these conformal theries.

Acknowledgementa

I would like to thank organizers of the Cambridge spring school 'Non-perturbative Aspects of Quantum Field Theories' for their kind hospitality. I also would like to thank D. Hanawa for helpful discussions on the conversion of solitons into squarks.

References

[1] N. Seiberg and E. Witten, *Nucl. Phys.* **B426** (1994) 19.

[2] N. Seiberg and E. Witten, *Nucl. Phys.* **B431** (1994) 484.

[3] P. Argyres and A. Faraggi, *Phys. Rev. Lett.* **74** (1995) 3931.

[4] A. Klemm, W. Lerche, S. Theisen and S. Yankielowicz, *Phys. Lett.* **B344** (1995) 169.

[5] A. Hanany and Y. Oz, *Nucl. Phys.* **B452** (1995) 283; A. Hanany, *Nucl. Phys.* **B466** (1996) 85.

[6] A. Brandhuber and K. Landsteiner, *Phys. Lett.* **B358** (1995) 73.

[7] P.C. Argyres, R.N. Plesser and A.D. Shapere, *Phys. Rev. Lett.* **75** (1995) 1699.

[8] J.A. Minahan and D. Nemeschansky, *Nucl. Phys.* **B464** (1996) 3.

[9] P.C. Argyres and A.D. Shapere, *Nucl. Phys.* **B461** (1996) 437.

[10] U.H. Danielsson and B. Sundborg, *Phys. Lett.* **370B** (1996) 83.

[11] A. Bilal, *Duality in N = 2 SUSY SU(2) Yang–Mills theory: a pedagogical introduction to the work of Seiberg and Witten*, hep-th/9601007.

[12] S.V. Ketov, 'Solitons, monopoles and duality: from sine-Gordon to Seiberg–Witten', *Fortsch. Phys.* **45** (1997), 237, hep-th/9611209.

[13] L. Alvarez-Gaumé and S. F. Hassan 'An introduction to S-Duality in N = 2 Supersymmetric Gauge Theory, (a pedagogical review of the work of Seiberg and Witten)', *Fortsch. Phys.* **45** (1997), 159, hep-th/9701069.

[14] H. Aoyama, T. Harano, M. Sato and S. Wada, *Phys. Lett.* **B388** (1996) 331; T. Harano, M. Sato, *Nucl. Phys.* **B484** (1997) 167.

[15] K. Ito and N. Sasakura, *Phys. Lett.* **B382** (1996) 95; *Nucl. Phys.* **B484** (1997) 141.

[16] N. Dorey, V.V. Khoze and M.P. Mattis, *Phys. Rev.* **D54** (1996) 2921; *Phys. Lett.* **B388** (1996) 324; *Phys. Rev.* **D54** (1996) 7832.

[17] M. Matone, *Phys. Lett.* **B357** (1995) 342.

[18] J. Sonnenschein, S. Theisen and S. Yankielowicz, *Phys. Lett.* **B367** (1996) 145.

[19] T. Eguchi and S.-K. Yang, *Mod. Phys. Lett.* **A11** (1996) 131.

[20] D. Hanawa, Master's thesis, Univ. of Tokyo, 1996.

[21] A. Bilal and F. Ferrari, *Nucl. Phys.* **B480** (1996) 589.

[22] P.C. Argyres and M. Douglas, *Nucl. Phys.* **B448** (1995) 93.

[23] P.C. Argyres, R.N. Plesser, N. Seiberg and E. Witten, *Nucl. Phys.* **B461** (1996) 71.

[24] T. Masuda and H. Suzuki, *Nucl.Phys.* **B495** (1997) 149.

[25] T. Eguchi, K. Hori, K. Ito and S.K. Yang, *Nucl. Phys.* **B471** (1996) 430; T. Eguchi and K. Hori, in *The Mathematical Beauty of Physics*, J.M. Drouffe and J.B. Zuber (eds.), World Scienticfic 1997.

[26] S. Kachru and C. Vafa, *Nucl. Phys.* **B450** (1995) 69.

[27] A. Klemm, W. Lerche, P. Mayr, C. Vafa and N. Warner, *Nucl. Phys.* **B477** (1996) 746.

[28] A. Klemm, W. Lerche and P. Mayr, *Phys. Lett.* **357B** (1995) 313.

[29] P.S. Aspinwall and J. Louis, *Phys. Lett.* **369B** (1996) 233.

[30] A. Gorsky, I. Krichever, A. Marshakov, A. Mironov and A. Morozov, *Phys. Lett.* **355B** (1995) 466.

[31] E. Martinec and N. Warner, *Nucl. Phys.* **B459** (1996) 97.

Supergravity, Brane Dynamics and String Duality

P.C. West

Abstract

In this review we show that a Clifford algebra possesses a unique irreducible representation, the spinor representation. We discuss what types of spinors can exist in Minkowski space-times and we explain how to construct all the supersymmetry algebras that contain a given space-time Lie algebra. After deriving the irreducible representations of the superymmetry algebras, we explain how to use them to systematically construct supergravity theories. We give the maximally supersymmetric supergravity theories in ten and eleven dimensions and discuss their propeties. We find which superbranes can exist for a given supersymmetry algebra and we give the dynamics of the superbranes that occur in M-theory. Finally, we discuss how the properties of supergravity theories and superbranes provide evidence for string duality.

In effect, we present a continuous chain of argument that begins with Clifford algebras and leads via supersymmetry algebras and their irreducible representations to supergravity theories, string duality, brane dynamics and M-theory.

Contents

---

*This material is based on lectures presented at the EU conference on *Duality and Supersymmetric Theories*, the Issac Newton Institute, Cambridge, UK and at the TASI 1997 Summer School, Boulder, Colorado, USA.

Introduction

In this review we begin with an account of Clifford algebras and show that they each possess only one irreducible representation, the spinor representation. We discuss the types of spinors can exist in Minkowski space-times. Equipped with this knowledge, we show how to systematically construct the supersymmetry algebras that contain a given space-time Lie group, such as the Poincaré group. In the context of the 4-dimensional supersymmetry algebras, we illustrate how to find all irreducible representations of the supersymmetry algebras and so arrive at a listing of all possible supersymmetric theories.

We then turn to the construction of theories of local supersymmetry, that is supergravity theories and show how to derive these theories from a knowledge of the corresponding irreducible representation of the supersymmetry algebra, (i.e. on-shell states). Of the three methods given for the construction of supergravity theories two are rather systematic in that they always lead to the desired theory with very little additional information. In Section 6 and we give the unique supergravity theory in eleven dimensions while in Section 7 we describe that two maximally supersymmetric theories in ten dimensions, the IIA and IIB supergravity theories. In addition to explaining how one constructs these theories according to the methods given in Section 5, we describe the properties of these theories. These include the $SL(2, \mathbf{R})$ invariance of the IIB theory and the derivation of the IIA supergravity theory from the 11-dimensional supergravity theory by compactification on a circle. The 10-dimensional IIA and IIB supergravity theories are the low-energy effective actions of IIA and IIB string theories and we discuss the consequences of this relationship for string theories. The 11-dimensional supergravity theory is thought to be the low-energy effective action of a yet to be clearly defined theory called M-theory.

The supergravity theories admit solitonic solutions to their classical field equations that correspond to static p-branes. A p-brane is an object which sweeps out a $(p + 1)$-dimensional surface as it moves through space-time. They generalise strings which are 1-branes. However, from the string theory perspective p-branes for $p > 1$ are non-perturbative objects. In Section 8, we give the dynamics of p-branes and discuss which p-branes can occur in the IIA, IIB and M-theories. In particular, we find that in M-theory we can have only 2-branes and 5-branes and we give the equations of motion of these branes.

The supergravity theories in ten and eleven dimensions form the basis for most discussions of string duality and in Section 9 we outline some of these arguments. In particular we discuss the non-perturbative string duality symmetries and the way they rotate perturbative string states into the *p*-branes discussed in Section 8.

In effect this review traces a continuous chain of argument that begins with Clifford algebras and leads via supersymmetry algebras and their representations to supergravity theories, superbranes and then to string dualities and *M*-theory.

1 Clifford Algebras and Spinors

In this section we define a Clifford algebra in an arbitrary dimension and find its irreducible representations and their properties. This enables us to find which types of spinors are allowed in a given Minkowski space-time.

The starting point for the construction of supersymmetric theories is the supersymmetry algebra which underlies it. Supersymmetry algebras contain supercharges which transform as spinors under the appropriate Lorentz group. Hence, even to construct the supersymmetry algebras, as we do in Section 2, we must first find out what types of spinors are possible in a given dimension and what are their properties. We will find in subsequent sections that supersymmetric algebras and the supersymmetric theories on which they are based rely for their existence in an essential way on the detailed properties of Clifford algebras, that we will derive in this section.

As far as I am aware the first discussion of spinors in arbitrary dimensions was given in [100] and many of the steps in this article are taken from that paper. Use has also been made of the reviews of reference [101], [102] and [194].

1.1 Clifford Algebras

A Clifford algebra in D dimensions is defined as a set containing D elements γ_m which satisfy the relation

$$\{\gamma_m, \gamma_n\} = 2\eta_{mn}, \tag{1.1}$$

where the labels $m, n, \ldots$ take D values, and η_{mn} is the flat metric in $\mathbf{R}^{s,t}$ $(s + t = D)$; that is the metric η_{mn} is a diagonal matrix whose first t entries down the diagonal are -1 and whose last s entries are $+1$. We can raise and lower the $m, n, \ldots$ indices using the metric $\eta_{mn} = \eta^{mn}$ in the usual way.

Under multiplication the D elements γ_n of the Clifford algebra generate a finite group denoted C_D which consists of the elements

$$C_D = \{\pm 1, \pm\gamma_m, \pm\gamma_{m_1,m_2}, \dots, \pm\gamma_{m_1\dots m_D}\}. \tag{1.2}$$

The $\gamma_{m_1 m_2\dots}$ is non-vanishing only if all indices $m_1, m_2, \dots$ are different, in which case it equals

$$\gamma_{m_1 m_2\dots} = \gamma_{m_1}\gamma_{m_2}\cdots .$$

The set of matrices $\gamma_{m_1 m_2\dots m_p}$ for all possible different values of the ms contains

$$\frac{D!}{(D-p)!p!} = \binom{D}{p}$$

different elements. Hence the group C_D generated by the γ_m has order

$$2\sum_{p=0}^{D}\binom{D}{p} = 2(1+1)^D = 2^{D+1}.$$

1.2   Clifford algebras in even dimensions

To find the representations of C_D is a standard exercise in representation theory of finite groups [116]. We will first consider the case of even D.

The number of irreducible representations of any finite-dimensional group, G equals the number of its conjugacy classes. We recall that the conjugacy class $[a]$ of $a \in G$ is given by

$$[a] = \{gag^{-1} \quad \forall\, g \in G\}.$$

For even D it is straightforward to show, using equation (1.1), that the conjugacy classes of C_D are given by

$$[+1], [-1], [\gamma_m], [\gamma_{m_1 m_2}], \dots, [\gamma_{m_1\dots m_D}]. \tag{1.3}$$

Hence for D even there are $2^D + 1$ inequivalent irreducible representations of C_D.

Next we use the fact that the number of inequivalent 1-dimensional representations of any finite group G is equal to the order of G divided by the order of the commutator group of G. We denote the commutator group of G by $\mathrm{Com}\,(G)$. It is defined to be the group $\mathrm{Com}\,(G) = aba^{-1}b^{-1}$, $\forall a, b \in G$. For D even the commutant of C_D is just the elements ± 1 and so has order 2. As a result, the number of inequivalent irreducible 1-dimensional representations of C_D is 2^D. Since the total number of irreducible representations is $2^D + 1$, we conclude that there is only one irreducible representation whose dimension is greater that 1.

Finally, we make use of the theorem that if we denote the order of any finite group by $\operatorname{ord} G$ and it has p irreducible inequivalent representations of dimension n_p then

$$\operatorname{ord} G = \sum_p (n_p)^2.$$

Applying this theorem to C_D we find that

$$2^{D+1} = 1^2 2^D + n^2,$$

where n is the dimension of the single irreducible representation whose dimension is greater than 1. We therefore conclude that $n = 2^{\frac{D}{2}}$. These results are summarised in the following theorem.

Theorem *For D even the group C_D has $2^D + 1$ inequivalent irreducible representations. Of these irreducible representations 2^D are 1-dimensional and the remaining representation has dimension $2^{\frac{D}{2}}$.*

This means that we can represent the γ_m as $2^{\frac{D}{2}}$ by $2^{\frac{D}{2}}$ matrices for the irreducible representation with dimension greater than 1. Our next task is to find the properties of this representation under complex conjugation and transpose. In fact, the 1-dimensional representations are not faithful representations of C_D and we shall not consider them in what follows.

That the above are irreducible representations of the group C_D means that they provide a representation of the group which consists of the elements given in equation (1.2) together with a group composition law which is derived from the Clifford algebra relations using only the operation of multiplication. In particular, the group operations do not include those of addition and subtraction which also occur in the Clifford algebra defining condition of equation 1.1. Hence the irreducible representations of C_D are not necessarily irreducible representations of the Clifford algebra itself. In fact, all the 1-dimensional irreducible representations of C_D do not extend to be also representations of the Clifford algebra, as they do not obey the rules for addition and subtraction. As such, the only representation of C_D and the Clifford algebra is the unique irreducible representation of dimension greater than 1, described above. It follows that the Clifford algebra itself has only one irreducible representation and this has dimension $2^{D/2}$. It is of course the well-known spinor representation.

Given an irreducible representation of the Clifford algebra, also denoted γ_m, with dimension greater than 1 we can take its complex conjugate. Denoting that complex conjugate by γ_m^*, it is obvious that γ_m^* also satisfies equation (1.1) and so forms a representation of the same Clifford algebra. It follows that it also forms a representation of C_D. However, there is only one irreducible representation of C_D of dimension greater than 1 and as a result the original representation and its complex conjugate must be equivalent. Consequently, there exists a matrix B such that

$$\gamma_m^* = B\gamma_m B^{-1}. \tag{1.4}$$

We can choose the scale of B such that $|\det B| = 1$. Taking the complex conjugate of equation (1.4) we find that

$$\gamma_m = (\gamma_m^*)^* = +B^* B\gamma^m B^{-1} B^{-1*}.$$

Hence, we conclude that $B^* B$ commutes with the irreducible representation and by Schur's Lemma must be a constant multiple of the identity matrix, i.e.

$$B^* B = \epsilon I.$$

Taking the complex conjugate of the above relation we find that $BB^* = \epsilon^* I$ and so $BB^* BB^{-1} = \epsilon^* I = \epsilon I$ thus $\epsilon = \epsilon^*$ i.e. ϵ is real. Since we have chosen $|\det B| = 1$ we conclude that that $|\epsilon| = 1$ and so $\epsilon = \pm 1$.

We can also consider the transpose of the irreducible representation γ_m that has dimension greater than one. Denoting the transpose of γ_m by γ_m^T, we find, using a very similar argument, that γ_m and γ_m^T are equivalent representations and so there exists a matrix C, called the *charge conjugation matrix*, such that

$$\gamma_m^T = -C\gamma_m C^{-1}. \tag{1.5}$$

We denote the Hermitian conjugate of γ_m by $\gamma_m^\dagger = \gamma_m^{*T}$. We can relate C to B if we know the Hermiticity properties of the γ_m. For simplicity, and because this is the case of most interest to us, from now on we assume that we are in a Lorentzian space-time whose metric η_{mn} is given by $\eta = \mathrm{diag}(-1, +1, +1, \ldots, +1)$. Any finite-dimensional representation of a finite group G can be chosen to be unitary. Making this choice for our group C_D we have $\gamma_m \gamma_m^\dagger = 1$. Taking into account the relationship $\gamma_n \gamma_n = \eta_{nn}$ we conclude that

$$\gamma_0^\dagger = -\gamma_0, \quad \gamma_m^\dagger = \gamma_m; \quad m = 1, \ldots, D - 1. \tag{1.6}$$

We could, as some texts do, regard this equation as part of the definition of the Clifford algebra. We may rewrite equation (1.6) as

$$\gamma_m^\dagger = \gamma_0 \gamma_m \gamma_0.$$

We may take C to be given by $C = -B^T \gamma_0$ as then

$$\begin{aligned}
C\gamma_m C^{-1} &= B^T \gamma_0 \gamma_m (-\gamma_0((B)^T)^{-1}) = -B^T \gamma_m^\dagger (B)^{T-1} = -(B^{-1}\gamma_m^* B)^T \\
&= -\gamma_m^T, \tag{1.7}
\end{aligned}$$

as required.

Further restrictions on B can be found by computing γ_m^T in two ways: we see that $\gamma_m^T = (\gamma_m^*)^\dagger = (\gamma_m^\dagger)^*$ implies

$$(B^{-1})^\dagger \gamma_0 \gamma_m \gamma_0 B^\dagger = B\gamma_0 \gamma_m \gamma_0 B^{-1}.$$

Using Schur's Lemma we deduce that $-\gamma_0 B^\dagger B \gamma_0$ is proportional to the unit matrix and as a result so is $B^\dagger B$, i.e. $B^\dagger B = \mu I$. Since $|\det B| = 1$ we have $|\mu| = 1$, but taking the matrix element of $B^\dagger B = \mu I$ with any vector we conclude that μ is real and positive. Hence $\mu = 1$ and consequently B is unitary, i.e. $B^\dagger B = I$. This result and the previously derived equation $BB^* = \epsilon I$ imply that

$$B^T = \epsilon B, \quad C^T = -\epsilon C. \tag{1.8}$$

We now wish to determine ϵ in terms of the space-time dimension D. Consider the set of matrices

$$I, \ \gamma_m, \ \gamma_{m_1 m_2 m_3}, \ \gamma_{m_1 m_2 m_3}, \ \gamma_{m_1 \cdots m_D}. \tag{1.9}$$

There are $2^D = \sum_p \binom{D}{p} = (1+1)^D$ such matrices and as they are linearly independent they form a basis for the space of all $2^{\frac{D}{2}} \times 2^{\frac{D}{2}}$ matrices. Using equation (1.1) we can relate $\gamma_{m_1 \cdots m_p}$ to $\gamma_{m_p \cdots m_1}$ to find that the sign change required to reverse the order of the indices is given by

$$\gamma_{m_1 \dots m_p} = (-1)^{p\frac{(p-1)}{2}} \gamma_{m_p \dots m_1}.$$

This, together with equation (1.7), imply that

$$C\gamma_{m_1 \cdots m_p} C^{-1} = (-1)^p (-1)^{p\frac{(p-1)}{2}} \gamma_{m_1 \cdots m_p}{}^T, \tag{1.10}$$

or equivalently

$$(C\gamma_{m_1 \cdots m_p}) = \epsilon(-1)^{\frac{(p-1)(p-2)}{2}} (C\gamma_{m_1 \cdots m_p})^T. \tag{1.11}$$

Using this result we can calculate the number of anti-symmetric matrices in the complete set of equation (1.9) when multiplied from the left by C; It is given by is given by

$$\sum_{p=0}^{D} \frac{1}{2}\left(1 - \epsilon(-1)^{\frac{(p-1)(p-2)}{2}}\right) \binom{D}{p}. \tag{1.12}$$

Using the relationship

$$(-1)^{\frac{(p-1)(p-2)}{2}} = -\frac{1}{2}[(1+i)i^n + (1-i)(-i)^n],$$

we can carry out the sum in equation (1.12). We know, however, that the number of anti-symmetric matrices is $2^{\frac{D}{2}}(2^{\frac{D}{2}}-1)\frac{1}{2}$. Equating these two methods of evaluating the number of anti-symmetric matrices we find that

$$\epsilon = -\sqrt{2}\cos\frac{\pi}{4}(D+1).$$

Put another way $\epsilon = +1$ for $D = 2, 4$; mod 8 and $\epsilon = -1$ for $D = 6, 8$; mod 8. It follows from equation (1.8) that for $D = 2, 4$; mod 8, B is a symmetric unitary matrix. Writing B in terms of its real and imaginary parts, $B = B_1 + iB_2$, where B_1 and B_2 are symmetric and real, the unitarity condition becomes $B_1^2 + B_2^2 = 1$ and $[B_1, B_2] = 0$. Under a change of basis of the γ_m matrices; $\gamma^{m\prime} = A\gamma^n A^{-1}$ we find that the matrix B changes as $B' = A^* B A^{-1}$. In fact, we can use A to diagonalize B_1 and B_2 which, still being unitary, must be of the form $B = \text{diag}(e^{i\alpha_1}, \ldots, e^{i\alpha_D})$. Carrying out another A transformation of the form $A = \text{diag}(e^{i\frac{\alpha_1}{2}}, \ldots, e^{i\frac{\alpha_D}{2}})$ we find the new B equals 1. Hence if $D = 2, 4$; mod 8 the γ_m matrices can be chosen to be real and $C = \gamma^0$.

1.3   Spinors in even dimensions

By definition a spinor λ transforms under Spin $(1, D-1)$ as

$$\delta\lambda = \frac{1}{4}w^{mn}\gamma_{mn}\lambda,$$

where $w^{mn} = -w^{nm}$ are the parameters of the Lorentz transformation. The group Spin $(1, D-1)$ is by definition the group generated by $\frac{1}{4}w^{mn}\gamma_{mn}$ and it is the covering group of $SO(1, D-1)$. The Dirac conjugate, denoted $\bar{\lambda}^D$, must transform so that $\bar{\lambda}^D\lambda \equiv \bar{\lambda}^{D\alpha}\lambda_\alpha$ is invariant and so transforms under a Lorentz transformation as

$$\delta\bar{\lambda}^D = \bar{\lambda}^D\left(-\frac{1}{4}w^{mn}\gamma_{mn}\right).$$

Using the relation $\gamma_{mn}^\dagger = -\gamma_0\gamma_{nm}\gamma_0 = \gamma_0\gamma_{mn}\gamma_0$ we find that

$$\delta(\lambda^+\gamma^0) = (\lambda^+\gamma^0)\left(-\frac{1}{4}w^{mn}\gamma_{mn}\right).$$

Consequently, we can take the Dirac conjugate to be defined by

$$\bar{\lambda}^D \equiv \lambda^\dagger\gamma^0.$$

The Majorana conjugate, denoted $\bar{\lambda}^M$, is defined by

$$\bar{\lambda}^m = \lambda^T C.$$

Using the relationship $\gamma_{mn}^T = C\gamma_n\gamma_m C^{-1} = -C\gamma_{mn}C^{-1}$ we find that

$$\delta\bar{\lambda}^M = \lambda^T \frac{1}{4} w^{mn} \gamma_{mn}^T C = -\lambda^T C \left(\frac{1}{4} w^{mn} \gamma_{mn}\right) = \bar{\lambda}^M \left(-\frac{1}{4} w^{mn} \gamma_{mn}\right).$$

Hence the Majorana conjugate transforms like the Dirac conjugate under Spin $(1, D-1)$ transformations and as a result we can define a Majorana spinor to be one is whose Dirac and Majorana conjugates are equal:

$$\bar{\lambda}^D = \bar{\lambda}^M.$$

The above condition can be rewritten as $\lambda^* = -(\gamma^0)^T C^T \lambda$ and, using the relation $C = B^T \gamma^0$, it becomes

$$\lambda^* = B\lambda.$$

We could have directly verified that $B^{-1}\lambda^*$ transforms under Spin $(1, D-1)$ in the same way as λ by using the equation $B\gamma_{mn}B^{-1} = (\gamma_{mn})^*$ and as a result have imposed this Majorana condition without any mention of the Dirac conjugate.

We are finally in a position to discover the dimensions in which Majorana spinors exist. If we impose the relationship $\lambda^* = B\lambda$, then, taking the complex conjugate we find that it implies the relationship $\lambda = B^*\lambda^*$. Substituting this condition into the first relation we find that

$$\lambda = B^* B\lambda = \epsilon\lambda.$$

since $B^*B = \epsilon I$. Consequently, Majorana spinors can only exist if $\epsilon = +1$, which is the case only in the dimensions $D = 2, 4$; mod 8, i.e. $D = 2, 4, 10, 12, \ldots$.

In an even-dimensional space-time we can construct the matrix

$$\gamma^{D+1} = \gamma_0\gamma_1 \cdots \gamma_{D-1} = \gamma_{01\cdots D-1}.$$

This matrix anti-commutes with γ_m and so commutes with the generators $\left(-\frac{1}{4}\gamma_{mn}\right)$ of spin$(1, D-1)$, the covering group of $SO(1, D-1)$. Hence $\gamma^{D+1}\chi$ transforms like a spinor if χ does. A straightforward calculation shows that

$$(\gamma^{D+1})^2 = (-1)^{\frac{D(D-1)}{2}}(-1) = (-1)^{\frac{D}{2}-1}. \tag{1.13}$$

Hence $(\gamma^{D+1})^2 = 1$ for $D = 2 \bmod 4$ while $(\gamma^{D+1})^2 = -1$ for $D = 4 \bmod 4$. In either case we can define Weyl spinors

$$\gamma^{D+1}\chi = \pm\chi \qquad \text{if } D = 2 \bmod 4$$

and

$$i\gamma^{D+1}\chi = \pm\chi \qquad \text{if } D = 4 \bmod 4.$$

We can now consider when Majorana–Weyl spinors exist. We found that Majorana spinors (i.e. $\chi^* = B\chi$) exist if $\epsilon = 1$, i.e. when $D = 2, 4$; mod 8. Taking the complex conjugate of the above Weyl conditions and using the relationship $(\gamma^{D+1})^* = B\gamma^{D+1}B^{-1}$ we find we get a non-vanishing solution only if $D = 2 \bmod 4$. Hence Majorana–Weyl spinors only exist if $D = 2 \bmod 8$ i.e. $D = 2, 10, 18, 26, \dots$. The factor of i is necessary for $D = 4 \bmod 4$ as the chirality condition must have an operator that squares to 1, however it is this same factor of i that gets a minus sign under complex conjugation and so rules out the possibility of having Majorana–Weyl spinors in these dimensions. We note that these are the dimensions in which self-dual Lorentzian lattices exist and, except for 18 dimensions, these are the dimensions in which critical strings exist.

Corresponding to the above chiral conditions we can defined projectors onto the spaces of positive and negative chiral spinors. These projectors are given by $P_\pm = \frac{1}{2}(1 \mp a\gamma^{D+1})$ where $a = 1$ if $D = 2 \bmod 4$ and $a = i$ if $D = 4 \bmod 4$. It is easy to verify that they are indeed projectors; i.e. $P_\pm P_\mp = 0$, $P_\pm^2 = P_\pm$, $P_\mp^2 = P_\mp$ and $P_\pm + P_\mp = 1$. Under complex conjugation the projectors transform as

$$P_\pm^* = \begin{cases} BP_\pm B^{-1}, & \text{if } D = 2 \bmod 4 \\ BP_\mp B^{-1}, & \text{if } D = 4 \bmod 4. \end{cases} \tag{1.14}$$

This equation places restrictions on the form that the matrices B and the chiral projectors can take. For example, let us write the $2^{\frac{D}{2}} \times 2^{\frac{D}{2}}$ γ-matrices in terms of $2^{\frac{D}{2}-1} \times 2^{\frac{D}{2}-1}$ blocks. We also choose our spinor basis such that the projection operators are diagonal and such that P_+ has only its upper diagonal block non-vanishing and equal to the identity matrix, and P_- with only its lower diagonal block non-zero and equal to the identity matrix space. Applying equation (1.14), we find that if $D = 2 \bmod 4$, the only non-zero blocks of the matrix B are its two diagonal ones and if $D = 0 \bmod 4$, the only non-zero blocks are its off-diagonal ones.

Under complex conjugation the chiral spinors transform as

$$B^{-1}(P_\pm \lambda)^* = \begin{cases} P_\pm B^{-1}\lambda^*, & \text{if } D = 2 \bmod 4 \\ P_\mp B^{-1}\lambda^*, & \text{if } D = 4 \bmod 4 \end{cases}.$$

Hence, complex conjugation and multiplication by B^{-1} relates the same chirality spinors if $D = 2 \bmod 4$, and opposite chirality spinors if $D = 4 \bmod 4$. As such, in the dimensions $D = 2, 4$; mod 8 where we can define the Majorana spinors, the Majorana condition relates same chirality spinors if $D = 2 \bmod 4$ and opposite chirality spinors if $D = 4 \bmod 4$.

We now investigate how a matrix transformation on λ acts on its chiral components. Under the matrix transformation $\lambda \to A\lambda$ we find that $B^{-1}\lambda^* \to (B^{-1}A^*B)B^{-1}\lambda^*$ and so the equivalent transformation on $B^{-1}\lambda^*$

is $B^{-1}A^*B$. Clearly if A is a polynomial in the γ-matrices with real coefficients then this transformation is the same on λ and λ^*. As we have already
discussed this is the case with Lorentz transformations which are generated
by $J^{mn} = \frac{1}{2}\gamma^{mn}$. However, if $A = EP_\pm$ where E is a polynomial in the
γ-matrices with real coefficients then the transformation becomes

$$B^{-1}(EP_\pm)^*B = \begin{cases} CP_\pm, & \text{if } D = 2\,\text{mod}\,4 \\ P_\mp, & \text{if } D = 4\,\text{mod}\,4 \end{cases}.$$

As an example of the latter let us consider the chiral projections of Lorentz
transformations which are given by $J^{mn}_\pm \equiv \frac{1}{2}\gamma^{mn}P_\pm$. In this case the above
equation becomes $B^{-1}J^{mn*}_\pm B = J^{mn}_\pm$ for $D = 2\,\text{mod}\,4$ and $B^{-1}J^{mn*}_\pm B = J^{mn}_\mp$
for $D = 4\,\text{mod}\,4$. Hence for $D = 2\,\text{mod}\,4$ we find that the representation
generated by $J^{mn}_\pm$ generate a representation which for a given chirality is
conjugate to the complex conjugate representation of the same chirality. On
the other hand, for $D = 4\,\text{mod}\,4$ we find that the complex conjugate of the
chiral Lorentz transformations of a given chirality is conjugate to that for the
opposite chirality. In dimensions $D = 2, 4$; mod 8 we can choose $B = 1$ and
it is straightforward to interpret the corresponding constraints. For example,
if $D = 2\,\text{mod}\,8$ then $(J^{mn}_\pm)^* = J^{mn}_\pm$ and the representation they generate is
contained in the group $SL(2^{\frac{D}{2}-1}, \mathbf{R})$.

However, for $D = 6, 8$; mod 8 the matrix B is anti-symmetric and so
can not be chosen to be one. If in addition we take $D = 2\,\text{mod}\,4$ that
is $D = 6\,\text{mod}\,8$ then we conclude that the chiral Lorentz transformations
are contained in the group $SU^*(2^{\frac{D}{2}-1})$. The group $SU^*(N)$ is the group of
$N \times N$ complex matrices of determinant 1 that commute with the operation
of complex conjugation and multiplication by an anti-symmetric matrix B
which obeys $B^\dagger B = 1$; see [103]. That is if $A \in SU^*(N)$ then $BAB^{-1} = A^*$.
Taking such an infinitesimal transformation $A = I + K$ we find that $SU^*(N)$
has real dimension $N^2 - 1$. The first such case is $D = 6$, since $SU^*(4)$ and
Spin$(1,5)$ both have dimension 15 we must conclude that the group of 6-
dimensional chiral Lorentz transformations generated by $J^{mn}_\pm$ is isomorphic
to the group $SU^*(4)$.

If the spinors carry internal spinor indices that transform under a pseudo-
real representation of an internal group, then we can also define a kind of
Majorana spinor, when $D = 6, 8$; mod 8, by the condition

$$(\lambda_i)^* = \Omega^{ij}B\lambda_j. \tag{1.15}$$

In this equation Ω^{ij} is a real anti-symmetric matrix which also obeys the
relations $\Omega^{ij} = \Omega_{ij}$ and $\Omega^{ij}\Omega_{jk} = -\delta^i_k$. We call spinors that satisfy this type
of Majorana condition *symplectic Majorana spinors*. Taking the complex
conjugate of this symplectic Majorana condition, using the above relations
and the fact that $B^*B = -1$ we find that it is indeed a consistent condition.

The symplectic Majorana condition should also be such that the internal group, which acts on the internal indices $i, j, \ldots$, acts in the same way on the left- and right-hand sides of the symplectic Majorana condition. This requires λ_i to carry a pseudo-real representation of the internal group.

The vector representation of the group $USp(N)$ provides one of the most important examples of a pseudo-real representation. The group $USp(N)$ consists of unitary matrices that in addition preserve Ω^{ij}; that is matrices A which satisfy $A^\dagger A = 1$ and $A^T \Omega A = \Omega$. Taking such an infinitesimal transformation we find that this group has dimension $\frac{1}{2}N(N + 1)$. Under $\lambda \to A\lambda$, we find that $\lambda^* \to A^*\lambda^*$, but $\Omega\lambda \to \Omega A\lambda = -\Omega A\Omega\Omega\lambda$. However, using the defining conditions of $USp(N)$ we can show that

$$-\Omega A\Omega = -(A^T)^{-1}\Omega\Omega = (A^T)^{-1} = (A^\dagger)^T = A^* \tag{1.16}$$

which means that the symplectic Majorana condition preserves $USp(N)$. We note that $USp(2) = SU(2)$.

In addition to the symplectic Majorana condition of equation (1.15) which requires $D = 6, 8; \mod 8$ we can also impose a Weyl constraint if $D = 2 \mod 4$. That is dimensions $D = 6 \mod 8$ symplectic Majorana–Weyl spinors exist.

An important example of a symplectic Majorana–Weyl spinors are those in 6 dimensions that transform under $USp(4)$. These spinors naturally arise when we reduce that Majorana 11-dimensional spinors to 6 dimensions. The 11-dimensional spinors transform under $\text{Spin}(1, 10)$ which under the reduction to 6 dimensions becomes $\text{Spin}(1, 5) \times \text{Spin}(5)$. The $\text{Spin}(5)$ which is isomorphic to $USp(4)$ and becomes the internal group in 6 dimensions. In fact we get symplectic Majorana–Weyl spinors of both chirality each of which has $4 \times 4 = 16$ components.

1.4 Clifford algebras in odd dimensions

We now take the dimension of space-time D to be odd. The group C_D of equation (1.2) is generated by the γ_n and has order 2^{D+1}. The irreducible representations can be found using the same arguments as we did for the case of an even-dimensional space-time. However, there are some differences which are a consequence of the fact that the conjugacy classes are not given by the obvious generalisation of those for even-dimensional case which were listed in equation (1.3). From all the γ-matrices, γ_m, $m = 0, 1, \ldots, D - 1$ we can form the matrix $\gamma_D \equiv \gamma_0\gamma_1 \cdots \gamma_{D-1}$. This matrix commutes with all the γ_m $m = 0, 1, \ldots, D - 1$ and so all products of the γ_ms. As such, $\pm\gamma_D$ form conjugacy classes by themselves and as a result the full list of conjugacy classes is given by

$$[1], [-1], [\gamma_m], [\gamma_{m_1 m_2}], \ldots, [\gamma_{m_1 \ldots m_D}], [-\gamma_{m_1 \ldots m_D}].$$

There are $2^D + 2$ conjugacy classes and so $2^D + 2$ inequivalent irreducible representations of C_D.

The commutator group of C_D is given by $\{\pm 1\}$ and so has order 2. As such, the number of inequivalent irreducible 1-dimensional representations of C_D is 2^D. Hence, in an odd-dimensional space-time we have two inequivalent irreducible representations of C_D of dimension greater than 1. In either of these two irreducible representations, the matrix γ_D commutes with the entire representation and so by Schur's Lemma must be a multiple of the identity i.e. $\gamma_D = a^{-1}I$ where a is a constant. Multiplying both sides by γ_{D-1} we find the result

$$\gamma_{D-1} = a\gamma_0\gamma_1 \cdots \gamma_{D-2} = a\gamma_{01\cdots D-2}.$$

Using equation (1.13), we conclude that $(\gamma_{01\cdots D-2})^2 = -(-1)^{\frac{(D-1)}{2}}$ as the matrix $\gamma_{01\cdots D-2}$ is the same as that denoted by γ^{D+1} for the even-dimensional space-time with one dimension lower. However, as $\gamma_{D-1}^2 = +1$ we must conclude that $a = \pm 1$ for $D = 3 \bmod 4$ and $a = \pm i$ for $D = 5 \bmod 4$. The γ_m, $m = 0, 1, \ldots, D - 2$, generate an even-dimensional Clifford algebra and we recall that the corresponding subgroup C_{D-1} has a unique irreducible representation of dimension greater than 1, the dimension being $2^{\frac{(D-1)}{2}}$. It follows that the two irreducible representations for D odd which have dimension greater than 1 must coincide with this irreducible representation when restricted to C_{D-1}. Hence, the two inequivalent irreducible representations for odd D are generated by the unique irreducible representation for the γ_m, $m = 0, 1, \ldots, D - 2$, with the remaining γ-matrix being given by $\gamma_{D-1} = a\gamma_0\gamma_1 \cdots \gamma_{D-2}$. The two possible choices of a, given above, correspond to the two inequivalent irreducible representations. Clearly, these two inequivalent irreducible representations both have dimension $2^{\frac{(D-1)}{2}}$ as this is the dimension of the unique irreducible representation with dimension greater than 1 in the space-time with one dimension less. We can check that this is consistent with the relationship between the order, 2^{D+1}, of the group and the sum of the square of the dimension of all irreducible representations. The latter is given by $1^2 . 2^D + \left(2^{\frac{(D-1)}{2}}\right)^2 + \left(2^{\frac{(D-1)}{2}}\right)^2 = 2^{D+1}$, as required.

We now extend the complex conjugation and transpose properties discussed previously for even-dimensional space-time to the case of an odd-dimensional space-time. Clearly, for the matrices γ_m, $m = 0, 1, \ldots, D - 2$, these properties are the same and are given in equations (1.4) and (1.5). It only remains to consider $\gamma_{D-1} = a\gamma_0\gamma_1 \cdots \gamma_{D-2} \equiv a\gamma_{01\cdots D-2}$. It follows from the previous section that

$$\gamma_{01\cdots D-2}^* = B\gamma_{01\cdots D-2}B^{-1}$$

and

$$C\gamma_{01\cdots D-2}C^{-1} = (-1)^{\frac{(D-1)}{2}}\gamma_{01\cdots D-2}^T.$$

We may also write this last equation as

$$C\gamma_{01\cdots D-2} = -\epsilon(-1)^{\frac{(D-1)}{2}} (C\gamma_{01\cdots D-2})^T$$

as $C^T = -\epsilon C$. Taking into account the different possible values of a discussed above we conclude that

$$\gamma_{D-1}^* = -(-1)^{\frac{(D-1)}{2}} B\gamma_{D-1}B^{-1} \tag{1.17}$$

and

$$\gamma_{D-1}^T = (-1)^{\frac{(D-1)}{2}} C\gamma_{D-1}C^{-1}. \tag{1.18}$$

As we did for the even-dimensional case we can adopt the choice $C = B^T\gamma^0$, whereupon we find that $\gamma_{D-1}^\dagger = \gamma^0\gamma_{D-1}\gamma^0$. The representation is automatically unitary as a consequence of being unitary on the C_{D-1} subgroup.

For $D = 3 \bmod 4$, γ_{D-1} has the same relationships under complex conjugation and transpose as do the γ_m, $m = 0, 1, \ldots, D-2$, and as a result, for $D = 3 \bmod 4$,

$$(C\gamma_{m_1\ldots m_p})^T = \epsilon(-1)^{\frac{(p-1)(p-2)}{2}} (C\gamma_{m_1\ldots m_p}) \quad m_1, \ldots, m_p = 0, \ldots, D-1. \tag{1.19}$$

For $D = 5 \bmod 4$, we get an additional minus sign in this relationship if one of the $m_1, \ldots, m_p$ takes the value $D - 1$.

Let us now consider which types of spinors can exist in odd-dimensional space-times. Clearly there the Weyl condition is not a Lorentz-invariant condition and so one cannot define such spinors. However, we can ask which odd-dimensional space-times have Majorana spinors, which we take to be defined by $\chi^* = B\chi$. Since either of the two inequivalent irreducible representations coincides with the unique irreducible representation of dimension greater than 1 when restricted to the subgroup C_{D-1}, the matrix B is the same as in the even-dimensional case. It follows that we require $\epsilon = +1$ which is the case for $D = 3, 5$; mod 8. We must, however, verify that the Majorana condition is preserved by all Lorentz transformations. Those that are generated by $\frac{1}{2}\gamma_{mn}$, $m, n = 0, 1\ldots, D - 2$ are guaranteed to work; however, carrying out the Lorentz transformation $\delta\chi = \frac{1}{4}\gamma_{mD-1}\chi, m = 0, 1\ldots, D - 2$ we find it preserves the Majorana constraint only if $D = 3 \bmod 4$. Hence, Majorana spinors exist in odd D-dimensional space-time if $D = 3 \bmod 8$. We note that these odd dimensions are precisely one dimension higher than those where Majorana–Weyl spinors exist. This is not a coincidence as the reduction of a Majorana spinor in $D = 3 \bmod 8$ dimensions leads to two Majorana Weyl spinors of opposite chirality and, since the resulting matrix γ_{D+1}, is real we may Weyl project to find a Majorana–Weyl spinor.

1.5 Central charges

One important application of the above theory is to find what central charges can appear in a supersymmetry algebra. That is what generators can appear in the anti-commutator $\{Q_\alpha,\ Q_\beta\}$ where Q_α is the generator of supersymmetry transformations. As we shall see, the result depends on the dimension of space-time and on whether the spinor Q_α is Weyl or Majorana–Weyl. To begin with we take the dimension of space-time to be even.

The right-hand side of the anti-commutator of the supercharges takes the form (see [117])

$$\{Q_\alpha,\ Q_\beta\} = (\gamma_m C^{-1})_{\alpha\beta} P^m + \sum (\gamma_{m_1 \cdots m_p} C^{-1})_{\alpha\beta} Z^{m_1 \cdots m_p},$$

where $Z^{m_1 \cdots m_p}$ are the central charges and P^m is the generator of translations. The sum is over all possible central terms. Clearly, the matrix $(\gamma_{m_1 \ldots m_p} C^{-1})_{\alpha\beta}$ must be symmetric in α and β. Examining equation (1.11) we find that this will be the case if

$$\epsilon(-1)^{\frac{(p-1)(p-2)}{2}} = 1. \tag{1.20}$$

For $D = 2, 4;\ \mathrm{mod}\,8,\ \epsilon = 1$ and so we find central charges for $p = 1, 2;\ \mathrm{mod}\,4$. In these dimensions we can define Majorana spinors and adopting this constraint still allows these central charges although the Majorana condition will place reality conditions on them. For $D = 6, 8;\ \mathrm{mod}\,8,\ \epsilon = -1$ and so we find central charges of rank p for $p = 3, 4;\ \mathrm{mod}\,4$.

Let us now consider the case when the spinors are Weyl spinors, that is satisfy $(P_\pm Q)_\alpha = 0$. in this case we must modify the terms on the right-hand side of the anti-commutator to be given by

$$\{Q_\alpha, Q_\beta\} = (\gamma_{m_1 \cdots m_p} P_\pm C^{-1})_{\alpha\beta} Z^{m_1 \cdots m_p}.$$

In this case $(\gamma_{m_1 \cdots m_p} C^{-1})_{\alpha\beta}$ and $(\gamma_{m_1 \cdots m_p} \gamma^{D+1} C^{-1})_{\alpha\beta}$ must be symmetric. However, the latter matrix is equal to the product of ϵ and $(\gamma_{m_1 \ldots m_{D-p}} C^{-1})_{\alpha\beta}$. Hence central charges of rank p are possible if $p = 1, 2;\ \mathrm{mod}\,4$ and $D - p = 1, 2;\ \mathrm{mod}\,4$.

Finally, we can consider Majorana–Weyl spinors which only exist in $D = 2\,\mathrm{mod}\,8$. In this case we have that the condition on p of the Weyl case above, which must be taken with $D = 2 + 8n$ for $n \in \mathbf{Z}$, allows only central charges of rank $p = 1\,\mathrm{mod}\,4$. An example of this latter case is provided by $N = 1\ D = 10$ which, if we take Majorana–Weyl spinors, is the algebra that underlies the $N = 1$ Yang–Mills theory and the type I supergravity theory that exists in 10 dimensions. The algebra is then given by

$$\begin{aligned} \{Q_\alpha, Q_\beta\} &= (\gamma_m P_\pm C^{-1})_{\alpha\beta} P^m + (\gamma_{m_1 \cdots m_5} P_\pm C^{-1})_{\alpha\beta} Z^{m_1 \ldots m_5} \\ &\quad + (\gamma_{m_1 \cdots m_9} P_\pm C^{-1})_{\alpha\beta} Z^{m_1 \cdots m_9}. \end{aligned} \tag{1.21}$$

Clearly P^m is the usual generator of translations.

The above discussion can be generalised to the case of odd-dimensional space-times although in this case we do not have Weyl spinors. For the case of $D = 3 \bmod 4$, γ_{D-1} behaves exactly like the other γ-matrices under complex conjugation and transpose and as a result we find from equation (1.19) precisely the same condition for the existence of central charges i.e. equation (1.20), with the value of ϵ being the same as that in one dimension less. The case of $D = 5 \bmod 4$ can be deduced in a similar way by taking into account the discussion below equation (1.19).

A very important example is the supersymmetry algebra in 11 dimensions for Majorana supercharges. This algebra underlies the supergravity theory in this dimension. For this algebra we can have central charges of rank p with $p = 1, 2; \bmod 4$ and so the supersymmetry algebra takes the form (see [117]).

$$\{Q_\alpha, Q_\beta\} = (\gamma^m C^{-1})_{\alpha\beta} P_m + (\Gamma^{mn} C^{-1})_{\alpha\beta} Z_{mn} + (\Gamma^{mnpqr} C^{-1})_{\alpha\beta} Z_{mnpqr}. \tag{1.22}$$

We need only go up to rank 5 thanks to the identity

$$\epsilon^{m_1 \cdots m_p m_{p+1} \cdots m_{D-1}} \gamma_{n_{p+1} \cdots n_{D-1}} \propto \gamma^{m_1 \cdots m_p}$$

which is true in all odd-dimensional spaces.

To find the possible central charges is the obvious part of constructing the supersymmetry algebra we must also find their relations with the rest of the generators of the algebra. This involves enforcing the super Jacobi identities as explained in the next section.

It is straightforward to extend the discussion to supersymmetry algebras which contain supersymmetry generators Q_α^i that carry a representation of an internal algebra. The anti-commutator $\{Q_\alpha^i, Q_\beta^j\}$ is now symmetric under interchange in α, i and β, j.

For an account of Clifford algebras and spinors in non-Lorentzian space-times the reader is encouraged to consult reference [102].

2   The Supersymmetry Algebra in 4 Dimensions

The starting point for the construction of any supersymmetric theory which is invariant under either rigid or local supersymmetry is the supersymmetry algebra which underlies it. In this section we demonstate how to systematically construct supersymmetry algebras from some very mild assumptions.

Supersymmetry algebras contain generators, called *supercharges*, that are Grassmann odd, transform under the Lorentz group as spinors and obey anti-commutation relations. It is a remarkable fact that supersymmetric algebras can be constructed from very little information. We must specify the spinorial character of the supercharges and what space-time Lie algebra is contained in the supersymmetry algebra. In Section 1, we deduced the possible spinors that can exist in a given dimension and we shall see that the different choices lead to different superysymmetry algebras. The most important of the space-time Lie subalgebras is the Poincaré algebra, but other possible choices include the de Sitter algebra or the conformal algebra. Given a choice of the spinorial character of the supercharges and the space-time Lie subalgebra, the deduction of the corresponding supersymmetry algebra relies on a generalisation of the Jacobi identites that occur in Lie algebras.

In this section, we show how to systematically construct the supersymmetry algebras in four dimensions. In particular, we demonstrate how to construct all supersymmetry algebras that contain the Poincaré Lie algebra. We also state the supersymmetry algebras that contain the conformal and de Sitter Lie algebras. The generalisation to the sytematic construction of supersymmetry algebras in higher dimensions is straightforward. The only difference is that the no-go theorem must be used with more caution as its proof contains a number of assumptions that are not valid for higher-dimensional theories.

This material is essentially the same as that given in reference [0]. We thank World Scientific publishing for their kind permission to reproduce this material.

In the 1960s, with the growing awareness of the significance of internal symmetries such as $SU(2)$ and larger groups, physicists attempted to find a symmetry which would combine in a non-trivial way the space-time Poincaré group with an internal symmetry group. After much effort it was shown that such an attempt was impossible within the context of a Lie group. Coleman and Mandula [4] showed, on very general assumptions, that any Lie group which contained the Poincaré group P, whose generators P_a and J_{ab} satisfy the relations

$$
\begin{aligned}
[P_a, P_b] &= 0 \\
[P_a, J_{bc}] &= (\eta_{ab}P_c - \eta_{ac}P_b) \\
[J_{ab}, J_{cd}] &= -(\eta_{ac}J_{bd} + \eta_{bd}J_{ac} - \eta_{ad}J_{bc} - \eta_{bc}J_{ad}),
\end{aligned}
\tag{2.1}
$$

and an internal symmetry group G with generators T_s such that

$$
[T_r, T_s] = f_{rst}T_t,
\tag{2.2}
$$

must be a direct product of P and G; or in other words

$$
[P_a, T_s] = 0 = [J_{ab}, T_s].
\tag{2.3}
$$

They also showed that G must be of the form of a semisimple group with additional $U(1)$ groups.

It is worthwhile to make some remarks concerning the status of this no-go theorem. Clearly there are Lie groups that contain the Poincaré group and internal symmetry groups in a non-trivial manner; however, the theorem states that these groups lead to trivial physics. Consider, for example, two-body scattering; once we have imposed conservation of angular momentum and momentum the scattering angle is the only unknown quantity. If there were a Lie group that had a non-trivial mixing with the Poincaré group then there would be further generators associated with space-time. The resulting conservation laws will further constrain, for example, two-body scattering, and so the scattering angle can only take on discrete values. However, the scattering process is expected to be analytic in the scattering angle, θ, and hence we must conclude that the process does not depend on θ at all.

Essentially the theorem shows that if one used a Lie group that contained an internal group which mixed in a non-trivial manner with the Poincaré group then the S-matrix for all processes would be zero. The theorem assumes among other things, that the S-matrix exists and is non-trivial, the vacuum is non-degenerate and that there are no massless particles. It is important to realise that the theorem only applies to symmetries that act on S-matrix elements and not on all the other many symmetries that occur in quantum field theory. Indeed it is not uncommon to find examples of the latter symmetries. Of course, no-go theorems are only as strong as the assumptions required to prove them.

In a remarkable paper, [1], Gol'fand and Likhtman showed that provided one generalised the concept of a Lie group one could indeed find a symmetry that included the Poincaré group and an internal symmetry group in a non-trivial way. In this section we will discuss this approach to the supersymmetry group; having adopted a more general notion of a group, we will show that one is led, with the aid of the Coleman–Mandula theorem, and a few assumptions, to the known supersymmetry group. Since the structure of a Lie group, at least in some local region of the identity, is determined entirely by its Lie algebra, it is necessary to adopt a more general notion than a Lie algebra. The vital step in discovering the supersymmetry algebra is to introduce generators Q_α^i, which satisfy anti-commutation relations, i.e.

$$\{Q_\alpha^i, Q_\beta^j\} = Q_\alpha^i Q_\beta^j + Q_\beta^j Q_\alpha^i \qquad (2.4)$$
$$=\text{some other generator.}$$

The significance of the i and α indices will become apparent shortly. Let us therefore assume that the supersymmetry group involves generators P_a, J_{ab}, T_s and possibly some other generators which satisfy commutation relations, as well as the generators Q_α^i ($i = 1, 2, \ldots, N$). We will call the former generators which satisfy equations (2.1), (2.2) and (2.3) *even*, and those satisfying equation (2.4) *odd* generators.

Having let the genie out of the bottle we promptly replace the stopper and demand that the supersymmetry algebras have a Z_2 graded structure. This simply means that the even and odd generators must satisfy the rules:

$$\begin{aligned} [\text{even, even}] &= \text{even} \\ \{\text{odd, odd}\} &= \text{even} \\ [\text{even, odd}] &= \text{odd.} \end{aligned} \quad (2.5)$$

We must still have the relations

$$[P_a, T_s] = 0 = [J_{ab}, T_s] \quad (2.6)$$

since the even (bosonic) subgroup must obey the Coleman–Mandula theorem.

Let us now investigate the commutator between J_{ab} and Q_α^i. As a result of equation (2.5) it must be of the form

$$[Q_\alpha^i, J_{ab}] = (b_{ab})_\alpha^\beta Q_\beta^i, \quad (2.7)$$

since by definition the Q_α^i are the only odd generators. We take the α indices to be those rotated by J_{ab}. As in a Lie algebra we have some generalised Jacobi identities. If we denote an even generator by B and an odd generator by F, we find that

$$\begin{aligned}{} [[B_1, B_2], B_3] + [[B_3, B_1], B_2] + [[B_2, B_3], B_1] &= 0 \\ [[B_1, B_2], F_3] + [[F_3, B_1], B_2] + [[B_2, F_3], B_1] &= 0 \\ \{[B_1, F_2], F_3\} + \{[B_1, F_3], F_2\} + [\{F_2, F_3\}, B_1] &= 0 \\ [\{F_1, F_2\}, F_3] + [\{F_1, F_3\}, F_2] + [\{F_2, F_3\}, F_1] &= 0. \end{aligned}$$

The reader may verify, by expanding each bracket, that these relations are indeed identically true.

The identity

$$[[J_{ab}, J_{cd}], Q_\alpha^i] + [[Q_\alpha^i, J_{ab}], J_{cd}] + [[J_{cd}, Q_\alpha^i], J_{ab}] = 0,$$

upon use of equation (2.7), implies that

$$[b_{ab}, b_{cd}]_\alpha^\beta = -\eta_{ac}(b_{bd})_\alpha^\beta - \eta_{bd}(b_{ac})_\alpha^\beta + \eta_{ad}(b_{bc})_\alpha^\beta + \eta_{bc}(b_{ad})_\alpha^\beta.$$

This means that the $(b_{cd})_\alpha^\beta$ form a representation of the Lorentz algebra or in other words that the Q_α^i carry a representation of the Lorentz group. We will select Q_α^i to be in the $\left(0, \frac{1}{2}\right) \oplus \left(\frac{1}{2}, 0\right)$ representation of the Lorentz group, i.e.

$$[Q_\alpha^i, J_{ab}] = \frac{1}{2}(\sigma_{ab})_\alpha^\beta Q_\beta^i. \quad (2.8)$$

We can choose Q_α^i to be a Majorana spinor, i.e.

$$Q_\alpha^i = C_{\alpha\beta}\bar{Q}^{\beta i}, \tag{2.9}$$

where $C_{\alpha\beta} = -C_{\beta\alpha}$ is the charge conjugation matrix (see Appendix A of [0]). This does not represent a loss of generality since, if the algebra admits complex conjugation as an involution we can always redefine the supercharges so as to satisfy equation (2.9) (see Note 1 at the end of this section).

The above calculation reflects the more general result that the Q_α^i must belong to a realization of the even (bosonic) subalgebras of the supersymmetry group. This is a simple consequence of demanding that the algebra be Z_2 graded. The commutator of any even generator B_1, with Q_α^i is of the form

$$[Q_\alpha^i, B_1] = (h_1)_{\alpha j}^{i\beta} Q_\beta^j.$$

The generalised Jacobi identity

$$[[Q_\alpha^i, B_1], B_2] + [[B_1, B_2], Q_\alpha^i] + [[B_2, Q_\alpha^i], B_1] = 0$$

implies that

$$[h_1, h_2]_{\alpha j}^{i\beta} Q_\beta^j = [Q_\alpha^i [B_1, B_2]],$$

or, in other words, the matrices h represent the Lie algebra of the even generators.

The above remarks imply that

$$[Q_\alpha^i, T_r] = (l_r)_j^i Q_\alpha^j + (t_r)_j^i (i\gamma_5)_\alpha^\beta Q_\beta^j, \tag{2.10}$$

where $(l_r)_j^i + i\gamma_5(t_r)_j^i$ represent the Lie algebra of the internal symmetry group. This results from the fact that δ_β^α and $(\gamma_5)_\alpha^\beta$ are the only invariant tensors which are scalar and pseudoscalar.

The remaining odd-even commutator is $[Q_\alpha^i, P_a]$. A possibility that is allowed by the generalised Jacobi identities that involve the internal symmetry group and the Lorentz group is

$$[Q_\alpha^i, P_a] = c(\gamma_a)_\alpha^\beta Q_\beta^i. \tag{2.11}$$

However, the $[[Q_\alpha^i, P_a], P_b] + \cdots$ identity implies that the constant $c = 0$, i.e.

$$[Q_\alpha^i, P_a] = 0. \tag{2.12}$$

More generally we could have considered $(c\gamma_a + d\gamma_a\gamma_5)Q$ on the right-hand side of equation (2.11); however the above Jacobi identity and the Majorana condition would then imply that $c = d = 0$. (See Note 2 at the end of this section.) Let us finally consider the $\{Q_\alpha^i, Q_\beta^j\}$ anti-commutator. This object must be composed of even generators and must be symmetric under interchange of $\alpha \leftrightarrow \beta$ and $i \leftrightarrow j$. The even generators are those of the Poincaré group, the internal symmetry group and other even generators which, from the Coleman–Mandula theorem, commute with the Poincaré group, i.e. they are scalar and pseudoscalar. Hence the most general possibility is of the form

$$\{Q_\alpha^i, Q_\beta^j\} = r(\gamma^a C)_{\alpha\beta} P_a \delta^{ij} + s(\sigma^{ab} C)_{\alpha\beta} J_{ab} \delta^{ij} + C_{\alpha\beta} U^{ij} + (\gamma_5 C)_{\alpha\beta} V^{ij}. \quad (2.13)$$

We have not included a $(\gamma^b \gamma_5 C)_{\alpha\beta} L_b^{ij}$ term as the (Q, Q, J_{ab}) Jacobi identity implies that L_b^{ij} mixes nontrivially with the Poincaré group and so is excluded by the no-go theorem.

The fact that we have only used numerically invariant tensors under the Poincaré group is a consequence of the generalised Jacobi identities between two odd and one even generators.

To illustrate the argument more clearly, let us temporarily specialise to the case $N = 1$ where there is only one supercharge Q_α. Equation (2.13) then reads

$$\{Q_\alpha, Q_\beta\} = r(\gamma^a C)_{\alpha\beta} P_a + s(\sigma^{ab} C)_{\alpha\beta} J_{ab}.$$

Using the Jacobi identity

$$\{[P_a, Q_\alpha], Q_\beta\} + \{[P_a, Q_\beta], Q_\alpha\} + [\{Q_\alpha, Q_\beta\}, P_a] = 0,$$

we find that

$$0 = s(\sigma^{cd} C)_{\alpha\beta} [J_{cd}, P_a] = s(\sigma^{cd} C)_{\alpha\beta} (-\eta_{ac} P_d + \eta_{ad} P_c),$$

and, consequently, $s = 0$. We are free to scale the generator P_a in order to bring $r = 2$.

Let us now consider the commutator of the generator of the internal group and the supercharge. For only one supercharge, equation (2.10) reduces to

$$[Q_\alpha, T_r] = l_r Q_\alpha + i(\gamma_5)_\alpha^\beta t_r Q_\beta.$$

Taking the adjoint of this equation, multiplying by $(i\gamma^0)$ and using the definition of the Dirac conjugate given in Appendix A of [0], we find that

$$[\bar{Q}^\alpha, T_r] = l_r^* \bar{Q}^\alpha + \bar{Q}^\beta (it_r^*)(\gamma_5)_\beta^\alpha.$$

Multiplying by $C_{\gamma\alpha}$ and using equation (2.9), we arrive at the equation

$$[Q_\alpha, T_r] = l_r^* Q_\alpha + i t_r^* (\gamma_5)_\alpha^\beta Q_\beta.$$

Comparing this equation with the one we started from, we therefore conclude that

$$l_r^* = l_r, \qquad t_r^* = t_r.$$

The Jacobi identity

$$[\{Q_\alpha, Q_\beta\}, T_r] + [[T_r, Q_\alpha], Q_\beta] + [[T_r, Q_\beta], Q_\alpha] = 0$$

results in the equation

$$[0 + (l_r\delta_\alpha^\gamma + it_r(\gamma_5)\alpha^\gamma)2(\gamma_a C)_{\gamma\beta}P_a] + (\alpha \leftrightarrow \beta)$$
$$= 2P_a\{l_r(\gamma_a C)_{\alpha\beta} + it_r(\gamma_5\gamma_a C)_{\alpha\beta}\} + (\alpha \leftrightarrow \beta) = 0.$$

Since $(\gamma_a C)_{\alpha\beta}$ and $(\gamma_5\gamma_a C)_{\alpha\beta}$ are symmetric and anti-symmetric in α, β respectively, we conclude that $l_r = 0$ but t_r has no constant placed on it. Consequently, we find that we have only one internal generator R and we may scale it such that

$$[Q_\alpha, R] = i(\gamma_5)_\alpha^\beta Q_\beta.$$

The $N = 1$ supersymmetry algebra is summarised in equation (2.18).

Let us now return to the extended supersymmetry algebra. The even generators $U^{ij} = -U^{ji}$ and $V^{ij} = -V^{ji}$ are called central charges [5] and are often also denoted by Z. It is a consequence of the generalised Jacobi identities $((Q, Q, Q)$ and $(Q, Q, Z))$ that they commute with all other generators including themselves, i.e.

$$[U^{ij}, \text{anything}] = 0 = [V^{ij}, \text{anything}]. \tag{2.14}$$

We note that the Coleman–Mandula theorem allowed a semisimple group plus $U(1)$ factors. The details of the calculation are given in Note 5 at the end of the section. Their role in supersymmetric theories will emerge later.

In general, we should write, on the right-hand side of (2.13),

$$(\gamma^a C)_{\alpha\beta}\omega^{ij}P_a + \cdots,$$

where ω^{ij} is an arbitrary real symmetric matrix. However, one can show that it is possible to redefine (rotate and rescale) the supercharges, whilst preserving the Majorana condition, in such a way as to bring ω^{ij} to the form $\omega^{ij} = r\delta^{ij}$ (see Note 3 at the end of this section). The $[P_a, \{Q_\alpha^i, Q_\beta^j\}]+\cdots = 0$ identity implies that $s = 0$ and we can normalise P_a by setting $r = 2$ yielding the final result

$$\{Q_\alpha^i, Q_\beta^j\} = 2(\gamma_a C)_{\alpha\beta}\delta^{ij}P^a + C_{\alpha\beta}U^{ij} + (\gamma_5 C)_{\alpha\beta}V^{ij}. \tag{2.15}$$

In any case r and s have different dimensions and so it would require the introduction of a dimensional parameter in order that they were both non-zero.

Had we chosen another irreducible Lorentz representation for Q_α^i other than $\left(j + \frac{1}{2}, j\right) \oplus \left(j, j + \frac{1}{2}\right)$ we would not have been able to put P_a, i.e. a

$\left(\frac{1}{2}, \frac{1}{2}\right)$ representation, on the right-hand side of equation (2.15). The simplest choice is $\left(0, \frac{1}{2}\right) \oplus \left(\frac{1}{2}, 0\right)$. In fact this is the only possible choice (see Note 4).

Finally, we must discuss the constraints placed on the internal symmetry group by the generalised Jacobi identity. This discussion is complicated by the particular way the Majorana constraint of equation (2.9) is written. A 2-component version of this constraint is

$$\overline{Q}_{\dot{A}i} = (Q_A^i)^*; \quad A, \dot{A} = 1, 2$$

(see Appendix A of [0] for 2-component notation). Equations (2.13) and (2.10) then become

$$\begin{aligned} \{Q_A^i, \overline{Q}_{\dot{B}j}\} &= -2i(\sigma^a)_{A\dot{B}} \delta_j^i P_a \\ \{Q_A^i, Q_B^j\} &= \varepsilon_{AB}(U^{ij} + iV^{ij}) \\ [Q_A^i, J_{ab}] &= +\frac{1}{2}(\sigma_{ab})_A^B Q_B^i \end{aligned} \tag{2.16}$$

and

$$[Q_A^i, T_r] = (l_r + it_r)_j^i Q_A^j. \tag{2.17}$$

Taking the complex conjugate of the last equation and using the Majorana condition we find that

$$[Q_{\dot{A}i}, T_r] = Q_{\dot{A}k}(U_r^\dagger)_i^k,$$

where $(U_r)_j^i = (l_r + it_r)_j^i$. The $(Q, \overline{Q}, T)$ Jacobi identity then implies that δ_j^i be an invariant tensor of G, i.e.

$$U_r + U_r^\dagger = 0.$$

Hence U_r is an anti-Hermitian matrix and so represents the generators of the unitary group $U(N)$. However, taking account of the central charge terms in the (Q, Q, T) Jacobi identity one finds that there is for every central charge an invariant anti-symmetric tensor of the internal group and so the possible internal symmetry group is further reduced. If there is only one central charge, the internal group is $Sp(N)$ while if there are no central charges it is $U(N)$.

To summarise, once we have adopted the rule that the algebra be Z_2 graded and contain the Poincaré group and an internal symmetry group then the generalised Jacobi identities place very strong constraints on any possible algebra. In fact, once one makes the further assumption that Q_α^i are spinors under the Lorentz group then the algebra is determined to be of the form of equations (2.1), (2.6), (2.8), (2.10), (2.12) and (2.15).

The simplest algebra is for $N = 1$ and takes the form

$$\begin{aligned} \{Q_\alpha, Q_\beta\} &= 2(\gamma_a C)_{\alpha\beta} P^a \\ [Q_\alpha, P_a] &= 0 \\ [Q_\alpha, J_{cd}] &= \frac{1}{2}(\sigma_{cd})_\alpha^\beta Q_\beta \\ [Q_\alpha, R] &= i(\gamma_5)_\alpha^\beta Q_\beta \end{aligned} \tag{2.18}$$

as well as the commutation relations of the Poincaré group. We note that there are no central charges (i.e. $U^{11} = V^{11} = 0$), and the internal symmetry group becomes just a chiral rotation with generator R.

We now wish to prove three of the statements above. This is done here rather than in the above text, in order that the main line of argument should not become obscured by technical points. These points are best clarified in 2-component notation.

Note 1 Suppose we have an algebra that admits a complex conjugation as an involution; for the supercharges this means that

$$(Q_A^i)^* = b_i{}^j Q_{\dot A j}; \quad (Q_{\dot A j})^* = d^j{}_k Q_A^k$$

There is no mixing of the Lorentz indices since $(Q_A^i)^*$ transforms like $Q_{\dot A i}$, namely in the $(0, \frac{1}{2})$ representation of the Lorentz group, and not like Q_A^i which is in the $(\frac{1}{2}, 0)$ representation. The lowering of the i index under $*$ is at this point purely a notational device. Two successive $*$ operations yield the unit operation and this implies that

$$(b_i{}^j)^* d^j{}_k = \delta_k^i \tag{2.19}$$

and in particular that $b_i{}^j$ is an invertible matrix. We now make the redefinitions

$$Q'^i_A = Q_A^i, \quad Q'_{\dot A i} = b_i{}^j Q_{\dot A j}.$$

Taking the complex conjugate of $Q'_{\dot A i}$, we find

$$(Q'^i_A)^* = (Q_A^i)^* = b_i{}^j Q_{\dot A j} = Q'_{\dot A j}$$

while

$$(Q'_{\dot A i})^* = (b_i{}^j)^* (Q_{\dot A j})^* = (b_i{}^j)^* d^j{}_k Q_A^k = Q_A^i,$$

using equation (2.19).

Thus the Q'^i_A satisfy the Majorana condition, as required. If the Qs do not initially satisfy the Majorana condition, we may simply redefine them so that they do.

Note 2 Suppose the $[Q_A, P_a]$ commutator were of the form

$$[Q_A, P_a] = e(\sigma_a)_{A\dot B} Q^{\dot B},$$

where e is a complex number and for simplicity we have suppressed the i index. Taking the complex conjugate (see Appendix A of [0]), we find that

$$[Q_{\dot A}, P_a] = -e^*(\sigma_a)_{B\dot A} Q^B.$$

Consideration of the $[[Q_A, P_a], P_b] + \ldots = 0$ Jacobi identity yields the result

$$-|e|^2(\sigma_a)_{A\dot{B}}(\sigma^b)^{C\dot{B}} - (a \leftrightarrow b) = 0.$$

Consequently $e = 0$ and we recover the result

$$[Q_A, P_a] = 0.$$

Note 3 The most general form of the $Q^{Ai}, Q_j^{\dot{B}}$ anti-commutator is

$$\{Q^{Ai}, Q_j^{\dot{B}}\} = -2iU_j^i(\sigma^m)^{A\dot{B}}P_m + \text{terms involving other Dirac matrices.}$$
(2.20)

Taking the complex conjugate of this equation and comparing it with itself, we find that U is a Hermitian matrix

$$(U_j^i)^* = U_i^j.$$

We now make a field redefinition of the supercharge

$$Q'^{Ai} = B_j^i Q^{Aj}$$

and its complex conjugate

$$Q'^{\dot{A}}_i = (B_j^i)^* Q_j^{\dot{A}}.$$

Upon making this redefinition in equation (2.20), the U matrix becomes replaced by

$$U'^i_j = B_k^i U_l^k (B_l^j)^* \quad \text{or} \quad U' = BUB^\dagger.$$

Since U is a Hermitian matrix, we may diagonalise it in the form $c_i\delta_j^i$ using a unitarity matrix B. We note that this preserves the Majorana condition on Q^{Ai}. Finally, we may scale $Q^i \to (1/\sqrt{c^i})Q^i$ to bring U to the form $U = d_i\delta_j^i$, where $d_i = \pm 1$. In fact, taking $A = B = 1$ and $i = j = k$, we realise that the right-hand side of equation (2.20) is a positive-definite operator and since the energy $-iP_0$ is assumed positive-definite, we can only find $d_i = +1$. The final result is

$$\{Q^{Ai}, Q_j^{\dot{B}}\} = -2i\delta_j^i(\sigma^m)^{A\dot{B}}P_m.$$

Note 4 Let us suppose that the supercharge Q contains an irreducible representation of the Lorentz group other than $\left(0, \frac{1}{2}\right) \oplus \left(\frac{1}{2}, 0\right)$, say, the representation $Q_{A_1 \cdots A_n, \dot{B}_1 \cdots \dot{B}_m}$ where the A and B indices are understood to be separately symmetrised and $n + m$ is odd in order that Q be odd and $n + m > 1$. By projecting the $\{Q, Q^\dagger\}$ anti-commutator we may find the anti-commutator involving $Q_{A_1 \cdots A_n, \dot{B}_1 \cdots \dot{B}_m}$ and its Hermitian conjugate. Let us consider in particular the anti-commutator involving $Q = Q_{11\cdots 1, \dot{i}\dot{i}\cdots \dot{i}}$, this must result in an object of spin $n + m > 1$. However, by the Coleman–Mandula no-go theorem no such generator can occur in the algebra and so the anti-commutator must vanish, i.e. $QQ^\dagger + Q^\dagger Q = 0$.

Assuming the space on which Q acts has a positive-definite norm, one such example being the space of on-shell states, we must conclude that Q vanishes. However if $Q_{11\cdots 1, 1\dot{1}\cdots \dot{1}}$ vanishes, so must $Q_{A_1\cdots A_n, \dot{B}_1\cdots \dot{B}_m}$ by its Lorentz properties, and we are left only with the $\left(0, \frac{1}{2}\right) \oplus \left(\frac{1}{2}, 0\right)$ representation.

Note 5 We now return to the proof of equation (2.14). Using the (Q, Q, Z) Jocobi identity it is straightforward to show that the supercharges Q commute with the central charges Z. The (Q, Q, U) Jacobi identity then implies that the central charges commute with themselves. Finally, one considers the (Q, Q, T_r) Jacobi identity, which shows that the commutator of T_r and Z takes the generic form $[T_r, Z] = \cdots Z$. However, the generators T_r and Z form the internal symmetry group of the supersymmetry algebra and from the no-go theorem we know that this group must be a semisimple Lie group multiplied by $U(1)$ factors. We recall that a semisimple Lie group is one that has no normal Abelian subgoups other that the group itself and the identity element. As such, we must conclude that T_r and Z commute, and hence our final result that the central charges commute with all generators, that is, they really are central.

Although the above discussion started with the Poincaré group, one could equally well have started with the conformal or (anti-)de Sitter groups and obtained the superconformal and super (anti-)de Sitter algebras. For completeness, we now list these algebras. The superconformal algebra which has the generators $P_n, J_{mn}, D, K_n, A, Q^{\alpha i}, S^{\alpha i}$ and the internal symmetry generators T_r and A is given by the Lorentz group plus:

$$
\begin{aligned}
[J_{mn}, P_k] &= \eta_{nk} P_m - \eta_{mk} P_n \\
[J_{mn}, K_k] &= \eta_{nk} K_m - \eta_{mk} K_n \\
[D, P_K] &= -P_K, \quad [D, K_K] = +K_K \\
[P_m, K_n] &= -2J_{mn} + 2\eta_{mn} D, \quad [K_n, K_m] = 0, \quad [P_n, P_m] = 0 \\
[Q_\alpha^i, J_{mn}] &= \frac{1}{2}(\gamma_{mn})_\alpha^\beta Q_\beta^i, \quad [S_\alpha^i, J_{mn}] = \frac{1}{2}(\gamma_{mn})_\alpha^\beta S^{\beta i} \\
\{Q_\alpha^i, Q_\beta^j\} &= -2(\gamma^n C^{-1})_{\alpha\beta} P_n \delta^{ij} \\
\{S_\alpha^i, S_\beta^j\} &= +2(\gamma^n C^{-1})_{\alpha\beta} K_n \delta^{ij} \\
[Q_\alpha^i, D] &= \frac{1}{2} Q_\alpha^i, \quad [S_\alpha^i, D] = -\frac{1}{2} S_\alpha^i \\
[Q_\alpha^i, K_n] &= -(\gamma_n)_\alpha^\beta S_\beta^i, \quad [S_\alpha^i, P_n] = (\gamma_n)_\alpha^\beta Q_\beta^i \\
[Q_\alpha^i, T_r] &= (\delta_\alpha^\beta (\tau_{r_1})_j^i + (\gamma_5)_\alpha^\beta (\tau_{r_2})_j^i) Q_\beta^j
\end{aligned}
\tag{2.21}
$$

$$\left[S^i_\alpha, T_r\right] = (\delta^\beta_\alpha(\tau_{r_1})^i_j - (\gamma_5)^\beta_\alpha(\tau_{r_2})^i_j)Q^j_\beta$$

$$\left[Q^i_\alpha, A\right] = -i(\gamma_5)^\beta_\alpha Q^i_\beta \left(\frac{4-N}{4N}\right)$$

$$\left[S^i_\alpha, A\right] = \frac{4-N}{4N}i(\gamma_5)^\beta_\alpha S^i_\beta$$

$$\{Q^i_\alpha, S^j_\beta\} = -2(C^{-1}_{\alpha\beta})D\delta^{ij} + (\gamma^{mn}C^{-1})_{\alpha\beta}J_{mn}\delta^{ij} + 4i(\gamma_5 C^{-1}_{\alpha\beta})A\delta^{ij}$$
$$-2(\tau_{r_1})^{ij}(C^{-1})_{\alpha\beta} + ((\tau_{r_2})^{ij}(\gamma_5 C^{-1})_{\alpha\beta})T_r.$$

The T_r and A generate $U(N)$, and $\tau_1 + \gamma_5\tau_2$ are in the fundamental representation of $SU(N)$.

The case of $N = 4$ is singular and one can have either

$$[Q^i_\alpha, A] = 0 \quad \text{or} \quad [Q^i_\alpha, A] = -i(\gamma_5)^\beta_\alpha Q^i_\beta$$

and similarly for S^i_α and A. One may verify that both possibilities are allowed by the $N = 4$ Jacobi identities and so form acceptable superalgebras.

The anti-de Sitter superalgebra has generators M_{mn}, $T_{ij} = -T_{ji}$ and $Q^{\alpha i}$, and is given by

$$[M_{mn}, M_{pq}] = \eta_{np}M_{mq} + 3 \text{ terms}$$

$$[M_{mn}, T_{ij}] = 0 \quad [Q^i_\alpha, M_{mn}] = \frac{1}{2}(\gamma_{mn})^\beta_\alpha Q^i_\beta$$

$$\left[Q^i_\alpha, T^{jk}\right] = -2i(\delta^{ij}Q^k_\alpha - \delta^{ik}Q^j_\alpha)$$

$$\{Q^i_\alpha, Q^j_\beta\} = \delta^{ij}(\gamma_{mn}C^{-1})_{\alpha\beta}iM_{mn} + (C^{-1})_{\alpha\beta}T^{ij}$$

$$\left[T^{ij}, T^{kl}\right] = -2i(\delta^{jk}T^{il} + 3 \text{ terms}).$$

3 Irreducible Representations of 4-dimensional Supersymmetry

It is a relatively straightforward proceedure to find the irreducible representations of any supersymmetry algebra. These representations tell us which supersymmetric theories are possible for a given supersymmetry algebra. To be precise, they provide a list of all the particles that occur in each of the theories that have a given supersymmetric algebra as a symmetry. In this section, we carry out this proceedure for the 4-dimensional supersymmetry algebras. Using a very similar procedure one can find the irreducible representations of the supersymmetry algebras in other dimensions. The reader is referred to reference [160] where this is procedure is sketched in an arbitrary dimension and where lists of irreducible representations are given.

One of the most interesting features of the construction of these irreducible representations is when the central charges of the supersymmetry algebra are non-trivial. In this case, and when the central charges take particular values, the massive representations contain many fewer states that those contained in the generic massive representation. When this occurs the states in the representation are called BPS states and they play an important role in discussions of string duality.

The first part of this section is taken from reference [0].

In this chapter we wish to find the irreducible representations of supersymmetry [11], or, put another way, we want to know what is the possible particle content of supersymmetric theories. As is well-known, the irreducible representations of the Poincaré group are found by the Wigner method of induced representations [12]. This method consists of finding a representation of a subgroup of the Poincaré group and boosting it up to a representation of the full group. In practice, one adopts the following recipe: we choose a given momentum q^μ which satisfies $q^\mu q_\mu = 0$ or $q^\mu q_\mu = -m^2$ depending which case we are considering. We find the subgroup H which leaves q^μ intact and find a representation of H on the $|q^\mu\rangle$ states. We then induce this representation to the whole of the Poincaré group P, in the usual way. In this construction there is a one-to-one correspondence between points of P/H and four-momentum which satisfies $P_\mu P^\mu = 0$ or $P_\mu P^\mu = -m^2$. One can show that the result is independent of the choice of momentum q^μ one starts with.

In what follows we will not discuss the irreducible representations in general, but only that part applicable to the rest frame, i.e. the representations of H in the states at rest. We can do this safely in the knowledge that once the representation of H on the rest-frame states is known then the representation of P is uniquely given and that every irreducible representation of the Poincaré group can be obtained by considering every irreducible representation of H.

In terms of physics, the procedure has a simple interpretation, namely, the properties of a particle are determined entirely by its behaviour in a given frame (i.e. for given q^μ). The general behaviour is obtained from the given q^μ by boosting either the observer or the frame with momentum q^μ to one with arbitrary momentum.

The procedure outlined above for the Poincaré group can be generalised to any group of the form $S \otimes_s T$ where the symbol $\otimes_s$ denotes the semi-direct product of the groups S and T where T is Abelian. It also applies to the supersymmetry group and we shall take it for granted that the above recipe is the correct procedure and does in fact yield all irreducible representations of the supersymmetry group.

Let us first consider the massless case $q_\mu q^\mu = 0$, for which we choose the standard momentum $q_s^\mu = (m, 0, 0, m)$ for our 'rest frame'. We must now find H whose group elements leave $q_s^\mu = (m, 0, 0, m)$ intact. Clearly this contains

Q^i_α, P_μ and T_s, since these generators all commute with P_μ and so rotate the states with q^μ_s into themselves. As we will see in the final section one cannot have non-vanishing central charges for the massless case.

Under the Lorentz group the action of the generator $\frac{1}{2}\Lambda^{\mu\nu}J_{\mu\nu}$ creates an infinitesimal transformation $q^\mu \to \Lambda^\mu_\nu q^\nu + q^\mu$. Hence q^μ_s is left-invariant provided the parameters obey the relations

$$\Lambda_{30} = 0, \quad \Lambda_{10} + \Lambda_{13} = 0, \quad \Lambda_{20} + \Lambda_{23} = 0.$$

Thus the Lorentz generators in H are

$$T_1 = J_{10} + J_{13}, \quad T_2 = J_{20} + J_{23}, \quad J = J_{12}.$$

These generators form the algebra

$$
\begin{aligned}
[T_1, J] &= -T_2 \\
[T_2, J] &= +T_1 \\
[T_1, T_2] &= 0
\end{aligned}
\tag{3.1}
$$

The reader will recognise this to be the Lie algebra of E_2, the group of translations and rotations in a 2-dimensional plane.

Now the only unitary representations of E_2 which are finite-dimensional have T_1 and T_2 trivially realised, i.e.

$$T_1|q^\mu_s\rangle = T_2|q^\mu_s\rangle = 0.$$

This results from the theorem that all non-trivial unitary representations of noncompact groups are infinite-dimensional. We will assume we require finite-dimensional representations of H.

Hence for the Poincaré group, in the case of massless particles, finding representations of H results in finding representations of E_2 and consequently for the generator J alone. We choose our states so that

$$J|\lambda\rangle = i\lambda|\lambda\rangle.$$

Our generators are anti-Hermitian. In fact, J is the helicity operator and we select λ to be integer or half-integer (i.e. $J = \underline{q} \cdot \underline{J}/|\underline{q}|$ evaluated at $\underline{q} = (0,0,m)$ where $J_i = \varepsilon_{ijk}J_{jk}$, $i,j = 1,2,3$).

Let us now consider the action of the supercharges Q^i_α on the rest-frame states, $|q^\mu_s\rangle$. The calculation is easiest when performed using the 2-component formulation of the supersymmetry algebra of equation (2.16). On rest-frame states we find that

$$\{Q^{Ai}, Q^{\dot B}_j\} = -2\delta^i_j(\sigma_\mu)^{A\dot B}q^\mu_s = -2\delta^i_j(\sigma_0 + \sigma_3)^{A\dot B}m = +4m\delta^i_j\begin{pmatrix} 0 & 0 \\ 0 & 1 \end{pmatrix}^{A\dot B}$$
$$\tag{3.2}$$

In particular these imply the relations

$$\{Q^{1i}, Q^1_j\} = 0$$
$$\{Q^{2i}, Q^2_j\} = 4m\delta^i_j \qquad (3.3)$$
$$\{Q^i_i, Q^{2j}\} = \{Q^1_i, Q^{2j}\} = 0.$$

The first relation implies that

$$\langle q^\mu_s | (Q^{1i}(Q^{1i})^* + (Q^{1i})^* Q^{1i}) | q^\mu_s \rangle = 0.$$

Demanding that the norm on physical states be positive-definite and vanishes only if the state vanishes, yields

$$Q^i_2 | q^\mu_s \rangle = Q_{\dot{2}i} | q^\mu_s \rangle = 0.$$

Hence, all generators in H have zero action on rest-frame states except J, T_s, P_μ, Q^i_1 and Q_{1i}. Using equation (2.16) we find that

$$[Q^i_1, J] = \frac{1}{2}(\sigma_{12})^1_1 Q^i_1 = -\frac{i}{2}Q^i_1. \qquad (3.4)$$

Similarly, we find that complex conjugation implies

$$[(Q^i_1)^*, J] = +\frac{i}{2}(Q^i_1)^*. \qquad (3.5)$$

The relations between the remaining generators, summarised in equations (3.3), (3.4), (3.5) and (2.17), can be summarised by the statement that Q^i_1 and $(Q^i_1)^*$ form a Clifford algebra, act as raising and lowering operators for the helicity operator J, and transform under the N and $\overline{N}$ representation of $SU(N)$.

We find the representations of this algebra in the usual way; we choose a state of given helicity, say λ, and let it be the vacuum state for the operator $(Q^i_1)^*$, i.e.

$$Q^i_1 | \lambda \rangle = 0, \qquad J | \lambda \rangle = i\lambda | \lambda \rangle.$$

The states of this representation are then

$$|\lambda\rangle = |\lambda\rangle$$
$$|\lambda - \frac{1}{2}, i\rangle = (Q^i_1)^* |\lambda\rangle \qquad (3.6)$$
$$|\lambda - 1, [ij]\rangle = (Q^i_1)^* (Q^j_1)^* |\lambda\rangle$$

etc. These states have the helicities indicated and belong to the $[ijk\cdots]$ anti-symmetric representation of $SU(N)$. The series will terminate after the helicity $\lambda - (N/2)$, as the next state will be an object anti-symmetric in $N+1$

indices. Since there are only N labels, this object vanishes identically. The states have helicities from λ to $\lambda - (N/2)$, there being $N!/(m!(N-m)!)$ states with helicity $\lambda - (m/2)$.

To obtain a set of states which represent particles of both helicities we must add to the above set the representations with helicities from $-\lambda$ to $-\lambda + (N/2)$. The exception is the so-called CPT self-conjugate sets of states which automatically contain both helicity states.

The representations of the full supersymmetry group are obtained by boosting the above states in accordance with the Wigner method of induced representations.

Hence the massless irreducible representation of $N = 1$ supersymmetry comprises only the two states

$$|\lambda\rangle; \qquad |\lambda - \tfrac{1}{2}\rangle = (Q_1)|\lambda\rangle$$

with helicities λ and $\lambda - \tfrac{1}{2}$ and since

$$Q_1 Q_1 |\lambda\rangle = 0$$

there are no more states.

To obtain a CPT-invariant theory we must add states of the opposite helicities, i.e. $-\lambda$ and $-\lambda + \tfrac{1}{2}$. For example, if $\lambda = \tfrac{1}{2}$ we get on-shell helicity states 0 and $\tfrac{1}{2}$ and their CPT conjugates with helicities $-\tfrac{1}{2}, 0$, giving a theory with two spin 0 and one Majorana spin $\tfrac{1}{2}$. Alternatively, if $\lambda = 2$ then we get on-shell helicity states $\tfrac{3}{2}$ and 2 and their CPT self-conjugates with helicity $\tfrac{-3}{2}$ and -2; this results in a theory with one spin 2 and one spin $\tfrac{3}{2}$ particles. These on-shell states are those of the Wess–Zumino model and $N = 1$ supergravity respectively. Later in this discussion we will give a complete account of these theories.

For $N = 4$ with $\lambda = 1$ we get the massless states

$$|1\rangle, \ |\tfrac{1}{2}, i\rangle, \ |0, [ij]\rangle, \ |-\tfrac{1}{2}, [ijk]\rangle, \ |-1, [ijkl]\rangle.$$

This is a CPT self-conjugate theory with one spin 1, four spin $\tfrac{1}{2}$ and six spin 0 particles.

Table 1 gives the multiplicity for massless irreducible representations which have maximal helicity 1 or less. We see that as N increases, the multiplicities of each spin and the number of different types of spin increases. The simplest theories are those for $N = 1$. The 1 in the first column in the Wess–Zumino model and the 1 in the second column is the $N = 1$ supersymmetric Yang–Mills theory. The latter contains one spin 1 and one spin $\tfrac{1}{2}$, consistent with the formula for the lowest helicity $\lambda - \tfrac{N}{2}$, which in this case gives $1 - \tfrac{1}{2} = \tfrac{1}{2}$. The $N = 4$ multiplet is CPT self-conjugate, since in this case we have

$\lambda - \frac{N}{2} = 1 - \frac{4}{2} = -1$. The table stops at $N = 4$ since when $N > 4$ we must have particles of spin greater than 1. Clearly, $N > 4$ implies that $\lambda - \frac{N}{2} = 1 - \frac{N}{2} < -1$. This leads us to the well-known statement that the $N = 4$ supersymmetric theory is the maximally extended Yang–Mills theory.

Table 1: Multiplicities for massless irreducible representations with maximal helicity 1 or less

	$N = 1$	$N = 1$	$N = 2$	$N = 2$	$N = 4$
Spin 1	–	$N = 1$	$N = 1$	–	$N = 1$
Spin $\frac{1}{2}$	$N = 1$	$N = 1$	$N = 2$	$N = 2$	$N = 4$
Spin 0	$N = 2$	–	$N = 2$	$N = 4$	$N = 6$

The content for massless on-shell representations with a maximum helicity 2 is given in Table 2.

The $N = 1$ supergravity theory contains only one spin 2 graviton and one spin $\frac{3}{2}$ gravitino. It is often referred to as simple supergravity theory. For the $N = 8$ supergravity theory, $\lambda - \frac{N}{2} = 2 - \frac{8}{2} = -2$. Consequently it is CPT self-conjugate and contains all particles from spin 2 to spin 0. Clearly, for theories in which $N > 8$, particles of spin higher than 2 will occur. Thus, the $N = 8$ theory is the maximally extended supergravity theory.

It has sometimes been claimed that this theory is in fact the largest possible consistent supersymmetric theory. This contention rests on the widely held belief that it is impossible to couple consistently massless particles of spin $\frac{5}{2}$ to other particles. In fact superstring theories do include spin $\frac{5}{2}$ particles, but these are massive.

We now consider the massive irreducible representations of supersymmetry. We take our rest-frame momentum to be

$$q_s^\mu = (m, 0, 0, 0).$$

Table 2: Multiplicity for massless on-shell representations with maximal helicity 2.

	$N = 1$	$N = 2$	$N = 3$	$N = 4$	$N = 5$	$N = 6$	$N = 7$	$N = 8$
Spin 2	1	1	1	1	1	1	1	1
Spin $\frac{3}{2}$	1	2	3	4	5	6	8	8
Spin 1		1	3	6	10	16	28	28
Spin $\frac{1}{2}$			1	4	11	26	56	56
Spin 0				2	10	30	70	70

The corresponding little group is then generated by

$$P_m, Q^{\alpha i}, T^r, Z_1^{ij}, Z_2^{ij}, J_m \equiv \frac{1}{2}\varepsilon_{mnr}J^{nr},$$

where $m, n, r = 1, 2, 3$ for the present discussion. The J_m generate the group $SU(2)$. Let us first consider the case where the central charges are trivially realised.

When acting on the rest-frame states the supercharges obey the algebra

$$\{Q^{Ai}, (Q^{Bj})^*\} = 2\delta_B^A \delta_j^i m; \qquad \{Q^{Ai}, Q^{Bj}\} = 0. \tag{3.7}$$

The action of the T^r is that of $U(N)$ with the $SU(2)$ rotation generators satisfy

$$[J_m, J_n] = \varepsilon_{mnr}J_r; \qquad [Q^{Ai}, J_m] = i(\sigma_m)^A_B Q^{Bi}, \tag{3.8}$$

where (σ_m) are the Pauli matrices. We note that as far as $SU(2)$ is concerned the dotted spinor $Q^{\dot{A}i}$ behaves like the undotted one Q_{Ai}.

We observe that, unlike the massless case, none of the supercharges are trivially realised and so the Clifford algebra they form has $4N$ elements, that is, twice as many as those for the massless case. The unique irreducible representation of the Clifford algebra is found in the usual way. We define a Clifford vacuum

$$Q_A^i |q_s^\mu\rangle = 0, \quad A = 1, 2, \quad i = 1, \dots, N,$$

and the representation is carried by the states

$$|q_s^\mu\rangle, \ (Q_A^i)^*|q_s^\mu\rangle, \ (Q_A^i)^*(Q_B^j)^*|q_s^\mu\rangle, \dots \ . \tag{3.9}$$

Thanks to the anti-commuting nature of the $(Q_A^i)^*$ this series terminates when Q^* is applied $2N + 1$ times.

The structure of the above representation is not particularly apparent since it is not clear how many particles of a given spin it contains. The properties of the Clifford algebra are more easily displayed by defining the real generators

$$\Gamma^i_{2A-1} = \frac{1}{2m}(Q^{Ai} + (Q^{Ai})^*); \qquad \Gamma^i_{2A} = \frac{i}{2m}(Q^{Ai} - (Q^{Ai})^*),$$

where the

$$\Gamma^i_p = (\Gamma^i_1, \Gamma^i_2, \Gamma^i_3, \Gamma^i_4)$$

are Hermitian. The Clifford algebra of equation (3.7) now becomes

$$\{\Gamma^i_p, \Gamma^j_q\} = \delta^{ij}\delta_{pq}.$$

The $4N$ elements of the Clifford algebra carry the group $SO(4N)$ in the standard manner, the $4N(4N-1)/2$ generators of $SO(4N)$ being

$$O^{ij}_{mn} = \frac{1}{2}[\Gamma^i_m, \Gamma^j_n].$$

As there are an even number of elements in the basis of the Clifford algebra, we may define a 'parity' (γ_5) operator by

$$\Gamma_{4N+1} = \prod_{p=1}^{4} \prod_{i=1}^{N} \Gamma_p^i,$$

which obeys the relations

$$(\Gamma_{4N+1})^2 = +1; \qquad \{\Gamma_{4N+1}, \Gamma_p^i\} = 0 \tag{3.10}$$

Indeed, the irreducible representation of equation (3.9) is of dimension 2^{2N} and transforms according to an irreducible representation of $SO(4N)$ of dimension 2^{2N-1}, with $\Gamma_{4N+1} = -1$, and another of dimension 2^{2N-1}, with $\Gamma_{4N+1} = +1$. Now any linear transformation of the Q, Q^* (for example $\delta Q = rQ$) can be represented by a generator formed from the commutator of the Q and Q^* (for example, $r[Q, Q^*]$). In particular the $SU(2)$ rotation generators are given by

$$s_k = -\frac{i}{4m}(\sigma_k)_B^A[Q^{jB}, (Q^{jA})^*].$$

One may easily verify that

$$[Q^{jA}, s_k] = i(\sigma_k)_B^A Q^{Bj}.$$

The states of a given spin will be classified by that subgroup of $SO(4N)$ which commutes with the appropriate $SU(2)$ rotation subgroup of $SO(4N)$. This will be the group generated by all generators bilinear in Q, Q^* that have their 2-component index contracted, i.e.

$$\Lambda_j^i = \frac{i}{2m}[Q^{Ai}, (Q_A^j)^*]; \qquad k^{ij} = \frac{i}{2m}[Q^{Ai}, Q_A^j],$$

with $(k^{ij})^\dagger = k_{ij}$. It is easy to verify that the Λ_j^i, k^{ij} and k_{ij} generate the group $USp(2N)$ and so the states of a given spin are labelled by representations of $USp(2N)$. That the group is $USp(2N)$ is most easily seen by defining

$$Q_A^a = \begin{cases} Q_A^i \delta_i^a & a = 1, \ldots, N \\ \varepsilon_{AB}(Q^{Bi})^* & a = N+1, \ldots, 2N \end{cases}$$

for then the generators Λ_j^i, k^{ij} and k_{ij} are given by

$$s^{ab} = \frac{i}{2m}[Q^{Aa}, Q_A^b].$$

Using the fact that

$$\{Q_A^a, Q_B^b\} = \varepsilon_{AB}\Omega^{ab},$$

where

$$\Omega^{ab} = \begin{pmatrix} 0 & 1 \\ -1 & 0 \end{pmatrix}$$

we can verify that

$$[s^{ab}, s^{cd}] = \Omega^{ac} s^{bd} + \Omega^{ad} s^{bc} + \Omega^{bc} s^{ad} + \Omega^{bd} s^{ac},$$

which is the algebra of $USp(2N)$.

The particle content of a massive irreducible representation is given by the following (see [21])

Theorem *If our Clifford vacuum is a scalar under the $SU(2)$ spin group and the internal symmetry group, then the irreducible massive representation of supersymmetry has the following content*

$$2^{2N} = \left[\frac{N}{2}, (0)\right] + \left[\frac{N-1}{2}, (1)\right] + \cdots + \left[\frac{N-\kappa}{2}, (\kappa)\right] + \cdots + [0, (N)],$$

where the first entry in the bracket denotes the spin and the last entry, say (k), denotes which kth-fold anti-symmetric traceless irreducible representation of $USp(2N)$ to which this spin belongs.

Consider an example with two supercharges. The classifying group is $USp(4)$ and the 2^4 states are one spin 1, four spin $\frac{1}{2}$, and five spin 0 corresponding to the $\underline{1}$-, $\underline{4}$- and $\underline{5}$-dimensional representations of $USp(4)$. For more examples see Table 3.

Should the Clifford vacuum carry spin and belong to a non-trivial representation of the internal group $U(N)$, then the irreducible representation would be found by taking the tensor product of the vacuum and the representation given in the above theorem.

Table 3: Some massive representations (without central charges) labelled in terms of the $USp(2N)$ representations.

	$N = 1$			$N = 2$			$N = 3$		$N = 4$
Spin 2		1			1			1	1
Spin $\frac{3}{2}$		1	2		1	4	1	6	8
Spin 1	1	2	1	1	4	$5+1$	6	$14+1$	27
Spin $\frac{1}{2}$	1	2	1	4	$5+1$	4	14	$14'+6$	48
Spin 0	2	1		5	4	1	$14'$	14	42

4 Massive Representations with a Central Charge

We now consider the case of particles that are massive, but which also possess a central charge. We take the particles to be in their rest frame with momentum $q^\mu \equiv (M, 0, 0, 0)$. The isotropy group, H contains $(P^a, Q_A^i, Q_i^A,$ $\underline{J}, T_r$ and $Z^{ij})$. In the rest-frame of the particles, that is for the momentum q^μ, the algebra of the supercharges is given by

$$\left\{ Q^{Ai}, (Q^{Bj})^* \right\} = 2\delta_B^A \delta_j^i M \tag{4.1}$$

and

$$\{Q_A^i,\ Q_B^j\} = \epsilon_{AB} Z^{ij}.$$

To discover what is the particle content in a supermultiplet we would like to rewrite the above algebra as a Clifford algebra. The first step in this procedure is to carry out a unitary transformation on the internal symmetry index of the supercharges i.e. $Q_A^i \to U_j^i Q_A^j$ or $Q_A \to U Q_A^i$ with $U^\dagger U = 1$. Such a transformation preserves the form of the first relation of equation (4.1). However, the unitary transformation can be chosen (see [104]) such that the central charge, which transforms as $Z \to U Z U^T$, can be brought to the form of a matrix which has all its entries zero except for the 2×2 matices down its diagonal. These 2×2 matrices are anti-symmetric as a consequence of the anti-symmetric nature of Z^{ij} which is preserved by the unitary transformation. This is the closest one can come to diagonalising an anti-symmetric matrix. Let us for simplicity restrict our attention to even N. We can best write down this matrix by replacing the $i, j = 1, 2, \ldots N$ internal indices by $i = (a, m)$, $j = (b, n)$, $a, b = 1, 2$, $m, n = 1, \ldots, \frac{N}{2}$, whereupon

$$Z^{(a,m)(b,n)} = 2\epsilon^{ab} \delta^{mn} Z_n. \tag{4.2}$$

In fact one also show that $Z_n \geq 0$. The supercharges in the rest-frame satisfy the relations

$$\left\{ Q^{A(am)}, (Q^{B(bn)})^* \right\} = 2\delta_B^A \delta_b^a \delta_n^m M \text{ and } \left\{ Q_A^{(am)}, Q_B^{(bn)} \right\} = 2\epsilon_{AB} \epsilon^{ab} \delta^{mn} Z_n.$$

We now define the supercharges

$$S_1^{Am} = \frac{1}{\sqrt{2}}(Q^{A1m} + (Q^{B2m} \epsilon_{BA})^*); \qquad S_2^{Am} = \frac{1}{\sqrt{2}}(Q^{A1m} - (Q^{B2m} \epsilon_{BA})^*)$$

in terms of which all the anti-commutators vanish except for

$$\left\{ S_1^{Am},\ (S_1^{Bn})^* \right\} = 2\delta^{AB} \delta^{mn} (M - Z_n) \tag{4.3}$$

$$\left\{ S_2^{Am}, \ (S_2^{Bn})^* \right\} = 2\delta^{AB}\delta^{mn}(M + Z_n). \tag{4.4}$$

This algebra is a Clifford algebra formed from the $2N$ operators S_1^{Am} and S_2^{Am} and their complex conjugates. It follows from equation (4.4) that if we take the same indices on each supercharge that the right-hand side is positive-definite and hence $Z_n \le M$.

To find the irreducible representation of supersymmetry we follow a similar procedure to that which we followed for massive and massless particles. The result crucially depends on whether $Z_n < M$, $\forall \ n$, or if one or more values of n we saturate the bound $Z_n = M$.

Let us first consider $Z_n < M$, $\forall \ n$. In this case, the right-hand sides of both equations (4.3) and (4.4) are non-zero. Taking S_1^{Am} and S_2^{Am} to annihilate the vacuum the physical states are given by the creation opperators $(S_1^{Am})^*$ and $(S_2^{Am})^*$ acting on the vacuum. The resulting representation has 2^{2N} states and has the same structure as for the massive case in the absence of a central charge. The states are classified by $USP(2N)$ as for the massive case.

Let us now suppose that q of the Z_n saturate the bound i.e., $Z_n = M$. For these values of n the right-hand side of equation (4.3) vanishes; taking the expectation value of this relation for any physical state we find that

$$\langle \text{phys} \mid S_1^{An}(S_1^{An})^* \mid \text{phys} \rangle + \langle \text{phys} \mid (S_1^{An})^*(S_1^{An}) \mid \text{phys} \rangle = 0 \tag{4.5}$$

The scalar product on the space of physical states satisfies all the axioms of a scalar product and hence we conclude that both of the above terms vanish and as a result

$$(S_1^{Bn})^* \mid \text{phys} \rangle = 0 = (S_1^{An}) \mid \text{phys} \rangle.$$

This argument is the same as that used to eliminate half of the supercharges and their complex conjugates in the massless case; however in the case under consideration it only eliminates q of the supercharges and their complex conjugates. There remain the $\frac{N}{2}$ supercharges (S_2^{Bm}) and the $\frac{N}{2} - q$ supercharges (S_1^{Bm}) for the values of m for which Z_m do not saturate the bound, as well as their complex conjugates. These supercharges form a Clifford algebra and we can take the $\frac{N}{2}$ supercharges (S_2^{Bm}) and the $\frac{N}{2} - q$ supercharges (S_1^{Bm}) to annihilate the vacuum and their complex conjugates to be creation opperators. The resulting massive irreducible representation of supersymmetry has $2^{2(N-q)}$ states and it has the same form as a massive representation of $N - q$ extended supersymmetry. The states will be classified by $USp(2N - 2q)$.

Clearly, a representation in which some or all of the central charges are equal to their mass has fewer states than the massive representation formed when none of the central charges saturate the mass, or a massive representation for which all the central charges vanish. This is a consequence of the fact that the latter Clifford algebra has more of its supercharges active in the irreducible representation. In almost all cases, the representation with some of its central charges saturated contains a smaller range of spins than the massive representation with no central charges. This feature plays a very important role in discussions of duality in supersymmetric theories.

Let us consider the irreducible representations of $N = 4$ supersymmetry which has both of its two possible central charges saturated. These representations are like the corresponding $N = 2$ massive representations. An important example has a $\underline{1}$ of spin 1, a $\underline{4}$ of spin $\frac{1}{2}$ and $\underline{5}$ of spin 0. The underlined numbers are their $USp(4)$ representations. This representation arises when the $N = 4$ Yang–Mills theory is spontaneously broken by one of its scalars acquiring a vaccum expectation value. The theory, before being spontaneously broken, has a massless representation with one of spin 1, two of spin $\frac{1}{2}$, and six of spin 0. Examining the massive representations for $N = 4$ in the absence of a central charge one finds that the representation with the smallest spins has all spins from spin 2 to spin 0. Hence the spontaneously broken theory can only be supersymmetric if the representation has a central charge. Another way to get the count in the above representation is to take the massless representation and recall that when the theory is spontaneously broken one of the scalars has been eaten by the vector as a result of the Higgs mechanism.

We close this section by answering a question which may have arisen in the mind of the reader. For the N extended supersymmetry algebra the supersymmetry algebra in the rest frame of equation (4.2) representation has $\frac{N}{2}$ possible central charges. This makes one central charge for the case of $N = 2$. However, this number conflicts with our understanding that a particle in $N = 2$ supersymmetric Yang–Mills theory can have two central charges corresponding to its electric and magnetic fields. The resolution of this conundrum is that although one can use a unitary transformation to bring the central charge of a given irreducible representation (i.e. particle), to have only one independent component one cannot do this simultaneously for all irreducible multiplets or particles.

Some examples of massive representations with central charges are given in Table 4.

We have seen that when the central charges take on special values the irreducible representation contain fewer states than the generic irreducible massive representation. When this occurs the states in the representation are called BPS states and the multiplet of states is called a BPS or short multiplet. Quantum corrections do not, except in very unusual circumstances, alter the number of states in a theory, and as such one expects the number and type of BPS multiplets to be the same in the classical and quantum theories. This follows from the observation that for a BPS multiplet to become a generic massive multiple,t the presence of more states would be required in the theory and one does not expect suddenly to find new states in the theory as one

Table 4: Some massive representations with one central charge ($|Z| = m$). All states are complex.

	$N = 2$		$N = 4$		$N = 6$		8
Spin 2					1	1	
Spin $\frac{3}{2}$			1	1	6	8	
Spin 1	1	1	4	6	14+1	27	
Spin $\frac{1}{2}$	1	2	4	5+1	14	$14' + 6$	48
Spin 0	2	1	5	4	$14'$	14	42

smoothly alters the parameters of the theory. As such, we can expect this result to hold regardless of the value of the coupling constant of the quantum theory.

This property is very useful since one can verify the existence of certain BPS states when the coupling constant of the theory is small by conventional techniques and it then follows that these BPS states will be present in the same theory when the coupling constant is large. In fact, the presence of BPS states is one of the few things that one can reliably establish in the strong-coupling regime of a supersymmetric theory. BPS multiplets play an important role in discussions on duality; if one suspects that two theories are dual one of the reliable signs is the correspondence between the BPS states in each theory.

The account of the massive irreducible representations of supersymmetry given here is along similar lines to the review by Ferrara and Savoy in [21].

5 Three Ways to Construct a Supergravity Theory

In this section it is explained how to construct a supergravity theory from the knowledge of its on-shell states.

There are three main ways. The one which has been used most often is the Noether method. It was used to construct the 4-dimensional $N = 1$ supergravity in its on-shell (see [14], [15]) and off-shell formulations (see [16], [17]). This method was also used in the construction of the $d = 11$ supergravity theory (see [106]) and a variant of it was used to find many of the properties of the IIB theory (see [111]). Although the method does not make use of any sophistocated mathematics and can be rather lengthy, it is very powerful. Starting from the linearised theory for the relevant supermultiplet, it gives a systematic way of finding the final nonlinear theory. To illustrate the method clearly and without undue technical complexity, we explain, in

Subsection 5.1.1, how to construct the Yang–Mills theory from the linearised theory. In Subsection 5.1.2 we find the linearised supergravity theory in 4 dimensions and apply the Noether procedure to find the full nonlinear theory. In fact, we will only carry out the first steps in this Noether procedure, but these clearly illustrate how to find the final result, most of whose features are already apparent at an early stage of the process. We then give the $N = 1$ supergravity theory and show it is invariant.

The second method uses the superspace description of supergravity theories and in Subection 5.2 we begin by summarising this approach. Supergravity theories in superspace share a number of similarities with the usual theory of general relativity. They are built from a supervierbein and a spin-connection and are invariant under superdiffeomorphisms. Using the supervierbein and spin-connection we can construct covariant derivatives and then define torsions and curvatures in the usual way. The superspace formulation differs from the usual formulation of general relativity in the nature of the tangent space group. The tangent space of the superspace formulation of supergravity contains fermionic and bosonic subspaces; however, the tangent space group is only the Lorentz group which rotates the odd (bosonic) and even (fermionic) subspaces seperately. In general relativity, the tangent space group is also the Lorentz group but in this case it rotates all vectors of a given length into each other.

The use of a restricted tangent space group in the superspace formulation allows us to take some of the torsions and curvatures to be zero since these constraints are respected by the Lorentz tangent space group. In fact the torsions and curvatures form a highly reducible representation of the Lorentz group. Indeed, the imposition of such constraints is precisely what is required to find the correct theory of supergravity. Hence the problem of finding the superspace formulation of supergravity is to find which of the torsions and curvatures are zero. We require different sets of constriants for the on-shell and off-shell supergravity theory. Clearly one gets from the latter to the former by imposing more constraints. To find the constraints for the off-shell theory, when this exists, can be rather difficult; however it turns out that to find the constraints for the on-shell theory is very straightforward and, remarkably, requires only dimensional analysis. Given a knowledge of the on-shell states in x-space, which are determined in a straightforward way from the supersymmetry algebra as explained in Section 3, we can deduce the dimensions of all the gauge-invariant quantities. By introducing a notion of geometric dimension, which in effect absorbs all factors of Newton's constant κ into the fields, one finds that for sufficently low dimensions and certain Lorentz character there are no gauge-invariant tensors in x-space. The superspace torsions and curvatures are gauge-invariant, so the superspace tensors with these dimensions and Lorentz character must then vanish as there is no available

x-space tensor that their lowest component could equal. We can substitute these superspace constraints into the the Bianchi idenitites satisfied by the torsions and curvatures to find constraints on the higher-dimensional torsions and curvatures. From this set of constraints one can find an x-space theory and it turns out that in all known cases this theory is none other than the corresponding on-shell supergravity theory. In other words, the constraints deduced from dimensional analysis and the use of the Bianchi identities are sufficient to find the on-shell supergravity in superspace and hence also in x-space. In Subsection 5.2.2 we explicity carry out this programme for the 4-dimensional $N = 1$ supergravity theory and recover the theory we found by the Noether method. This procedure was used to find the full IIB supergravity in superspace and in x-space (see [111]). We refer the reader to reference [111] for the details of this construction.

Finally, we briefly mention in Subsection 5.3 the third method of finding supergravity theories by the gauging certain space-time groups.

Several parts of this section are taken from reference [0].

5.1   The Noether method

5.1.1   Yang–Mills theory and the Noether technique

Any theory whose nonlinear form is determined by a gauge principle can be constructed by a Noether procedure; see [9]. Because of the importance of the Noether technique in constructing theories of supergravity we will take this opportunity to illustrate the technique within the framework of the simpler supersymmetric Yang–Mills theory; see [10].

Let us begin by considering the construction of the Yang–Mills theory itself from the linearised (free) theory. At the free level the theory is invariant under two distinct transformations: rigid and local Abelian transformations. Rigid transformations belong to a group S with generators R_i which satisfy

$$[R_i, R_j] = s_{ij}{}^k R_k.$$

The structure constants $s_{ij}{}^k$ may be chosen to be totally anti-symmetric and the indices $i, j, k, \ldots$ can be raised and lowered with the Kronecker delta δ_i^j. Under these rigid transformations the vector fields A_a^i transform as

$$\delta A_a^i = s_{jk}{}^i T^j A_a^k \tag{5.1}$$

where T_j are the infinitesimal group parameters. The other type of transformations is local Abelian transformations

$$\delta A_a^i = \partial_a \Lambda^i \tag{5.2}$$

188 West

Clearly both these transformations form a closed algebra*. The linearised theory which is invariant under the transformations of equations (5.1) and (5.2) is given by

$$A^{(0)} = \int d^4x \left\{ -\frac{1}{4} f^i_{ab} f^i_{ab} \right\},$$

where $f^i_{ab} = \partial_a A^i_b - \partial_b A^i_a$ The nonlinear theory is found in a series of steps, the first of which is to make the rigid transformations local, i.e. $T^j = T^j(x)$. Now, $A^{(0)}$ is no longer invariant under

$$\delta A^i_a = s_{jk}{}^i T^j(x) A^k_a \tag{5.3}$$

but its variation may be written in the form

$$\delta A^{(0)} = \int d^4x \{ (\partial_a T^k(x)) j^{ak} \},$$

where

$$j^{ak} = -s_{ij}{}^k A^i_b f^{abj}.$$

Now consider the action A_1

$$A_1 = A^{(0)} - \frac{1}{2} g \int d^4x (A^i_a j^{ai}),$$

where g is the gauge coupling constant; it is invariant to order g^0 provided we combine the local transformation $T^i(x)$ with the local transformation $\Lambda^i(x)$ with the identification $\Lambda^i(x) = (1/g)T^i(x)$. That is, the initially separate local and rigid transformations of the linearised theory become knitted together into a single local transformation given by

$$\delta A^i_a = \frac{1}{g} \partial_a T^i(x) + s^{ijk} T_j(x) A^k_a(x). \tag{5.4}$$

The first term in the transformation of δA^i_a yields in the variation of the last term in A_1, a term which cancels the unwanted variation of $A^{(0)}$.

We now continue with this process of amending the Lagrangian and transformations order-by-order in g until we obtain an invariant Lagrangian. The variation of A_1 under the second term of the transformation of equation (5.4) is of order g and is given by

$$\delta A_1 = \int d^4x \{ -g(A^i_a A^j_b s_{ij}{}^k)(A^l_b \partial_a T^m s_{lm}{}^k) \}.$$

An action invariant to order g is

$$A_2 = A_1 + \int d^4x \frac{g^2}{4}(A^i_a A^j_b s_{ij}{}^k)(A^{bl} A^{am} s_{lm}{}^k) = \frac{1}{4} \int d^4x (F^i_{ab})^2,$$

*In particular one finds $[\delta_\Lambda, \delta_T] A^i_a = \partial_a(s_{jk}{}^i T^j \Lambda^k)$.

where

$$F^i_{ab} = \partial_a A^i_b - \partial_b A^i_a - g s_{jk}{}^i A^j_a A^k_b.$$

In fact the action A_2 is invariant under the transformations of equation (5.4) to all orders in g and so represents the final answer, and is, of course, the well-known action of Yang–Mills theory. The commutator of two transformations on A^i_a is

$$[\delta_{T_1}, \delta_{T_2}]A^i_a = s_{ij}{}^k T^j_2 \left(\frac{1}{g}\partial_a T^k_1 + s_{lm}{}^k T^l_1 A^m_a \right) - (1 \leftrightarrow 2)$$

$$= \frac{1}{g}\partial_a T^i_{12} + s_{ij}{}^k T^j_{12} A^k_a,$$

where

$$T^i_{12} = s_{jk}{}^i T^j_2 T^k_1$$

and so the transformations form a closed algebra. In the last step we used the Jacobi identity in terms of the structure constants.

For supergravity and other local theories the procedure is similar, although somewhat more complicated. The essential steps are to first make the rigid transformations local and find invariant Lagrangians order-by-order in the appropriate gauge coupling constant. This is achieved in general not only by adding terms to the action, but also adding terms to the transformation laws of the field. If the latter process occurs one must also check the closure order-by-order in the gauge coupling constant.

Although one can use a Noether procedure which relies on the existence of an action, one can also use one which uses the transformation laws alone. This works, in the Yang–Mills case, as follows: upon making the rigid transformation local as in equation (5.3) one finds that the algebra no longer closes, i.e.

$$[\delta_\Lambda, \delta_T]A^i_a = \partial_a(s_{ij}{}^i(T^j \Lambda^k)) - s_{jk}{}^i(\partial_a T^j)\Lambda^k.$$

The cure for this is to regard the two transformations as simultaneous and knit them together as explained above. Using the new transformation for A^i_a of equation (5.4) we then test the closure to order g^0. In fact, in this case the closure works to all orders in g and the process stops here; in general however, one must close the algebra order-by-order in the coupling constant modifying the transformation laws and the closure relations for the algebra. Having the full transformations it is then easy to find the full action when that exists.

Let us apply the above Noether procedure to find the $N = 1$ supersymmetric Yang–Mills theory from its linearised theory. In the linearised theory the fields A^i_a, λ^i, D^i have the rigid transformations T_i and local transformations $\Lambda^i(x)$ given by

$$\delta A^i_a = s_{jk}{}^i T^j A^k_a, \quad \delta \lambda^i = s_{jk}{}^i T^j \lambda^k \quad \delta D^i = s_{jk}{}^i T^j D^k$$

and

$$\delta A_a^i = \partial_a \Lambda^i \quad \delta D^i = 0, \quad \delta \lambda^i = 0.$$

The supersymmetry transformations of the linearised theory are given by

$$\left.\begin{array}{rcl} \delta A_a^i &=& \bar{\varepsilon}\gamma_a\lambda^i \\ \delta\lambda^i &=& -\dfrac{1}{2}\sigma^{cd}f_{cd}^i\varepsilon + i\gamma_5 D^i\varepsilon \\ \delta D^i &=& i\bar{\varepsilon}\gamma_5\partial\!\!\!/\lambda^i. \end{array}\right\} \tag{5.5}$$

These transformations form a closed algebra, and leave invariant the following linearised Lagrangian

$$L = -\frac{1}{4}(f_{cd}^i)^2 - \frac{1}{2}\bar{\lambda}^i\partial\!\!\!/\lambda^i + \frac{1}{2}(D^i)^2.$$

Let us use the Noether method on the algebra to find the nonlinear theory. Making the rigid transformation on A_a^i local we must, as in the Yang–Mills case, knit the rigid and local transformations together (i.e. $\Lambda^i(x) = (1/g)T^i(x)$) to gain closure of gauge transformations on A_μ^i. Closure of supersymmetry and gauge transformations implies that the rigid transformations on λ^i and D^i also become local. This particular closure also requires that all the supersymmetry transformations are modified to involve covariant quantities. For example, we find that on D^i

$$[\delta_T, \delta_\varepsilon]D^i = i\bar{\varepsilon}\gamma_5\gamma^a s_{jk}{}^i(\partial_a T^j)\lambda^k.$$

and as a result we must replace $\partial_a\lambda^i$ by $\mathcal{D}_a\lambda^i = \partial_a\lambda^i - gs_{jk}{}^i A_a^j\lambda^k$ in the δD^i of equation (7.17) and then the commutator $[\delta_\Lambda, \delta_\varepsilon]$ is zero to all orders in g. The algebra then takes the form

$$\left.\begin{array}{rcl} \delta A_a^i &=& \bar{\varepsilon}\gamma_a\lambda^i \\ \delta\lambda^i &=& \left(-\dfrac{1}{2}\sigma^{cd}F_{cd}^i + i\gamma_5 D^i\right)\varepsilon \\ \delta D^i &=& i\bar{\varepsilon}\gamma_5\mathcal{D}\!\!\!/\lambda^i, \end{array}\right\} \tag{5.6}$$

where

$$F_{ab}^i = \partial_a A_b^i - \partial_b A_a^i - gs_{jk}{}^i A_a^j A_b^k.$$

We must now verify that the above supersymmetry transformations close. For other supersymmetric gauge theories we must add further terms to the supersymmetry transformations in order to regain closure. However, in this case gauge covariance and dimensional analysis ensure that there are no possible terms that one can add to these supersymmetry transformations and so the transformations of equations (5.6) must be the complete laws for the full theory. The reader may verify that there are no inconsistencies by showing that the algebra does indeed close.

The action invariant under these transformations is

$$A = \int d^4x \left\{ -\frac{1}{4}(F^i_{ab})^2 - \frac{1}{2}\bar{\lambda}^i \not{D}\lambda^i + \frac{1}{2}(D^i)^2 \right\}. \tag{5.7}$$

One could also have used the Noether procedure on the action. Gauge invariance implies that the action is that given in equation (5.7). Demanding that this gauge invariant action be supersymmetric requires us to modify the supersymmetry transformations to those of equation (5.6).

5.1.2 $N = 1$ $D = 4$ supergravity

We will now construct $N = 1$ $D = 4$ supergravity using the Noether method. The starting point is the linearised theory which we now construct along similar lines to the method used to find the Wess–Zumino model and $N = 1$ super QED in Chapters 5 and 6 in reference [0].

The Linearised Theory We will start with the on-shell states and construct the linearized theory, without and then with auxiliary fields. The $N = 1$ irreducible representations of supersymmetry which include a spin 2 graviton contain either a spin $\frac{3}{2}$ or a spin $\frac{5}{2}$ fermion. The spin $\frac{5}{2}$ particle would seem to have considerable problems in coupling to other fields and so we will choose the spin $\frac{3}{2}$ particle.

As in the Yang–Mills case, the linearized theory possesses rigid supersymmetry and local Abelian gauge invariances. The latter are required, in order that the fields do describe the massless on-shell states alone without involving ghosts. We recall that a rigid symmetry is one whose parameters are space-time independent while a local symmetry has space-time dependent parameters.

These on-shell states are represented by a symmetric second-rank tensor field, $h_{\mu\nu}$, ($h_{\mu\nu} = h_{\nu\mu}$) and a Majorana vector spinor, $\psi_{\mu\alpha}$. For these fields to represent a spin 2 particle and a spin $\frac{3}{2}$ particle they must possess the infinitesimal gauge transformations

$$\delta h_{\mu\nu} = \partial_\mu \xi_\nu(x) + \partial_\nu \xi_\mu(x); \qquad \partial\psi_{\mu\alpha} = \partial_\mu \eta_\alpha(x). \tag{5.8}$$

The unique ghost-free gauge-invariant, free-field equations are

$$E_{\mu\nu} = 0, \qquad R^\mu = 0, \tag{5.9}$$

where $E_{\mu\nu} = R^L_{\mu\nu} - \frac{1}{2}\eta_{\mu\nu}R^L$, and $R^{Lab}_{\mu\nu}$ is the linearized Riemann tensor given by

$$R^{Lab}_{\mu\nu} = -\partial_a \partial_\mu h_{b\nu} + \partial_b \partial_\mu h_{a\nu} + \partial_a \partial_\nu h_{b\mu} - \partial_b \partial_\nu h_{a\mu}$$

and

$$R^\mu = \varepsilon^{\mu\nu\rho\kappa} i\gamma_5\gamma_\nu\partial_\rho\psi_\kappa$$
$$R^{Lb}{}_\mu = R^{Lab}{}_{\mu\nu}\delta^\mu_a$$
$$R^L = R^{La}{}_\mu \delta^\mu_a.$$

For an explanation of this point see van Nieuwenhuizen, [13].

We must now search for the supersymmetry transformations that form an invariance of these field equations and represent the supersymmetry algebra on-shell. On dimensional grounds the most general transformation is

$$\left.\begin{array}{l} \delta h_{\mu\nu} = \dfrac{1}{2}(\bar\varepsilon\gamma_\mu\psi_\nu + \bar\varepsilon\gamma_\nu\psi_\mu) + \delta_1\eta_{\mu\nu}\bar\varepsilon\gamma^\kappa\psi_\kappa \\ \delta\psi_\mu = +\delta_2\sigma^{ab}\partial_a h_{b\mu}\varepsilon + \delta_3\partial_\nu h^\nu{}_\mu\varepsilon. \end{array}\right\} \quad (5.10)$$

The parameters δ_1, δ_2 and δ_3 will be determined by the demanding that the set of transformations which comprise the supersymmetry transformations of equation (5.10) and the gauge transformations of equation (5.8) should form a closing algebra when the field equations of equation (5.9) hold. At the linearized level the supersymmetry transformations are linear rigid transformations, that is, they are *first order* in the fields $h_{\mu\nu}$ and $\psi_{\mu\alpha}$ and parametrized by *constant* parameters ε^α.

Carrying out the commutator of a Rarita–Schwinger gauge transformation, $\eta_\alpha(x)$ of equation (5.8) and a supersymmetry transformation, ε of equation (5.10) on $h_{\mu\nu}$, we get:

$$[\delta_\eta, \delta_\varepsilon]h_{\mu\nu} = \frac{1}{2}(\bar\varepsilon\gamma_\mu\partial_\nu\eta + \bar\varepsilon\gamma_\nu\partial_\mu\eta) + \delta_1\eta_{\mu\nu}\bar\varepsilon\partial\!\!\!/\eta.$$

This is a gauge transformation with parameter $\frac{1}{2}\bar\varepsilon\gamma_\mu\eta$ on $h_{\mu\nu}$ provided $\delta_1 = 0$. Similarly, calculating the commutator of a gauge transformation of $h_{\mu\nu}$ and a supersymmetry transformation on $h_{\mu\nu}$ automatically yields the correct result, i.e., zero. However, carrying out the commutator of a supersymmetry transformation and an Einstein gauge transformation on $\psi_{\mu\alpha}$ yields

$$[\delta_{\xi_\mu}, \delta_\varepsilon]\psi_\mu = +\delta_2\sigma^{ab}\partial_a(\partial_\mu\xi_b)\varepsilon + \delta_3\partial_\nu\partial^\nu\xi_\mu\varepsilon + \delta_3\partial_\nu\partial_\mu\xi^\mu\varepsilon,$$

which is a Rarita–Schwinger gauge transformation on ψ_μ provided $\delta_3 = 0$. Hence we take $\delta_1 = \delta_3 = 0$.

We must test the commutator of two supersymmetries. On $h_{\mu\nu}$ we find the commutator of two supersymmetries to give

$$\begin{aligned} [\delta_{\varepsilon_1}, \delta_{\varepsilon_2}]h_{\mu\nu} &= +\frac{1}{2}\{\bar\varepsilon_2\gamma_\mu\delta_2\sigma^{ab}\partial_a h_{b\nu}\varepsilon_2 + (\mu \leftrightarrow \nu)\} - (1 \leftrightarrow 2) \\ &= \delta_2\{\bar\varepsilon_2\gamma^b\varepsilon_1\partial_\mu h_{b\nu} - \bar\varepsilon_2\gamma^a\varepsilon_1\partial_a h_{\mu\nu} - (\mu \leftrightarrow \nu)\}. \quad (5.11) \end{aligned}$$

This is a gauge transformation on $h_{\mu\nu}$ with parameter $\delta_2 \bar{\varepsilon}_2 \gamma^b \varepsilon_1 h_{b\nu}$ as well as a space-time translation. The magnitude of this translation coincides with that dictated by the supersymmetry group provided $\delta_2 = -1$, which is the value we now adopt.

It is important to stress that linearized supergravity differs from the Wess–Zumino model in that one must take into account the gauge transformations of equations (5.8) as well as the rigid supersymmetry transformations of equation (5.10) in order to obtain a closed algebra. The resulting algebra is the $N = 1$ supersymmetry algebra when supplemented by gauge transformations. This algebra reduces to the $N = 1$ supersymmetry algebra only on gauge-invariant states.

For the commutator of two supersymmetries on ψ_μ we find

$$
\begin{aligned}
[\delta_{\varepsilon_1}, \delta_{\varepsilon_2}]\psi_\mu &= -\sigma^{ab}\partial_a \varepsilon_2 \frac{1}{2}(\varepsilon_1 \gamma_b \psi_\mu + \bar{\varepsilon}_1 \gamma_\mu \psi_b) - (1 \leftrightarrow 2) \\
&= +\frac{1}{2.4}\sum_R c_R \bar{\varepsilon}_1 \gamma_R \varepsilon_2 \sigma^{ab}\partial_a \gamma^R(\gamma_b \psi_\mu + \gamma_\mu \psi_b) - (1 \leftrightarrow 2) \\
&= \frac{1}{8}\bar{\varepsilon}_1 \gamma_R \varepsilon_2 \sigma^{ab}\gamma^R\left(\gamma_b \psi_{a\mu} + \frac{1}{2}\gamma_\mu \psi_{ab}\right) \\
&\quad +\partial_\mu\left(\frac{1}{8}\bar{\varepsilon}_1 \gamma_R \varepsilon_2 \sigma^{ab}\gamma^R \gamma_b \psi_a\right) - (1 \leftrightarrow 2)
\end{aligned}
$$

where $\psi_{\mu\nu} = \partial_\mu \psi_\nu - \partial_\nu \psi_\mu$. Using the different forms of the Rarita–Schwinger equation of motion, given by

$$
R^\mu = 0 \Leftrightarrow \gamma^\mu \psi_{\mu\nu} = 0 \Leftrightarrow \psi_{\mu\nu} + \frac{1}{2}i\gamma_5 \varepsilon_{\mu\nu\rho\kappa}\psi^{\rho\kappa} = 0,
$$

we find the final result to be

$$
\begin{aligned}
&[\delta_{\varepsilon_1}, \delta_{\varepsilon_2}]\psi_\mu \\
&= 2\bar{\varepsilon}_2 \gamma^c \varepsilon_1 \partial_c \psi_\mu + \partial_\mu\left(-\bar{\varepsilon}_2 \gamma^c \varepsilon_1 \psi_c + \sum_R c_R \frac{1}{8}\bar{\varepsilon}_1 \gamma_R \varepsilon_2 \sigma^{ab}\gamma^R \gamma_b \psi_a - (1 \leftrightarrow 2)\right).
\end{aligned}
$$

This is the required result: a translation and a gauge transformation on ψ_μ.

The reader can verify that the transformations of equation (5.10) with the values of $\delta_1 = \delta_3 = 0$, $\delta_2 = -1$ do indeed leave invariant the equations of motion of $h_{\mu\nu}$ and $\psi_{\mu\alpha}$.

Having obtained an irreducible representation of supersymmetry carried by the fields $h_{\mu\nu}$ and $\psi_{\mu\alpha}$ when subject to their field equations we can now find the algebraically on-shell Lagrangian. The action (Freedman, van Nieuwenhuizen and Ferrara [14], Deser and Zumino [15]) from which the field equations of equation (5.9) follow, is

$$
A = \int d^4x \left\{-\frac{1}{2}h^{\mu\nu}E_{\mu\nu} - \frac{1}{2}\bar{\psi}_\mu R^\mu\right\}. \tag{5.12}
$$

It is invariant under the transformations of equation (5.10) provided we adopt the values for the parameters δ_1, δ_2 and δ_3 found above. This invariance holds without use of the field equations, as it did in the Wess–Zumino case.

We now wish to find a linearized formulation which is built from fields which carry a representation of supersymmetry without imposing any restrictions (i.e., equations of motion), namely, we find the auxiliary fields. As a guide to their number we can apply our Fermi-Bose counting rule which, since the algebra contains gauge transformations, applies only to the gauge-invariant states. On-shell, $h_{\mu\nu}$ has two helicities and so does $\psi_{\mu\alpha}$; however off-shell, $h_{\mu\nu}$ contributes $(5 \times 4)/2 = 10$ degrees of freedom minus 4 gauge degrees of freedom giving 6 bosonic degrees of freedom. On the other hand, off-shell, $\psi_{\mu\alpha}$ contributes $4 \times 4 = 16$ degrees of freedom minus 4 gauge degrees of freedom, giving 12 fermionic degrees of freedom. Hence the auxiliary fields must contribute 6 bosonic degrees of freedom. If there are n auxiliary fermions there must be $4n + 6$ bosonic auxiliary fields.

Let us assume that a minimal formulation exists, that is, there are no auxiliary spinors. Let us also assume that the bosonic auxiliary fields occur in the Lagrangian as squares without derivatives (like F and G) and so are of dimension 2. Hence we have 6 bosonic auxiliary fields; it only remains to find their Lorentz character and transformations. We will assume that they consist of a scalar M, a pseudoscalar N and a pseudovector b_μ, rather than an anti-symmetric tensor or 6 spin-0 fields. We will give the motivating arguments for this later.

Another possibility is the two fields A_μ and $a_{\kappa\lambda}$ which possess the gauge transformations $\delta A_\mu = \partial_\mu \Lambda$ and $\delta a_{\kappa\lambda} = \partial_\kappa \Lambda_\lambda - \partial_\lambda \Lambda_\kappa$. A contribution $\varepsilon_{\mu\nu\rho\kappa} A^\mu \partial^\nu a^{\rho\kappa}$ to the action would not lead to propagating degrees of freedom.

The transformation of the fields $h_{\mu\nu}$, $\psi_{\mu\alpha}$, M, N and b_μ must reduce on-shell to the on-shell transformations found above. This restriction, dimensional arguments and the fact that if the auxiliary fields are to vanish on-shell they must vary into field equations, gives the transformations to be (see [16], [17])

$$\left. \begin{aligned}
\delta h_{\mu\nu} &= \frac{1}{2}(\bar{\varepsilon}\gamma_\mu\phi_\nu + \bar{\varepsilon}\gamma_\nu\psi_\mu) \\
\delta\psi_{\mu\alpha} &= -\sigma^{ab}\partial_a h_{b\mu}\varepsilon - \frac{1}{3}\gamma_\mu(M + i\gamma_5 N)\varepsilon + b_\mu i\gamma_5\varepsilon + \delta_6\gamma_\mu \slashed{b} i\gamma_5\varepsilon \\
\delta M &= \delta_4 \bar{\varepsilon}\gamma \cdot R \\
\delta N &= \delta_5 i\bar{\varepsilon}\gamma_5\gamma \cdot R \\
\delta b_\mu &= +\delta_7 i\bar{\varepsilon}\gamma_5 R_\mu + \delta_8 i\bar{\varepsilon}\gamma_5\gamma_\mu\gamma \cdot R.
\end{aligned} \right\} \quad (5.13)$$

The parameters δ_4, δ_5, δ_6, δ_7 and δ_8 are determined by the restriction that the above transformations of equation (5.13) and the gauge transformations of equation (5.8) should form a closed algebra. For example, the commutator of two supersymmetries on M gives

$$[\delta_{\varepsilon_1}, \delta_{\varepsilon_2}]M = \delta_4\{-\bar{\varepsilon}_2\gamma^\mu\varepsilon_1\partial_\mu M + 16\bar{\varepsilon}_2\sigma^{\mu\nu}i\gamma_5\varepsilon_1(1 + 3\delta_6)\partial_\mu b_\nu\},$$

which is the required result provided $\delta_4 = -\frac{1}{2}$ and $\delta_6 = -\frac{1}{3}$. Carrying out the commutator of two supersymmetries on all fields we find a closing algebra provided

$$\delta_4 = -\frac{1}{2}, \quad \delta_5 = -\frac{1}{2}, \quad \delta_6 = -\frac{1}{3}, \quad \delta_7 = \frac{3}{2} \quad \text{and} \quad \delta_8 = -\frac{1}{2}.$$

We henceforth adopt these values for the parameters. An action which is constructed from the fields $h_{\mu\nu}$, $\psi_{\mu\alpha}$, M, N and b_μ and is invariant under the transformations of equation (5.13) with the above values of the parameters is (see [16], [17])

$$A = \int d^4x \left\{ -\frac{1}{2}h_{\mu\nu}E^{\mu\nu} - \frac{1}{2}\bar{\psi}_\mu R^\mu - \frac{1}{3}(M^2 + N^2 - b^\mu b_\mu) \right\}. \qquad (5.14)$$

This is the action of linearized $N = 1$ supergravity and upon elimination of the auxiliary field M, N and b_μ it reduces to the algebraically on-shell Lagrangian of equation (5.12).

The Nonlinear Theory　The full nonlinear theory of supergravity can be found from the linearized theory discussed above by applying the Noether technique discsuued at the beginning of this section. Just as in the case of Yang–Mills the reader will observe that the linearized theory possesses the local Abelian invariances of equation (5.8) as well as the rigid (i.e., constant parameter) supersymmetry transformations of equation (5.10).

We proceed just as in the case of the Yang–Mills theory and make the parameter of rigid transformations space-time dependent, i.e., set $\varepsilon = \varepsilon(x)$ in equation (5.10). The linearized action of equation (5.14) is then no longer invariant, but its variation must be of the form

$$\delta A_0 = \int d^4x\partial_\mu\bar{\varepsilon}^\alpha j^\mu_\alpha$$

since it is invariant when $\bar{\varepsilon}^\alpha$ is a constant. The object j^μ_α is proportional to $\psi_{\mu\alpha}$ and linear in the bosonic fields $h_{\mu\nu}$, M or N and b_μ. As such, on dimensional grounds, it must be of the form

$$j_{\mu\alpha} \propto \partial_\tau h_{\rho\mu}\psi_{\nu\beta} + \cdots .$$

Consider now the action, A, given by

$$A_1 = A_0 - \frac{\kappa}{4}\int d^4x\bar{\psi}^\mu j_\mu,$$

where κ is the gravitational constant. The action A is invariant to order κ^0 *provided* we combine the now local supersymmetry transformation of equation (5.13) with a local Abelian Rarita–Schwinger gauge transformation of equation (5.8) with parameter $\eta(x) = (2/\kappa)\varepsilon(x)$. That is, we make a transformation

$$\delta\psi_\mu = \frac{2}{\kappa}\partial_\mu\varepsilon(x) - \partial_a h_{b\mu}\sigma^{ab}\varepsilon(x) - \frac{1}{3}\gamma_\mu(M + i\gamma_5 N)\varepsilon(x) + i\gamma_5\left(b_\mu - \frac{1}{3}\gamma_\mu\gamma^\nu b_\nu\right)\varepsilon(x),$$

the remaining fields transforming as before except that ε is now space-time dependent.

As in the Yang–Mills case the two invariances of the linearized action become knitted together to form one transformation, the role of gauge coupling being played by the gravitational constant, κ. The addition of the term $(-\kappa/4)\bar\psi^\mu j_\mu$ to A_0 does the required job; its variation is

$$-\frac{\kappa}{4}\cdot 2\cdot\left(\frac{2}{\kappa}\right)(\partial_\mu\bar\varepsilon)j^\mu + \text{ terms of order } \kappa^1.$$

The order κ^1 terms do not concern us at the moment. We note that $j_{\mu\alpha}$ is linear in $\psi_{\mu\alpha}$ and so we get a factor of 2 from $\delta\psi_{\mu\alpha}$.

In fact, one can carry out the Noether procedure in the context of pure gravity where one finds, at the linearized level, the rigid translation

$$\delta h_{\mu\nu} = \zeta^\lambda\partial_\lambda h_{\mu\nu}$$

and the local gauge transformation

$$\delta h_{\mu\nu} = \partial_\mu\xi_\nu + \partial_\nu\xi_\mu.$$

These become knitted together at the first stage of the Noether procedure to give

$$\delta h_{\mu\nu} = \frac{1}{\kappa}\partial_\mu\zeta_\nu + \frac{1}{\kappa}\partial_\nu\zeta_\mu + \zeta^\lambda\partial_\lambda h_{\mu\nu},$$

since $\xi_\nu = (1/\kappa)\zeta_\nu$. This variation of $h_{\mu\nu}$ contains the first few terms of an Einstein general coordinate transformation of the vierbein which is given in terms of $h_{\mu\nu}$ by

$$e_\mu{}^a = \eta_\mu{}^a + \kappa h_\mu{}^a.$$

We proceed in a similar way to the Yang–Mills case. We obtain order-by-order in κ an invariant Lagrangian by adding terms to the Lagrangian and in this case also adding terms to the transformations of the fields. For example, if we added a term to $\delta\psi_\mu$ say, $\delta\bar\psi_\mu = \cdots + \bar\varepsilon X_\mu\kappa$, then from the linearized action we receive a contribution $-\kappa\bar\varepsilon X_\mu R^\mu$ upon variation of ψ_μ. It is necessary at each step (order of κ) to check that the transformations of the fields form a closed algebra. In fact, any ambiguities that arise in the procedure are resolved by demanding that the algebra closes.

The final set of transformations (see [16], [17]) is

$$\delta e_\mu{}^a = \kappa \bar{\varepsilon} \gamma^a \psi_\mu$$

$$\delta \psi_\mu = 2\kappa^{-1} D_\mu(w(e,\psi))\varepsilon + i\gamma_5 \left(b_\mu - \frac{1}{3}\gamma_\mu \slashed{b}\right)\varepsilon - \frac{1}{3}\gamma_\mu(M + i\gamma_5 N)\varepsilon$$

$$\delta M = -\frac{1}{2}e^{-1}\bar{\varepsilon}\gamma_\mu R^\mu - \frac{\kappa}{2}i\bar{\varepsilon}\gamma_5\psi_\nu b^\nu - \kappa\bar{\varepsilon}\gamma^\nu\psi_\nu M + \frac{\kappa}{2}\bar{\varepsilon}(M + i\gamma_5 N)\gamma^\mu\psi_\mu$$

$$\delta N = -\frac{e^{-1}}{2}i\bar{\varepsilon}\gamma_5\gamma_\mu R^\mu + \frac{\kappa}{2}\bar{\varepsilon}\psi_\nu b^\nu - \kappa\bar{\varepsilon}\gamma^\nu\psi_\nu N - \frac{\kappa}{2}i\bar{\varepsilon}\gamma_5(M + i\gamma_5 N)\gamma^\mu\psi_\mu$$

$$\delta b_\mu = \frac{3i}{2}e^{-1}\bar{\varepsilon}\gamma_5\left(g_{\mu\nu} - \frac{1}{3}\gamma_\mu\gamma_\nu\right)R^\nu + \kappa\bar{\varepsilon}\gamma^\nu b_\nu\psi_\mu - \frac{\kappa}{2}\bar{\varepsilon}\gamma^\nu\psi_\nu b_\mu$$

$$- \frac{\kappa}{2}i\bar{\psi}_\mu\gamma_5(M + i\gamma_5 N)\varepsilon - \frac{i\kappa}{4}\varepsilon_\mu{}^{bcd}b_b\bar{\varepsilon}\gamma_5\gamma_c\psi_d, \tag{5.15}$$

where

$$R^\mu = \varepsilon^{\mu\nu\rho\kappa}i\gamma_5\gamma_\nu D_\rho(w(e,\psi))\psi_\kappa$$

$$D_\mu(w(e,\psi)) = \partial_\mu + w_{\mu ab}\frac{\sigma^{ab}}{4},$$

and

$$w_{\mu ab} = \frac{1}{2}e^\nu{}_a(\partial_\mu e_{b\nu} - \partial_\nu e_{b\mu}) - \frac{1}{2}e_b{}^\nu(\partial_\mu e_{a\nu} - \partial_\nu e_{a\mu})$$

$$- \frac{1}{2}e_a{}^\rho e_b{}^\sigma(\partial_\rho e_{\sigma c} - \partial_\sigma a_{\rho c})e_\mu{}^c$$

$$+ \frac{\kappa^2}{4}(\bar{\psi}_\mu\gamma_a\psi_b + \bar{\psi}_a\gamma_\mu\psi_b - \bar{\psi}_\mu\gamma_b\psi_a).$$

They form a closed algebra, the commutator of two supersymmetries on any field being

$$[\delta_{\varepsilon_1}, \delta_{\varepsilon_2}] = \delta_{\text{supersymmetry}}(-\kappa\xi^\nu\psi_\nu) + \delta_{\text{general coordinate}}(2\xi_\mu)$$

$$+\delta_{\text{Local Lorentz}}\left(-\frac{2\kappa}{3}\varepsilon_{ab\lambda\rho}b^\lambda\xi^\rho\right.$$

$$\left. - \frac{2\kappa}{3}\bar{\varepsilon}_2\sigma_{ab}(M + i\gamma_5 N)\varepsilon_1 + 2\xi^d w_d{}^{ab}\right),$$

where

$$\xi_\mu = \bar{\varepsilon}_2\gamma_\mu\varepsilon_1.$$

The transformations of equation (5.15) leave invariant the action (see [16], [17])

$$A = \int d^4x \left\{\frac{e}{2\kappa^2}R - \frac{1}{2}\bar{\psi}_\mu R^\mu - \frac{1}{3}e(M^2 + N^2 - b_\mu b^\mu)\right\},$$

where

$$R = R_{\mu\nu}{}^{ab}e_a{}^\mu e_b{}^\nu \quad \text{and} \quad R_{\mu\nu}{}^{ab}\frac{\sigma_{ab}}{4} = [D_\mu, D_\nu].$$

The auxiliary fields M, N and b_μ may be eliminated to obtain the nonlinear algebraically on-shell Lagrangian which was the form in which supergravity was originally found in references [14] and [15].

As discussed at the beginning of this section one could also build up the non-local theory by working with the algebra of field transformations alone.

Invariance of $N = 1$ Supergravity   We refer the reader to Chapter 10 of reference [0] for a demonstration of the invariance of the supergravity under local supersymmetry transformations.   This proof (see [18], [19]) uses the 1.5-order formalism.

5.2   On-shell $N = 1$ $D = 4$ superspace

In this subsection we will construct the on-shell superspace formulation of $N = 1$ $D =$ supergravity, from which we recover the equations of motion in x-space.

5.2.1   Geometry of local superspace

The geometrical framework [69] of superspace supergravity has many of the constructions of general relativity, but also requires additional input. A useful guide in the construction of local superspace is that it should admit rigid superspace as a limit.

We begin with an 8-dimensional manifold $z^\Pi = (x^u, \theta^{\underline{\alpha}})$ where x^u is a commuting coordinate and $\theta^{\underline{\alpha}}$ is a Grassmann odd coordinate. On this manifold a super-general coordinate reparametrization has the form

$$z^\Pi \rightarrow z'^\Pi = z^\Pi + \xi^\Pi,$$

where $\xi^\Pi = (\xi^u, \xi^{\underline{\alpha}})$ are arbitrary functions of z^Π.

Just as in general relativity we can consider scalar superfields, that is, fields for which

$$\phi'(z') = \phi(z)$$

and superfields with superspace world indices φ_Λ; for example

$$\varphi_\Lambda = \frac{\partial \phi}{\partial z^\Lambda}.$$

The latter transform as

$$\varphi'_\Lambda(z') = \frac{\partial z^\Pi}{\partial z'^\Lambda} \varphi_\Pi(z).$$

The transformation properties of higher-order tensors is obvious.

We must now specify the geometrical structure of the manifold. For reasons that will become apparent, the superspace formulation is essentially a vierbein formulation. We introduce supervierbeins $E_\Pi{}^N$ which transform under the supergeneral coordinate transformations as

$$\delta E_\Pi{}^N = \xi^\Lambda \partial_\Lambda E_\Pi{}^N + \partial_\Pi \xi^\Lambda E_\Lambda{}^N.$$

The N-index transforms under the tangent space group which is taken to be just the Lorentz group; and so $\delta E_\Pi{}^N = E_\Pi{}^M \Lambda_M{}^N$ where

$$\Lambda_M{}^N = \begin{pmatrix} \lambda_m{}^n & 0 & 0 \\ 0 & -\frac{1}{4}(\sigma_{mn})_A{}^B \Lambda^{mn} & 0 \\ 0 & 0 & +\frac{1}{4}(\sigma_{mn})_{\dot{A}}{}^{\dot{B}} \Lambda^{mn} \end{pmatrix}. \tag{5.16}$$

The matrix $\Lambda_m{}^n$ is an arbitrary function on superspace and it governs not only the rotation of the vector index, but also the rotation of the spinorial indices. Since we are dealing with an 8-dimensional manifold one could choose a much larger tangent space group. For example, $\Lambda_M{}^N$ could be an arbitrary matrix that preserves the metric

$$g_{NM} = \text{diag}\,(a_1 \eta_{mn}, a_2 \varepsilon_{AB}, a_3 \varepsilon_{\dot{A}\dot{B}}), \tag{5.17}$$

where a_1, a_2 and a_3 are non-zero arbitrary constants. Demanding reality of the metric implies $a_2^* = a_3$ and we may scale away one factor. This corresponds to taking the tangent space group to be $OSp(4,1)$. In such a formulation one could introduce a metric $g_{\Pi\Lambda} = E_\Pi{}^N g_{NM} E_\Lambda{}^M$ and one would have a formulation which mimicked Einstein's general relativity at every step; see [70].

Such a formulation, however, would not lead to the x-space component $N = 1$ supergravity given earlier. One way to see this is to observe that the above tangent space group does not coincide with that of rigid superspace (super Poincaré/Lorentz), which has the Lorentz group, as given in equation (5.16) with $\Lambda_m{}^n$ a constant matrix, as its tangent space group. As linearized superspace supergravity must admit a rigid superspace formulation, any formulation based on an $OSp(4,1)$ tangent group will not coincide with linearized supergravity. In fact the $OSp(4,1)$ formulation has a higher derivative action.

An important consequence of this restricted tangent space group is that tangent supervectors $V^N = V^\Pi E_\Pi{}^N$ belong to a reducible representation of the Lorentz group. This allows one to write down many more invariants. The objects $V^m V_m$, $V^A V^B \varepsilon_{AB}, V^{\dot{A}} V^{\dot{B}} \varepsilon_{\dot{B}\dot{A}}$ are all separately invariant.

In other words, in the choice of metric in equation (5.17) the constants a_1, a_2 and a_3 can have any value, including zero.

We define a Lorentz-valued spin connection

$$\Omega_{\Lambda M}{}^{N} = \begin{pmatrix} \Omega_{\Lambda m}{}^{n} & 0 & 0 \\ 0 & -\frac{1}{4}\Omega_{\Lambda}{}^{mn}(\sigma_{mn})_{A}{}^{B} & 0 \\ 0 & 0 & \frac{1}{4}\Omega_{\Lambda}{}^{mn}(\bar{\sigma}_{mn})_{\dot{A}}{}^{\dot{B}} \end{pmatrix}.$$

This object transforms under super general coordinate transformations as

$$\delta\Omega_{\Lambda M}{}^{N} = \xi^{\Pi}\partial_{\Pi}\Omega_{\Lambda M}{}^{N} + \partial_{\Lambda}\xi^{\Pi}\Omega_{\Pi M}{}^{N},$$

and under tangent space rotation as

$$\delta\Omega_{\Lambda M}{}^{N} = -\partial_{\Lambda}\Omega_{M}{}^{N} + \Omega_{\Lambda M}{}^{S}\Lambda_{S}{}^{N} + \Omega_{\Lambda R}{}^{N}\Lambda_{M}{}^{R}(-1)^{(M+R)(N+R)}.$$

The covariant derivatives are then defined by

$$D_{\Lambda} = \partial_{\Lambda} + \frac{1}{2}\Omega_{\Lambda}{}^{mn}J_{mn}, \tag{5.18}$$

where J_{mn} are the appropriate Lorentz generators (see Appendix A of reference [0]). The covariant derivative with tangent indices is

$$D_{N} = E_{N}{}^{\Lambda}D_{\Lambda}, \tag{5.19}$$

where $E_{N}{}^{\Lambda}$ is the inverse vierbein defined by

$$E_{N}{}^{\Lambda}E_{\Lambda}{}^{M} = \delta_{N}{}^{M} \quad \text{or} \quad E_{\Lambda}{}^{M}E_{M}{}^{\Pi} = \delta_{\Lambda}{}^{\Pi}.$$

Equipped with super-vierbein and spin-connection we define the torsion and curvature tensors as usual

$$[D_{N}, D_{M}\} = T_{NM}{}^{R}D_{R} + \frac{1}{2}R_{NM}{}^{mn}J_{mn}.$$

Using equations (5.18) and (5.19) we find that

$$\begin{aligned} T_{NM}{}^{R} &= E_{M}{}^{\Lambda}\partial_{\Lambda}E_{N}{}^{\Pi}E_{\Pi}{}^{R} + \Omega_{MN}{}^{R} - (-1)^{MN}(M \leftrightarrow N) \\ R_{MN}{}^{rs} &= E_{M}{}^{\Lambda}E_{N}{}^{\Pi}(-1)^{\Lambda(N+\Pi)}\{\partial_{\Lambda}\Omega_{\Pi}{}^{rs} + \Omega_{\Lambda}{}^{rk}\Omega_{\Pi k}{}^{s} - (-1)^{\Lambda\Pi}(\Lambda \leftrightarrow \Pi)\}. \end{aligned}$$

The supergeneral coordinate transformations can be rewritten using these tensors

$$\begin{aligned} \delta E_{\Lambda}{}^{M} &= -E_{\Lambda}{}^{R}\xi^{N}T_{NR}{}^{M} + D_{\Lambda}\xi^{M} \\ \delta\Omega_{\Lambda M}{}^{N} &= E_{\Lambda}{}^{R}\xi^{P}R_{PRM}{}^{N}, \end{aligned}$$

where $\xi^{N} = \xi^{\Lambda}E_{\Lambda}{}^{N}$ and we have discarded a Lorentz transformation.

The torsion and curvature tensors satisfy Bianchi identities which follow from the identity

$$[D_{M}, [D_{N}, D_{R}\}\} - [[D_{M}, D_{N}\}, D_{R}\} + (-1)^{RN}[[D_{M}, D_{R}\}, D_{N}\} = 0.$$

They read

$$
\begin{aligned}
0 \; = \; & I^{(1)}_{RMN}{}^F = [-(-1)^{(M+N)R} D_R T_{MN}{}^F + T_{MN}{}^S T_{SR}{}^F + R_{MNR}{}^F] \\
& + [+(-1)^{MN} D_N T_{MR}{}^F - (-1)^{NR} T_{MR}{}^S T_{SN}{}^F - (-1)^{NR} R_{MRN}{}^F] \\
& + [-D_M T_{NR}{}^F + (-1)^{(N+R)M} T_{NR}{}^S T_{SM}{}^F + (-1)^{(N+R)M} R_{NRM}{}^F]
\end{aligned}
$$

and

$$
\begin{aligned}
& I^{(2)}_{RMN}{}^{mn} \\
& = [(-1)^{(M+N)R} D_R R_{MN}{}^{mn} + T_{MN}{}^S R_{SR}{}^{mn}] \\
& \quad -(-1)^{NR}(R \to N, \; N \to R, \; M \to M \text{ in the first bracket}) \\
& \quad +(-1)^{(N+R)M}(M \to N, \; N \to R, \; R \to M \text{ in the first bracket}) \\
& = 0.
\end{aligned}
$$

It can be shown that if $I^{(1)}_{MNR}{}^F$ holds then $I^{(2)}_{RMN}{}^{mn}$ is automatically satisfied. This result holds in the presence of constraints on $T_{MN}{}^R$ and $R_{MN}{}^{mn}$ and is a consequence of the restricted tangent space choice. We refer to this as *Dragon's theorem*, [71]. For all fermionic indices we find that

$$
\begin{aligned}
I_{ABC}{}^N \; = \; & -D_A T_{BC}{}^N + T_{AB}{}^S T_{SC}{}^N + R_{ABC}{}^N - D_C T_{AB}{}^N + T_{CA}{}^S T_{SB}{}^N \\
& + R_{CAB}{}^N - D_B T_{CA}{}^N + T_{BC}{}^S T_{SA}{}^N + R_{BCA}{}^N \\
= \; & 0,
\end{aligned}
$$

and for fermionic indices with one bosonic index

$$
\begin{aligned}
I_{ABr}{}^N \; = \; & -D_A T_{Br}{}^N + T_{AB}{}^S T_{Sr}{}^N + R_{ABr}{}^N - D_r T_{AB}{}^N + T_{rA}{}^S T_{SB}{}^N \\
& + R_{rAB}{}^N + D_B T_{rA}{}^N - T_{Br}{}^S T_{SA}{}^N - R_{BrA}{}^N \\
= \; & 0,
\end{aligned}
$$

while

$$
\begin{aligned}
I_{Anr}{}^N \; = \; & -D_A T_{nr}{}^N + T_{An}{}^s T_{sr}{}^N + R_{Anr}{}^N - D_r T_{An}{}^N + T_{rA}{}^s T_{sn}{}^N \\
& + R_{rAn}{}^N - D_n T_{rA}{}^N + T_{nr}{}^s T_{sA}{}^N + R_{nrA}{}^N \\
= \; & 0.
\end{aligned}
$$

Clearly one can replace any undotted index by a dotted index and the signs remain the same. We recall that for rigid superspace all the torsions and curvatures vanish except for $T_{A\dot{B}}{}^n = -2i(\sigma^n)_{A\dot{B}}$. Clearly this is inconsistent with an $Osp(4,1)$ tangent space group.

The dimensions of the torsions and curvature can be deduced from the dimensions of D_N. If F and B denote fermionic and bosonic indices respectively then

$$
[D_F] = \frac{1}{2}; \quad [D_B] = 1
$$

and

$$[T_{FF}{}^B] = 0; \quad [T_{FF}{}^F] = [T_{FB}{}^B] = \frac{1}{2}; \quad [T_{FB}{}^F] = [T_{BB}{}^B] = 1; \quad [T_{BB}{}^F] = \frac{3}{2},$$

while

$$[R_{FF}{}^{mn}] = 1; \quad [R_{FB}{}^{mn}] = \frac{3}{2}; \quad [R_{BB}{}^{mn}] = 2.$$

It is useful to consider the notion of the geometric dimension of fields. This is the dimension of the field as it appears in the torsions and curvature. Such expressions never involve κ and as they are nonlinear in certain bosonic fields, such as the vierbein $e_\mu{}^n$, these fields must have zero geometric dimensions. The dimensions of the other fields are determined in relation to $e_\mu{}^n$ to be given by

$$[e_\mu{}^n] = 0; \quad [\psi_\mu{}^\alpha] = \frac{1}{2}; \quad [M] = [N] = [b_\mu] = 1.$$

These dimensions can,for example, be read off the supersymmetry transformations. These dimensions differ from the canonical assignment of dimension by one unit. The difference comes about as we have absorbed factors of κ into the fields.

5.2.2   On-shell derivation of $N = 1$ $D = 4$ superspace supergravity

Having set up the appropriate geometry of superspace we are now in a position to derive on-shell $N = 1$ $D = 4$ supergravity using its superspace setting and solely from a knowledge of the on-shell states of a given spin in the irreducible representation. The result is derived by using dimensional analysis and the Bianchi identities in superspace discussed above. We now illustrate this procedure for $N = 1$ supergravity. It was this method that was used in [111] to find the full equations of motion in superspace and in x-space of IIB supergravity.

The on-shell states are represented by $h_{\mu\nu}$ ($h_{\mu\nu} = h_{\nu\mu}$) and $\psi_\mu{}^\alpha$ which have the gauge transformations

$$\delta h_{\mu\nu} = \partial_\mu\xi_\nu + \partial_\nu\xi_\mu; \quad \delta\psi_{\mu\alpha} = \partial_\mu\eta_\alpha.$$

We have omitted the nonlinear terms, as only the general form is important. The geometric dimension of $h_{\mu\nu}$ is zero while that of $\psi_\mu{}^\alpha$ is $\frac{1}{2}$. The lowest dimension gauge covariant objects are of the form

$$\partial\psi \quad \text{and} \quad \partial\partial h,$$

which have dimensions $\frac{3}{2}$ and 2 respectively.

Consider now the supertorsion and curvature; these objects at $\theta = 0$ must correspond to covariant x-space objects. If there is no such object then the corresponding tensor must vanish at $\theta = 0$ and hence to all orders in θ. The only 0-dimensional tensors are $T_{AB}{}^{n}$ and $T_{A\dot{B}}{}^{n}$. There are no 0-dimensional covariant objects except the numerically invariant tensor $(\sigma^{n})_{A\dot{B}}$. Hence, we must conclude that

$$T_{AB}{}^{n} = 0; \quad T_{A\dot{B}}{}^{n} = c(\sigma^{n})_{A\dot{B}}, \tag{5.20}$$

where c is a constant. We choose $c \neq 0$ in order to agree with rigid superspace. The reality properties of $T_{A\dot{B}}{}^{n}$ imply that c is imaginary and we can normalize it to take the value $c = -2i$.

There are no $\frac{1}{2}$-dimensional covariant tensors in x-space and so

$$T_{A\dot{B}}{}^{\dot{C}} = T_{AB}{}^{C} = T_{Am}{}^{n} = 0. \tag{5.21}$$

There are no 1-dimensional covariant objects in x-space. This would not be the case if one had an independent spin connection, $w_{\mu}{}^{rs}$, for $\partial e + w + \cdots$ would be a covariant quantity. When $w_{\mu}{}^{rs}$ is not an independent quantity it must be given in terms of $e_{\mu}{}^{n}$ and $\psi_{\mu}{}^{\alpha}$ in such a way as to render the above 1-dimensional covariant quantity zero. Hence, for a dependent spin connection, i.e., in second-order formalism, we have

$$T_{nA}{}^{\dot{B}} = T_{nA}{}^{B} = R_{AB}{}^{mn} = 0 = R_{A\dot{B}}{}^{mn}. \tag{5.22}$$

In other words, every 0-, $\frac{1}{2}$-, and 1-dimensional torsion and curvature vanishes with the exception of $T_{A\dot{B}}{}^{n} = -2i(\sigma^{n})_{A\dot{B}}$.

The reader who is familiar with the off-shell constraints for $N = 1$ supergravity can compare them with the on-shell constraints found here. The set of constraints of off-shell supergravity is given in Section 16.2 of reference [0]. We find that that the extra constraints are $T_{nA}{}^{B} = T_{nA}{}^{\dot{B}} = 0 = R_{AB}{}^{AB}$. In terms of the superfields R, $W_{(ABG)}$ and $G_{A\dot{B}}$ this is equivalent to

$$R = G_{A\dot{B}} = 0.$$

Returning to the on-shell theory. The $\frac{3}{2}$-dimensional tensors can involve at $\theta = 0$ the spin-$\frac{3}{2}$ object $\partial \psi$ and so these will not all be zero. The only remaining non-zero tensors are $T_{mn}{}^{A}$, $R_{Ar}{}^{mn}$ and $R_{st}{}^{mn}$ and of course $T_{A\dot{B}}{}^{n} = -2i(\sigma^{n})_{A\dot{B}}$. However, the previous constraints of equations (5.20)–(5.22) are sufficient to specify the entire theory, as we will now demonstrate. The first nontrivial Bianchi identity is $\frac{3}{2}$-dimensional and is

$$\begin{aligned} I_{nB\dot{D}}{}^{\dot{C}} &= -D_{n}T_{B\dot{D}}{}^{\dot{C}} + T_{nB}{}^{F}T_{F\dot{D}}{}^{\dot{C}} + R_{nB\dot{D}}{}^{\dot{C}} + D_{\dot{D}}T_{nB}{}^{\dot{C}} - T_{\dot{D}n}{}^{F}T_{FB}{}^{\dot{C}} \\ &\quad - R_{\dot{D}nB}{}^{\dot{C}} - D_{B}T_{\dot{D}n}{}^{\dot{C}} + T_{B\dot{D}}{}^{\dot{F}}T_{Fn}{}^{\dot{C}} + R_{B\dot{D}n}{}^{\dot{C}} \\ &= 0. \end{aligned}$$

Using the above constraints this reduces

$$-2i(\sigma^m)_{B\dot{D}}T_{mn}{}^{\dot{C}} - R_{nB\dot{D}}{}^{\dot{C}} = 0.$$

Tracing on $\dot{D}$ and $\dot{C}$ then yields

$$(\sigma^m)_{B\dot{D}}T_{mn}{}^{\dot{D}} = 0.$$

This is the Rarita–Schwinger equation as we will demonstrate shortly.

The spin-2 equation must have dimension 2 and is contained in the $I_{Bmn}{}^A$ Bianchi identity.

$$\begin{aligned}
I_{Bmn}{}^A &= -D_B T_{mn}{}^A + T_{Bm}{}^F T_{Fn}{}^A + R_{Bmn}{}^A - D_n T_{Bm}{}^A + T_{nB}{}^F T_{Fm}{}^A \\
&\quad + R_{nBm}{}^A - D_m T_{nB}{}^A + T_{mn}{}^F T_{FB}{}^A + R_{mnB}{}^A \\
&= 0.
\end{aligned}$$

Application of the constraints gives

$$-D_B T_{mn}{}^A + T_{mnB}{}^A = 0 \tag{16.124}$$

On contracting with $(\sigma^m)_{\dot{B}A}$ we find

$$(\sigma^m)_{\dot{B}A} D_B T_{mn}{}^A = 0 = R_{mnB}{}^A (\sigma^m)_{\dot{B}A}.$$

Using the fact that $R_{mnB}{}^A = -\frac{1}{4} R_{mn}{}^{pq} (\sigma_{pq})_B{}^A$ yields the result

$$R_{mn} - \frac{1}{2}\eta_{mn} R = 0$$

or

$$R_{mn} = 0 \quad \text{where} \quad R_{mn} = R_{msn}{}^s. \tag{5.23}$$

We now wish to demonstrate that these are the spin-$\frac{3}{2}$ and spin-2 equations. The $\theta = 0$ components of $E_\mu{}^n$ and $E_\mu{}^A$ are denoted as follows:

$$E_\mu{}^n(\theta = 0) = e_\mu{}^n, \quad E_\mu{}^A(\theta = 0) = \frac{1}{2}\psi_\mu{}^A.$$

At this stage the above equation is simply a definition of the fields $e_\mu{}^n$ and $\psi_\mu{}^A$. The $\theta = 0$ components of $E_A{}^n$ may be gauged away by an appropriate supergeneral coordinate transformation. As

$$\delta E_A{}^n(\theta = 0) = \xi^\Pi \partial_\Pi E_A{}^n|_{\theta=0} + \partial_A \xi^\Pi E_\Pi{}^n|_{\theta=0} = \cdots + \partial_A \xi^\mu e_\mu{}^n + \cdots,$$

we may clearly choose $\partial_A \xi^\mu$ so that $E_A{}^n = 0$. Similarly we may choose

$$E_{\underline{A}}{}^{\dot{B}} = \delta_{\dot{A}}{}^{\dot{B}}, \quad E_{\underline{A}}{}^B = \delta_A{}^B, \quad E_{\underline{A}}{}^{\dot{B}} = 0.$$

To summarize,

$$E_\Pi{}^m(\theta = 0) = \begin{pmatrix} e_\mu{}^n & \frac{1}{2}\psi_\mu{}^A & \frac{1}{2}\psi_\mu{}^{\dot{A}} \\ 0 & \delta_B^A & 0 \\ 0 & 0 & \delta_{\dot{B}}^{\dot{A}} \end{pmatrix}.$$

For the spin connection $\Omega_\Pi{}^m$ we define

$$\Omega_\mu{}^{mn}(\theta = 0) = w_\mu{}^{mn},$$

and we use a Lorentz transformation to gauge

$$\Omega_\alpha{}^{mn}(\theta = 0) = 0.$$

At $\theta = 0$ we then find

$$T_{\mu\nu}{}^{\dot{A}} = -\frac{1}{2}\partial_\mu\psi_\nu{}^{\dot{A}} + \Omega_{\mu\nu}{}^{\dot{A}} - (\mu \leftrightarrow \nu) = -\frac{1}{2}\psi_{\mu\nu}{}^{\dot{A}},$$

where

$$\psi_{\mu\nu}{}^{\dot{A}} \equiv D_\mu\psi_\nu{}^{\dot{A}} - (\mu \leftrightarrow \nu)$$

and

$$D_\mu\psi_\nu{}^{\dot{A}} = \partial_\mu\psi_\nu{}^{\dot{A}} - \psi_\nu{}^{\dot{B}}w_{\mu\dot{B}}{}^{\dot{A}}.$$

Here we have used the results

$$\Omega_{\mu\nu}{}^{\dot{A}} = E_\nu{}^N w_{\mu N}{}^{\dot{A}} = \frac{1}{2}\psi_\nu{}^{\dot{B}}w_{\mu\dot{B}}{}^{\dot{A}}.$$

The torsion with all tangent indices is given in terms of $T_{\mu\nu}{}^{\dot{A}}$ by the relation

$$\begin{aligned} T_{\mu\nu}{}^{\dot{A}}(\theta = 0) &= E_\mu{}^N(\theta = 0)E_\nu{}^M(\theta = 0)T_{NM}^{\dot{A}}(\theta = 0)(-1)^{NM} \\ &= e_\mu{}^n e_\nu{}^m T_{nm}{}^{\dot{A}}(\theta = 0), \end{aligned}$$

where we have used the constraints $T_{Bn}{}^{\dot{A}} = T_{\dot{B}C}{}^{\dot{A}} = 0$. Consequently

$$0 = (\sigma^m)_{A\dot{B}}T_{mn}^{\dot{B}}(\theta = 0) = -\frac{1}{2}(\sigma^m)_{A\dot{B}}e_m{}^\mu e_n{}^\nu\psi_{\mu\nu}{}^{\dot{B}} \tag{5.24}$$

and we recognize the Rarita–Schwinger equation on the right-hand side.

Actually to be strictly rigorous we must also show that $w_\mu{}^{mn}$ is the spin connection given in terms of $e_\mu{}^n$ and $\psi_\mu{}^A$. In fact, this follows from the constraint $T_{nm}{}^r = 0$. We note that

$$T_{\mu\nu}{}^r(\theta = 0) = -\partial_\mu e_\nu{}^r + w_{\mu\nu}{}^r - (\mu \leftrightarrow \nu).$$

However,

$$
\begin{aligned}
T_{\mu\nu}{}^{r}(\theta = 0) &= E_{\mu}{}^{N}(\theta = 0)E_{\nu}{}^{M}(\theta = 0)T_{NM}{}^{r}(\theta = 0)(-1)^{MN} \\
&\quad - \frac{1}{4}\psi_{\mu}{}^{A}\psi_{\nu}{}^{\dot{B}}T_{A\dot{B}}{}^{r}(\theta = 0) - \psi_{\mu}{}^{\dot{B}}\psi_{\nu}{}^{A}T_{\dot{B}A}{}^{r}(\theta = 0) \\
&= +\frac{1}{2}i\psi_{\nu}{}^{\dot{B}}(\sigma^{r})_{A\dot{B}}\psi_{\mu}{}^{A} - (\mu \leftrightarrow \nu).
\end{aligned}
$$

Consequently we find that

$$
w_{\mu n}{}^{m}e_{\nu}{}^{n} - \partial_{\nu}e_{\mu}{}^{m} - (\mu \leftrightarrow \nu) = +\frac{i}{2}\psi_{\nu}{}^{\dot{B}}(\sigma^{m})_{A\dot{B}}\psi_{\mu}{}^{A} - (\mu \leftrightarrow \nu),
$$

which can be solved in the usual way to yield the correct expression for $w_{\mu n}{}^{m}$.

The spin-2 equation is handled in the same way:

$$
R_{\mu\nu}{}^{mn}(\theta = 0) = \partial_{\mu}w_{\nu}{}^{mn} + w_{\mu}{}^{mr}w_{\nu r}{}^{n} - (\mu \leftrightarrow \nu).
$$

However

$$
\begin{aligned}
R_{\mu\nu}{}^{nm}&(\theta = 0) \\
&= E_{\mu}{}^{N}(\theta = 0)E_{\nu}{}^{M}(\theta = 0)R_{NM}{}^{mn}(\theta = 0)(-1)^{mN} \\
&= e_{\mu}{}^{p}e_{\nu}{}^{q}R_{pq}{}^{nm}(\theta = 0) \\
&\quad + \frac{1}{2}(\psi_{\mu}{}^{A}e_{\nu}{}^{p}R_{\dot{A}p}{}^{nm} + \psi_{\mu}{}^{A}e_{\nu}{}^{p}R_{Ap}{}^{nm}(\theta = 0) - (\mu \leftrightarrow \nu)). \quad (5.25)
\end{aligned}
$$

The object $R_{Ap}{}^{nm}$ can be found from the Bianchi identity $I_{Anr}{}^{s}$

$$
\begin{aligned}
0 = I_{Anr}{}^{s} &= -D_{A}T_{nr}{}^{s} + T_{An}{}^{F}T_{Fr}{}^{s} + R_{Anr}{}^{s} - D_{r}T_{An}{}^{s} + T_{rA}{}^{F}T_{Fn}{}^{s} \\
&\quad + R_{rAn}{}^{s} - D_{n}T_{rA}{}^{s} + T_{nr}{}^{F}T_{FA}{}^{s} + R_{nrA}{}^{s}.
\end{aligned}
$$

Using the constraints we find that

$$
R_{Anr}{}^{s} + R_{rAn}{}^{s} = +2iT_{nr}{}^{\dot{B}}(\sigma^{s})_{A\dot{B}}.
$$

From equation (5.24) we find that

$$
R_{Anr}{}^{s} + R_{rAn}{}^{s} = -i(\sigma^{s})_{A\dot{B}}e_{n}{}^{\mu}e_{r}{}^{\nu}\psi_{\mu\nu}{}^{\dot{B}}. \quad (5.26)
$$

Contracting equation (5.25) with e_{m}^{ν} we find

$$
e_{m}^{\nu}R_{\mu\nu}{}^{nm}(\theta = 0) \equiv R_{\mu}{}^{n} = e_{\mu}{}^{p}R_{pm}{}^{nm} + \frac{1}{2}(\psi_{\mu}{}^{A}R_{Am}{}^{nm} + \psi_{\mu}{}^{A}R_{\dot{A}m}{}^{nm}).
$$

Equation (5.26) then gives

$$
e_{\mu}{}^{p}R_{pm}{}^{nm} = R_{\mu}{}^{n} - \left(\frac{i}{2}\psi_{\mu}{}^{A}(\sigma^{m})_{A\dot{B}}e_{m}^{\lambda}e^{n\tau}\psi_{\lambda\tau}{}^{\dot{B}} + \text{h.c.}\right).
$$

Equation (5.23), $R_{mn} = 0$, then yields the spin-2 equation of $N = 1$ supergravity, which is the left-hand side of the above equation.

At first sight it appears that the task is not finished; one should also analyze all the remaining Bianchi identities and show that they do not lead to any inconsistencies. However, it can be shown that the other Bianchi identities are now automatically satisfied (see [71], [82]).

5.3 Gauging of space-time groups

It has been know for many years that Einstein's theory of general relativity contains a local Lorentz symmetry. When the action is expressed in first-order formalism the spin-connection is the gauge field and the Riemann tensor the field strength for Lorentz group, [114]. It was only in reference [18] that a space-time group was gauged and general relativity theory was deduced from a gauge theory viewpoint. In fact, reference [18] gauged the super Poincaré group and deduced the supergravity theory from this view point. It is straightforward to restrict the calculation to that for the Poincaré group and deduce just general relativity. Of course at that time supergravity had been constructed ([14], [15]), but the gauging procedure provided the first analytical proof (see [18]) of its invariance under local supersymmetry. The theory of supergravity with a cosmological constant by gauging the the super de Sitter group was independently found in reference [113].

Let us consider gauging the $N = 1$ super Poincaré group; corresponding to the generators J_{ab}, P_a and Q^α we introduce the gauge fields w_μ^{ab}, e_μ^a and $Q_{\mu\alpha}$ which will become the spin-connection, the vierbein and gravitino. It is straightforward to calculate the field strengths $R_{\mu\nu}^{ab}$, $C_{\mu\nu}^a$ and $\phi_{\mu\nu\alpha}$ and the gauge transformations of the gauge fields. Supergravity is not the gauge theory of the super Poincaré group in an obvious way and we must proceed by setting the field strength $C_{\mu\nu}^a$ associated to translations to zero. We now construct an action to the appropriate order in derivatives that is invariant under the gauge transformations of the super Poincaré group subject to the constraint $C_{\mu\nu}^a$. In particular, we start from the most general action which is first-order in the field strengths i.e.

$$\int d^4x \; \epsilon^{\mu\nu\rho\lambda} (\epsilon_{abcd} e_\mu^a e_\nu^b R_{\rho\lambda}^{cd} + i f \bar\psi_\mu \gamma_5 \gamma_\nu \phi_{\rho\lambda}).$$

The constant f is readily fixed by demanding the invariance of this action. Hence, we rapidly arrive at the supergravity action. In fact the constraint $C_{\mu\nu}^a = 0$ is just that required to correctly express the spin-connection in terms of the vierbein and gravitino, that is to go from first- to second-order formalism. The constraint is also just that required to convert gauge transformations associated with translations into general coordinate trasformations.

When carrying out the variation of the action subject to the constraint the variation of the spin-connection is irrelevant, since its variation multiplies its equation of motion which vanishes due to the constraint $C_{\mu\nu}^a = 0$. However, this constraint is none other than the condition for expressing the spin-connection in terms of the vierbein, that is the transition to second-order formalism. Thus the invariance of the above action subject to the constraint provides an analytical proof of the invariance of supergravity under local supersymmetry; see [18], [19]. This way of proceeding became known as the 1.5-order formalism and it is reviewed in Chapter 10 of reference [0].

The construction of the theories of conformal supergravity were carried out using this gauging method ([118]). The key to getting the gauge method to work is to guess the appropriate constraints. However, since these constraints break the original gauge transformations they are not always easy to find. Much effort has been devoted to developing the method discussed into something systematic. One such work was that of reference [115] where the full gauge symmetry was realised, but was spontaneously broken.

The gauge techninque has not been used to construct the theories of supergravity in 10 and 11 dimensions and it may be instructive to derive them using this method. It cannot be a coincidence that gravity and supergravity admit such simple formulations as a gauge theories and this connection suggests that there is something to be understood at a deeper level.

6   11-dimensional Supergravity

11 is the highest number of dimensions in which a supergravity theory can exist; see [105]. In this section, we give the 11-dimensional supergravity theory and describe its properties. 11-dimensional supergravity is thought to be the low-energy effective action of a new kind of theory called M-theory which is believed to underlie string theory. Little is known about M-theory apart from its relation to 11-dimensional supergravity.

The non-trivial representation of the Clifford algebra in 11 dimensions is inherited from that in 10, and so has dimension $2^{\frac{10}{2}} = 32$. We also inherit the properties $\epsilon = 1$ and so $B^T = B$ and $C = -C^T$ which were discussed in Section 1. The resulting properties of the γ matrices are given in equations (1.17) and (1.19).

11-dimensional supergravity is based on the $D = 11$ supersymmetry algebra with Majorana spinor Q_α which has 32 real components. As we discussed in Chapter 2 of [0] the algebra takes the form (see [117])

$$\{Q_\alpha, Q_\beta\} = (\gamma^m C^{-1})_{\alpha\beta} P_m + (\Gamma^{mn} C^{-1})_{\alpha\beta} Z_{mn} + (\Gamma^{mnpqr} C^{-1})_{\alpha\beta} Z_{mnpqr},$$
$$(6.1)$$

where P_m is the translation operator and Z_{mn} and Z_{mnpqr} are central charges. Although these play little apparent role in the construction of the supergravity theory they are very important in M-theory.

By a supergravity theory we mean a theory with spins 2 and less. The observation at the start of this section follows from the study of the massless irreducible representations of supersymmetry, which can be deduced in a straightforward way from the relevant supersymmetry algebras. The irreducible representations of the 4-dimensional supersymmetry algebras were given in Section 3. We found that the maximal supergravity theory in 4 dimensions corresponded to $N = 8$ supersymmetry. This is the algebra with 8

Majorana supercharges, each of which has 4 real components. In fact, this theory can be obtained the 11-dimensional supergravity theory by dimensional reduction.

We now explain this result and show that it implies that a supergravity theory can live in at most 11 dimensions. The 4-dimensional result follows in an obvious way from the fact that in the massless case from each 4-component supercharge only 2 of the components act non-trivially on the physical states. Furthermore, these 2 components form a Clifford algebra, one of which raises, and one of which lowers the helicity by $\frac{1}{2}$. Choosing the supercharges that raises the helicity to annihilate the vacuum, the physical states are given by the action of the remaining supercharges. If the supersymmetry algebra has N Majorana supercharges, the physical states are given by the action of N creation operators each of which lowers the helicity by $\frac{1}{2}$. Consequently, if we take the vacuum to have helicity 2, the lowest helicity state in the representation will be $2 - \frac{N}{2}$. To have a supergravity theory we cannot have less than helicity -2 and hence the limit $N \leq 8$. Given a supergravity theory in a dimension greater than 4 we can reduce it in a trivial way, by taking all the fields to be independent of the extra dimensions, to find a supergravity theory in 4 dimensions. However, the number of supercharges is unchanged in the reduction and so the maximal number is $4 \times 8 = 32$. Hence the supergravity in the higher dimension must arise in a dimension whose spinor representation has dimension 32 or less. Thus we find the desired result; 11 dimensions is the highest dimension in which a supergravity theory can exist. It also follows that any supergravity theory must have 32, or fewer supercharges and that the maximal, or largest, supergravity theory in a given dimension has 32 supercharges.

The irreducible representation, or particle content, of 11-dimensional supergravity was found in reference [105] by analysing the irreducible representations of the supersymmetry algebra of equation (6.1). One could also deduce it by requiring that it reduce to the irreducible representation of the 4-dimensional $N = 8$ supergravity theory given in Section 3. We now give a more intuitive argument for the particle content.

11-dimensional supergravity must be invariant under general coordinate transformations (i.e. local translations) and local supersymmetry transformations. To achieve these symmetries it must possess the equivalent 'local gauge' fields, the vielbein e_μ^a and the gravitino $\psi_{\mu\alpha}$. The latter transforms as $\delta\psi_{\mu\alpha} = \partial_\mu\eta_\alpha + \ldots$ and so must be the same type of spinor as the supersymmetry parameter which in this case is a Majorana spinor.

For future use we now give the on-shell count of degrees of freedom of the graviton and gravitino in D dimensions. The relevant bosonic part of the little group which classifies the irreducible representation is $SO(D-2)$. The graviton encoded in e_μ^a is a second-rank symmetric traceless tensor of

$SO(D-2)$ and as such has $\frac{1}{2}(D-2)(D-1)-1$ degrees of freedom on-shell. The gravitino has $(D-3)cr$ real components. Here c is the dimension of the Clifford algebra in dimension $D-2$ and so is given by $c = 2^{\frac{D}{2}-1}$ if D is even and $c = 2^{\frac{(D-1)}{2}-1}$ if D is odd. The quantity r is 2, 1 or $\frac{1}{2}$ if $\psi_{\mu\alpha}$ is a Dirac, Majorana or Majorana–Weyl spinor respectively. In terms of little group representations, the gravitino is a vector spinor ϕ_i, $i = 1,\ldots,D-2$, which is γ-traceless $\gamma^i\phi_i = 0$,. A general vector spinor in $D-2$ dimensions has $(D-2)cr$ components, but the γ-trace subtracts another spinor's worth of components (i.e. cr).

For 11 dimensions we find that the graviton and gravitino have 44 and 128 degrees of freedom on-shell. However, in any supermultiplet of on-shell physical states the fermionic and bosonic degrees of freedom must be equal. Assuming that there are no further fermionic degrees of freedom we require another 84 bosonic on-shell degrees of freedom. If we take these to belong to an irreducible representation of $SO(9)$ then the unique solution would be a third-rank anti-symmetric tensor. This can only arise from a third-rank gauge field $A_{\mu_1\mu_2\mu_3}$ whose fourth-rank gauge field $F_{\mu_1\mu_2\mu_3\mu_4} \equiv 4\partial_{[\mu_1}A_{\mu_2\mu_3\mu_4]}$, the anti-symmetry being with weight 1. We note that in D dimensions a rank p anti-symmetric gauge field belongs to the anti-symmetric rank p tensor representation of $SO(D-2)$ and so has $\frac{(D-2)\ldots(D-p-1)}{p!}$ degrees of freedom on-shell.

The 11-dimensional supergravity Lagrangian was constructed in reference [106] and is given by

$$L = -\frac{e}{4\kappa^2}R(\Omega(e,\psi)) - \frac{e}{48}F_{\mu_1\ldots\mu_4}F^{\mu_1\ldots\mu_4} - \frac{ie}{2}\bar{\psi}_\mu\Gamma^{\mu\nu\varrho}D_\nu\left(\frac{1}{2}(\Omega+\hat{\Omega})\right)\psi_\varrho$$
$$+\frac{1}{192}e\kappa(\bar{\psi}_{\mu_1}\Gamma^{\mu_1\ldots\mu_6}\psi_{\mu_2} + 12\bar{\psi}^{\mu_3}\Gamma^{\mu_4\mu_5}\psi^{\mu_6})(F_{\mu_3\ldots\mu_6} + \hat{F}_{\mu_3\ldots\mu_6})$$
$$+\frac{2\kappa}{(12)^4}\epsilon^{\mu_1\ldots\mu_{11}}F_{\mu_1\ldots\mu_4}F_{\mu_5\ldots\mu_8}A_{\mu_9\mu_{10}\mu_{11}} \qquad (6.2)$$

where

$$F_{\mu_1\ldots\mu_4} = 4\partial_{[\mu_1}A_{\mu_2\mu_3\mu_4]}, \quad \hat{F}_{\mu_1\ldots\mu_4} = F_{\mu_1\ldots\mu_4} - 3\bar{\psi}_{[\mu_1}\Gamma_{\mu_2\mu_3}\psi_{\mu_4]},$$

and

$$\Omega_{\mu mn} = \hat{\Omega}_{\mu mn} - \frac{i}{4}\bar{\psi}_\nu\Gamma_{\mu mn}{}^{\nu\lambda}\psi_\lambda,$$

$$\hat{\Omega}_{\mu mn} = \Omega^0_{\mu mn}(e) + \frac{i}{2}(\bar{\psi}_\nu\Gamma_n\psi_m - \bar{\psi}_\nu\Gamma_m\psi_n + \bar{\psi}_n\Gamma_\mu\psi_m).$$

The symbol $\Omega^0_{\mu mn}(e)$ is the usual expression for the spin-connection in terms of the vielbein e^n_μ which can be found in Section 5.

It is invariant under the local supersymmetry transformations

$$
\left.\begin{aligned}
\delta e_\mu^m &= -i\kappa\bar{\epsilon}\Gamma^m\psi_\mu \\
\delta\psi_\mu &= \frac{1}{\kappa}D_\mu(\hat{\Omega})\epsilon + \frac{i}{12^2}(\Gamma_\mu{}^{\nu_1...\nu_4} - 8\delta_\mu^{\nu_1}\Gamma^{\nu_2\nu_3\nu_4})F_{\nu_1...\nu_4}\epsilon \\
\delta A_{\mu_1\mu_2\mu_3} &= \frac{3}{2}\bar{\epsilon}\Gamma_{[\mu_1\mu_2}\psi_{\mu_3]}.
\end{aligned}\right\} \qquad (6.3)
$$

Although the result may at first sight look complicated most of the terms can be understood if one were to consider constructing the action using the Noether method. In this method, which we discussed in Section 5, we start from the linearized theory for the graviton, the gravitino and the gauge field $A_{\mu_1\mu_2\mu_3}$. The linearized action is bilinear in the fields and is invariant under a set of rigid supersymmetry transformations which are linear in the fields as well as the local Abelian transformations $\delta\psi_{\mu\alpha} = \partial_\mu\eta_\alpha$, $A_{\mu_1\mu_2\mu_3} = \partial_{[\mu_1}\Lambda_{\mu_2\mu_3]}$ as well as an appropriate analogous transformation for the linearised vielbein. The linearised supersymmetry transformations are found by using dimensional analysis and closure of the linearised supersymmetry algebra. We now let the rigid supersymmetry parameter ϵ depend on space-time and identify the two spinor parameters by $\eta_\alpha = \frac{1}{\kappa}\epsilon_\alpha$. We know that the final result will be invariant under general coordinate transformations and so we may at each step in the Noether procedure insert the vielbein in all terms so as to ensure this invariance. Even at this stage in the Noether procedure we recover all the terms in the transformations in the fields given above in equation (6.3) except for some of the terms in the spin-connection $\hat{\Omega}$. For the action, we find all the terms except the last two terms and again some terms in the spin-connection of the gravitino. The action at this stage is not invariant under the now local supersymmetry transformations and as explained in Section 5 we can gain invariance at order κ^0 by adding a term of the form $\bar{\psi}^{\mu\alpha}j_{\mu\alpha}$ where $j_{\mu\alpha}$ is the Noether current for the supersymmetry of the linearised theory. This term is none other than the penultimate term in the above action. To gain invariance to order κ^1 we must cancel the variations of this penultimate term under the supersymmetry transformation. This is achieved if we add the last term to the action. Hence even at this stage in the procedure we have accounted for essentially all the terms in the action and transformation laws. While one can pursue the Noether procedure to the end, to find the final form of the action and transformations laws, it is perhaps best to guess the final form of the connection and verify that the action is invariant and the local supersymmetry transformations close.

The 11-dimensional action contains only one coupling constant, Newton's constant κ, which has the dimensions of mass$^{-\frac{9}{2}}$ and so defines a Planck mass m_p by $\kappa = m_p^{-\frac{9}{2}}$. We note that there are no scalars in the action whose expectation value could be used to define another coupling constant. If we scale the fields by $\psi_\mu \to \kappa^{-1}\psi_\mu$ and $A_{\mu_1\mu_2\mu_3} \to \kappa^{-1}A_{\mu_1\mu_2\mu_3}$ we find that all factors of κ drop out of the action except for a prefactor of κ^{-2} and all factors of κ drop out of the supersymmetry transformation laws. As such, when expressed in terms of these variables, the classical field equations do not contain κ. In fact, the value of the coupling constant κ has no physical meaning. One way to see this fact is to observe that if, after carrying out the above redefinitions, we Weyl-scale the fields by

$$e_\mu^m \to e^{-\alpha}e_\mu^m, \quad \psi_\mu \to e^{-\frac{\alpha}{2}}\psi_\mu \quad \text{and} \quad A_{\mu_1\mu_2\mu_3} \to e^{-3\alpha}A_{\mu_1\mu_2\mu_3} \qquad (6.4)$$

as well as scale the coupling constant by $\kappa \to e^{-\frac{9\alpha}{2}}\kappa$; we then find that the action is invariant. In deriving this result we used the equation

$$R(e^\tau e_\mu^m) = e^{-2\tau}(R(e_\mu^m) + 2(D-1)D^\mu D_\mu\tau + (D-1)(D-2)g^{\mu\nu}\partial_\mu\tau\partial_\nu\tau),$$

where R is the Ricci tensor in dimension D. Of course this is not a symmetry of the action in the usual sense as we have rescaled the coupling constant. However, as the coupling constant only occurs as a prefactor multiplying the entire action, it is a symmetry of the classical equations of motion. Hence, we can only specify the value of the constant κ with respect to a particular metric. These transformations are not a symmetry of the quantum theory where, in the path integral, the prefactor of κ^{-2} which multiplies all terms in the action becomes $\kappa^{-2}\hbar$ where $\hbar$ is Planck's constant. In this case we can absorb the rescaling either in κ or $\hbar$. The above Weyl-scaling of the vielbein implies that the proper distance d^2s scales as $d^2s \to e^{-\alpha}d^2s$. Taking the scaling to be absorbed by $\hbar$ we find that $\hbar \to e^{-9\alpha}\hbar$ and so small proper distance corresponds to small $\hbar$. Put another small κ, or equivalently $\hbar$, is the same as working at small proper distance.

Although we have constructed the supergravity theories in 10 and 11 dimensions in this section we have omitted many of the significant formulae, such as the supersymmetry transformations. Since these are contained in the original papers, [105], [106], [107], [108], [109], on the subject, we have used the same metric as in these papers, that is the tangent space metric $\eta_{mn} = \text{diag}(+1, -1, \ldots, -1)$. Since many practitioners nowadays prefer the other signature we now give the rules to change to the tangent space metric, which is mainly positive, i.e. $\eta_{nm} = \text{diag}(-1, 1, \ldots, 1)$. To go to the latter metric we must take $\eta_{nm} \to -\eta_{nm}$, $\gamma^a \to i\gamma^a$, $e_\mu^n \to e_\mu^n$. Using this rule it is easy to carry out the change. One finds, for example, that $g^{\mu\nu}\partial_\mu\sigma\partial_\nu\sigma \to -g^{\mu\nu}\partial_\mu\sigma\partial_\nu\sigma$ and $R \to -R$.

7   IIA and IIB Supergravity

In this section we give the supergravity theories in 10 dimensions which have the maximal supersymmetry. There are two such theories, called IIA and

IIB, and they are the effective low-energy actions for the IIA and IIB string theories respectively. We describe the properties of these supergravity theories that are relevant for the string theories and play an important role in string duality.

7.1 Supergravity theories in 10 dimensions

In 10 dimensions the non-trivial representation of the Clifford algebra has dimension $2^{\frac{10}{2}} = 32$. As we found in Section 1, the matrices B and C, associated with the complex conjugation and transpose of the γ-matrices, obey the properties $B^T = B$ and $C = -C^T$ (i.e. $\epsilon = 1$). The properties of the γ matrices are given in equations (1.10) and (1.11). In 10 dimensions a Majorana spinor has 32 real components; however, we can also have Majorana–Weyl spinors and these only have 16 real components.

The supersymmetry algebra for a single Majorana–Weyl supercharge which has 16 real components is given in equation (1.21). There are two supersymmetric theories that are based on this algebra: the so-called $N = 1$ Yang–Mills theory ([119]) and the $N = 1$ supergravity theory ([120]) which is more often called type I supergravity. The coupling between the two theories was given in references [121].

In the discussion at the beginning of Section 6 we found that a supergravity theory can be based on a supersymmetry algebra with 32 or fewer supercharges. If we consider the supersymmetry algebra with a 32-component Majorana spinor we find the IIA supergravity theory which was constructed in references [107], [108] and [109]. Clearly, when we decomposed the Majorana spinor into Majorana–Weyl spinors we get two such spinors – one of each chirality. The other 10-dimensional supersymmetry algebra with 32 supercharges has two Majorana–Weyl spinors of the same chirality and the corresponding supergravity is IIB supergravity. This theory was constructed in references [110], [112] and [111]. Unlike the other supergravity theories in 10 dimensions IIB supergravity has an internal symmetry which is the group $SL(2, \mathbf{R})$; see [110].

Upon reduction of 11-dimensional supergravity to 10 dimensions by taking the 11th dimension to be a circle, we will obtain a 10-dimensional theory that possess a supersymmetry algebra based on a 32-component Majorana supercharge which decomposes into two Majorana–Weyl spinors of opposite Weyl chiralities. Thus the resulting theory can only be IIA supergravity. Indeed, this was how IIA supergravity was found; see [107], [108] and [109].

The importance of the IIA and IIB supergravity theories, which was the main motivation for their construction, is that they are the low-energy effective theories of the corresponding IIA and IIB closed string theories in 10 dimensions. Type I supergravity coupled to $N = 1$ Yang–Mills theory is the effective action for the low-energy limit of the $E_8 \otimes E_8$ or $SO(32)$ heterotic string. All these supergravity theories were constructed at a time when string theory was deeply unpopular, and when it did become fashionable, little interest was taken in supergravity theories. However, they now form the basis for many of the discussions of duality in string theories.

7.2 IIA supergravity

As we have mentioned, this theory is based on a supersymmetry algebra which contains one Majorana spinor Q_α, $\alpha = 1, \ldots, 32$. Following the discussion of Section 2, we conclude that the anti-commutator of two supersymmetry generators can have central charges of rank p where $p = 1, 2$; mod 4 and so the corresponding anti-commutator is given by (see [117])

$$
\begin{aligned}
\{Q_\alpha, \, Q_\beta\} \\
= (\gamma^m C^{-1})_{\alpha\beta} P_m + (\gamma^{mn} C^{-1})_{\alpha\beta} Z_{mn} + (\gamma^{m_1 \ldots m_5} C^{-1})_{\alpha\beta} Z_{m_1 \ldots m_5} \\
+ (\gamma^{m_1 \ldots m_4} \gamma_{11} C^{-1})_{\alpha\beta} Z_{m_1 \ldots m_4} + (\gamma^m \gamma_{11} C^{-1})_{\alpha\beta} Z_m + (\gamma_{11} C^{-1})_{\alpha\beta} Z
\end{aligned}
$$

We need only take $p \le 5$ since we may use

$$
\gamma^{m_1 \ldots m_s} \gamma_{11} = \frac{1}{(10 - s - 1)!} \eta \epsilon^{m_1 \ldots m_s m_{s+1} \ldots m_{10}} \gamma_{m_{s+1} \ldots m_{10}},
$$

where $\eta = \pm 1$, to eliminate terms with $p \ge 6$.

The IIA theory was obtained in references [107], [108] and [109] from the $D = 11$ theory by compactification and this is the method of construction we now follow. We consider the 11-dimensional supergravity of equation (6.2) and take the 11th dimension to be a circle S^1 of radius R. To be precise, we take the 11th coordinate x^{10} to be such that $x^{10'} \sim x^{10}$ if $x^{10'} = x^{10} + 2\pi n R$, $n \in \mathbf{Z}$ where $2\pi R$ parametrizes the range of x^{10}. We will also write x^{10} as $x^{10} = \theta R$ for $0 \le \theta < 2\pi$. We adopt the convention that hatted indices run over all 11-dimensional indices, but unhatted indices only run over the 10-dimensional indices, for example $\hat\mu, \hat\nu = 0, 1, \ldots, 10$ while $\mu, \nu = 0, 1, \ldots, 9$. Given any of the fields in the 11-dimensional supergravity we can take its Fourier transform on x^{10}. In particular, if ϕ represents any of these fields whose Lorentz and possible spinor indices are suppressed, we find that

$$
\phi(x^\mu, x^{10}) = \phi(x^\mu) + \sum_{n, n \ne 0} e^{in\theta} \phi_n(x^\mu). \tag{7.1}
$$

Thus from each particle in 10 dimensions we find an infinite number of particles in 10 dimensions. The non-zero modes (i.e. $n \ne 0$), however, will lead to massive particles whose masses are given by their momentum in the 11th direction. Such massive particles are called *Kaluza–Klein* particles. In the limit when the radius of the circle is large these particles become infinitely massive and can be neglected, whereupon one is left with a finite set of massless particles which form a supergravity theory. Discarding the massive particles can be achieved by taking all the 11-dimensional fields to be independent of x^{10} and this we now do.

This reduction proceeds in the following generic manner

$$
\begin{array}{cccc}
D = 11 & e_{\hat{\mu}}{}^{\hat{m}} & A_{\hat{\mu}_1 \hat{\mu}_2 \hat{\mu}_3} & \psi_{\hat{\mu}\alpha} \\
 & \downarrow & \downarrow & \downarrow \\
D = 10 & e_\mu{}^m, B_\mu, \phi & A_{\mu_1 \mu_2 \mu_3}, A_{\mu_1 \mu_2} & \psi_{\mu\alpha}, \lambda_\alpha.
\end{array}
\tag{7.2}
$$

While one can reduce the fields in many ways only some definitions of the 10-dimensional fields will lead to a final result which is in the generic form in which a supergravity theory is usually written. The three-form gauge field reduces in an obvious way $A_{\mu_1 \mu_2 \mu_3} = A_{\mu_1 \mu_2 \mu_3}$, $A_{\mu_1 \mu_2} = A_{\mu_1 \mu_2 10}$ where $\mu_1, \mu_2 \mu_3 = 0, 1, \ldots, 9$. However, the useful reductions for the other fields are more subtle; the vielbein takes the form

$$
e_{\hat{\mu}}{}^{\hat{m}} = \begin{pmatrix} e^{-\frac{1}{12}\sigma} e_\mu{}^m & 2e^{\frac{2}{3}\sigma} B_\mu \\ 0 & e^{\frac{2}{3}\sigma} \end{pmatrix}, \quad (e^{-1})_{\hat{m}}{}^{\hat{\mu}} = \begin{pmatrix} e^{\frac{1}{12}\sigma} e_m{}^\mu & -2e^{\frac{1}{12}\sigma} B_\nu (e^{-1})_m{}^\nu \\ 0 & e^{-\frac{2}{3}\sigma} \end{pmatrix}
\tag{7.3}
$$

while the 11-dimensional gravitino becomes

$$
\psi_{\hat{m}} = (e^{-\frac{1}{24}\sigma} e_m{}^\mu \psi'_\mu, \frac{2}{3}\sqrt{2} e^{\frac{17}{24}\sigma} \lambda),
$$

where

$$
\psi'_\mu = e^{-\frac{1}{24}\sigma} \left(\psi_\mu - \sqrt{\frac{1}{72}} \Gamma_\mu \Gamma^{11} \lambda \right) - \sqrt{\frac{32}{9}} e^{\frac{3}{4}\sigma} B_\mu \lambda \quad \text{and} \quad \Gamma^{11} = i\Gamma^1 \cdots \Gamma^{10}.
$$

The above formulae and those below are related to those of reference [107] by carrying out the transformation $\sigma \to \frac{2}{3}\sigma$ on the latter.

The field B_μ is a gauge field whose gauge transformation has a parameter that came from the general coordinate transformations with parameter ξ^{10} in the 11-dimensional theory. The component $e_{\hat{\mu}=10}{}^m$ of the vielbein can be chosen to be zero as a result of a local Lorentz transformation $w^m{}_{10}$. The strange factors involving e^σ and other redefinitions are required in order to get the usual Einstein and spinor kinetic energy terms. The 10-dimensional Newtonian coupling constant κ is given in terms of the 11-dimensional Newtonian constant κ_{11} by $\kappa^2 = \frac{(\kappa_{11})^2}{2\pi R}$, and has the dimensions of $(\text{mass})^{-8}$. This equation follows from examining the κ that results from the dimensional reduction. We have set $\kappa = 1$ in this section.

The resulting 10-dimensional IIA supergravity theory is given by

$$
L = L^B + L^F,
$$

where the first term contains all the terms which are independent of the fermions and the second term is the remainder. The bosonic part is given by (see [107], [108], [109])

$$
\begin{aligned}
L^B &= -eR(w(e)) - \frac{1}{12}ee^{\frac{5}{2}}F'_{\mu_1\ldots\mu_4}F'^{\mu_1\ldots\mu_4} + \frac{1}{3}ee^{-\sigma}F_{\mu_1\ldots\mu_3}F^{\mu_1\ldots\mu_3} \\
&\quad -ee^{\frac{3}{2}\sigma}F_{\mu_1\mu_2}F^{\mu_1\mu_2} + \frac{1}{2}\partial_\mu\sigma\partial^\mu\sigma + \frac{1}{2\cdot(12)^2}\epsilon^{\mu_1\ldots\mu_{10}}F_{\mu_1\ldots\mu_4}F_{\mu_5\ldots\mu_8}A_{\mu_9\mu_{10}}(7.4)
\end{aligned}
$$

where

$$
F_{\mu_1\mu_2} = 2\partial_{[\mu_1}B_{\mu_2]} \tag{7.5}
$$

$$
F_{\mu_1\mu_2\mu_3} = 3\partial_{[\mu_1}A_{\mu_2\mu_3]} \tag{7.6}
$$

$$
F'_{\mu_1\ldots\mu_4} = 4\partial_{[\mu_1}A'_{\ldots\mu_4]} + 12A_{[\mu_1\mu_2}G_{\mu_3\mu_4]}. \tag{7.7}
$$

In the last definition we have used the field $A'_{\mu_1\mu-2\mu_3} = A_{\mu_1\mu-2\mu_3} - 6B_{[\mu_1}A_{\mu_2\mu_3]}$ which is invariant under the gauge transformation with parameter ξ^{10}. The fermionic part of the Lagrangian is much more complicated and the first two terms are

$$
L^F = -\frac{i}{2}e\psi_{\mu_1}\Gamma^{\mu_1\mu_2\mu_3}D_{\mu_2}\psi_{\mu_3} + \frac{i}{2}e\bar{\lambda}\Gamma^\mu D_\mu\lambda + \cdots .
$$

The transformations of the fields can be deduced in a similar fashion from the transformation of 11-dimensional fields of equation (6.1). For example, the vielbein and dilaton transform as

$$
\delta e_\mu{}^m = -i\bar{\epsilon}\Gamma^m\psi_\mu, \quad \delta\sigma = \sqrt{2}i\bar{\lambda}\Gamma^{11}\epsilon,
$$

where ϵ is the suitably defined parameter of local supersymmetry transformations. We refer the reader to references [107], [108] and [109] for the transformations of the other fields and the terms in the fermionic part of the action.

The IIA action has an $SO(1,1)$ invariance with parameter c that transforms the fields as

$$
\sigma \to \sigma + c, \quad B_\mu \to e^{-\frac{3}{4}c}B_\mu, \quad A_{\mu\nu} \to e^{\frac{1}{2}c}A_{\mu\nu}, \quad A_{\mu\nu\rho} \to e^{-\frac{1}{4}c}A_{\mu\nu\rho},
$$

while the 10-dimensional vielbein is inert. This symmetry has its origin in the 11-dimensional theory and in particular the Weyl-scalings of equation (6.4) given in the previous section. Although these are not a symmetry of the 11-dimensional action, we can convert them into a symmetry of the 10-dimensional action provided we combine them with a diffeomorphism on x^{10}, in particular the diffeomorphism $x^{10} \to e^{-9\alpha}x^{10}$. To keep the range of x^{10} the same we also scale R by $R \to e^{-9\alpha}R$. This diffeomorphism is a symmetry of the 11-dimensional theory and from the active viewpoint transforms $\int dx^{10} \to$ $e^{-9\alpha} \int dx^{10}$ and the 11-dimensional Lagrangian L as $L \to e^{9\alpha} L$. The 10-dimensional theory is obtained by substituting the field expansion of equation (7.1) into the 11-dimensional action. However, the effect of the $\int dx^{10}$ is just to extract the part of the 11-dimensional Lagrangian L that is independent of x^{10}, which then becomes the 10-dimensional Lagrangian and gives a factor of $2\pi R$. The latter is then combined with the factor κ_{11}^{-2} to define the 10-dimensional Newtonian coupling constant $\kappa_{10}^{-2} = 2\pi R\kappa_{11}^{-2}$. Clearly, the 10-dimensional coupling constant κ_{10} is inert under the combined transformation as $R \to e^{-9\alpha} R$ under the diffeomorphism, and $\kappa_{11}^{-2} \to e^{9\alpha} \kappa_{11}^{-2}$ under the Weyl-scaling. Similarly the 10-dimensional action in inert as it scales by $e^{9\alpha}$ under the diffeomorphism and $e^{-9\alpha}$ under the Weyl-scaling. Thus we have found that dimensional reduction has transformed a symmetry of the equations of motion into a symmetry of the action.

An alternative way of writing the IIA Lagrangian is to use the so-called string metric $g_{\mu\nu}^s$. This metric is the one that occurs in the sigma model approach to string theory which starts with the 2-dimensional sigma model action

$$-\frac{1}{4\pi\alpha'} \int d^2\xi \left(\sqrt{-g} g^{\alpha\beta} \partial_\alpha x^\mu \partial_\beta x^\nu g_{\mu\nu}^s + \epsilon^{\alpha\beta} \partial_\alpha x^\mu \partial_\beta x^\nu A_{\mu\nu} \right) + \frac{1}{4\pi} \int d^2\xi \sigma R^{(2)},$$
(7.8)

where $R^{(2)}$ is the 2-dimensional curvature scalar. The fields $A_{\mu\nu}$ and σ are the anti-symmetric tensor gauge field and dilaton which appear in the massless string spectrum and occur in the IIA action given above. The constant α' has the dimensions of $(\text{mass})^{-2}$ and defines the mass scale m_s of the string by $m_s^2 = \frac{1}{4\pi\alpha'}$. The combination in front of the first two terms of the string action is often called the *string tension* T (i.e. $T = \frac{1}{4\pi\alpha'}$). One recovers the tree-level string equations and thus at lowest order in α' we find the supergravity equations of motion, by demanding conformal invariance. The corresponding string vielbein $e_\mu^{s\ m}$ is related to the above vielbein by $e_\mu^{s\ m} = e^{\frac{1}{4}\sigma} e_\mu^{\ m}$.

The last term in the above sigma model action, for constant σ, takes the form $\langle\sigma\rangle\chi$ where χ is the Euler number and is given by $\chi = \frac{1}{4\pi} \int d^2\xi R^{(2)}$. For a closed Riemann surface of genus g it is given by $\chi = 2 - 2g$.

Making this change in the bosonic part of the IIA Lagrangian of equation (7.4) and dropping the 's' superscript on the string vielbein we find that the Lagrangian becomes

$$L_B = ee^{-2\sigma}\{-R + 4\partial_\mu\sigma\partial^\mu\sigma - \frac{1}{3}F_{\mu_1\mu_2\mu_3}F^{\mu_1\mu_2\mu_3}\}$$
$$+ \left\{ -\frac{1}{12}eF'_{\mu_1...\mu_4}F'^{\mu_1...\mu_4} - eF_{\mu_1\mu_2}F^{\mu_1\mu_2} \right\}$$
$$+ \frac{2}{12^2}\epsilon^{\mu_1...\mu_{10}}F_{\mu_1...\mu_4}F_{\mu_5...\mu_8}A_{\mu_9\mu_{10}}.$$
(7.9)

218   *West*

As we have mentioned, the IIA action is the lower energy limit of the
IIA string theory, which is obtained as the string tension goes to zero (i.e.
$\alpha' \to \infty$). In this limit one is left with only the massless particles of the
IIA supergravity theory. It will be very useful to know how these particles
arise in the IIA string. This closed string theory in its formulation with
manifest world surface supersymmetry, that is the Neveu–Schwarz–Rammond
formulation [7], has four sectors; the $NS \otimes NS$, the $R \otimes R$, $R \otimes NS$ and
the $NS \otimes R$ corresponding to the different boundary conditions that can be
adopted for the 2-dimensional spinor in the theory. Clearly, the $NS \otimes R$ and
$R \otimes NS$ sectors contain the fermions, while the $NS \otimes NS$ and the $R \otimes R$
sectors contain the bosons. It is straightforward to solve the physical state
conditions in these sectors and one finds that the bosonic fields of the IIA
supergravity arise as

$$\underbrace{e^a_\mu, A_{\mu\nu}, \sigma;}_{NS \otimes NS} \quad \underbrace{A_{\mu\nu\varrho}, B_\mu;}_{R \otimes R} \quad \underbrace{\psi_{\mu\alpha}, \lambda_\alpha}_{NS \otimes R \; and \; R \otimes NS} .$$

Looking at the IIA Lagrangian in the string frame of equation (7.9) we
find that all the fields that arise in the $NS \otimes NS$ sector occur in a different
way from those in the $R \otimes R$ sector; While the former have a factor of $e^{-2\sigma}$
the latter do not have such a factor. We will find the same phenomenon for
the IIB string.

The IIA supergravity has two parameters. It has the Newtonian coupling
constant κ, which we have suppressed, and the parameter $\langle e^\sigma \rangle$. The IIA
string also has two parameters the string tension T, or equivalently the string
mass scale m_s, and the string coupling constant g_s. Since the low-energy
effective action of the string is IIA supergravity, these two sets of parameters
must be related.

We first consider how the parameters arise in the string theory. In a
second quantized formulation of string theory, one finds that the action can
be written in a way where the string coupling only occurs as a prefactor of g_s^{-2}.
Examples of such formulations are the light-cone gauge action or the gauge
covariant action. The parameter α' only occurs in these formulations through
the masses of the the particles or equivalently the L_n operators that occur
in these formulations. In the path integral formulation, the action becomes
multiplied by $\hbar^{-1}$ and so we find that $\hbar$ and g_s only occur in the combination
$\hbar g_s^2$. This situation is identical to the way in which the gauge coupling occurs
in Yang–Mills theory. The power of Planck's constant measures the number
of loops. Indeed if we have any Feynman graph with n loops, I propagators
and V vertices, the power of $\hbar$ associated with an n-loop graph is given by
$\hbar^{(I-V)}$, since each vertex carries a power of $\hbar$ and each propagator an inverse
power of $\hbar$. Using the topological relation $n = I - V + 1$, the power of $\hbar$ for
a n-loop diagram becomes $\hbar^{(n-1)}$. Our previous discussion then implies that
each n-loop diagram has a power $g_s^{(2n-2)}$ associated with it.

Now let us examine how the parameters arise in the effective action. The 2-dimensional first-quantized string-action of equation (7.8) contains the dilaton σ multiplied by the Euler number χ of the Riemann surface. A surface of genus n corresponds to a n-loop string amplitude and so in the path integral of this action one finds a factor of $e^{(2n-2)\langle\sigma\rangle}$. As a result, we conclude that the IIA string coupling and the expectation value of the dilaton are related by $g_s = \langle e^\sigma \rangle$ Since κ has the dimension of $(\text{mass})^{-4}$ it must be proportional to $(\alpha')^2$ and is given by the relation $\kappa = (\alpha')^2 e^{\langle\sigma\rangle}$.

We now finish our discussion of the IIA supergravity by reiterating some of the above features that will be useful for discussions of string duality. The IIA supergravity theory has the gauge fields σ, B_μ, $A_{\mu\nu}$ and $A_{\mu\nu\varrho}$. This means the IIA theory has gauge fields of rank q where $q = 1, 2, 3$ and these have corresponding field strengths of rank $q + 1 = 2, 3, 4$. Given a field strength $F_{\mu_1\ldots\mu_q}$ of rank $q + 1$ we can define a dual field strength by $F_{\mu_1\ldots\mu_{(D-1-q)}} = \frac{1}{(q+1)!}\epsilon_{\mu_1\ldots\mu_{10}}F^{\mu_{D-q}\ldots\mu_{10}}$. Hence the duals of the above field strengths are forms of rank $q = 6, 7, 8$. When the original field strengths are on-shell we can, at least at the linearized level, write the dual field strengths in terms of dual gauge fields of ranks 5, 6 and 7. Hence the IIA theory has gauge fields of ranks $p = 1, 2, 3, 5, 6, 7$ if we include the dual gauge fields as well as the original ones. It is instructive to list the above gauge fields according to the string sector in which they arise. In the $NS \otimes NS$ sector we find gauge fields of ranks 2 and 6 while in the $R \otimes R$ sector we find gauge fields of ranks 1, 3, 5 and 7. We observe that classifying the gauge fields according to the different sectors splits them into fields of odd and even rank.

From equation (7.3) we read off the component of the vielbein associated with the circle to be $e_{10}{}^{10} = e^{\frac{2}{3}\langle\sigma\rangle}$, with corresponding metric $g_{10\,10} = e^{\frac{4}{3}\langle\sigma\rangle}$. The parameter R introduced into the defining range of the variable x^{10} has no physical meaning as it only parametrizes the range of x^{10}. However, from the metric we can compute the radius R_{11} of the circle in the 11th dimension. We find that $R_{11} = Re^{\frac{2}{3}\langle\sigma\rangle}$. We recall that the string coupling is given by $g_s = e^{\langle\sigma\rangle}$ and as a result we find that

$$g_s = \left(\frac{R_{11}}{R}\right)^{\frac{3}{2}}. \tag{7.10}$$

The above relationship between the radius R_{11} of compactification of the 11-dimensional theory and the IIA string coupling constant implies in particular that as $R_{11} \to \infty$ we find that $g_s \to \infty$. Thus in the strong-coupling limit of the IIA string, the radius of the circle of compactification becomes infinite suggesting that the theory decompactifies; see [124].

We now consider some properties of the Kaluza–Klein modes which we have so far ignored in the reduction from 11 dimensions. Their masses are given by the action of $(e^{-1})_{11}^{\hat{\mu}}\partial_{\hat{\mu}} = e^{-\frac{2}{3}\sigma}\partial_{11}$ where we have used equation

(7.3). Examining the expansion of equation (7.1) we find that the masses of Kaluza–Klein particles are given by $n\frac{e^{-\frac{2}{3}(\sigma)}}{R}$ for integer n. However, using the relationship between R and R_{11} given just above, we find that the masses of the Kaluza–Klein particles are given by $\frac{n}{R_{11}}$. The gauge field B_μ which originated from the 11-dimensional metric couples to the Kaluza–Klein particles in a way which is governed by the derivative

$$(e^{-1})_m^{\hat\mu}\partial_{\hat\mu} = e^{\frac{1}{12}\sigma}(e^{-1})_m^{\mu}(\partial_\mu - 2B_\mu\partial_{11})$$

where we have again used equation (7.3). As such, we find that the Kaluza–Klein particles have charges given by $\frac{2n}{R}$ for integer n. From the IIA string perspective, the B_μ field is in the $R \otimes R$ sector and so the Kaluza–Klein particles couple with these charges to the $R \otimes R$ sector. It turns out that the IIA supergravity possess solitonic particle solutions that have precisely the masses and charges of the Kaluza–Klein particles; see [123]. Thus the IIA supergravity, and so in effect the IIA string, knows about all the particle content of the 11-dimensional theory and not only the massless modes that arise after the compactification.

It is these observations that underlie the conjecture, [123], [124], that the strong-coupling limit of the IIA string theory is an 11-dimensional theory, called M-theory, whose low-energy limit is 11-dimensional supergravity.

7.3 IIB supergravity

The IIB supergravity was found in references [110], [111] and [112] using two different methods. In reference [110], a variant, [126], of the Noether method was used: rather than working with an action and transformation laws, one can just use the transformation laws. One begins with the rigid supersymmetry transformations and local Abelian transformations of the linearised theory. Letting the supersymmetry parameter become space-time dependent, the transformations laws no longer close; however we may still close the supersymmetry algebra order-by-order in κ by adding terms to the transformation laws, provided we also identify the now local spinor parameter of the supersymmetry transformation with the spinor parameter that occurs in the local Abelian transformation of the gravitino. In this way the supersymmetry transformations laws of the IIB theory and the fact that the scalars belong to the coset $SU(1,1)/U(1)$ were found; see [110]. Using the fact that the transformations laws only close on-shell, this work was extended in reference [112] to find the equations of motion in the absence of fermions. In the independent work of reference [111], the full equations of motion in superspace and x-space were found using the on-shell superspace techniques of Subsection 5.2. The third-order terms of the IIB theory were constructed for the light-cone gauge Hamiltonian in reference [125].

The strategy behind these calculations is explained in Chapter 4 of [0] and although the ideas are straightforward the calculations themselves are technically complicated to the extend that they will not be reproduced here. Nonetheless we will describe the essential features of the IIB theory so that the reader will grasp some of the ideas involved and gain a feel for the IIB theory itself.

7.3.1 The algebra

The IIB supergravity is based on a supersymmetry algebra whose two super-charges Q_α^i, $i = 1, 2$ and $\alpha = 1, \ldots, 32$ are Majorana–Weyl spinors of the same chirality. They therefore obey the conditions

$$(Q_\alpha^i)^* = Q_\alpha^i, \quad \Gamma_{11} Q^i = Q^i.$$

The supersymmetry algebra is given by

$$\{Q_\alpha^i, Q_\beta^j\} = (\gamma^\mu C^{-1})_{\alpha\beta} \delta^{ij} P_\mu + \cdots,$$

where $+\cdots$ denote terms with central charges whose form the reader may readily find by following the discussion at the end of Subsection 1.5.

It is more useful to work instead with the complex Weyl supercharges $Q_\alpha = Q_\alpha^1 + iQ_\alpha^2$, $\overline{Q}_\alpha = Q_\alpha^1 - iQ_\alpha^2 = (Q_\alpha)^*$. The supersymmetry algebra also contains a $U(1)$ generator denoted R, $(R^\dagger = -R)$, which acts on the supercharges as

$$[Q_\alpha, R] = iQ_\alpha, \quad [\overline{Q}_\alpha, R] = -i\overline{Q}_\alpha.$$

7.3.2 The particle content

The field content of the IIB theory is

$$e_\mu{}^m, \ A_{\mu\nu}, \ a, \ B_{\mu\nu\rho\kappa}, \ \psi_{\mu\alpha}, \ \lambda_\alpha. \tag{7.11}$$

The fields $A_{\mu\nu}$ and a are complex, while the gauge field $B_{\mu\nu\rho\kappa}$ is real. The spinors are complex Weyl spinors; the graviton $\psi_{\mu\alpha}$ is of the opposite chirality to λ_α, but has the same chirality as the supersymmetry parameter ϵ_α. Recalling the discussion above equation (6.2) we find that these fields lead to 35, 56, 2, 35, 112 and 16 on-shell degrees of freedom respectively. The gauge field $B_{\mu_1\mu_2\mu_3\mu_4}$ defines the linearised five-rank field strength $g_{\mu_1\mu_2\mu_3\mu_4\mu_5} \equiv 5\partial_{[\mu_1} B_{\mu_2\mu_3\mu_4\mu_5]}$ which satisfies a self-duality condition. At the linearized level this self-duality condition is given by

$$g_{\mu_1\mu_2\mu_3\mu_4\mu_5} = \frac{1}{5!} \epsilon_{\mu_1\mu_2\mu_3\mu_4\mu_5\nu_1\nu_2\nu_3\nu_4\nu_5} g^{\nu_1\nu_2\nu_3\nu_4\nu_5} \equiv {}^* g_{\mu_1\mu_2\mu_3\mu_4\mu_5}.$$

Without the self-duality condition this gauge field corresponds to a particle that belongs to the fourth-rank totally anti-symmetric representation of the little group $SO(8)$. The self-duality condition above corresponds to the constraint that this representation is self-dual and hence the 35 degrees of freedom given above. Thus the supermultiplet of IIB supergravity has 128 bosonic degrees of freedom and 128 fermionic degrees of freedom on-shell.

Most of the fields of equation (7.11) transform under the $U(1)$ transformations; their R weights are 0, 2, 4, 0, 1 and 3 respectively. Clearly, real fields must have R weight zero and the gravitino must have the opposite R weight as the supercharge Q_α since $\delta\psi_\mu = \partial_\mu\epsilon$.

Following the pattern of the standard action for gauge fields one may be tempted to use the linearised action

$$\int d^{10}x \ g_{\nu_1\nu_2\nu_3\nu_4\nu_5}g^{\nu_1\nu_2\nu_3\nu_4\nu_5}$$

for the fourth-rank gauge field. However, if g is the five-form associated with the gauge field, the above action is given by $\int g \wedge {}^*g = \int g \wedge g = 0$. This discussion can be rephrased without using forms as follows: using the self-duality condition we can rewrite one of the field strengths in terms of ${}^*g$; swopping the indices on the ϵ symbol such that the last five indices are at the beginning, we incur a minus sign; using the self-duality condition once more we again find that the above action vanishes as it is equal to its negative.

Clearly this result holds for any rank $2n + 1$ self-dual gauge field strength in a space-time of dimension $4n + 2$, $n \in \mathbf{Z}$. Indeed there is no simple action for the fourth-rank gauge field and so for the IIB theory itself. Although some actions have been suggested there are reasons to believe that they do not correctly capture all the physics of the theory. As a result we will content ourselves with deriving the equations of motion.

The IIB supergravity is the theory that describes the effective action of the low-energy limit of the IIB string. The massless fields in the IIB string being those that occur in the IIB supergravity. To find the massless fields we must examine the physical state conditions for the IIB string which, being a closed superstring, has in the Neveu–Schwarz–Rammond formulation, the usual $NS \otimes NS$, $R \otimes R$, $R \otimes NS$ and $NS \otimes R$ sectors corresponding to the possible boundary conditions for the 2-dimensional spinor in the theory. Clearly, the last two sectors contains the fermions while the $NS \otimes NS$ and the $R \otimes R$ sectors contain the bosons. The bosons arise as

$$\underbrace{e_\mu^a, \ A^1_{\mu\nu}, \ \sigma}_{NS \otimes NS} \quad \underbrace{A^2_{\mu\nu}, \ l, \ B_{\mu\nu\rho\tau},}_{R \otimes R}$$

where $A^1_{\mu\nu}$ and $A^2_{\mu\nu}$ are real fields which make up the complex fields $A_{\mu\nu}$, and we replace the complex scalar a by two real fields l and σ. The precise way in which these decompositions are defined will be specified later. The fact that the two rank-2 gauge fields are split between the $NS \otimes NS$ and $R \otimes R$ sectors has important consequences for discussions of string duality. Since the physical state condition in the $NS \otimes NS$ sector are exactly the same as in the IIA string we should not be surprised to find that this sector contains exactly the same bosonic field content.

7.3.3   The scalars

The gauge fields, the graviton and the gravitini possess gauge transformations as a result of which they can only occur in gauge-invariant quantities, which in the sense of Section 5, have geometric dimensions greater than zero. At first sight, this is not the case for the scalars of the theory since they have geometric dimension zero. As we explained in Section 5, dimensional analysis plays an important role in the construction of supergravity theories and as such it might seem that the role of the scalars in the theory is difficult to determine. Fortunately, however, the scalars belong (see [110]) to the coset space $SU(1,1)/U(1)$ and consequently the way they can can occur in the theory is strongly constrained.

The use of coset spaces to describe scalar fields was described in reference [127]. Since it is just as simple to describe the general theory (see [127]) we will give the construction for a general coset space. Let G be any group, H one of its subgroups and denote the coset space by G/H. Let us consider any space-time dependent $\mathcal{V} \in G$ which is taken to transform as

$$\mathcal{V} \to g\mathcal{V}h, \tag{7.12}$$

where $h \in H$ is a local (i.e. space-time dependent) transformation and $g \in G$ is a rigid transformation. We may use the local H transformations to gauge away $\dim H$ scalar fields leaving $\dim G - \dim H$ scalar fields in $\mathcal{V}$. The object $\Omega_\mu \equiv \mathcal{V}^{-1}\partial_\mu\mathcal{V}$ belongs to the Lie algebra of G and so can be written in the form

$$\Omega_\mu = \mathcal{V}^{-1}\partial_\mu\mathcal{V} = f_\mu^a T_a + w_\mu^i H_i, \tag{7.13}$$

where H_i are generators of H, and T_a are the remaining generators of G. The object ω_μ is invariant under the rigid transformations $g \in G$, but transforms under local H transformations as

$$\omega_\mu \to h^{-1}\partial_\mu h + h^{-1}\omega_\mu h. \tag{7.14}$$

The theory simplifies if we restrict ourselves to reductive cosets which are those for which the commutator $[T_a, H_i]$ can be written in terms of only the coset generators T_a. In this case the above transformation rule implies that $f_\mu \equiv f_\mu^a T_a \to h^{-1}f_\mu h$ and $w_\mu \equiv w_\mu^i H_i \to h^{-1}w_\mu h + h^{-1}\partial_\mu h$. We can think of f_μ^a as a vielbein on G/H defining a set of preferred frames and w_μ^i as the connection associated with local H transformations.

An invariant Lagrangian is given by

$$\eta^{\mu\nu}\text{Tr}\,(f_\mu f_\nu).$$

The corresponding equation of motion is given by

$$D_\mu f_a^\mu = 0, \tag{7.15}$$

where we have introduced the covariant derivative $D_\mu f_\nu \equiv \partial_\mu f_\nu + [w_\mu,\,f_\nu]$. It is straightforward to verify that the above Lagrangian and equation of motion are invariant under both local H transformations and rigid G transformations.

Let us work out the above expressions for the case of interest, namely for $G = SU(1,1)$ and $H = U(1)$. The group $G = SU(1,1)$ is the set of 2×2 matrices g of determinant 1 which acts on the column vector $\begin{pmatrix} z_1 \\ z_2 \end{pmatrix}$ by $\begin{pmatrix} z_1 \\ z_2 \end{pmatrix} \to g \begin{pmatrix} z_1 \\ z_2 \end{pmatrix}$ in such a way as to preserve $|z_1|^2 - |z_2|^2$. The most general element of $SU(1,1)$ can be written in the form

$$U = \begin{pmatrix} u & v \\ v^* & u^* \end{pmatrix}, \tag{7.16}$$

subject to $uu^* - vv^* = 1$. An infinitesimal element of $G = SU(1,1)$ can therefore be written in the form $g = I + A$ where A is given by

$$A = -2i\hat{a}\sigma_3 + b_1\sigma_1 - b_2\sigma_2 = \begin{pmatrix} -2i\hat{a} & b \\ b^* & +2i\hat{a} \end{pmatrix},$$

where $\hat{a}$, b_1 and b_2 are real, $b = b_1 + ib_2$, and σ_i, $i = 1,2,3$, are the Pauli matrices. An alternative parameterization of elements of $G = SU(1,1)$ is given by exponentiating the above infinitesimal element;

$$U = e^A = \begin{pmatrix} \cosh\rho - 2i\hat{a}\frac{\sinh\rho}{\rho} & b\frac{\sinh\rho}{\rho} \\ b^*\frac{\sinh\rho}{\rho} & \cosh\rho + 2i\hat{a}\frac{\sinh\rho}{\rho} \end{pmatrix}, \tag{7.17}$$

where $\rho^2 = b^*b - 4\hat{a}^2$. The $U(1)$ subgroup is generated by $i\sigma_3$ and its elements take the form

$$h = \begin{pmatrix} e^{-2i\hat{a}} & 0 \\ 0 & e^{+2i\hat{a}} \end{pmatrix}. \tag{7.18}$$

Taking $\mathcal{V}$ to be a general elements of $SU(1,1)$ we find that ω_μ takes the form

$$\omega_\mu = \begin{pmatrix} 2iQ_\mu & P_\mu \\ \bar{P}_\mu & -2iQ_\mu \end{pmatrix}, \tag{7.19}$$

where $Q_\mu = (Q_\mu)^*$. Under an infinitesimal local $U(1)$ transformation of equation (7.18), we find from equation (7.14) that the vielbein and $U(1)$-connection transform as

$$\delta Q_\mu = -\partial_\mu \hat{a}, \quad \delta P_\mu = 4i\hat{a}P_\mu.$$

The last equation corresponds to the $U(1)$ weight 4 assignment given earlier to the scalars. If we write $\mathcal{V} = (C_+, C_-)$, where C_+ and C_- are two column vectors, then $C_\pm$ transforms as $C_\pm \to gC_\pm e^{\mp 2i\hat{a}}$ under local and rigid transformations. If we further take the ratio of the top and bottom components of the column vectors $C_\pm$ and denote it by $c_\pm$, then it follows that $c_\pm$ is inert under local H transformations but transforms as

$$c_\pm \to \frac{uc_\pm + v}{v^*c_\pm + u^*}$$

under a rigid transformation of the form of equation (7.16).

We can use a local $U(1)$ transformation to bring $\mathcal{V}$ to be of the form

$$\mathcal{V} = \frac{1}{\sqrt{1 - \phi\phi^*}} \begin{pmatrix} 1 & \phi \\ \phi^* & 1 \end{pmatrix}. \tag{7.20}$$

This choice is most easily achieved using the form of $SU(1,1)$ elements given in equation (7.17). Examining the second column vector in $\mathcal{V}$ we find that $c_- = \phi$ and so ϕ transforms as under G as

$$\phi \to \frac{u\phi + v}{v^*\phi + u^*}. \tag{7.21}$$

For the choice of $\mathcal{V}$ of equation (7.20) we find that

$$Q_\mu = -\frac{i}{4} \frac{(-\phi\partial_\mu\phi^* + \phi^*\partial_\mu\phi)}{(1 - \phi\phi^*)}, \quad P_\mu = \frac{\partial_\mu\phi}{(1 - \phi\phi^*)}.$$

The corresponding equation of motion is found by substituting equation (7.19) into equation (7.15) to find

$$D_\mu P^\mu \equiv \partial_\mu P^\mu + 4iQ_\mu P^\mu = 0. \tag{7.22}$$

The actual IIB equations of motion of the scalars must be of this form, but will also include other terms containing the superpartners of the scalars. We observe that the scalars only occur through P_μ or, for derivatives of other fields with non-zero $U(1)$ weight, through Q_μ. Both these fields are given by equation (7.13) which contains the space-time derivatives acting on the group element and so they have geometric dimension 1. This fact plays a crucial role in the way the scalars are encoded into the IIB theory as it allows the

Table 5: Coset spaces of the maximal supergravities

D	G	H
11	1	1
10, IIB	$SL(2)$	$SO(2)$
10, IIA	$SO(1,1)/Z_2$	1
9	$GL(2)$	$SO(2)$
8	$E_3 \sim SL(3) \times SL(2)$	$U(2)$
7	$E_4 \sim SL(5)$	$USp(4)$
6	$E_5 \sim SO(5,5)$	$USp(4) \times USp(4)$
5	E_6	$USp(8)$
4	E_7	$SU(8)$
3	E_8	$SO(16)$

use of dimensional analysis to restrict the way the scalars can occur in the theory.

It has been found that the scalar fields that occur in supergravity theories always belong to a coset space; see [169], [170], [171]. Some of the other interesting cases are the $N = 4$ and $N = 8$ supergravity theories in four dimensions where the coset spaces are $SU(1,1)/U(1)$[169] and $E_7/SU(8)$; see [170]. In Table 5 we give the coset spaces associated with the maximal supergravities. This table was taken from references [167] and [181]. All these theories, except for IIB, arise from the dimensional reduction of the 11-dimensional supergravity theory.

7.3.4 The gauge fields

Let us now examine how the gauge fields associated with the rank-2 tensor gauge field $A_{\mu\nu}$ can occur in the theory. We define the field strength of $A_{\mu\nu}$ to be Im $_{\mu_1\mu_2\mu_3} = 3\partial_{[\mu_1} A_{\mu_2\mu_3]}$. This gauge field must be inert under the local $U(1)$ transformations associated with the coset $SU(1,1)/U(1)$. If it were to have a non-trivial $U(1)$ transformation then it could only transform covariantly, but in this case the field strength would not transform covariantly. One could attempt to avoid this latter conclusion by including the $U(1)$ connection Q_μ in the definition of the field strength; however then the corresponding field strength would not be invariant under the standard $U(1)$ gauge transformations of the gauge field. Under rigid $g \in SU(1,1)$ it transforms as

$$(\overline{\text{Im}}, \text{Im}) \to (\overline{\text{Im}}, \text{Im})g^{-1}. \tag{7.23}$$

We can define a $SU(1,1)$-invariant field strength by

$$(\bar{F}, F) = (\overline{\text{Im}}, \text{Im})\mathcal{V}. \tag{7.24}$$

Using equations (7.12) and (7.18) we find that under a local $U(1)$ transformation the new fields $(\overline{F}, F)$ have weights $(-2, 2)$ and so transform as

$$(\overline{F}, F) \rightarrow (\overline{F}, F)h.$$

Using the form of $\mathcal{V}$ given in equation (7.20) we can readily find $(\overline{\mathrm{Im}}, \mathrm{Im})$ in terms of $(\overline{F}, F)$.

We note that the coset space description of the scalar fields plays an important part in the formulation of the gauge field which will occur in the equations of motion. It will follow that the equations of motion will admit duality transformations. Since almost all maximal supergravity have gauge fields and coset space scalars this is a general feature of supergravity theories. We refer the reader to the lectures of M.K. Gaillard and B. Zumino in this volume for a complete discussion of this topic.

7.3.5 The equations of motion

As we have discussed, the derivation of the equations of motion was carried out using a variant of the Noether method (see Subsection 5.1) and the on-shell superspace technique discussed in Subsection 5.2. These calculations are too involved to reproduce here; however, many features of the equations can be deduced using the features of the IIB theory discussed above. For example, the equation of motion must reduce to the correct linearised equations, obey the requirements of dimensional analysis and contain terms of the same $U(1)$ weight. In addition, the gauge fields can only occur in terms of their field strengths $F^{\mu_1\mu_2\mu_3}$ and
$overline{F}^{\mu_1\mu_2\mu_3}$ as well as a five-rank field strength $G_{\mu_1\ldots\mu_5}$. These field strengths all have geometric dimension 1. The vielbein must occur through the usual Ricci tensor $R_{\mu\nu}$ which has geometric dimension 2 and as we discussed above the scalars belong to the coset $SU(1,1)/U(1)$ and hence are only contained in the objects of geometric dimension 1, namely, P_μ and Q_μ. The latter can only occur as part of a covariant derivative.

Let us begin with the scalars; their equation of motion must generalise equation (7.22) and hence the terms in the equation must have $U()$ weight 4 and geometric dimension 2. The only possible candidate is $F_{\mu_1\mu_2\mu_3}F^{\mu_1\mu_2\mu_3}$. The equation for $A_{\mu\nu}$ must contain $D^{\mu_3}F_{\mu_1\mu_2\mu_3}$ and so it has geometric dimension 2 and $U(1)$ weight 2. The only such terms we can add are $F_{\mu_1\mu_2\mu_3}P^{\mu_3}$ and $G_{\mu_1\ldots\mu_5}F^{\mu_3\cdots\mu_5}$. The equation of motion for the fourth-rank gauge field is just the self-duality condition on the five-rank field strength and so this equation has geometric dimension 1 and $U(1)$ weight zero. There are no terms one can add to this equation other than the duality condition on the field strength itself. Analysing the vielbein equation in the same way we essentially determine the field equations up to constants.

The equations of motion of IIB supergravity in the absence of fermions is given by (see [111], [112])

$$D^\mu P_\mu = \frac{1}{6} F_{\mu_1\mu_2\mu_3} F^{\mu_1\mu_2\mu_3} \tag{7.25}$$

$$D^{\mu_3} F_{\mu_1\mu_2\mu_3} = \overline{F}_{\mu_1\mu_2\mu_3} P^{\mu_3} - \frac{i}{6} G_{\mu_1...\mu_5} F^{\mu_3...\mu_5} \tag{7.26}$$

$$R_{\mu\nu} = -2\overline{P}_{(\mu} P_{\nu)} - F_{(\mu}{}^{\nu_1\nu_2} F_{\nu)\nu_1\nu_2} + \frac{1}{12} g_{\mu\nu} \overline{F}_{\mu_1\mu_2\mu_3} F^{\mu_1\mu_2\mu_3} - \frac{1}{96} G_\mu{}^{\mu_1...\mu_4} G_{\nu\mu_1...\mu_4} \tag{7.27}$$

$$G_{\mu_1...\mu_5} = {}^*G_{\mu_1...\mu_5}, \tag{7.28}$$

where

$$G_{\mu_1...\mu_5} = 5\partial_{[\mu_1} A_{\mu_2...\mu_5]} + 20i(A_{[\mu_1\mu_2} \text{Im}{}^*_{\mu_3...\mu_5]} - A^*_{[\mu_1\mu_2} \text{Im}_{\mu_3...\mu_5]}). \tag{7.29}$$

The reader is referred to reference [111] for the fermionic contribution. They are invariant (see [111], [112]) under the local supersymmetry and $U(1)$ transformations of reference [110].

7.3.6 The $SL(2, \mathbf{R})$ version

The group $SU(1,1)$ is isomorphic to the group $SL(2, \mathbf{R})$. For some purposes it is better to formulate the theory in a manner where the $SL(2, \mathbf{R})$ form of the invariance is manifest, rather than as above, where the $SU(1,1)$ symmetry is apparent. As we explained above, $g \in SU(1,1)$ acts on the column vector $\begin{pmatrix} z_1 \\ z_2 \end{pmatrix}$ by $\begin{pmatrix} z_1 \\ z_2 \end{pmatrix} \to g \begin{pmatrix} z_1 \\ z_2 \end{pmatrix}$. If we denote the ratios of the column vector by $z = z_1/z_2$ then the action of $SU(1,1)$ becomes

$$z \to \frac{uz + v}{v^* z + u^*}. \tag{7.30}$$

This action is such that it takes the unit disc $|z| \le 1$ to itself. We can map the unit desk to the upper half-plane $H = \{w : \text{Im } w \ge 0\}$ by the transformation

$$z \to w = i\left(\frac{1-z}{1+z}\right).$$

The action induced by the transformation of equation (7.30) on H is given by

$$w \to \frac{aw + b}{cw + d},$$

where $ad - bc = 1$ and a, b, c, d are real. In this last transformation we recognise the action of the group $SL(2, \mathbf{R})$, corresponding to the element

$$\hat{g} = \begin{pmatrix} a & b \\ c & d \end{pmatrix} \in SL(2, \mathbf{R}).$$

It is well-known that $SL(2, \mathbf{R})$ is the largest group that maps the upper half-plane to itself and so we should not be surprised that in mapping from the unit disc onto the upper half-plane, the action of $SU(1, 1)$ becomes that of $SL(2, \mathbf{R})$. The precise relationship between the parameters of the two groups is given by

$$
\left.
\begin{aligned}
a &= \frac{1}{2}(u + u^* - v - v^*), \; b = \frac{i}{2}(-u + u^* - v + v^*), \\
c &= -\frac{i}{2}(-u + u^* + v - v^*), \; d = \frac{1}{2}(u + u^* + v + v^*).
\end{aligned}
\right\}
\tag{7.31}
$$

For the $SU(1, 1)$ formulation of the IIB theory given above, we found that the scalar field ϕ of equation (7.20) transformed under $SU(1, 1)$ in the same way as the variable z of equation (7.21). Hence if we make the transformation from ϕ to the variable φ by

$$
\phi \to \varphi = i \left(\frac{1 - \phi}{1 + \phi} \right),
$$

then φ transforms under $SL(2, \mathbf{R})$ just like w, that is

$$
\varphi \to \frac{a\varphi + b}{c\varphi + d}.
$$

It remains to find new variables for the rank-2 gauge field that transform in a recognisable way under $SL(2, \mathbf{R})$. Let us write the field strength $\mathfrak{F}_{\mu\nu\rho}$ as $\mathfrak{F}_{\mu\nu\rho} = \mathfrak{F}^2_{\mu\nu\rho} + i\mathfrak{F}^1_{\mu\nu\rho}$ then an explicit calculation shows that the transformation law of equation (7.23) for $\mathfrak{F}_{\mu\nu\rho}$ becomes

$$
\begin{pmatrix} \mathfrak{F}^1_{\mu\nu\rho} \\ \mathfrak{F}^2_{\mu\nu\rho} \end{pmatrix} \to \hat{g} \begin{pmatrix} \mathfrak{F}^1_{\mu\nu\rho} \\ \mathfrak{F}^2_{\mu\nu\rho} \end{pmatrix}.
\tag{7.32}
$$

It is straightforward to substitute for the new variables into the equations of motion (7.25)–(7.29) to find a formulation that is manifestly $SL(2, \mathbf{R})$-invariant. Carrying out this transformation and also making the substitution $\varphi = l + ie^\sigma$, one finds that the $NS \otimes NS$ fields, (i.e. $e_\mu{}^a$, $A^1_{\mu\nu}$ and σ) have identical equations of motion as the $NS \otimes NS$ sector of the IIA supergravity. In fact, since these fields do not include the rank-4 gauge field, we can formulate the dynamics of the $NS \otimes NS$ sector of the theory in terms of an action, and this action will have a Lagrangian which is none other than the first term of equation (7.9), if we choose to work with the string metric.

We close with some comments on some of the features of the IIB theory discussed that are most relevant to our discussion on string duality. Like the IIA theory, the IIB theory has two coupling constants; the Newtonian constant κ, whose dependence we have suppressed, and the expectation value of $\langle e^\sigma \rangle$. As we have explained in the previous section, the latter plays the role of the IIB string coupling constant, i.e. $g_s = \langle e^\sigma \rangle$. In general, an $SL(2,\mathbf{R})$-transformation changes from weak to strong string coupling. For example, the transformation $\varphi = l + ie^\sigma \rightarrow \varphi' = l' + ie^{\sigma'} = -\frac{1}{\varphi}$ implies that

$$g_s' = \langle e^{\sigma'} \rangle = \frac{1}{\langle e^\sigma \rangle} = \frac{1}{g_s} \tag{7.33}$$

maps from the weak to the strong regime.

The $SL(2,\mathbf{R})$-transformation also mixes the two real rank-3 field strengths $\mathfrak{F}^i_{\mu\nu\rho}$, $i = 1, 2$, one of which arises from the $NS \otimes NS$ sector and one from the $R \otimes R$ sector of the string theory. Hence a generic $SL(2,\mathbf{R})$ transformation transforms fields in the $NS \otimes NS$ sector into those in the $R \otimes R$ sector and vice versa.

The IIB theory cannot be obtained from the 11-dimensional supergravity by a reduction; however if we were to reduce the IIB theory on a circle to 9 dimensions then the resulting supergravity theory would have an underlying supersymmetry algebra with 32 supercharges which, being an odd dimension, would be of no fixed chirality. This is precisely the same supersymmetry algebra that would result if we were to reduce the IIA theory on a circle to 9 dimensions. In particular, they form two 16-component Majorana spinors. Since this algebra uniquely determines the maximal 9-dimensional supergravity theory we must conclude that reducing the IIA and IIB supergravity theories to 9 dimensions leads to the same supergravity theory; see [107].

7.4   Type I supergravity

The type I Supergravity was in fact the first supergravity theory in 10 dimensions to be constructed; see [120]. Its underlying algebra contains a Majorana–Weyl spinor supercharge whose anti-commutator was given in equation (1.21). The field content is given by

$$\underbrace{e^a_\mu, \phi}_{NS \otimes NS} \; ; \quad \underbrace{A_{\mu\nu}}_{R \otimes R}; \quad \text{plus} \quad \underbrace{\psi_{\mu\alpha}, \lambda_\alpha}_{NS \otimes R},$$

where we have also indicated the sectors in the type I string from which they come. The gravitino $\psi_{\mu\alpha}$ and the spinor λ_α are Majorana–Weyl spinors of opposite chirality and all the bosonic fields are real.

This theory can be obtained form the either the IIA or the IIB theory by truncation. From the IIA theory we impose the obvious Weyl conditions on the gravitino and spinor in the IIA theory, and to be consistent with supersymmetry we must also set $A_{\mu\nu\rho} = 0 = B_\mu$. It is straightforward to find the Lagrangian for the bosonic fields by truncating the corresponding action for the IIA theory of equation (7.4).

We can also obtain the type I theory from the IIB theory by a truncation. In the latter we consider the operator Ω that changes the sign of $\mathfrak{F}^2_{\mu\nu\rho}$, l and $B_{\mu\nu\rho\sigma}$, which are in the $NS \otimes NS$, $R \otimes R$ and $R \otimes R$ sectors respectively, but leaves inert e^a_μ, $\mathfrak{F}^1_{\mu\nu\rho}$ and σ, which are in $NS{\otimes}NS$, $R{\otimes}R$ and $NS{\otimes}NS$ sectors respectively. To recover the type I supergravity, we keep only fields which are left-inert by Ω and so we find only the latter fields. On the spinors we impose a Majorana condition which leads to a gravitini and another fermion which are both Majorana–Weyl although of opposite chirality. In fact Ω corresponds in string theory to world sheet parity, that is it exchanges left- and right-moving modes.

The $N = 1$ Yang–Mills theory is also based on the supersymmetry algebra with one Majorana–Weyl supercharge. It consists of a gauge field A_μ and one Majorana–Weyl spinor λ_α. This theory (see [119]) is easily derived. We first write down the linearised transformation laws that are determined up to two constants by dimensional analysis. The constants are then fixed by demanding that the supersymmetry transformations and the linearised gauge transformations form a closed algebra. The full theory is uniquely found by demanding that the algebra closes and that the action be invariant under the usual non-Abelian gauge transformation of the gauge field. The result is given by

$$\int d^{10}x \left(-\frac{1}{4}F^i_{\mu\nu}F^{\mu\nu i} - \frac{i}{2}\bar{\lambda}^i\gamma^\mu(D_\mu\lambda)^i \right),$$

which is invariant under

$$\delta A^i_\mu = i\bar{\epsilon}\gamma_\mu\lambda^i, \ \delta\lambda^i = -\frac{1}{2}F^i_{\mu\nu}\gamma^{\mu\nu}\epsilon,$$

where

$$F^i_{\mu\nu} = \partial_\mu A^i_\nu - \partial_\nu A^i_\mu - gf_{jk}{}^i A^j_\mu A^k_\nu$$

is the Yang–Mills field strength and

$$D_\mu\lambda^i = \partial_\mu\lambda^i - gf_{jk}{}^i A^j_\mu\lambda^k.$$

We can also consider the coupling between the $N = 1$ Yang–Mills theory and the type I supergravity. This was found in reference [121]. If we take the gauge group to be $SO(32)$ or $E_8 \otimes E_8$ we find the theory that results from the low-energy limit of the corresponding heterotic string theory or the type I string theory.

8   Brane Dynamics

8.1   Bosonic branes

Bosonic p-branes are extended objects that sweep out a $p + 1$-dimensional space-time manifold in a background superspace-time. A 0-brane is just a

particle and a 1-brane is a string. However p-branes for $p \geq 2$ also occur
in string theory as solitons and are thus non-perturbative objects. In this
section we find what possible superbranes can exist and give their dynamics.
Although p-branes with $p \geq 2$ are intrinsically non-perturbative objects, they
are related by duality symmetries to perturbative particle and string states.
As such, they play an important role in discussions of string duality.

We first consider a brane that has no supersymmetry. A p-brane sweeps out
a $p + 1$-dimensional world surface M, with coordinates ξ^m, $m = 0, 1, \ldots, p$ in
a D-dimensional target space $\underline{M}$ with coordinates $X^{\underline{n}}$, $\underline{n} = 0, 1, \ldots, D-1$. As
the symbols imply, we use $m, n, p, \ldots$ for the embedded surface world indices
and $\underline{m}, \underline{n}, \underline{p}, \ldots$ for target space world indices. The corresponding tangent
space indices are $a, b, c, \ldots$ for world surface indices and $\underline{a}, \underline{b}, \underline{c}, \ldots$ for target
space world indices. This notation is used extensively in the literature in this
subject. The reader should have no difficulty making the transition from the
$\mu, \nu \ldots$ and $m, n, p \ldots$ used in the previous section for the world and tangent
target space indices respectively. The surface M swept out by the p-brane in
the target space $\underline{M}$ is specified by the functions $X^{\underline{n}}(\xi^n)$ which extremise the
action

$$ - T \int d^{p+1}\xi \sqrt{-\det g_{mn}} \qquad (8.1)$$

where

$$ g_{mn} = \partial_n X^{\underline{n}} \partial_m X^{\underline{m}} g_{\underline{nm}} $$

and $g_{\underline{mm}}$ is the metric of the target space-time often referred to as the background metric. The constant T is the brane tension and has the dimensions
of $(\text{mass})^{p+1}$. The action of equation (8.1) is invariant under reparameterizations of both the target space $\underline{M}$ and the world surface M. The metric g_{mn}
is the metric induced on the world surface M by the background metric of
the target space. As such, we recognize the action in equation (8.1) as the
area swept out by the p-brane. Hence, like the string and point particle, a
p-brane moves so as to extremise the volume of the surface it sweeps out. If
the target space is flat the background metric is just the Minkowski metric
$g_{\underline{mm}} = \eta_{\underline{mm}}$. A 0-brane is just a point particle and if it has mass m then
$T = m$. A 1-brane is just the usual bosonic string and the action of equation
(7.1.1) is the Nambu action for the string if we take the background metric
to be flat. In this case we often write $T = 12\pi\alpha'$ where α' is the string Regge
slope parameter.

The bosonic brane does not have enough symmetry to determine its couplings to the fields in the target space. However, a p-brane naturally couples
to a $(p + 1)$-gauge field $A_{\underline{m}_1 \cdots \underline{m}_{p+1}}$ of the target space by a term of the form

$$ \int d^{p+1}\xi \, \epsilon^{n_1 \cdots n_{p+1}} \partial_{n_1} X^{\underline{m}_1} \cdots \partial_{n_{p+1}} X^{\underline{m}_{p+1}} A_{\underline{m}_1 \cdots \underline{m}_{p+1}}. \qquad (8.2)$$

So for example, the motion of a 0-brane, that is a point particle, is described by the functions $X^{\underline{n}}(\tau)$, where $\xi^0 = \tau$, and it naturally couples to a vector field $A_{\underline{n}}$ in the form

$$\int d\tau \frac{dX^{\underline{n}}}{d\tau} A_{\underline{n}}.$$

If we couple this expression to that in equation (8.1) for a flat target space then the equations of motion for $X^{\underline{n}}$ are nothing but the Lorentz force law for a charged particle in an electromagnetic field. A 1-brane, i.e. string, couples to a 2-form $A_{\underline{nm}}$ in the manner

$$\int d^2\xi \epsilon^{mn} \partial_m X^{\underline{m_1}} \partial_n X^{\underline{m_2}} A_{\underline{m_1 m_2}}.$$

We can split the target space indices $\underline{n}$, $\underline{m}$, into those associated with the directions longitudinal, and those with directions which are transverse to the brane. We denote the former by $n, m, \ldots = 0, \ldots, p$ and the latter by $n', m', \ldots = p+1, \ldots, D-1$. A useful gauge is the static gauge in which we use the reparameterization transformations of the world surface to identify the $p+1$ longitudinal coordinates $X^n(\xi)$, $n = 0, 1, \ldots, p$, with the coordinates ξ^n, $n = 0, 1, \ldots, p$, of the p-brane; in other words

$$X^n(\xi) = \xi^n, \quad n = 0, 1, \ldots, p \ . \tag{8.3}$$

This leaves the transverse coordinates $X^{n'}(\xi)$, $n' = p+1, \ldots, D-1$, to describe the dynamics of the brane. We can think of the $D - p - 1$ transverse coordinates as the Goldstone bosons or zero modes of the broken translations due to the presence of the p-brane.

As we have explained in Section 7, the low-energy effective action of a string theory is a supergravity theory. It has been found that p-brane solutions arise from this supergravity theory. For such a static solution the supergravity fields do not depend on the p spatial coordinates of the p-brane world surface, but do depend on the $D - p - 1$ coordinates transverse to the brane. It turns out that the supergravity fields usually depend on functions which are harmonic functions of the transverse coordinates.

It will be instructive to consider a simpler and better understood example of solitons, namely the monopoles (i.e. a 0-brane) that occur in 4 dimensions. Monopoles arise as static solutions in the $N = 2$ supersymmetric Yang–Mills field theory. The space of parameters required to specify the solution is called the *moduli space*. If there are K monopoles then the moduli space has dimension $4K$. For one monopole this 4-dimensional space is $\mathbf{R}^3 \otimes S^1$. It is made up of the three spatial coordinates $\mathbf{R}^3$ which specify the position of the monopole and a further parameter associated with gauge transformations which have a non-trivial behaviour at infinity. For $K \geq 2$ the moduli space is more complicated, but an explicit metric is known for the case of $K =$

2; see [128]. Since there exist static monopoles solutions, it follows that monopoles do not experience any forces when at rest. However, when they are set in motion they do experience velocity-dependent forces. In general the behaviour of the monopoles is described by the quantum field theory in which they arise. However, if the monopoles have only a small amount of energy above their rest masses then we can approximate their motion in terms of the coordinates of their moduli space; see [129]. The motion is then described by an action whose fields are the coordinates of the moduli space which are given a time dependence corresponding to the monopole motion. If we had only one monopole then three of the moduli X^i, $i = 1, 2, 3$, could be interpreted as the Goldstone bosons (resulting from the breaking of the spatial translations due to the presence of the monopole) and the fourth moduli η would be related to the existence of non-trivial gauge transformations at infinity. As such, the motion is described by $X^i(\tau)$, $i = 1, 2, 3$, $\eta(\tau)$ where $\xi^0 = \tau$. A much more detailed account of these ideas can be found in the lectures of N. Manton in this volume.

In a similar spirit we can consider the moduli space and action which describes the low-energy behaviour of p-brane solitons. The moduli space of the p-brane solitons contains the positions of the p-branes and to describe their low-energy motion we let the moduli depend on the world-volume coordinates of the p-brane. Indeed, we can think of the action of equation (8.1), when we take only its terms to lowest order in derivatives, as the effective action which describes the low-energy behaviour of a single p-brane soliton.

There is an alternative interpretation of the action of equation (8.1). Were we to believe that a certain p-brane was a fundamental object then we might take the action of equation (8.1) to describe its dynamics completely. This was precisely the viewpoint of for the string for many years.

8.2   Types of superbranes

A super p-brane can be viewed as a $(p + 1)$-dimensional *bosonic* submanifold M, with coordinates ξ^n, $n = 0, 1, \ldots, p$, that moves through a target superspace $\underline{M}$ with coordinates

$$Z^{\underline{N}} = (X^{\underline{n}}, \Theta^{\underline{\alpha}}).$$

We use the superspace index convention that $\underline{N}, \underline{M}, \ldots$ and $\underline{A}, \underline{B}, \ldots$ represent the world and tangent space indices of the target space. Later, in Subsection 8.5, use the same symbols without the underlining to represent the corresponding indices of the world surface superspace.

In this section we wish to find which types of branes can exist in which space-time dimensions with particular emphasis on 10 and 11 dimensions. We will also discuss the features of brane dynamics which are generic to all branes, leaving to the next three subseections a more detailed discussion of the dynamics of the specific types of branes.

Superbranes come in various types. The simplest are those whose dynamics can be described entirely by specifying the superworld surface $Z^{\underline{N}}(\xi^n)$ that the p-brane sweeps out in the target space. We refer to these branes as *simple superbranes*; they were also previously called type I branes, not to be confused with type I strings. As we shall see there are other types of branes that have higher spin fields living on their world surfaces. In particular, we will discuss branes that have vectors and second-rank anti-symmetric tensor gauge fields on their world surface.

A superbrane has an action of the form

$$A = A_1 + A_2. \tag{8.4}$$

The first term is given by

$$A_1 = -T \int d^{p+1}\xi \sqrt{-\det g_{mn}} + \cdots \,,$$

where

$$g_{mn} = \partial_n Z^{\underline{N}} \partial_m Z^{\underline{M}} g_{\underline{NM}}, \quad g_{\underline{NM}} = E_{\underline{N}}{}^{\underline{a}} E_{\underline{M}}{}^{\underline{b}} \eta_{\underline{ab}},$$

and $E_{\underline{N}}{}^{\underline{a}}$ is the supervielbein on the target superspace. By abuse of notation we use the same symbol for the world surface metric as for the bosonic case; the reader will be able to distinguish between the two from the context. The $+ \cdots$ denotes terms involving possible world surface fields as well as those that depend on the other background fields. The constant T is the p-brane tension and has the dimensions of $(\text{mass})^{p+1}$ since the action is dimensionless and $X^{\underline{n}}$ has the dimension of $(mass)^{-1}$. If we consider a brane that is static we can think of the tension T as the mass per unit spatial volume of the brane.

The symbol $g_{\underline{NM}}$ is not really a background metric in the usual sense since the sum on the tangent space indices is restricted to be only over the bosonic part. Such a restricted summation is possible as a consequence of the fact that the superspace tangent space group is just the Lorentz group. In fact, as we discussed in Section 5, the theory of local superspace is formulated in terms of the supervielbein since the metric is not uniquely defined.

The second part of the action A_2 of equation (8.4) contains, in addition to others, the term

$$\int d^{p+1}\xi \, \epsilon^{n_1 \cdots n_{p+1}} \partial_{n_1} X^{\underline{m_1}} \cdots \partial_{n_{p+1}} X^{\underline{m_{p+1}}} A_{\underline{m_1} \cdots \underline{m_{p+1}}},$$

where $A_{\underline{m_1} \cdots \underline{m_{p+1}}}$ is a background gauge field.

The background space-time fields can only belong to a supermultiplet that exists in the target superspace. In this review, we will take these supermultiplets to be the supergravity theory that has the background supersymmetry algebra possessed by the p-brane. The field content of the possible supergravity theories can be deduced from the supersymmetry algebra using the methods given in Section 3. However, as we discussed in Section 6, the supergravity theory is essentially unique if the supersymmetry algebra has 32 supercharges and the type of spinors contained in the supersymmetry algebra are specified. Thus, unlike bosonic branes, the background fields are specified by the background supersymmetry of the brane, if that supersymmetry has 32 supercharges. If the brane has 16 supercharges then although one cannot generally specify the background fields uniquely, the possible supergravity theories are very limited. The general coupling of a p-brane to the background fields is complicated, but it always has a coupling to a $(p+1)$-gauge field in the form of equation (8.2). However, since this $(p+1)$-gauge field must be one of the background fields of the supergravity theory, this places an important restriction on which superbranes can arise for a given dimension and target space supersymmetry algebra.

If we consider the case of a super 1-brane, the action of equation (8.4) is just the Green–Schwarz action [130] for the superstring. The case $p = 2$ is often referred to as the *membrane* and the action for the supermembrane in 11 dimensions was found in [151, 157].

The action of equation (8.4) is invariant under super-reparameterizations of the target superspace $\underline{M}$, but only bosonic reparameterizations of the world surface M since the embedded manifold M is a bosonic manifold. It is also invariant under a Fermi-Bose symmetry called κ-supersymmetry. From the viewpoint adopted here this is a complicated symmetry that ensures that the fermions have the correct number of degrees of freedom on-shell and we will discuss it in more detail in the next section. This symmetry relates the terms in A_1 to those in A_2 and vice versa, and in fact fixes uniquely the form of A_2 given that of A_1.

Even if the target space is flat superspace, the supervielbein has a nontrivial dependence on the coordinates given by

$$\partial_m Z^{\underline{N}} E_{\underline{N}}{}^{\underline{a}} = \partial_m X^{\underline{a}} - \frac{i}{2} \bar{\Theta} \gamma^{\underline{a}} \partial_m \Theta, \qquad \partial_m Z^{\underline{N}} E_{\underline{N}}{}^{\underline{\alpha}} = \partial_m \Theta^{\underline{\alpha}}. \qquad (8.5)$$

In this case, the target space super-reparameterization invariance reduces to just rigid supersymmetry:

$$\delta X^{\underline{a}} = \frac{i}{2} \bar{\epsilon} \gamma^{\underline{a}} \Theta, \quad \delta \theta^{\underline{\alpha}} = \epsilon^{\underline{\alpha}}.$$

Note that the action of equation (8.4) does not appear to possess world-surface supersymmetry. We recall for the case of a 1-brane (that is a string) there are two formulations: the Green–Schwarz formulation given here, and the original Neveu–Schwarz–Ramond formulation. The latter is formulated in terms of $X^{\underline{n}}$ and a spinor which, in contrast to above, possess a target space vector index and is a spinor with respect to the 2-dimensional world sheet. This formulation is manifestly invariant under world surface reparameterizations, but not under target space super-reparameterizations. If one goes to light-cone gauge (i.e. in effect static gauge) then the two formulations become the same and so the Green–Schwarz formulation has a hidden world sheet supersymmetry and the Neveu–Schwarz–Ramond (see [172]) a hidden target space supersymmetry provided we carry out the GSO projection.

It is thought (see [173]) that all branes which in their Green–Schwarz formulation admit κ-supersymmetry actually have a hidden word surface supersymmetry. Although the analogue of a Neveu–Schwarz–Ramond formulation is not known for p-branes when $p > 1$, there does exist a superembedding formalism; see [131], [132], [133], [134], [137], [138], [139]. In this formulation the p-brane sweeps out a *supermanifold* which is embedded in the target superspace. Although this approach leads to equations of motions and not an action, it has the advantage that it possesses super-reparmeterisation invariance in both the world surface and the target superspaces. The κ-symmetry is then just part of the super reparameterizations of the world surface. Its particular form is a result of the gauge fixing required to get from the super-embedding formalism to the so-called Green–Schwarz formulation. This origin of κ-symmetry was first found in reference [178] within the context of the point particle. We will comment further on the superembedding formalism in Subsection 8.5.

The superbranes also possess a static gauge for which the bosonic coordinates take the form of equation (8.3) while κ-supersymmetry can be used to set half of the fermions $\Theta^{\underline{\alpha}} = (\Theta^{\alpha}, \Theta^{\alpha'})$ to vanish, i.e. $\Theta^{\alpha} = 0$. While the remaining $D - p - 1$ bosonic coordinates $X^{n'}$ correspond to the Goldstone bosons associated with the breaking of translations by the p-brane, the remaining $\Theta^{\alpha'}$ Goldstone fermions correspond to the breaking of half of the supersymmetries by the brane.

The fields of the p-brane belong to a supermultiplet of the world surface and so must have equal numbers of fermionic and bosonic degrees of freedom on-shell. Let us first consider a p-brane that arises in a theory that has the maximal supersymmetry. This would be the case if the brane describes the low-energy motion of a soliton of a maximal supergravity theory which breaks half the supersymmetry. In this case, if it breaks half of the 32 supersymmetries of the target space, it will, in static gauge, have only 16 $\Theta^{\alpha'}$ which will lead to 8 fermionic degrees of freedom on-shell. If we are dealing with a simple superbrane these must be matched by the coordinates $X^{n'}$ in static gauge which must therefore be 8 in number. Thus if we are in 11 dimensions the only simple superbrane is a 2-brane while if we are in 10 dimensions the only simple superbrane is a 1-brane.

There also exist D-branes whose dynamics requires a vector field A_n, $n = 0, 1, \ldots, p$, living on the brane in addition to the coordinates $X^{\underline{n}}$, $\Theta^{\underline{\alpha}}$ which describe the embedding of the brane in the target superspace. In this case, if we have a brane that arises in the background of a maximal supergravity theory and which breaks half of this supersymmetry, then we again have 8 fermionic degrees of freedom on-shell. This must be balanced by the vector which has $p - 1$ degrees of freedom on-shell and the remaining $D - p - 1$ transverse coordinates from which we deduce that $D = 10$. Hence such D-branes can only exist in 10 dimensions. For D-branes this simple counting argument allows branes for all p, however, as we shall see, not all values of p occur for a given target space supersymmetry. In addition, we will discuss other branes with higher-rank gauge fields living on their world surface.

To find further restrictions on which branes actually exist we can use the argument given above. A super p-brane couples to a $(p + 1)$-gauge field which must belong to the background supergravity theory. Hence to see which branes can exist we need only see which gauge fields are present in the corresponding supergravity theory. When doing this we must bear in mind that if we have a rank $p+1$ gauge field which has a rank-$(p+2)$ field strength $F_{(p+2)}$ we can take its dual to produce a rank-$(D - p - 2)$ field strength $^*F_{(D-p-2)}$ which, if the original field strength is on-shell, has a corresponding rank-$D - p - 3$ dual gauge field $^*A_{D-p-3}$. The dual field strength is defined by

$$^*F_{\underline{n}_1 \cdots \underline{n}_{(D-p-2)}} = \frac{1}{(p+2)!} \epsilon_{\underline{n}_1 \cdots \underline{n}_{(D-p-2)} \underline{n}_{(D-p-1)} \cdots \underline{n}_D} F^{\underline{n}_{(D-p-1)} \cdots \underline{n}_D}.$$

Hence, if the original rank-$(p + 1)$ gauge field couples to a p-brane, the dual gauge field is rank-$(D - p - 3)$ and couples to a $D - p - 4$-brane. In fact D-branes can only couple to the $R \otimes R$ sector of a string theory; see [135].

Let us begin with branes in 11 dimensions. This theory was described in Section 6 and possesses a third-rank gauge field $A_{(3)}$ which should couple to a 2-brane. The dual potential has rank 6, i.e. $^*A_6$, and this couples to a 5-brane. Thus in 11 dimensions we expect only a 2-brane and a 5-brane. The 2-brane is just the simple superbrane we discussed above. The 5-brane has 16 Goldstone fermions and these lead to the same 8 degrees of freedom on-shell; however it has only 5 transverse coordinates leading to only 5 degrees of freedom on-shell. Clearly, we require another 3 degrees of freedom on-shell. These must form a representation of the little group $SO(4)$. If they are belong to an irreducible representation of $SO(4)$, this can only be a second-rank self-dual anti-symmetric tensor. On-shell this corresponds to a second-rank anti-symmetric gauge field whose field strength obeys a self-duality condition. The dynamics of the 5-brane is significantly more complicated than that of simple branes and will be given in Subsection 8.4.

Let us now consider the branes that can couple to the IIA string. In Section 7 we found that the IIA theory has gauge fields and their duals of ranks 1, 2, 3, 5, 6, 7. Of these, ranks 2 and 6 arise in the $NS \otimes NS$ sector of the string while those of ranks 1, 3, 5 and 7 arise in the $R \otimes R$ sector of the string. This suggests the existence of p-branes, for $p = 1$ and 5, that can couple to the $NS \otimes NS$ sector of the IIA string, and p-branes, for $p = 0, 2, 4$ and 6, that can couple to the $R \otimes R$ sector of the IIA string. The latter are the D-branes. Of these only the 1-brane can be a simple brane and it is this brane that couples to the $A_{\mu\nu}$ gauge field of the IIA string. It is in fact the IIA string itself and it is sometimes therefore called the *fundamental string*. The 5-brane is the straightforward dimensional reduction of the 11-dimensional 5-brane that will be discussed in the next section. In the literature an 8-brane is also discussed. This brane is associated with the massive supergravity theory constructed in reference [174]. This theory necessarily contains a cosmological constant c and so has a term $\int d^{10}x\, c\sqrt{-\det g_{nm}}$ in the action. However, we can write this term as $\int d^{10}x\, c\sqrt{-\det g_{nm}} F^{\underline{n}_1 \cdots \underline{n}_{(10)}} F_{\underline{n}_1 \cdots \underline{n}_{(10)}}$ where $F_{\underline{n}_1 \cdots \underline{n}_{(10)}}$ is the curl of a rank-9 gauge field which suggests the existence of an 8-brane; see [185].

Let us now turn to the IIB theory and its branes. The original gauge fields in the IIB theory are of rank 2 in the $NS \otimes NS$ sector and ranks 0, 2 and 4 in the $R \otimes R$ sector. If we include their dual gauge fields we have gauge fields of ranks 2 and 6 in the $NS \otimes NS$ sector and ranks 0, 2, 4, 6, and 8 in the $R \otimes R$ sector. We do not include two rank-4 gauge fields as their field strength is the 5-rank self-dual field strength of the IIB theory. This suggests that there exist 1 and 5-branes which couple to the $NS \otimes NS$ sector of the IIB theory and $-1, 1, 3, 5$ and 7-branes which couple to the $R \otimes R$ sector. The latter are the D-branes. The 1-brane which couples to the rank-2 gauge field in the $NS \otimes NS$ sector is the IIB string itself. The 5-brane which couples to the $NS \otimes NS$ sector is a more complicated object. Note that it is the 3-brane that couples to the rank-4 gauge field whose field strength satisfies a self-duality property. As one might expect this D-brane possesses a self-duality symmetry; see [136]. The $p = -1$-brane which couples to the $R \otimes R$ sector occupies just a point in space-time and so is an instanton.

In fact, all the branes discussed above exist and the possible branes in 11 and 10 dimensions are listed in Table 6. In this table a 'D' or 's' subsrcipt denotes the brane to be a Dirichlet and simple brane respectively.

As we have mentioned above, one can also search for the p-brane solitons of the corresponding supergravity theories. The p-brane actions then correspond to the low-energy motions of these solitons and it has been shown that there exist p-brane solitons for all the above superbranes. We refer the reader to the lectures of G. Gibbons in this volume and the reviews of reference [176].

We now discuss one further guide to determining which superbranes occur

Table 6: Super brane scan

	p								
M-theory			2_S			5			
IIA	0_D	1_S	2_D		4_D	5	6_D		8_D
IIB		1_S+1_D		3_D		$5+5_D$		7_D	

in which theory. Given a p-brane one can construct the following current

$$j^{n \underline{m_1} \cdots \underline{m_p}} = \epsilon^{n n_1 \cdots n_p} \partial_{n_1} X^{\underline{m_1}} \cdots \partial_{n_p} X^{\underline{m_p}},$$

which is obviously conserved. The charge associated with this current is given by

$$Z^{\underline{m_1} \cdots \underline{m_p}} = \int d^p \xi j^{0 \underline{m_1} \cdots \underline{m_p}},$$

where the integral is over the p spatial coordinates of the brane. Our previous discussion on the coupling of the p-brane to a $p + 1$-form gauge background gauge field can be restated as the $p + 1$-form background gauge field couples to the current $j^n \partial_n X^{\underline{m}}$. When computing the space-time supersymmetry algebra in the presence of the p-brane it turns out (see [180]) that the above charge occurs as a central charge. Hence, if a p-brane arises in a particular theory we expect to find its p-form central charge in the corresponding supersymmetry algebra. Clearly, if the supersymmetry algebra does not admit a p-form central charge, the p-brane cannot occur unless further supersymmetry is broken. However, we can turn this argument around and find out which central charges can occur in the appropriate supersymmetry algebra and then postulate the existence of a p-brane for each corresponding p-form central charge. Which central charges can occur for a specified type of supercharge was discussed in Section 1. For example, in equation (1.22), we found that the 11-dimensional supersymmetry algebra with one Majorana spinor, which is the supersymmetry algebra appropriate to 11-dimensional supergravity, had a 2-form and a 3-form central charge. Hence in M-theory we expect from the above argument to find a 2-brane and a 5-brane. This agrees with our previous considerations.

We close this subsection by giving some simple dimensional arguments concerning the tensions of various branes. In 11 dimensions, that is M-theory, we have only one scale, namely the Planck scale m_p. As a result, up to constants, the tensions T_2 and T_5 of the 2-brane and 5-brane can only be given by $T_2 = (m_p)^3$ and $T_5 = (m_p)^6$ respectively. As we discussed in Section 7, we can obtain the IIA supergravity by dimensional reduction of 11-dimensional supergravity on a circle and it is conjectured that IIA string theory can be obtained by reduction of M-theory in the same way. Hence for the IIA theory in 10 dimensions we have two scales, the Planck mass m_p and the radius of compactification of the circle R_{11}. Since the string coupling is dimensionless it must be a function of $m_p R_{11}$ and we found in Section 7 that it is given by $g_s = e^{\langle \sigma \rangle} = (m_p R_{11})^{\frac{3}{2}}$.

From the 2-brane in 11 dimensions we can obtain a 2-brane in 10 dimensions by simply ignoring the dependence of the fields of the 11-dimensional 2-brane on x^{10}. We can also find a 1-brane, that is a string, if we simultaneously reduce and wrap the 2-brane on the circle; see [175]. The tensions of these two objects in 10 dimensions are therefore given by $T_2 = (m_p)^3$ and $T_1 = R_{11}(m_p)^3$ respectively. The factor of R_{11} occurs because the energy per unit length of the 1-brane arises from the energy of the 2-brane on the circle. Similarly, from the 5-brane in 11 dimensions we can obtain a 5-brane and a 4-brane in 10 dimensions with tensions $T_5 = (m_p)^5$ and $T_4 = R_{11}(m_p)^5$. The 1-brane in 10 dimensions is just the fundamental string (i.e. the IIA string) and its tension T_1 can be identified as the square of the string mass; that is $T_1 = R_{11}(m_p)^3 = \frac{1}{4\pi\alpha'} \equiv (m_s)^2$. It is instructive to express the tensions in terms of the variables g_s and m_s appropriate for the IIA string. Using our relationship $R_{11} = (m_p)^{-1}(g_s)^{\frac{2}{3}}$, we find that $R_{11} = g_s/m_s$ and $m_p^3 = m_s^3/g_s$. Substituting into the above tensions we find that $T_2 = (m_s)^3/g_s$, $T_4 = (m_s)^5/g_s$ and $T_2 = (m_s)^6/g_s^2$. We observe the characteristic inverse coupling constant dependence that is typical of nonperturbative solitons. The 2- and 4-branes are Dirichlet branes and have a $(1/g_s)$-dependence while the 5-brane has a $(1/g_s^2)$-dependence. This dependence of the coupling constant on the tension of Dirichlet branes is universal and is related to the occurance of open strings in the theory.

In fact, one can account from M-theory for two of the remaining branes that are associated with the IIA string. The 0-brane and the 6-brane arise from the pp-wave and Kaluza–Klein monopole solutions in 11-dimensional supergravity respectively. The 8-brane is associated with massive supergravity in 10 dimensions and its connection to 11 dimensions is unclear.

8.3 Simple superbranes

In this section we give the complete dynamics of simple superbranes (see [151], [157], [173]) which, we recall, depend only on the embedding coordinates $Z^{\underline{N}} = (X^{\underline{n}}, \Theta^{\underline{\alpha}})$. The action for a simple super p-brane is given by

$$A = A_1 + A_2. \tag{8.6}$$

The first term is given by

$$A_1 = -T \int d^{p+1}\xi \sqrt{-\det g_{mn}},$$

where

$$g_{mn} = \partial_n Z^{\underline{N}} \partial_m Z^{\underline{M}} g_{\underline{NM}}, \quad \text{and} \quad g_{\underline{NM}} = E_{\underline{N}}{}^{\underline{a}} E_{\underline{M}}{}^{\underline{b}} \eta_{\underline{ab}},$$

where $E_{\underline{N}}{}^{\underline{A}}$ is the supervielbein of the background superspace. The second term in the action of equation (8.6) is given by

$$A_2 = -\frac{T}{(p+1)!} \int d^{p+1}\xi \epsilon^{n_1 \cdots n_{p+1}} \partial_{n_1} Z^{\underline{M_1}} E_{\underline{M_1}}^{\underline{A_1}} \cdots \partial_{n_{p+1}} Z^{\underline{M_{p+1}}} E_{\underline{M_{p+1}}}^{\underline{A_{p+1}}} B_{\underline{A_1} \cdots \underline{A_{p+1}}}.$$

In this expression, $B_{\underline{A_1} \cdots \underline{A_{p+1}}}$ is a background superspace $(p+1)$-gauge field referred to superspace tangent indices. Its corresponding superspace $(p+1)$-form is

$$B = \frac{1}{(p+1)!} E^{\underline{A_1}} \cdots E^{\underline{A_{p+1}}} B_{\underline{A_1} \cdots \underline{A_{p+1}}} = dZ^{\underline{N_{p+1}}} \cdots dZ^{\underline{N_1}} B_{\underline{N_1} \cdots \underline{N_{p+1}}},$$

where $E^{\underline{A}} = dZ^{\underline{N}} E_{\underline{N}}{}^{\underline{A}}$.

The geometry of the target superspace is described as in Section 5 using supervielbeins, but now also with the addition of the gauge field B. The covariant objects are the torsions and curvatures as before, but now we have in addition the gauge field strength of the gauge field B. The corresponding superfield strength H is the exterior derivative, in superspace of course, acting on the superspace gauge field; i.e. $H = dB$. The superspace torsions and curvatures satisfy Bianchi identities as in Section 5, but in addition we have the identity

$$D_{\underline{A_1}} H_{\underline{A_2} \cdots \underline{A_{p+2}}} + T_{\underline{A_1} \underline{A_2}}{}^{\underline{B}} H_{\underline{B} \underline{A_3} \cdots \underline{A_{p+1}}} + \text{super cyclic permutations} = 0.$$

In the case that the background superspace is flat then the supervielbein are given as in equation (8.5). However, the gauge field B also has a nontrivial form. The resulting torsions all vanish except for

$$T_{\underline{\alpha}\underline{\beta}}{}^{\underline{a}} = i(\gamma^{\underline{a}} C^{-1})_{\underline{\alpha}\underline{\beta}},$$

and

$$H_{\underline{\alpha}\underline{\beta}\underline{a}_1 \cdots \underline{a}_p} = -i(-1)^{\frac{1}{4}p(p-1)} (\gamma_{\underline{a}_1 \cdots \underline{a}_p} C^{-1})_{\underline{\alpha}\underline{\beta}}.$$

In terms of forms we may express these results as

$$T^{\underline{a}} = \frac{1}{2} i d\overline{\Theta} \gamma^{\underline{a}} d\Theta$$

and

$$H = \frac{i}{2p!} E^{a_p} \cdots E^{\underline{a}_1} d\overline{\Theta} \gamma_{\underline{a}_1 \cdots \underline{a}_p} d\Theta. \tag{8.7}$$

Since H is an exact form $dH = 0$. Using this on equation (8.7) and from the form of $E^{\underline{a}}$ in equation (8.5) we find (see [173]) that $dH = 0$ is equivalent to the condition

$$d\overline{\Theta}\gamma_{\underline{a}} d\Theta d\overline{\Theta} \gamma^{\underline{ab}_1 \cdots \underline{b}_{(p-1)}} d\Theta = 0.$$

Since Θ is Grassmann-odd, $d\Theta$ is Grassmann-even and hence the above condition must hold for a Grassmann-even spinor of the appropriate type that is complex, Weyl, Majorana or Majorana–Weyl, but is otherwise arbitrary. This means in effect that we can discard Θ from the above equation provided we include a projector if the spinor is Weyl, and symmetrise on all the free spinor indices if it is Majorana, and on the first and third and second and fourth indices separately if it is complex. For example, if Θ is Majorana then we find the identity

$$(\gamma_{\underline{a}}C^{-1})_{(\underline{\alpha}\underline{\beta}}(\gamma^{\underline{ab_1}\cdots\underline{b_{p-1}}}C^{-1})_{\underline{\delta}\underline{\epsilon})} = 0.$$

These identities can be reformulated using the appropriate Fierz identity and it can then be shown (see [173]) that they hold if the number of fermion and boson degrees of freedom of the brane are equal. We have already used this condition to find the simple branes in 10 and 11 dimensions.

Finally it remains to discuss the κ-symmetry of the action of equation (8.6). The variation of the coordinates $Z^{\underline{N}}$ under this symmetry when referred to the tangent basis are given by

$$\delta Z^{\underline{N}}E_{\underline{N}}{}^{\underline{a}} = 0, \quad \delta Z^{\underline{N}}E_{\underline{N}}{}^{\underline{\alpha}} = (1 + \Gamma)^{\underline{\alpha}}_{\underline{\beta}}\kappa^{\underline{\beta}},$$

where the local parameter $\kappa^{\underline{\beta}}$ is a world surface scalar, but a tangent space spinor. The matrix Γ is given by

$$\Gamma = \frac{(-1)^{\frac{1}{4}(p-2)(p-1)}}{(p+1)!\sqrt{-\det g_{nm}}}\epsilon^{n_1\cdots n_{p+1}}\partial_{n_1}Z^{\underline{N_1}}E_{\underline{N_1}}^{\underline{a_1}}\cdots\partial_{n_{p+1}}Z^{\underline{N_{p+1}}}E_{\underline{N_{p+1}}}^{\underline{a_{p+1}}}\gamma_{\underline{a_1}\cdots\underline{a_{p+1}}}.$$

The matrix Γ obeys the remarkably simple property $\Gamma^2 = 1$ and as a result $\frac{1}{2}(1 \pm \Gamma)$ are projectors. Clearly a κ of the form $\kappa = \frac{1}{2}(1 - \Gamma)\kappa^1$ leads to no contribution to the symmetry to which therefore only half of the components of κ actually contribute. The remaining half allows us to gauge away the corresponding half of Θ. This property of κ-symmetry is essential for getting the correct number of on-shell fermion degrees of freedom, namely 8. In fact the action is only invariant under κ-symmetry if the background fields obey their equations of motion.

For the case when the background superspace is flat the above κ-symmetry reduces to

$$\delta X^{\underline{n}} = \frac{i}{2}\bar{\Theta}\gamma^{\underline{n}}\delta\Theta, \quad \delta\Theta = (1 + \Gamma)\kappa.$$

The form of the redundancy discussed above in the κ-invariance leads to very difficult problems when gauge-fixing the κ-symmetry and so quantizing the theory. Imposing a gauge condition on the spinor Θ to fix the κ-symmetry can only fix part of the symmetry and leaves unfixed that part of κ which is given by κ^1. A little thought shows that as a result the ghost action itself will have a local symmetry which must itself be fixed. In fact, this process continues and results in an infinite number of ghosts for ghosts corresponding to the infinite set of invariances. This would not in itself be a problem, but it has proved impossible to find a Lorentz-covariant gauge-fixed formulation. The problem of quantizing can also be seen from the viewpoint of the Hamiltonian approach where one finds that the system possesses first and second class constraints that cannot be separated in a Lorentz-covariant manner. The Green–Schwarz action for the string has so far defied attempts to be quantized in a truly covariant manner.

8.4 D-brane dynamics

As we discussed, D-branes have a vector field A_n living on their world surface in addition to the embedding coordinates $X^{\underline{n}}$, $\Theta^{\underline{\alpha}}$. The action for the D-branes is a generalization of that of the simple brane of equation (8.6) to include this vector field A_n. In the absence of background fields other than the background metric and when $\Theta^{\underline{\alpha}} = 0$, the action is just the Born–Infeld action

$$-T \int d^{p+1}\xi \sqrt{-\det(g_{nm} + F_{nm})},$$

where $F_{mn} = 2\partial_{[m}A_{n]}$. The full action was found in references [147]–[150] by using κ-symmetry. Its precise form and the relationship to Dirichlet open strings can be found in the lectures of Costas Bachas in this volume.

8.5 Branes in M-theory

As we have discussed above, there exists a 2-brane and a 5-brane in 11 dimensions. The 2-brane is a simple brane and so its dymanics are those given in Subsection 8.3. The dynamics of the 5-brane is the subject of this section. As we discussed in Subsection 8.2, the 5-brane contains 5 scalar fields and a 16-component spinor corresponding to the breaking of translations and supersymmetry by the 5-brane. It also contains, living on its world surface, an anti-symmetric second-rank tensor gauge field whose field strength obeys a self-duality condition. As this condition is required to obtain the correct number of degrees of freedom, we only need it to hold at the linearized level and it is an important feature of the five brane dynamics that the self-duality condition in the full theory is a very nonlinear condition on the third-rank field strength.

As for other branes, the bosonic indices of the fields on the 5-brane can be decomposed into longitudinal and transverse indices (i.e. for the world target space indices $\underline{n} = (n, n')$) according to the decomposition of the 11-dimensional Lorentz group $SO(1, 10)$ into $SO(1, 5) \times SO(5)$. The corresponding decomposition of the 11-dimensional spin group $\mathrm{Spin}\,(1, 10)$ is into Spin $(1,5) \times$ Spin (5). These spin groups divided by Z_2 are isomorphic to their corresponding Lorentz groups in the usual way. In fact Spin (5) is isomorphic to the group $USp(4)$ which is defined below equation (1.16). Thus the 5-brane possess a Spin $(1,5) \times$ Spin (5) or Spin $(1,5) \times USp(4)$ symmetry.

We now assign the fields of the 5-brane to multiplets of this symmetry. The 5 real scalars are of course Spin $(1,5)$ singlets, but belong to the vector representation of $SO(5)$, $X^{n'}$, $n' = 6, \ldots, 10$. This representation corresponds to the second-rank anti-symmetric tensor representation ϕ_{ij}, $i, j = 1, \ldots, 4$, of the isomorphic $USp(4)$ group which is traceless with respect to the anti-symmetric metric Ω^{ij} of this group. The gauge field B_{mn} is a singlet under $USp(4)$. We recall from Section 1 that in 6 dimensions one cannot have Majorana spinors, but can have symplectic Majorana spinors and even symplectic Majorana–Weyl spinors. The fermions of the 5-brane belong to the 4-dimensional vector representation of $USp(4)$ and are $USp(4)$ symplectic Majorana–Weyl spinors and so obey equation (1.15). Since they are Weyl we can work with their Weyl projected components which take only 4 values as opposed to the usual $8 = 2^3$ components for a 6-dimensional spinor. The spinor indices of the groups Spin $(1,5)$ and $USp(4)$ are denoted by $\alpha, \beta, \ldots = 1, \ldots, 4$, and $i, j, \ldots = 1, \ldots, 4$, respectively and thus the spinors carry the indices Θ^i_α, $\alpha = 1, \ldots 4$, $i = 1, \ldots 4$. The spinors therefore have the required 16 real components.

It is instructive to examine how this spinor arises from the original 11-dimensional spinor $\Theta^{\underline{\alpha}}$ in terms of which the 5-brane dynamics was first formulated. Although we began with spinor indices $\underline{\alpha}$ that took 32-dimensional values, we can, as above for all branes, split these indices into two pairs of indices each taking 16 values $\underline{\alpha} = (\alpha, \alpha')$. In the final 6-dimensional expressions the spinor indices are further written according to the above decomposition of the spin groups and we take $\alpha \to \alpha i$ and $\alpha' \to {}^i_\alpha$ when appearing as superscripts and $\alpha \to \alpha i$ and $\alpha' \to {}^\alpha_i$ when appear as subscripts. It should be clear whether we mean α to be 16- or 4-dimensional depending on the absence or presence of $i, j, \ldots$, indices respectively. For example, we will write $\Theta^{\alpha'} \to \Theta^i_\alpha$.

The fields of the 5-brane belong to the so-called $(2,0)$ tensor multiplet which transforms under $(2,0)$ 6-dimensional supersymmetry. The $(2,0)$ notation means that the supersymmetry parameter is a $USp(4)$ symplectic Majorana–Weyl spinor. By contrast, we note that $(1,0)$ supersymmetry means that the supersymmetry parameter is a $USp(2)$ symplectic Majorana–Weyl spinor while, if the parameter is just a $USp(4)$ symplectic Majorana spinor, the supersymmetry would be denoted by $(2,2)$.

The classical equations of motion of the 5-brane in the absence of fermions and background fields are (see [137])

$$G^{mn}\nabla_m \nabla_n X^{a'} = 0 \, ,$$

and

$$G^{mn}\nabla_m H_{npq} = 0,$$

where the world surface indices are m, n, $p = 0, 1, \ldots, 5$, and a, b, $c = 0, 1, \ldots, 5$, for world and tangent indices respectively. The transverse indices are a', $b' = 6$, 7, 8, 9, 10. We now define the symbols that occur in the equation of motion. The usual induced metric for a p-brane is given, in static gauge and flat background superspace, by

$$g_{mn} = \eta_{mn} + \partial_m X^{a'} \partial_n X^{b'} \delta_{a'b'}.$$

The covariant derivative in the equations of motion is defined with the Levi–Civita connection with respect to the metric g_{mn}. Its action on a vector field T_n is given by

$$\nabla_m T_n = \partial_m T_n - \Gamma_{mn}{}^p T_p,$$

where

$$\Gamma_{mn}{}^p = \partial_m \partial_n X^{a'} \partial_r X^{b'} g^{rs} \delta_{a'b'}.$$

We define the world surface vielbein associated with the above metric in the usual way $g_{mn} = e_m{}^a \eta_{ab} e_n{}^b$. There is another inverse metric G^{mn} which occurs in the equations of motion and it is related to the usual induced metric given above by the equation

$$G^{mn} = (e^{-1}){}_c{}^m \eta^{ca} m_a{}^d m_d{}^b (e^{-1}){}_b{}^m,$$

where the matrix m is given by

$$m_a{}^b = \delta_a{}^b - 2h_{acd} h^{bcd}.$$

The field h_{abc} is an anti-symmetric 3-form which is self-dual:

$$h_{abc} = \frac{1}{3!} \varepsilon_{abcdef} h^{def},$$

but it is not the curl of a 3-form gauge field. It is related to the field $H_{mnp} = 3\partial_{[m} B_{np]}$ which appears in the equations of motion and is the curl of a gauge field, but H_{mnp} is not self-dual. The relationship between the two fields is given by

$$H_{mnp} = e_m{}^a e_n{}^b e_p{}^c (m^{-1}){}_c{}^d h_{abd}.$$

Clearly, the self-duality condition on h_{abd} transforms into a condition on H_{mnp} and vice versa for the Bianchi identity $dH = 0$.

The 5-brane equations of motion were found in reference [137] for arbitrary background fields and to first order in the fermions; we refer the reader to this reference for the construction.

The above 5-brane equations of motion were found using the so-called embedding formalism. This formalism, including the superspace embedding condition of equation (8.8), was first given within the context of the superparticle in reference [178]. Further work on the superparticle was carried out in references [195] and [196]. The superembedding formalism was first applied to p-branes in references [131], [132], [197] and [133], and to the M-theory 5-brane in [134]. The above 5-brane equations of motion were first derived in [137]. The form of the 5-brane equations of motion and the relationship to the D4-brane of the IIA theory was discussed in reference [137]. Reference [138] contains a review of the embedding formalism. The superembedding approach was also used to find p-brane actions and equations of motion in references [141] and [198].

We now briefly explain the simple idea that underlies this formalism. We consider a target or background superspace $\underline{M}$ in which a brane sweeps out a superspace M. On each of these superspaces we have a set of preferred frames or supervielbeins. The frame vector fields on the target manifold $\underline{M}$ and the 5-brane world surface M are denoted by $E_{\underline{A}} = E_{\underline{A}}{}^{\underline{M}} \partial_{\underline{M}}$ and $E_A = E_A{}^M \partial_M$ respectively. We recall that we use the superspace index convention that $\underline{N}, \underline{M}, \ldots$, and $\underline{A}, \underline{B}, \ldots$, represent the world and tangent indices of the target superspace $\underline{M}$ while $N, M, \ldots$, and $A, B, \ldots$, represent the world and tangent space indices of the embedded superspace M.

Since the supermanifold M is embedded in the supermanifold $\underline{M}$, the frame vector fields of M must point somewhere in $\underline{M}$. Exactly where they point is encoded in the coefficients $E_A{}^{\underline{A}}$ which relate the vector fields E_A and $E_{\underline{A}}$, i.e.,

$$E_A = E_A{}^{\underline{A}} E_{\underline{A}}.$$

Applying this relationship to the coordinate $Z^{\underline{M}}$ we find the equation

$$E_A{}^{\underline{A}} = E_A{}^N \partial_N Z^{\underline{M}} E_{\underline{M}}{}^{\underline{A}}.$$

It is now straightforward to express the torsion and curvature tensors of M in terms of those of $\underline{M}$ plus terms involving a suitable covariant derivative of $E_A{}^{\underline{A}}$; one finds that

$$\nabla_A E_B{}^{\underline{C}} - (-1)^{AB} \nabla_B E_A{}^{\underline{C}} + T_{AB}{}^C E_C{}^{\underline{C}} = (-1)^{A(B+\underline{B})} E_B{}^{\underline{B}} E_A{}^{\underline{A}} T_{\underline{AB}}{}^{\underline{C}}, \quad (8.8)$$

where the derivative ∇_A is covariant with respect to both embedded and target superspaces, that is, it has connections which act on both underlined and non-underlined indices.

The tangent space of the superspaces $\underline{M}$ and M can be divided into their Grassmann-odd and -even sectors, that is the odd and even pieces of M are spanned by the vector fields E_α and E_a respectively and similarly for $\underline{M}$. The superembedding formalism has only one assumption; the odd tangent space of M should lie in the odd tangent space of $\underline{M}$. This means that

$$E_\alpha{}^{\underline{a}} = 0.$$

To proceed one substitutes this condition into the relationships of equation (8.8) between the torsions and curvatures of M in terms of $\underline{M}$ and analyses the resulting equations in order of increasing dimension. For example at dimension 0 one finds the equation

$$E_{\underline{a}}{}^{a}E_{\underline{b}}{}^{b}T_{\underline{ab}}{}^{\underline{c}} = T_{\underline{ab}}{}^{c}E_{c}{}^{\underline{c}} \ .$$

This procedure is much the same as for the usual superspace Bianchi identities for super Yang–Mills and supergravity theories. For the 2-brane and 5-brane of M-theory this procedure yields:

(a) the equations of motion for the fields of the brane,

(b) the equations of motion for the background fields,

(c) the geometry of the embedded manifold, that is its torsions and curvatures.

Finding the equations of motion of the supergravity background fields means finding the superspace constraints on the target superspace torsions and curvatures.

Although the embedding condition of equation (8.8) is very natural in that, as we saw in Section 5, all the geometry of superspace is contained in the odd sectors of the tangent space of supermanifolds, its deeper geometrical significance is unclear. However, the power of this approach became evident once it was shown to lead to the correct dynamics for the most sophisticated brane, the M-theory 5-brane. Although the above results hold for many branes, they do not hold for all unless the embedding condition is in general supplemented by a further condition.

The appearance of a metric in the equations of motion that is different to the usual induced metric has its origins in the fact that the natural metric which appears on the world surface of the 5-brane has an associated inverse vielbein denoted by $(E^{-1})_{a}{}^{m}$, which is related in the usual way through $G^{mn} = (E^{-1})_{a}{}^{m}(E^{-1})_{b}{}^{n}\eta^{ab}$. The relationship between the two inverse vielbeins is $(e^{-1})_{a}{}^{m} = (m^{-1})_{a}{}^{b}(E^{-1})_{b}{}^{m}$.

There is another formulation of the dynamics of the 5-brane given in references [141] and [142]. Although this formulation involves an action, this is not necessarily an advantage, as has been pointed out in reference [143]. Reference [141] used the superembedding formalism, but in a different way to that considered in this section.

The above equation of motion for the 5-brane admits two interesting solutions corresponding to 1-branes (self-dual strings) – see [144] – and 3-branes – see [145]. The 3-brane solution has been used (see [146], [177]) to derive the complete low-energy effective action of the $N = 2$ Yang–Mills theory. Indeed, the 5-brane may self-intersect on a 4-manifold which can be considered as this 3-brane. Just like the monopoles discussed at the beginning of this section we may consider the moduli space of 3-branes. The low-energy motion of several 3-branes can then be described in terms of their moduli which now depend on the world surface of the 3-brane. The corresponding action for the low-energy motion of these 3-branes is then a 4-dimensional quantum field theory with $N = 2$ supersymmetry which is none other that the complete chiral effective action of the spontaneously broken $N = 2$ supersymmetry gauge theory. Thus from the *classical* dynamics of the 5-brane we can deduce the chiral effective action of the spontaneously broken $N = 2$ Yang–Mills theory which includes an infinite number of instanton corrections, only the first two of which can be calculated using known instanton techniques.

9 String Duality

In this section we use our previous discussions on supergravity theories and brane dynamics to give some of the evidence for string dualities. We restrict our attention to the relationships between the IIA and IIB strings and M-theory and only aim to give some introductory remarks. A more complete discussion can be found in the lectures of Ashoke Sen in this volume.

Dualities in string theory can be of several different types. There are dualities that relate one string theory to another and dualities that relate a given string theory to itself. We refer to the latter as self-dualities. Dualities can also be classified by whether they map the perturbative regime of the theory to the same perturbative regime or whether they map the perturbative to the non-perturbative regime. Roughly speaking, the perturbative and non-perturbative regimes correspond to the small and large coupling constant regimes of the theory respectively. Currently, our ability to calculate systematically in quantum field theories is limited to the perturbative regime where one can extract meaningful answers through perturbation theory using the coupling constant as the parameter. In fact, in the absence of a special non-renormalization theorem, such expansions, which are derived from the path integral, are not convergent. However, for certain quantum field theories the first few terms give increasingly more accurate results and one can, by resumming the series, find the result to any desired accuracy as a matter of principle. To determine the range of values of the coupling constant for which the theory is in its pertubative regime is not always very straightforward and it is found by studying the behaviour of the amplitudes as a complex function of the coupling constant. In some theories there are no values of the coupling constant for which the perturbation series can be resummed and so these theories have in effect no perturbative regime.

In fact, many interesting phenomena are outside the range of perturbation theory. Such phenomena are often related to the presence of instantons and solitons in the theory. It is of course one of the outstanding problems of theoretical physics to understand properly non-perturbative effects, such as quark confinement.

Dualities which map from the perturbative to the same perturbative regime of a theory are mapping from a regime of the theory which is relatively well understood to the same region. As a result, such dualities are straightforward to check. The best example of this type of duality in string theory is so-called T-duality (see [183], [191], [192]) which occurs in string theories that are compactified. The simplest example of T-duality occurs for the closed bosonic string theory compactified on a circle of radius R. One finds that the theory is invariant under the transformation $R \to \alpha'/R$. To be more precise, the mass spectrum of the physical states and their scattering amplitudes are invariant under this transformation; see [184].

More generally, we can consider the T-duality which occurs in a string theory which has been compactified on d of its dimensions. In string theory we can make separate compactifications for the left- and right-modes of the string and as a result the relevant torus is $T^{(d,d)} = \mathbf{R}^{(d,d)}/\Lambda^{(d,d)}$ where $\Lambda^{(d,d)}$ is the Lorentzian lattice that generates the torus. The lattice is formed from the momenta in the left and right directions (p_L, p_R). This lattice is just the momenta on the torus, while the Lorentzian signature arises from the scalar product encoded in the constraint $(L_0 - \bar{L}_0)\psi = 0$. This constraint also implies the lattice is even. Modular invariance of this string theory requires that $\Lambda^{(d,d)}$ be an even self-dual Lorentzian lattice. Clearly, the resulting string theory depends on the torus, or equivalently the self-dual lattice used in the compactification. It turns out that all even self-dual Lorentzian lattices that can occur in such string compactifications are uniquely classified by the coset (see [190])

$$O(d; \mathbf{R}) \otimes O(d; \mathbf{R}) \backslash O(d, d; \mathbf{R}) / O(d, d; \mathbf{Z}).$$

Consequently, the resulting string theories are also classified by this coset. Clearly, we can act with the group $O(d, d; \mathbf{Z})$ on the basis vectors of a given lattice, and it will take that lattice to itself. Hence we must also divide out by this group. The d^2 parameters of the coset can be accounted for by the d^2 expectation values of the metric and anti-symmetric tensor of the string in the compactified directions; see [161]. The corresponding string theory is invariant under the duality symmetry $O(d, d; \mathbf{Z})$. The T-dualities map a given theory to itself and are therefore an example of a self-duality. The T-duality symmetries are well-understood and have been shown to hold to all orders in string perturbation theory; see [184].

Dualities which map from perturbative to non-perturbative regimes are not very well-understood. The problem with verifying such dualities is that they relate quantities in the perturbative regime, which in general can be reliably computed, to quantities in the non-perturbative regime which in general cannot be calculated. Therefore such dualities are in general difficult to test. However, for certain supersymmetric theories, some properties of the theory in the non-perturbative regime are reliably known. One example is the low-energy effective action of a string theory that possesses a supersymmetry with 32 supercharges. Such a low-energy effective action is none other than a supergravity theory. We recall that supergravity theories can have an underlying supersymmetry algebra that has at most 32 supercharges. Furthermore, for one of the maximal supergravity theories, the action is essentially uniquely determined by the underlying supersymmetry algebra. For example, if we are in 10 dimensions then a supersymmetry algebra with 32 supercharges has either two Majorana–Weyl spinors of the same chirality or two Majorana–Weyl spinors of the opposite chirality and the corresponding unique supergravity theories are the IIB and the IIA supergravities respectively. Therefore, given any string theory which is invariant under a space-time supersymmetry that has 32 supercharges, then the low-energy effective action of this string theory must be the unique maximal supergravity in that dimension for the particular maximal supersymmetry algebra involved. Hence if we suspect that two string theories are related by a duality symmetry, one test we can apply is to find if the suspected duality relates their low-energy effective actions which are the unique corresponding maximal supergravities. We can turn this approach around and consider which string theories have low-energy effective theories (i.e supergravity theories) that can be related by a duality symmetry and then consider if these symmetries can be promoted to string dualities.

A further consequence of these considerations concerns string theories that are invariant under a space-time supersymmetry with 32 supercharges and that are self-dual. Since their low-energy effective actions in both the strong and weak coupling limits are the same low-energy supergravity, it follows that this self-duality symmetry must be a symmetry of the maximal supergravity theory. Thus to look for self-duality symmetries we can examine the symmetries of the maximal supergravities and wonder which to promote to a string duality.

One other property that we can reliably establish in all coupling constant regimes of a supersymmetric theory is the existence of the BPS states discussed in Section 3. Recall from there that these supermultiplets have fewer states than a supermultiplet with a generic mass, and their existence relies on their mass being equal to one of the central charges that appear in the supersymmetry algebra. For these states to disappear as one changes the coupling constant would require the abrupt existence of additional degrees of freedom and this does not usually occur.

Equipped with this strategy we now examine the relationships between the maximal supergravity theories. For simplicity let us consider the relationships between the maximal supergravities in 11, 10 and 9 dimensions. These are the 11-dimensional supergravity and the IIA and IIB theories in 10 dimensions and the single 9-dimensional maximal supergravity. Their relationships are set out in Figure 1. We recall from Section 7 that IIA supergravity can be obtained from 11-dimensional supergravity by reduction on a circle, and that the radius R of this circle and the coupling constant g_s of the IIA string theory are related by equation (7.10). The IIB theory has a supersymmetry algebra with two Majorana–Weyl spinors of the same chirality and so cannot be obtained from 11-dimensional supergravity by dimensional reduction. It does, however, possess an $SL(2, \mathbf{R})$ symmetry. Finally, we found that if we reduce either the IIA or IIB theory to 9 dimensions then we obtain the same unique supergravity.

Let us now consider what these relationships suggest for the corresponding string theories. Let us begin by considering which of the theories in 10 dimensions could be self-dual. The obvious candidate is the IIB theory whose low-energy effective action has an $SL(2, \mathbf{R})$ symmetry; see [110]. It is natural to consider this group or one of its subgroups to be a symmetry of the IIB string theory. The IIB supergravity has two second-rank tensor gauge fields $B^1_{\mu\nu}$ and $B^2_{\mu\nu}$ which transform into each other under $SL(2, \mathbf{R})$ as in equation (7.32). From the viewpoint of the IIB string, these fields belongs to the $NS \otimes NS$ and $R \otimes R$ sectors respectively. As we explained in Subsection 8.2 these target space gauge fields couple to the 1-branes which are the IIB string itself and the Dirichlet 1-brane.

Let us consider an object which is charged with respect to both gauge fields $B^1_{\mu\nu}$ and $B^2_{\mu\nu}$ and denote the charges by q^1 and q^2 respectively. Since the charges they carry are proportional to their corresponding field strengths, they are rotated into each other under $SL(2, \mathbf{R})$. Using equation (7.32), we conclude that under an $SL(2, \mathbf{R})$ transformation the charges are changed (see [110]) according to

$$\begin{pmatrix} q^{1\prime} \\ q^{2\prime} \end{pmatrix} = \begin{pmatrix} a & b \\ c & d \end{pmatrix} \begin{pmatrix} q^1 \\ q^2 \end{pmatrix}. \qquad (9.1)$$

However, the analogue of the Dirac quantization condition for branes (see [191]) implies that the charges are always quantized in integers once we adopt suitable units. It is easy to see that the maximal subgroup of $SL(2, \mathbf{R})$ which preserves such a charge quantization is $SL(2, \mathbf{Z})$. The charge quantization condition can be expressed as the statement that one can only ever find whole fundamental strings and $D1$-branes and their bound states.

Thus we can at most choose the group $SL(2, \mathbf{Z})$ to be a symmetry of IIB string theory. This symmetry includes the transformation of equation (7.33) and so takes the weak-coupling regime theory to the strong coupling regime of the IIB string.

Clearly, in the IIB string theory there exist elementary string states of unit charge with respect to the gauge field $B^1_{\mu\nu}$, but no charge with respect to the other gauge field $B^2_{\mu\nu}$. One such state has charge $\begin{pmatrix} 1 \\ 0 \end{pmatrix}$. Acting with an $SL(2, \mathbf{Z})$ transformation we find, using equation (9.1), states with the charges $\begin{pmatrix} a \\ c \end{pmatrix}$ where a and c are integers and are the charges with respect $B^1_{\mu\nu}$ and $B^2_{\mu\nu}$ respectively. While the states charged with respect to just $B^1_{\mu\nu}$ are the elementary string states which occur in the perturbative domain, the $SL(2, \mathbf{Z})$ rotated states carry charge with respect to $B^1_{\mu\nu}$ and $B^2_{\mu\nu}$ and are non-perturbative in nature. Indeed, the elementary states of the IIB string cannot be charged with respect to $B^2_{\mu\nu}$. Hence, the existence of an $SL(2, \mathbf{Z})$ symmetry in the theory implies that there must exist states with the above charges for all integers a and c which are relatively prime, that is have no common factor. The latter condition follows from the $SL(2, \mathbf{Z})$ condition $ad - bc = 1$, since in this case $ad - bc$ would contain a common factor and so could not equal one. It was shown in [164], by considering parallel Dirichlet 1-branes, that these states actually exist and this provides us with an important consistency check on the conjectured $SL(2, \mathbf{Z})$ symmetry in the string theory.

It was in reference [152] that it was first suggested that string theory could possess a duality symmetry that transformed weak- to strong-coupling regimes. In particular, it was known that that the low-energy effective action for the heterotic string theory compactified on T^6 possessed an $SL(2, \mathbf{R})$ symmetry and the authors of references [152] and [153] suggested that the $SL(2, \mathbf{Z})$ subgroup be a symmetry of the the heterotic string theory compactified on T^6. By considering the action of this symmetry on BPS states, in references [152]–[154] and [155] evidence for this conjecture was given. In reference [156], it was suggested that the $SL(2, \mathbf{Z})$ subgroup of the known $SL(2, \mathbf{R})$ symmetry (see [110]) of the IIB low-energy effective action be a symmetry of the full IIB theory. Indeed, reference [156] conjectured that if we changed the field from $\mathbf{R}$ to $\mathbf{Z}$ in the known coset symmetry of the low-energy effective action (i.e. supergravity theory) then we will find a symmetry of the corresponding string theory. To be precise it was suggested that since the supergravity theories possessed scalars which were contained in coset spaces G/H that the corresponding string theory have a symmetry $G(\mathbf{Z})$ where $G(\mathbf{Z})$ is the group G with its field changed form $\mathbf{R}$ to $\mathbf{Z}$. The coset space symmetries of the maximal supergravity theories are listed in Table 5. For the IIA or IIB string compactified on a 6-torus, the low-energy effective action is the $N = 8$ supergravity in 4 dimensions and so we expect to find a $E_7(\mathbf{Z})$ symmetry. These references followed the pattern that occurs in the $N = 4$ Yang–Mills theory which was earlier given in reference [193].

Now we turn our attention to the relationship between the 11-dimensional

supergravity (see [106]) and the IIA supergravity [107–109]. As explained
in Subsection 7.2 the former theory results from the latter if we compactify
on a circle; see [107]–[109]. We found in equation (7.10) that the IIA string
coupling g_s and the radius of compactification R_{11} are related by $g_s \propto R_{11}^{\frac{3}{2}}$.
Clearly, in the strong-coupling limit of the IIA string i.e. as $g_s \to \infty$ the radius
$R_{11} \to \infty$. However, in this limit the circle of compactification becomes flat
and one expects to recover an 11-dimensional theory whose low-energy limit
is 11-dimensional supergravity. This realization has lead to the conjecture in
[123] and [124] that the strong-coupling limit of IIA theory defines a consistent
theory called M-theory which possesses 11-dimensional Poincaré invariance
and has 11-dimensional supergravity as its low-energy effective action.

    We now summarise some of the evidence for the existence of M-theory.
When compactifying the 11-dimensional supergravity in Subsection 7.2, in
addition to the massless modes of the IIA supergravity theory we found
Kaluza–Klein modes . These Kaluza–Klein modes have a mass n/R_{11} which
hve charge $n/R = (n/R_{11})g_s^{(2/3)}$ with respect to the $U(1)$ gauge field B_μ in
the IIA theory that arises from the graviton in 11 dimensions. Since the
low-energy effective action of M-theory is 11-dimensional supergravity, these
Kaluza–Klein modes are also present in M-theory. However, if we are to believe that the strong coupling limit of IIA string theory is M-theory then these
Kaluza–Klein modes should be also be present in the IIA string. In fact, the
Kaluza–Klein states belong to massive supermultiplets that have fewer states
than a supermultiplet with a generic mass and so they are the so-called BPS
states considered in Section 3 whose mass is equal to their central charge. As
we discussed at the end of that section, the existence of BPS states must be
present in the strong-coupling regime of a theory if they are present in the
weak-coupling regime and vice versa. As such, we should find the analogue
of the Kaluza–Klein states at all coupling constant regimes of the IIA string
theory.

    Hence an essential test of the conjecture is to find the analogues of the
Kaluza–Klein states in the IIA theory. The first evidence for the existence of
these states in the IIA string was the realization that the IIA supergravity
admitted solitonic states of the correct mass and charge; see [123]. The
elementary states, that is perturbative states of the IIA string, are not charged
with respect to the gauge field B_μ since it belongs to the $R \otimes R$ sector of the
IIA string. However, as we discussed in Section 7, Dirichlet 0-branes do
couple to the B_μ gauge field in the $R \otimes R$ sector of the IIA string. The
Dirichlet 0-branes are non-perturbatitve in nature, but this identification is
in accord with fact that the masses of the Kaluza–Klein states are given by
$(nR_{11}) = n(m_s/g_s)$ when expressed in terms of the string coupling constant.
These states becomes very large as the coupling constant becomes small, in a
manner typical of non-perturbative solitons. In the limit of strong-coupling

they become massless. The appearance of an infinite number of massless modes indicates the transition to a theory in which a dimension has become decompactified.

The dynamics of a single such Dirichlet 0-brane was outline in Section 8. In a flat background, it consists of a supersymmetric generalization of Born–Infeld action in 1 dimension. At low energy this theory in just the dimensional reduction of $D = 10$ $U(1)$ gauge theory to 1 dimension. Such a single Dirichlet 0-brane has the correct mass and charge to be identified with the lowest mass Kaluza–Klein state. It is thought that the higher mass Kaluza–Klein states can be identified with bound states of Dirichlet 0-branes at threshold, that is, bound states that have the same energy as the lowest energy state of two particles. The dynamics of several parallel Dirichlet 0-branes is described by a supersymmetric generalization of Born–Infeld theory which now carries a $U(N)$ gauge group if N is the number of parallel Dirichlet 0-branes. At low energy, this theory is just a dimensional reduction of $D = 10$ Yang–Mills theory to 1 dimension. Just as one quantizes the point particle to discover that it corresponds to the Klein–Gordon equation in quantum field theory and quantizes a 1-brane (i.e. string) to discover its particle spectrum, one must quantize this supersymmetric generalization of non-Abelian Born–Infeld theory to discover the states in the IIA string arising from the presence of the Dirichlet 0-branes. It is thought that this quantum mechanical system does indeed have bound states that have the correct mass and charge to be identified with the Kaluza–Klein states; see [182].

The last similarity between the maximal supergravity theories is the equivalence of the IIA and IIB supergravity theories when compactified on a circle; see [107]. Hence, if one compactifies the IIA and IIB string theories on circles of radius R_A and R_B respectively one obtains two 9-dimensional string theories which have the same low effective action, since the maximal supergravity theory in this dimension is unique. This suggests that these two 9-dimensional string theories are really the same theory. However, the variables in which the two theories are found after compactification may be related in a non-trivial way. In fact, the two IIA and IIB string theories compactified on circles of radius R_A and R_B respectively are the same theory, but they are related by a T-duality transformation such that $R_A \to \frac{\alpha'}{R_B}$; see [186]–[188]. In the limit $R_A \to \infty$ the theory decompactifies and one recovers the IIA string in 10 dimensions. Just as for the reduction of the IIA theory, the radius of the circle is related to a scalar field in the 9-dimensional theory that appears from the metric in 10 dimensions, and as a result each value of the radius corresponds to a point in the moduli space of the 9-dimensional theory. The limit in which $R_A \to \infty$, is just a limit to a point in moduli space. Similarly, in the limit $R_B \to \infty$, which is also the limit $R_A \to 0$, we recover the IIB string theory in 10 dimensions. Hence, different limits in the moduli space of the

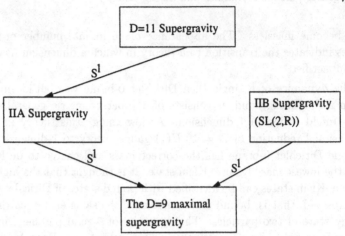

Figure 1: Relations between maximal supergravities

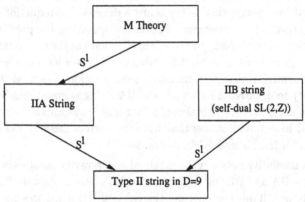

Figure 2: Relations between the IIA and IIB string theories and M-theory

9-dimensional theory lead to different 10-dimensional theories. Starting with say, the IIA theory in 10 dimensions, one can compactify on a circle, carry out a T-duality transformation and then take the appropriate limit to recover the IIB theory in 10 dimensions. Because of the need to take limits one cannot, in general, simply carry out the T-duality directly in the 10-dimensional theories.

Of course the IIA supergravity theory in 9 dimensions can be obtained from 11-dimensional supergravity by compactification on a torus and so one may expect that the IIA string in 9 dimensions can be obtained from compactification of M-theory on a torus. The resulting 9-dimensional theories will inherit the isometries of the torus as symmetries. In fact it can be shown that the $SL(2, \mathbf{Z})$, which acts on the basis vectors of the lattice which underlies the torus, can be identified with the $SL(2, \mathbf{Z})$ of the IIB theory after it has been compactified on a circle.

The relations between M-theory and the IIA and IIB string theories in 10 dimensions and the one string theory in 9 dimensions discussed above are summarized in Figure 2. The similarity, for the reasons explained above, between this table and Table 1 for the corresponding supergravity theories is obvious. It is relatively straightforward, using similar arguments, to incorporate the heterotic and type I string theories into these pictures. We refer to the lectures of Ashoke Sen contained in this volume for a more complete treatment of string duality.

Acknowledgement

I wish to thank World Scientific Publishing for their kind premission to reproduce some of the material from reference [0], Neil Lambert for suggesting many improvements, David Olive for discussions and Rachel George for help in preparing the manuscript.

References

[0] P. West, *Introduction to Supersymmetry and Supergravity*, (1990), Extended and Revised Second Edition, World Scientific Publishing.

[1] Y.A. Golfand and E.S. Likhtman, *JETP Lett.* **13** 323 (1971).

[2] D.V. Volkov and V.P. Akulov, *Pis'ma Zh. Eksp. Teor. Fiz.* **16** 621 (1972); *Phys. Lett.* **46B** 109 (1973).

[3] J. Wess and B. Zumino, *Nucl. Phys.* **B70** 139 (1974).

[4] S. Coleman and J. Mandula, *Phys. Rev.* **159** 1251 (1967).

[5] R. Hagg, J. Lopuszanski and M. Sohnius, *Nucl. Phys.* **B88** 61 (1975).

[6] P. van Nieuwenhuizen and P. West, *Principles of Supersymmetry and Supergravity*, forthcoming book to be published by Cambridge University Press.

[7] P. Ramond, *Phys. Rev.* **D3** 2415 (1971); A. Neveu and J.H. Schwarz, *Nucl. Phys.* **B31** 86 (1971); *Phys. Rev.* **D4** 1109 (1971); J.-L. Gervais and B. Sakita, *Nucl. Phys.* **B34** 477, 632 (1971); F. Gliozzi, J. Scherk and D.I. Olive, *Nucl. Phys.* **B122** 253 (1977).

[8] J. Wess and B. Zumino, *Nucl. Phys.* **B78** 1 (1974).

[9] For a discussion of the Noether procedure in the context of supergravity, see: S. Ferrara, D.Z. Freedman and P. van Nieuwenhuizen, *Phys. Rev.* **D13** 3214 (1976).

[10] S. Ferrara and B. Zumino, *Nucl. Phys.* **B79** 413 (1974); A. Salam and J. Strathdee, *Phys. Rev.* **D11** 1521 (1975).

[11] A. Salam and J. Strathdee, *Nucl. Phys.* **B80** 499 (1974); M. Gell-Mann and Y. Neeman, (1974) unpublished; W. Nahm, *Nucl. Phys.* **B135** 149 (1978). For

a review, see: D.Z. Freedman in *Recent Developments in Gravitation*, Cargèse (1978), M. Levy and S. Deser (eds.) (Gordon and Breach, 1979); S. Ferrara and C. Savoy in *Supergravity '81*, S. Ferrara and J. Taylor (eds.) (Cambridge University Press, 1982).

[12] E.P. Wigner, *Ann. of Math.* **40** 149 (1939).

[13] P. van Nieuwenhuizen, *Phys. Rep.* **68** 189 (198 1).

[14] D. Freedman, P. van Nieuwenhuizen and S. Ferrara, *Phys. Rev.* **D13** 3214 (1976); *Phys. Rev.* **D14** 912 (1976).

[15] S. Deser and B. Zumino, *Phys. Lett.* **62B** 335 (1976).

[16] K. Stelle and P. West, *Phys. Lett.* **B74** 330 (1978).

[17] S. Ferrara and P. van Nieuwenhuizen, *Phys. Lett.* **B74** 333 (1978).

[18] A. Chamseddine and P. West, *Nucl. Phys.* **B129** 39 (1977).

[19] P. Townsend and P. van Nieuwenhuizen, *Phys. Lett.* **B67** 439 (1977).

[20] J. Wess and B. Zumino, *Nucl. Phys.* **B78** 1 (1974).

[21] S. Ferrara in *Proceedings of the 9th International Conference on General Relativity and Gravitation* (1980), Ernst Schmutzer (ed.)

[22] M. Sohnius, K. Stelle and P. West, in *Superspace and Supergravity*, S.W. Hawking and M. Rocek (eds.), (Cambridge University Press, 1981).

[23] A. Salam and J. Strathdee, *Phys. Lett.* **51B** 353 (1974); P. Fayet, *Nucl. Phys.* **B113** 135 (1976).

[24] M. Sohnius, K. Stelle and P. West, *Nucl. Phys.* **B17** 727 (1980); *Phys. Lett.* **92B** 123 (1980).

[25] P. Fayet, *Nucl. Phys.* **B113** 135 (1976).

[26] P. Breitenlohner and M. Sohnius, *Nucl. Phys.* **B178** 151 (1981); M. Sohnius, K. Stelle and P. West, in *Superspace and Supergravity*, S.W. Hawking and M. Rocek (eds.), (Cambridge University Press, 1981).

[27] P. Howe and P. West, '$N = 1$, $d = 6$ Harmonic Superspace', in preparation.

[28] G. Sierra and P.K. Townsend, *Nucl. Phys.* **B233** 289 (1984); L. Mezincescu and Y.P. Yao, *Nucl. Phys.* **B241** 605 (1984).

[29] F. Gliozzi, J. Scherk and D. Olive, *Nucl. Phys.* **B122** 253 (1977); L. Brink, J. Schwarz and J. Scherk, *Nucl. Phys.* **B121** 77 (1977).

[30] In this context, see: M. Rocek and W. Siegel, *Phys. Lett.* **105B** 275 (1981); V.0. Rivelles and J.G. Taylor, *J. Phys. A. Math. Gen.* **15** 163 (1982).

[31] S. Ferrara, J. Scherk and P. van Nieuwenhuizen, *Phys. Rev. Lett.* **37** 1035 (1976); S. Ferrara, F. Gliozzi, J. Scherk and P. van Nieuwenhuizen, *Nucl. Phys.* **B117** 333 (1976); P. Breitenlohner, S. Ferrara, D.Z. Freedman, F. Gliozzi, J. Scherk and P. van Nieuwenhuizen, *Phys. Rev.* **D15** 1013 (1977); D.Z. Freedman, *Phys. Rev.* **D15** 1173 (1977).

[32] S. Ferrara and P. van Nieuwenhuizen, *Phys. Lett.* **76B** 404 (1978).

[33] K.S. Stelle and P. West, *Phys. Lett.* **77B** 376 (1978).

[34] S. Ferrara and P. van Nieuwenhuizen, *Phys. Lett.* **78B** 573 (1978).

[35] K.S. Stelle and P. West, *Nucl. Phys.* **B145** 175 (1978).

[36] M. Sohnius and P. West, *Nucl. Phys.* **B203** 179 (1982).

[37] R. Barbieri, S. Ferrara, D. Nanopoulos and K. Stelle, *Phys. Lett.* **113B** 219 (1982).

[38] E. Cremmer, S. Ferrara, B. Julia, J. Scherk and L. Girardello, *Phys. Lett.* **76B** 231 (1978).

[39] E. Cremmer, B. Julia, J. Scherk, S. Ferrara, L. Girardello and P. van Nieuwenhuizen, *Nucl. Phys.* **B147** 105 (1979).

[40] E. Cremmer, S. Ferrara, L. Girardello and A. Van Proeyen, *Nucl. Phys.* **B212** 413 (1983); *Phys. Lett.* **116B** 231 (1982).

[41] S. Deser, J. Kay and K. Stelle, *Phys. Rev. Lett.* **38** 527 (1977); S. Ferrara and B. Zumino, *Nucl. Phys.* **B134** 301 (1978).

[42] M. Sohnius and P. West, *Nucl. Phys.* **B198** 493 (1982).

[43] S. Ferrara, L. Girardello, T. Kugo and A. Van Proeyen, *Nucl. Phys.* **B223** 191 (1983).

[44] S. Ferrara, M. Grisaru and P. van Nieuwenhuizen, *Nucl. Phys.* **B138** 430 (1978).

[45] B. de Witt, J.W. van Holten and A. Van Proeyen, *Nucl. Phys.* **B184** 77 (1981); *Phys. Lett.* **95B** 51 (1980); *Nucl. Phys.* **B167** 186 (1980).

[46] A. Salam and J. Strathdee, *Phys. Rev.* **Dll** 1521 (1975); *Nucl. Phys.* **B86** 142 (1975).

[47] W. Siegel, *Phys. Lett.* **85B** 333 (1979).

[48] R. Arnowitt and P. Nath, *Phys. Lett.* **56B** 117 (1975); L. Brink, M. Gell-Mann, P. Ramond and J. Schwarz, *Phys. Lett.* **74B** 336 (1978); **76B** 417 (1978); S. Ferrara and P. van Nieuwenhuizen, *Ann. Phys.* **126** 111 (1980); P. van Nieuwenhuizen and P. West, *Nucl. Phys.* **B169** 501 (1980).

[49] M. Sohnius, *Nucl. Phys.* **B165** 483 (1980).

[50] P. Howe, K. Stelle and P. Townsend, *Nucl. Phys.* **B214** 519 (1983).

[51] M. Grisaru, M. Rocek and W. Siegel, *Nucl. Phys.* **B159** 429 (1979).

[52] E. Berezin, *The Method of Second Quantization* (Academic Press 1960).

[53] A. Salam and J. Strathdee, *Nucl. Phys.* **B76** 477 (1974); S. Ferrara, J. Wess and B. Zumino, *Phys. Lett.* **51B** 239 (1974).

[54] M. Grisaru, M. Rocek and W. Siegel, *Nucl. Phys.* **B159** 429 (1979).

[55] J. Wess, Lecture Notes in Physics 77 (Springer-Verlag, 1978).

[56] J. Gates and W. Siegel, *Nucl. Phys.* **B147** 77 (1979).

[57] J. Gates, K. Stelle and P. West, *Nucl. Phys.* **B169** 347 (1980).

[58] R. Grimm, M. Sohnius and J. Wess, *Nucl. Phys.* **B133** 275 (1978).

[59] P. Breitenlohner and M. Sohnius, *Nucl. Phys.* **B178** 151 (1981).

[60] P. Howe, K. Stelle and P. West, *Phys. Lett.* **124B** 55 (1983).

[61] P. Howe, K. Stelle and P. West, '$N = 1$, $d = 6$ Harmonic Superspace', Kings College preprint.

[62] M. Sohnius, K. Stelle and P. West, in *Superspace and Supergravity*, S.W. Hawking and M. Rocek (eds.) (Cambridge University Press 1981).

[63] A. Galperin, E. Ivanov, S. Kalitzin, V. Ogievetsky and E. Sokatchev, Trieste preprint.

[64] L. Mezincescu, JINR report P2-12572 (1979).

[65] J. Koller, *Nucl. Phys.* **B222** 319 (1983); *Phys. Lett.* **124B** 324 (1983).

[66] P. Howe, K. Stelle and P.K. Townsend, *Nucl. Phys.* **B236** 125 (1984).

[67] A. Salam and J. Strathdee, *Nucl. Phys.* **B80** 499 (1974).

[68] S. Ferrara, J. Wess and B. Zumino, *Phys. Lett.* **51B** 239 (1974).

[69] J. Wess and B. Zumino, *Phys. Lett.* **66B** 361 (1977); V.P. Akulov, D.V. Volkov and V.A. Soroka, *JETP Lett.* **22** 187 (1975).

[70] R. Arnowitt, P. Nath and B. Zumino, *Phys. Lett.* **56** 81 (1975); P. Nath and R. Arnowitt, *Phys. Lett.* **56B** 177 (1975); **78B** 581 (1978).

[71] N. Dragon, *Z. Phys.* **C2** 62 (1979).

[72] E.A. Ivanov and A.S. Sorin, *J. Phys. A. Math. Gen* **13** 1159 (1980).

[73] That some representations do not generalize to supergravity was noticed in M. Fischler, *Phys. Rev.* **D20** 1842 (1979).

[74] P. Howe and R. Tucker, *Phys. Lett.* **80B** 138 (1978).

[75] P. Breitenlohner, *Phys. Lett.* **76B** 49 (1977); **80B** 217 (1979).

[76] W. Siegel, *Phys. Lett.* **80B** 224 (1979).

[77] W. Siegel, 'Supergravity superfields without a supermetric', Harvard preprint HUTP-771 A068, *Nucl. Phys.* **B142** 301 (1978); S.J. Gates Jr. and W. Siegel, *Nucl. Phys.* **B147** 77 (1979).

[78] See also in this context: V. Ogievetsky and E. Sokatchev, *Phys. Lett.* **79B** 222 (1978).

[79] R. Grimm, J. Wess and B. Zumino, *Nucl. Phys.* **B152** 1255 (1979).

[80] These constraints were first given by: J. Wess and B. Zumino, *Phys. Lett.* **66B** 361 (1977).

[81] J. Wess and B. Zumino, *Phys. Lett.* **79B** 394 (1978).

[82] P. Howe and P. West, *Nucl. Phys.* **B238** 81 (1983).

[83] P. Howe, *Nucl. Phys.* **B199** 309 (1982).

[84] A. Salam and J. Strathdee, *Phys. Rev.* **D11** 1521 (1975).

[85] S. Ferrara and 0. Piguet, *Nucl. Phys.* **B93** 261 (1975).

[86] J. Wess and B. Zumino, *Phys. Lett.* **49B** 52 (1974).

[87] J. Iliopoulos and B. Zumino, *Nucl. Phys.* **B76** 310 (1974).

[88] S. Ferrara, J. Iliopoulos and B. Zumino, *Nucl. Phys.* **B77** 41 (1974).

[89] D.M. Capper, *Nuovo Cim.* **25A** 259 (1975); R. Delbourgo, *Nuovo Cim.* **25A** 646 (1975).

[90] P. West, *Nucl. Phys.* **B106** 219 (1976); D. Capper and M. Ramon Medrano, *J. Phys.* **62** 269 (1976); S. Weinberg, *Phys. Lett.* **62B** 111 (1976).

[91] M. Grisaru, M. Rocek and W. Siegel, *Nucl. Phys.* **B159** 429 (1979).

[92] B.W. Lee in *Methods in Field Theory*, Les Houches 1975, R. Balian and J. Zinn-Justin (eds.), (North Holland, and World Scientific, 1981).

[93] W. Siegel, *Phys. Lett.* **84B** 193 (1979); *94B* 37 (1980).

[94] L.V. Avdeev, G.V. Ghochia and A.A. Vladiminov, *Phys. Lett.* **105B** 272 (1981); L.V. Avdeev and A.A. Vladiminov, *Nucl. Phys.* **B219** 262 (1983).

[95] D.M. Capper, D.R.T. Jones and P. van Nieuwenhuizen, *Nucl. Phys.* **B167** 479 (1980).

[96] G. 't Hooft and M. Veltman, *Nucl. Phys.* **B44** 189 (1972); C. Bollini and J. Giambiagi, *Nuovo Cim.* **12B** 20 (1972); J. Ashmore, *Nuovo Cim. Lett.* **4** 37 (1972).

[97] P. Howe, A. Parkes and P. West, *Phys. Lett.* **147B** 409 (1984); *Phys. Lett.* **150B** 149 (1985).

[98] J.W. Juer and D. Storey, *Nucl. Phys.* **B216** 185 (1983); O. Piguet and K. Sibold, *Nucl. Phys.* **B248** 301 (1984).

[99] E. Witten, *Nucl. Phys.* **B188** 52 (1981).

[100] F. Gliozzi, D. Olive and J. Scherk, 'Supersymmetry, supergravity theories and the dual spinor model', *Nucl. Phys.* **B122** (1977) 253.

[101] P. van Niewenhuizen, 'Six lectures on supergravity', in *Supergravity '81*, S. Ferrara and J. Taylor (eds.), Cambridge University Press.

[102] T. Kugo and P. Townsend, *Supersymmetry and the division algebras*, *Nucl. Phys.* **B221** (1983) 357.

[103] Barut and Racka, *Theory of Group Representations*, World Scientific Publishing.

[104] B. Zumino, *J. Math. Phys.* **3** (1962) 1055.

[105] W. Nahm, 'Supersymmetries and their representations', *Nucl. Phys.* **B135** (1978) 149.

[106] E. Cremmer, B. Julia and J. Scherk, *Phys. Lett.* **76B** (1978) 409.

[107] C. Campbell and P. West, '$N = 2$ $D = 10$ nonchiral supergravity and its spontaneous compactification.' *Nucl. Phys.* **B243** (1984) 112.

[108] M. Huq and M. Namazie, 'Kaluza–Klein supergravity in ten dimensions', *Class. Quant. Grav.* **2** (1985).

[109] F. Giani and M. Pernici, '$N = 2$ supergravity in ten dimensions', *Phys. Rev.* **D30** (1984) 325.

[110] J, Schwarz and P. West, 'Symmetries and transformation of chiral $N = 2$ $D = 10$ supergravity', *Phys. Lett.* **126B** (1983) 301.

[111] P. Howe and P. West, 'The complete $N = 2$ $D = 10$ supergravity', *Nucl. Phys.* **B238** (1984) 181.

[112] J. Schwarz, 'Covariant field equations of chiral $N = 2$ $D = 10$ supergravity', *Nucl. Phys.* **B226** (1983) 269.

[113] S.W. MacDowell and F. Mansouri, *Phys. Rev. Lett.* **38** (1977) 739.

[114] T.W. B. Kibble, *J. Maths. Phys.* **2** (1961) 212.

[115] K.S. Stelle and P. West, 'Spontaneously broken de Sitter symmetry and the gravitational holonomy group', *Phys. Rev.* **D21** (1980) 1466.

[116] see for example M. Naimark and A. Stern, *Theory of Group Representations*, (Springer-Verlag, 1982).

[117] J.W. van Holten and A. van Proeyen, *J. Phys. Gen.* (1982) 3763.

[118] M. Kaku, P. van Niewenhuizen and P.K. Townsend, 'Properties of conformal supergravity', *Phys. Rev.* **D17** (1978) 3179.

[119] L. Brink, J. Scherk and J.H. Schwarz, 'Supersymmetric Yang–Mills theories', *Nucl. Phys.* **B121** (1977) 77; F. Gliozzi, J. Scherk and D. Olive, 'Supersymmetry, supergravity theories and the dual spinor model', *Nucl. Phys.* **B122** (1977) 253.

[120] A.H. Chamseddine, 'Interacting supergravity in ten dimensions: the role of the six-index gauge field', *Phys. Rev.* **D24** (1981) 3065; E. Bergshoeff, M. de Roo, B. de Witt and P. van Nieuwenhuizen, '10-dimensional Maxwell–Einstein supergravity, its currents, and the issue of its auxiliary fields', *Nucl. Phys.* **B195** (1982) 97; E. Bergshoeff, M. de Roo and B. de Witt, 'Conformal supergravity in ten dimensions', *Nucl. Phys.* **B217** (1983) 143.

[121] G. Chapline and N.S. Manton, 'Unification of Yang–Mills theory and supergravity in ten dimensions', *Phys. Lett.* **120B** (1983) 105.

[122] A.H. Chamseddine and P.C. West, 'Supergravity as a gauge theory of supersymmetry', *Nucl. Phys.* **B129** (1977) 39.

[123] P.K. Townsend, 'The 11-dimensional supermembrane revisited', *Phys. Lett.* **B350** (1995) 184, hep-th/9501068, 'D-branes from M-branes', *Phys. Lett.* **B373** (1996) 68, hep-th/9512062

[124] E. Witten, 'String theory dynamics in various dimensions', *Nucl. Phys.* **B443** (1995) 85, hep-th/9503124.

[125] M.B. Green and J.H. Schwarz, 'Superstring interactions', *Phys. Lett.* **122B** (1983) 143.

[126] P. West, 'Representations of supersymmetry', in *Supergravity '81*, S. Ferrara and J. Taylor (eds.), (Cambridge University Press, 1982).

[127] C. Callan, S. Coleman, J. Wess and B. Zumino, *Phys. Rev.* **177** (1969) 2247.

[128] N. Manton, *Phys. Lett.* **110B** (1982) 54.

[129] M. Atiyah and N. Hitchin, *The Geometry and Dynamics of Magnetic Monopoles*, (1988) Princeton University Press.

[130] M.B. Green and J.H. Schwarz, *Nucl. Phys.* **B243** (1984) 285.

[131] E. Bergshoff and E. Sezgin, 'Twistor-like formulation of super p-branes', *Nucl. Phys.* **B422** (1994) 329, hep-th/9312168; E. Sezgin, 'Space-time and worldvolume supersymmetric super p-brane actions', hep-th/9312168.

[132] I. Bandos, D. Sorokin and and D. Volkov, 'On the generalized action principle for superstrings and superbranes', *Phys. Lett.* **B352** (1995) 269, hep-th/9502141, I. Bandos, P. Pasti, D. Sorokin, M. Tonin and D. Volkov, 'Superstrings and supermembranes in the doubly supersymmetric geometrical approach', *Nucl. Phys.* **B446** (1995) 79, hep-th/9501113.

[133] P.S. Howe and E. Sezgin, 'Superbranes', hep-th/9607227

[134] P.S. Howe and E. Sezgin, '$D = 11$, $p = 5$',*Phys. Lett.* **B394** (1997) 62, hep-th/9611008

[135] J. Polchinski, *Phys. Rev. Lett.* **75** (1995) 4724.

[136] A. Tseytlin, 'Self-duality of the Born–Infeld action and the Dirichlet 3-brane of type IIB superstring theory', *Nucl. Phys.* **B469** (1996) 51; M.B. Green and M. Gutperle, 'Comment on three-branes', *Phys. Lett.* **B377** (1996) 28.

[137] P.S. Howe, E. Sezgin and P.C. West, 'Covariant field equations of the M-theory 5-brane',*Phys. Lett.* **B399** (1997) 49, hep-th/9702008.

[138] P.S. Howe, E. Sezgin and P.C. West, 'The 6-dimensional self-dual tensor', hep-th/9702111.

[139] P.S. Howe, E. Sezgin, and P.C. West, 'Aspects of superembeddings', Contribution to the D.V. Volkov memorial volume, hep-th/9705093

[140] M.K. Gaillard and B. Zumino, 'Duality rotations for interacting fields', *Nucl. Phys.* **B193** (1981) 221.

[141] I. Bandos, K. Lechner, A. Nurmagambetov, P. Pasti, D. Sorokin, and M. Tonin, 'Covariant action for the super 5-brane of M-theory', hep-th/9701149.

[142] M. Perry and J.H. Schwarz, *Nucl. Phys.* **B489** (1997) 47: hep-th/9611065; M. Aganagic, J. Park, C. Popescu, and J.H. Schwarz, 'Worldvolume action of the M-theory 5-brane', hep-th/9701166.

[143] E. Witten, 'The 5-brane effective action in M-theory', hep-th/9610234.

[144] P.S. Howe, N.D. Lambert and P.C. West, 'The selfdual string soliton', hep-th/9709014.

[145] P.S. Howe, N.D. Lambert and P.C. West, 'The 3-brane soliton of the M-5-brane', hep-th/9710033.

[146] P.S. Howe, N.D. Lambert and P.C. West, 'Classical M-5-brane dynamics and quantum $N = 2$ Yang–Mills', hep-th/9710034.

[147] M. Cederwall, A. von Gussich, B.E.W. Nilsson & A. Westerberg, 'The Dirichlet super-3-brane in 10-dimensional type IIB supergravity', hep-th/9610148.

[148] M. Aganagic, C. Popescu and J.H. Schwarz, 'D-brane actions with local kappa symmetry', hep-th/9610249.

[149] M. Cederwall, A. von Gussich, B.E.W. Nilsson, P. Sundell and A. Westerberg, 'The Dirichlet super p-branes in 10-dimensional type IIA and IIB supergravity', hep-th/9611159.

[150] M. Aganagic, C. Popescu and J.H. Schwarz, 'Gauge-invariant and gauge-fixed D-brane actions', hep-th/9612080.

[151] E. Bergshoeff, E. Sezgin and P.K. Townsend, 'Properties of 11-dimensional supermembrane theory', *Ann. Phys.* **185** (1988) 330.

[152] A. Font, L. Ibanez, D. Lust and F.D. Quevedo, *Phys. Lett.* **B249** (1990) 35.

[153] S.J. Rey, *Phys. Rev.* **D43** (1991) 526.

[154] A. Sen, *Phys. Lett.* **B303** (1993) 22; *Int. J. Mod. Phys.* **A9** (1994) 3707.

[155] J. Schwarz and A. Sen, *Nucl. Phys.* **B411** (1994) 35.

[156] C.M. Hull and P.K. Townsend, 'Unity of superstring dualities', *Nucl. Phys.* **B438** (1995) 109, hep-th/9410167.

[157] E. Bergshoeff, E. Sezgin and P.K. Townsend, *Phys. Lett.* **B189** (1987) 75.

[158] N. Seiberg and E. Witten, *Nucl. Phys.* **B426** (1994) 19, hep-th/9407087.

[159] E. Witten, *Nucl. Phys.* **B500** (1997) 3, hep-th/9703166.

[160] J. Strathdee, *Int. J. Mod. Phys.* **A2** (1987) 273.

[161] K. Narain, H. Sarmardi and E. Witten, *Nucl. Phys.* **B279** (1987) 369.

[162] C. Montonen and D. Olive, 'Magnetic monopoles as gauge particles?', *Phys. Lett.* **72B** (1977) 117.

[163] E. Witten and D. Olive, 'Supersymmetry algebras that include topological charges', *Phys. Lett.* **78B** (1978) 97.

[164] E. Witten, 'Bound states of strings and p-branes', *Nucl. Phys.* **B460** (1996) 335, hep-th/9510135.

[165] For a review see, J.A. Harvey 'Magnetic monopoles, duality and supersymmetry', hep-th/9603086.

[166] M. Green, J. Schwarz and E. Witten, *Superstring theory*, Vol. 1&2, Cambridge University Press, (1987).

[167] B. Julia, 'Group Disintegrations', in *Superspace & Supergravity*, S.W. Hawking and M. Roček, (eds.) Cambridge University Press (1981).

[168] S. Ferrara, J. Scherk and B. Zumino, 'Algebraic properties of extended supersymmetry', *Nucl. Phys.* **B121** (1977) 393; E. Cremmer, J. Scherk and S. Ferrara, 'SU(4) invariant supergravity theory', *Phys. Lett.* **74B** (1978) 61; B. de Witt, 'Properties of SO(8)-extended supergravity', *Nucl. Phys.* **B158** (1979) 189; B. de Witt and H. Nicolai, '$N = 8$ supergravity', *Nucl. Phys.* **B208** (1982) 323.

[169] A.A. Tseytlin, 'On the dilaton dependence of type II superstring action', hep-th/9601109.

[170] E. Cremmer, J. Scherk and S. Ferrara, 'SU(4) invariant supergravity theory', *Phys. Lett.* **74B** (1978) 61.

[171] E. Cremmer and B. Julia, 'The $N = 8$ supergravity theory. I. The Lagrangian', *Phys. Lett.* **80B** (1978) 48.

[172] P. Ramond, 'Dual theory for free fermions', *Phys. Rev.* **D3** (1971) 2415; A. Neveu and J.H. Schwarz, 'Factorizable dual model of pions', *Nucl. Phys.* **B31** (1971) 86.

[173] A. Achucarro, J.M. Evans, P.K. Townsend and D.L. Wiltshire, 'Super *p*-branes' *Phys. Lett.* **198** (1987) 441.

[174] L. Romans, *Phys. Lett.* **B169** (1986) 374.

[175] M.J. Duff, P.S. Howe, T. Inami and K.S. Stelle, 'Superstrings in $D = 10$ from supermembranes in $D = 11$', *Phys. Lett.* **191B** (1987) 70.

[176] C.G. Callan, J.A. Harvey and A. Strominger, 'Supersymmetric string solitons', hep-th/9112030; M.J. Duff, R.R. Khuri and J.X. Lü, 'String solitons', *Phys. Rept.* **259** (1995) 213, hep-th/9412184; M. Duff, 'Supermembranes', hep-th/9611203; K.S. Stelle, 'Lectures on supergravity *p*-branes', lectures given at the 1996 ICTP Summer School in High Energy Physics and Cosmology, Trieste, hep-th/9701088.

[177] N.D. Lambert and P.C. West, 'Gauge fields and M-5-brane dynamics', hep-th/9712040.

[178] D. Sorokin, V. Tkach and D.V. Volkov, 'Superparticles, twistors and Siegel symmetry', *Mod. Phys. Lett.* **A4** (1989) 901; D. Sorokin, V. Tkach, D.V. Volkov and A. Zheltukhin, 'From superparticle Siegel supersymmetry to the spinning particle proper-time supersymmetry', *Phys. Lett.* **B259** (1989) 302.

[179] W. Siegel, 'Hidden local supersymmetry in the supersymmetric particle action', *Phys. Lett.* **128B** (1983) 397.

[180] A. de Azcarraga, J.P. Gauntlett, J.M. Izquierdo and P.K. Townsend, 'Topological extensions of the supersymmetry algebra for extended objects, **63** (1989) 2443.

[181] E. Cremmer, 'Supergravities in 5 dimensions', in *Superspace & Supergravity*, S.W. Hawking and M. Roček (eds.), Cambridge University Press (1981).

[182] S. Sethi and M. Stern, hep-th/9705046.

[183] T.H. Buscher, *Phys. Lett.* **B194** (1987) 51; **B201** (1988) 466.

[184] M. Rocek and E. Verlinde, *Nucl. Phys.* **B373** (1992) 630.

[185] E. Bergshoff, M. de Roo, M.B. Green, G. Papadopoulos and P.K. Townsend, 'Duality of type II 7-branes and 8-branes', Hep-th/9601150

[186] M. Dine, P. Huet and N. Seiberg, 'Large and small radius in string theory', *Nucl. Phys.* **B322** (1989) 301.

[187] J. Dai, R.G. Leigh, J. Polchinski, 'New connections between string theories', *Mod. Phys. Lett.* **A4** (1989) 2073.

[188] E. Bergshoeff, C. Hull and T. Ortin, 'Duality in the type-II superstring effective action', *Nucl. Phys.* **B451** (1995) 547, hep-th/9504081.

[189] P.K. Townsend and P. van Nieuwenhuizen, 'Geometrical interpretation of extended supergravity', *Phys. Lett.* **67B** (1977) 439

[190] K.S. Narain, 'New heterotic string theories in uncompactified dimensions < 10', *Phys. Lett.* **169B** (1986) 41; K.S. Narain, M.H. Samadi and E. Witten, 'A note on toroidal compactification of heterotic string theory', *Nucl. Phys.* **B279** (1987) 369.

[191] R. Nepomechie, *Phys. Rev.* **D31** (1984) 1921; C. Teitelboim, *Phys. Lett.* **B167** (1986) 69.

[192] K. Kikkawa and M. Yamasaki, *Phys. Lett.* **B149** (1984) 357; N. Sakai and I. Senda, *Prog. Theor. Phys.* **75** (1986) 692.

[193] A. Sen, *Phys. Lett.* **B329** (1994) 217.

[194] H. Braden, 'N-dimensional spinors: their properties in terms of finite groups', *Journal of Mathematical Physics* **26** (1985) 613.

[195] P.S. Howe and P.K. Townsend, 'The massless superparticle as Chern–Simons mechanics', *Phys. lett.* **B259** (1991) 285; F. Deldue and E. Sokatchev, 'Superparticle with extended worldline supersymmetry', *Class. Quantum Grav.* **9** (1992) 361; A. Galperin, P. Howe and K. Stelle, 'The superparticle and the Lorentz group', *Nucl. Phys.* **B368** (1992) 248; A. Galperin and E. Sokatchev, 'A twistor-like $D = 10$ superparticle action with manifest $N = 8$ world-line supersymmetry', *Phys. Rev.* **D46** (1992) 714, hep-th/9203051.

[196] F. Deldue, A. Galperin, P.S. Howe and E. Sokatchev, ' A twistor formulation of the heterotic $D = 10$ superstring with manifest $(8,0)$ worldsheet supersymmetry', *Phys. Rev.* **D47** (1993) 578, hep-th/9207050; F. Deldue, E. Ivanov and E. Sokatchev, 'Twistor-like superstrings with $D = 3, 4, 6$ target-superspace and $N = (1,0), (2,0), (4,0)$ world-sheet supersymmetry', *Nucl. Phys.* **B384** (1992) 334, hep-th/9204071.

[197] P. Pasti and M. Tonin, 'Twistor-like formulation of the supermembrane in $D = 11$', *Nucl. Phys.* **B418** (1994) 337, hep-th/9303156.

[198] I. Bandos, 'Generalized action principle and geometrical approach for superstrings and super p-branes', hep-th/9608094; I. Bandos, P. Pasti, D. Sorokin and M. Tonin, 'Superbrane actions and geometrical approach', hep-th/9705064; I. Bandos, D. Sorokin and M. Tonin, 'Generalized action principle and superfield equations of motion for $D = 10$ Dp-branes', *Nucl. Phys.* **B497** (1997) 275, hep-th/9701127.

Supergravity Vacua and Solitons

G. W. Gibbons

1 Introduction

The enormous recent progress, detailed in other articles in this volume, on the non-perturbative structure of string/M-theory and its low energy approximation, quantum field theory, has been based on the recognition of the importance of p-branes, i.e. of extended objects with p spatial dimensions. Nowadays these can now be viewed in various ways. Historically, however, they first appeared as classical solutions of the low energy limits, of SUSY Yang-Mills theory or of the various supergravity theories. The purpose of these lectures is to provide a pedagogic introduction to the properties of solitons in supergravity theories and how one constructs the classical solutions. No claim is made for completeness; the subject is by now far too vast to survey the entire subject in just two lectures, and what follows is to some extent a rather personal account of the theory emphasising the features which are peculiar to the gravitational context. Thus lecture one is mainly concerned with exploring the question: what are the analogues of 'solitons' in gravity theories? The second is concerned with finding p-brane solutions. An important subsidiary technical theme is the use of sigma models or harmonic maps to solve Einstein's equations. Most, but not all, of the material is about four spacetime dimensions, partly because this is the best understood case and the most extensively studied and partly because my intuition (at least) is strongest in that dimension.

My own interest in this subject began in the early days of supergravity theories with the realization that since perturbatively such theories describe just a system of interacting massless particles (namely the graviton $g_{\mu\nu}$ N gravitini ψ^i_μ, n_v abelian vectors A^B_μ, $n_{\frac{1}{2}}$ spin-$\frac{1}{2}$ fermions λ and n_s scalars $\overline{\phi}^a$), then unless non-perturbative effects come into play these theories can have little connection with physics. It soon became apparent to me that the non-perturbative structure must involve gravity in an essential way, and, influenced by some ideas of Hájíček, which he mainly applied to gravity coupled to non-abelian gauge theory, I was led to propose extreme black holes as the appropriate analogue of solitons, what we now call BPS states. With this in mind I embarked in the early 1980s on a series of investigations aimed at uncovering the essential features of the soliton concept in gravity. What follows is largely a synopsis of those ideas updated to take into account recent developments. It has been both gratifying and a little surprising to see how well they have remained relevant over the intervening years.

Gibbons

2 The soliton concept in SUGRA theories

2.1 SUGRA in 4 dimensions

In what follows we shall concentrate on the theory in four spacetime dimensions, but much of what follows goes through with appropriate modifications in higher dimensions. Some examples of this will be given *en passant*.

2.1.1 The Lagrangians

If spacetime $\{\mathcal{M}, g_{\mu\nu}\}$ has four dimensions the Lagrangian $L = L_F + L_B$ of an ungauged supergravity theory with N supersymmetries has a fermionic L_F and bosonic L_B part. The latter is given by

$$
\begin{aligned}
L_B \;=\; & \frac{1}{16\pi}R - \frac{1}{8\pi}\overline{G}_{ab}g^{\mu\nu}\partial_\mu\overline{\phi}^a\partial_\nu\overline{\phi}^b - \frac{1}{16\pi}\mu(\overline{\phi})_{AB}F^A_{\mu\nu}F^{B\mu\nu} \\
& -\frac{1}{16\pi}\nu(\overline{\phi})_{AB}F^A_{\mu\nu}\star F^{B\mu\nu}
\end{aligned}
\tag{2.1}
$$

where R is the Ricci scalar of the spacetime metric $g_{\mu\nu}$ and $\star$ is the Hodge dual. As is usual when combining gravity and electromagnetism, I am using Gaussian or 'unrationalized' units in order to avoid extraneous factors of 4π in the formulae. We have also set Newton's constant $G = 1$. Occasionally I will, without further comment, reinstate it.

I have adopted the spacetime signature $(-1, +1, +1, +1)$. The particular convenience of that signature choice is that because the Clifford algebra $\mathrm{Cliff}_{\mathbb{R}}(3,1) \equiv \mathbb{R}(4)$ the algebra of 4×4 real matrices, *everything* in the purely classical theory, including all spinors and spinor Lagrangians, may be taken to be real. The same is of course true of classical M-theory since $\mathrm{Cliff}_{\mathbb{R}}(10,1) \equiv \mathbb{R}(32)$ and continues to hold if we descend to ten spacetime dimensions. It seems that it is only when we pass to the quantum theory that we need to introduce complex numbers. When dealing with Grassmann algebra valued spinors in classical supersymmetric theories one need only consider algebras over the field of real numbers. However the consistent adoption of this point of view entails some minor changes of the conventions, concerning for example 'conjugation' which are customary in the subject because i's never appear in any formulae. My own view is that the customary conventions introduce into the theory extraneous and unnatural elements which disguise the underlying mathematical simplicity. Fortunately, perhaps, the explicit details of the necessary changes in conventions will not be needed in what follows.

2.1.2 The scalar manifold

The scalar fields $\overline{\phi}^a$, $a = 1, \ldots, n_s$, take values in some target manifold $M_{\overline{\phi}}$ with metric $\overline{G}_{ab}$ which is typically a non-compact *symmetric space*, $M_{\overline{\phi}} = \overline{G}/\overline{H}$ where $\overline{H}$ is the maximum compact subgroup of $\overline{G}$ and the Lie algebra $\overline{\mathfrak{g}} = \overline{\mathfrak{h}} \oplus \overline{\mathfrak{k}}$ admits the involution $(\overline{\mathfrak{h}}, \overline{\mathfrak{k}}) \to (\overline{\mathfrak{h}}, -\overline{\mathfrak{k}})$. As a consequence

- topologically $M_{\overline{\phi}}$ is trivial, $M_{\overline{\phi}} \equiv \mathbb{R}^{n_s}$.

- The sectional curvatures $K(e_1, e_2) = R_{abcd}e_1^a e_2^b e_1^c e_2^d$ are non-positive.

These conditions play an important role in the theory of harmonic maps. For example, a well known theorem states that the only regular harmonic maps from a compact Riemannian manifold of positive Ricci curvature to a compact Riemannian manifold of negative sectional curvature are trivial. It is not difficult to adapt the proof of this theorem to rule out static asymptotically flat Skyrmion type solutions of the purely gravity-scalar sector of the theory. As a consequence we can anticipate that there are few non-perturbative features of the purely scalar sector. This situation changes when one passes to the gauged SUGRA theories. In that case there is a potential function for the scalars and various domain wall configurations exist. I shall not pursue this aspect of the theory further here.

2.1.3 Duality

The abelian vector fields A_μ^B, $B = 1, \ldots, n_v$, transform under a representation of G and and may be thought of as sections of the pull-back under the map $\overline{\phi} : \mathcal{M} \to M_{\overline{\phi}}$ of a vector bundle over $M_{\overline{\phi}}$ tensored with spacetime one-forms. In fact the structure is somewhat richer. As explained by Zumino elsewhere in this volume, one may define an electromagnetic induction 2-form by

$$G_A^{\mu\nu} = -8\pi \frac{\partial L_B}{\partial F_{\mu\nu}^A}. \tag{2.2}$$

The pair $(F^A, \star G_A)$ also carries a representation of G and in fact may be considered as the pull-back of an $Sp(2n_v, \mathbb{R})$ bundle over $M_{\overline{\phi}}$ tensored with spacetime two-forms.

To see what is going on more explicitly it is helpful to cast the vector Lagrangian in non-covariant form. In an orthonormal frame it becomes, up to a factor of 4π

$$\frac{1}{2}\mu_{AB}(\mathbf{E}^A \cdot \mathbf{E}^B - \mathbf{B}^A \cdot \mathbf{B}^B) + \nu_{AB}\mathbf{E}^A \cdot \mathbf{B}^B. \tag{2.3}$$

We define

$$\mathbf{D}_A = \mu_{AB}\mathbf{E}^B + \nu_{AB}\mathbf{B}^B \tag{2.4}$$

and
$$\mathbf{H}_A = \mu_{AB}\mathbf{B}^B - \nu_{AB}\mathbf{E}^B. \tag{2.5}$$
Thus
$$\begin{pmatrix} \mathbf{H}_A \\ \mathbf{E}^A \end{pmatrix} = \begin{pmatrix} \mu + \nu\mu^{-1}\nu & -\nu\mu^{-1} \\ -\mu^{-1}\nu & \mu^- - 1 \end{pmatrix} \begin{pmatrix} \mathbf{B}^A \\ \mathbf{D}_A \end{pmatrix}. \tag{2.6}$$

We define the matrix
$$\mathcal{M}(\overline{\phi}) = \begin{pmatrix} \mu + \nu\mu^{-1}\nu & -\nu\mu^{-1} \\ -\mu^{-1}\nu & \mu^- - 1 \end{pmatrix}, \tag{2.7}$$

and finds that
$$\det \mathcal{M}(\overline{\phi}) = 1. \tag{2.8}$$

The local density of energy due to the vectors is
$$\frac{1}{8\pi}(\mathbf{H}_A \cdot \mathbf{B}^A + \mathbf{D}_A \cdot \mathbf{E}^A). \tag{2.9}$$

The energy density and the equations of motion, will be invariant under the action of $S \in SL(2,\mathbb{R})$
$$\begin{pmatrix} \mathbf{B} \\ \mathbf{D} \end{pmatrix} \rightarrow S \begin{pmatrix} \mathbf{B} \\ \mathbf{D} \end{pmatrix} \tag{2.10}$$

provided we can find an action of $SL(2,\mathbb{R})$ on the scalar manifold $S : M_{\overline{\phi}} \rightarrow M_{\overline{\phi}}$ which leaves the metric $\overline{G}_{ab}$ invariant and under which the matrix $\mathcal{M}(\overline{\phi})$ pulls back as
$$\mathcal{M} \rightarrow (S^t)^{-1}\mathcal{M}S^{-1}. \tag{2.11}$$

As explained by Zumino this happy situation does indeed prevail in the SUGRA models under present consideration. We note *en passant* that electric-magnetic duality transformations of this type may extended to the *non-linear* electrodynamic theories of Born–Infeld type which one encounters in the world volume actions of Dirichlet 3-branes.

The set-up described above may be elaborated somewhat . For general $N = 2$ models, including supermatter, $M_{\overline{\phi}}$ is a Kähler manifold subject to 'Special Geometry'. The properties described in the present article do not require this extra structure and it will not be necessary to understand precisely what special geometry is in the sequel.

2.1.4 Scaling symmetry

As mentioned above, perturbatively theories of this kind describe massless particles. In fact, since the vectors are *abelian*, even more is true. The theory admits a global *scale-invariance*:
$$(g_{\mu\nu}, F_{\mu\nu}^A, \overline{\phi}^a) \rightarrow (\lambda^2 g_{\mu\nu}, \lambda F_{\mu\nu}^A, \overline{\phi}^a) \tag{2.12}$$

with $\lambda \in \mathbb{R}_+$ which takes classical solutions to classical solutions and takes the Lagrangian density $\sqrt{-g}L_B \rightarrow \lambda^2 \sqrt{-g}L_B$.

It is an easy exercise to check that our general Lagrangian has an energy-momentum tensor T_ν^μ which satisfies the *dominant energy condition*, i.e $T_\nu^\mu p^\nu$ lies in or on the future light cone for all vectors p^ν which themselves lie in or on the future light cone. Thus these theories have good local stability and causality properties.

Given the scaling symmetry and the positivity of energy one may construct arguments analogous to the well known theorem of Derrick in flat space to show that there are no non-trivial everywhere static or stationary solutions of the field equations. These types of theorems, are often referred to as Lichnerowicz theorems, though they go back to Serini, Einstein and Pauli. They may be summarized by the slogan: **No solitons without horizons** .

2.1.5 Examples

At this point some examples of $\overline{G}/\overline{H}$ are in order.

- $N = 1$. We only have a graviton, $n_s = 0 = n_v$, and thus bosonically this is just ordinary General Relativity.

- $N = 2$, $U(1)/U(1)$, there are no scalars, $n_s = 0$ and one vector, $n_v = 1$. This is just Einstein-Maxwell theory

- $N = 4$, $SO(2,1)/SO(2) \times SO(6)/SO(6)$. We have two scalars and six vectors.

- $N = 4$ plus supermatter, $SO(2,1)/SO(2) \times SO(6,6)/SO(6) \times SO(6)$. This is what you get if you dimensionally reduce $N = 1$ supergravity from 10 dimensions.

- The reduction of the Heterotic theory in 10 dimensions, $SO(2,1)/SO(2) \times SO(22,6)/SO(22) \times SO(6)$

- $N = 8$, $E_{7(7)}/SU(8)$.

Note that, except for $N = 8$, $G = S \times T$ where $S = SL(2, \mathbb{R})$ is what is now called the S-duality group, and T is now called the T-duality group. In the case of $N = 8$ both S and T are contained in the Hull–Townsend single U-duality group.

2.2 Vacua

Presumably the minimum requirement of a classical *vacuum* or *ground state* is that it be a homogeneous spacetime $M = G/H$ with constant scalars,

$\partial_\mu \overline{\phi}^a = 0$ and covariantly constant Maxwell fields $\nabla_\mu F_{\nu\rho}^A = 0$. Actually one sometimes wishes to consider a 'linear dilaton' but we shall not consider that possibility here. The ground state will thus be labelled in part by a point $p \in M_{\overline{\phi}}$. Thus one may consider $M_{\overline{\phi}}$, or possibly some submanifold of it, as parameterizing a *moduli space* of vacua. We usually demand that the ground state be classically stable, at least against small disturbances, and this typically translates into the requirement that it have least 'energy' among all nearby solutions with the same asymptotics. Unfortunately there is no space here to enter in detail into a general discussion of the how energy is defined in general relativity using an appropriate globally timelike Killing vector field K^μ. Later we shall outline an approach based on supersymmetry and Bogomol'nyi bounds. Suffice it to emphasise at present two important general principles.

- The dominant energy condition plays an essential role in establishing the positivity.

- Stability cannot be guaranteed if there is no globally timelike Killing field K^μ.

Thus, if spacetime M admits non-extreme Killing horizons across which a Killing field K^μ switches from being timelike to being spacelike, the local energy momentum vector $T_{\mu\nu}K^\nu$ *relative to* K^μ can become spacelike. This fact is behind the quantum instability of non-extreme Killing horizons due to Hawking radiation. In the context of homogeneous spacetimes this means that while globally static anti-de Sitter spacetime

$$AdS_n = SO(n-1,2)/SO(n-1,1)$$

has positive energy properties, and can be expected to be stable, de Sitter spacetime $dS_n = SO(n+1,1)/SO(n-1,1)$ with its cosmological horizons does not admit a global definition of positive energy.

For the class of Lagrangians we are considering in four spacetime dimensions, there are two important ground states. One is familiar Minkowski spacetime $\mathbb{E}^{3,1} = E(3,1)/SO(3,1)$, for which $F_{\mu\nu}^A = 0$ and $\overline{\phi}^a$ is arbitrary. The other, probably less familiar, is Bertotti–Robinson spacetime $AdS_2 \times S^2 = SO(1,2)/SO(1,1) \times SO(3)/SO(2)$. This represents a 'compactification' from four to two spacetime dimensions on the two-sphere S^2. One has

$$F^A = \frac{p^A}{A}\eta_{S^2} \tag{2.13}$$

and

$$\star G_A = \frac{q_A}{A}\eta_{S^2} \tag{2.14}$$

where η_{S^2} is the volume form on the S^2 factor, A is its area, (p^A, q_A) are constant magnetic and electric charges and the scalar field $\overline{\phi}^a$ is 'frozen' at a value $\overline{\phi}^{\text{frozen}}$ which extremizes a certain potential $V(\overline{\phi}, p, q)$ which may be read off from the Lagrangian. The potential is given by

$$V(\overline{\phi}, p, q) = (p \quad q)\, \mathcal{M}(\overline{\phi}) \begin{pmatrix} p \\ q \end{pmatrix} \tag{2.15}$$

and is invariant under S-duality which acts on the charges as

$$\begin{pmatrix} p \\ q \end{pmatrix} \to S \begin{pmatrix} p \\ q \end{pmatrix}. \tag{2.16}$$

The frozen values of $\overline{\phi}$ are given by

$$\frac{\partial V(\overline{\phi}, p, q)}{\overline{\phi}}\Big|_{\overline{\phi} = \overline{\phi}^{\text{frozen}}} = 0. \tag{2.17}$$

In fact

$$A = 4\pi V(\overline{\phi}^{\text{frozen}}, p, q). \tag{2.18}$$

Thus if $V(\overline{\phi}, p, q)$ has a unique minimum value then the vacuum solution is specified entirely by giving the charges (p, q).

As we shall see, it turns out that almost all of the black hole properties of the theory are determined entirely by the function $V(\overline{\phi}, p, q)$. In particular SUGRA theories, such as $N = 2$ theories based on Special Geometry, one may know some special facts about $V(\overline{\phi}, p, q)$ and one may read off much of the nature of the BPS configurations directly, without a detailed investigation of particular spacetime metrics.

2.3   Supersymmetry and Killing spinors

The supersymmetry transformations take the schematic form:

$$\delta_\epsilon B = F \tag{2.19}$$

$$\delta_\epsilon F = B \tag{2.20}$$

where (B, F) are bosonic and fermionic fields respectively and $\epsilon = \epsilon^i$, $i = 1, \ldots, N$, is an N-tuplet of spinor fields. A purely bosonic background $(B, 0)$ is said to admit a Majorana Killing spinor field ϵ, or to admit SUSY, if

$$\delta_\epsilon(B, 0) = 0. \tag{2.21}$$

Because the Killing spinor condition is linear in spinor fields, we may take the Grassmann algebra valued spinor fields ϵ to be commuting spacetime dependent spinor fields (which by an abuse of notation we also call ϵ) multiplied by a constant Grassmann coefficient. The Killing spinor condition reduces to $\delta_\epsilon F = 0$, or in components:

$$\delta_\epsilon \psi^i{}_\mu = (\hat{\nabla}_\mu \epsilon)^i \tag{2.22}$$

$$\delta_\epsilon \lambda = M\epsilon \tag{2.23}$$

where ψ^i_μ are the N gravitini and λ are the spin-$\frac{1}{2}$ fields. The operator $\hat{\nabla}_\mu = \nabla_\mu + E_\mu$, where ∇_μ is the Levi-Civita covariant derivative acting on spinors, and E_μ are endomorphisms depending on the bosonic field and whose precise form depends upon the particular SUGRA theory under consideration.

It follows from the supersymmetry algebra that

$$K_\epsilon = \bar{\epsilon}\gamma^\mu \epsilon \tag{2.24}$$

is a Killing vector field which necessarily lies in or on the future light cone. The solution is said to admit maximal SUSY if the dimension of the space of Killing spinors is N. If it is less, we speak of having BPS states with $\frac{N}{k}$ SUSY. From the point of view of SUSY representation theory BPS solutions correspond to short multiplets.

Typically *both* Minkowski spacetime $\mathbb{E}^{3,2}$ *and* Bertotti-Robinson spacetime $AdS_s \times S^2$ admit maximal SUSY. In gauged supergravity one has either a negative cosmological constant or a negative potential for the scalars. It then turns out that anti-de Sitter spacetime is a ground state with maximal SUSY. Because it does not admit an everywhere causal Killing field, de Sitter space, even if it were a solution, could never admit SUSY.

2.3.1 Remark on 5 dimensions

For the purposes of discussing black hole entropy it is often simpler to treat five-dimensional supergravity theories. I will not discuss them here in detail. I will simply remark that much of the present theory goes through. After dualization the bosonic Lagrangian is

$$L_B = \frac{1}{16\pi}R - \frac{1}{8\pi}G_{ab}g^{\mu\nu}\partial_\mu\phi^a\partial_\nu\phi^b - \frac{1}{16\pi}\mu(\phi)_{AB}F^A_{\mu\nu}F^{B\mu\nu}$$
$$+\text{Chern–Simons term}. \tag{2.25}$$

and $AdS_2 \times S^2$ is replaced by $AdS_2 \times S^3$. For example if $N = 8$, $G/H = E_{6,(+6)}/USp(8)$.

2.4 pp-waves

The perturbative, massless, states of the theory correspond to wave solutions. For example

$$ds^2 = -dt^2 + dz^2 + dx^2 + dy^2 + K(t-z,x,y)(dt-dz)^2 \tag{2.26}$$

with $K(t - z, x, y)$ being a harmonic function of (x, y)

$$(\partial_x^2 + \partial_y^2)K = 0 \qquad (2.27)$$

with arbitrary dependence on on $t - z$, is a vacuum solution representing a classical gravitational wave propagating in the positive z direction. Such solutions are called 'pp-waves' . As one would expect from basic supersymmetry representation theory they admit $N = \frac{1}{2}$ SUSY. If

$$(\gamma^t - \gamma^z)\epsilon = 0 \qquad (2.28)$$

then ϵ is a Killing spinor and by suitably scaling it we have

$$\bar{\epsilon}\gamma^\mu\epsilon = K^\mu \qquad (2.29)$$

where

$$K = \frac{\partial}{\partial t} + \frac{\partial}{\partial z} \qquad (2.30)$$

is the lightlike Killing field of the metric.

This example admits an obvious generalization to arbitrary spacetime dimension n. One simply replaces the two transverse coordinates (x, y) by $n-2$ transverse coordinates $(x^1, \ldots, x^{n-2})$. If K is taken to be independent of $t-z$ and one sets $H = 1 + K$ one may dimensionally reduce to $n - 1$ spacetime dimensions:

$$ds^2 = H\left(dz - \frac{dt}{H}\right)^2 - \frac{1}{H^{\frac{1}{n-3}}}\left(-\frac{dt^2}{H^{n-3}} + H^{\frac{1}{n-3}}dx_{n-2}^2\right). \qquad (2.31)$$

This gives a Kaluza–Klein 0-brane, with $A = \frac{dt}{2H}$. For example if $n = 5$ we get a so-called $a = \sqrt{3}$ extreme black hole which is S-dual to a the Kaluza–Klein monopole based on the Taub–NUT metric. If $n = 11$ we get the Dirichlet 0-brane of 10-dimensional type IIA theory.

2.5 Asymptotically flat solutions

We suppose that $g_{00} \sim -1 + \frac{2GM}{r}$ and $g_{ij} \sim (1 + \frac{2GM}{r})\delta_{ij}$, where M is the ADM mass. The scalar field $\bar{\phi} \sim \bar{\phi}_\infty + (\Sigma^a/r)$ where the the scalar charge $\Sigma^a \in TM_{\bar{\phi}_\infty}$. Because the scalar manifold $M_{\bar{\phi}}$ is topologically trivial and because its sectional curvature is non-positive it is not difficult to prove, using standard techniques from the theory of harmonic maps, that there are no non-singular solutions without horizons with vanishing vector fields. Thus there are no analogues of Skyrmions.

We define the total electric charge q_A and magnetic charge p^A by the usual 2-surface integrals at infinity,

$$p^A = \frac{1}{4\pi}\int_{S_\infty^2} F^A \qquad (2.32)$$

and

$$q_A = \frac{1}{4\pi} \int_{S^2_\infty} \star G_A. \qquad (2.33)$$

We now arrive at some absolutely crucial points.

- The fundamental fields carry neither electric nor magnetic charges. Thus perturbative states cannot carry them. In fact, by Maxwell's equations, a solution can only carry non-vanishing charges if it is in some way singular or topologically non-trivial or both. In fact we encounter here the phenomenon of 'charge without charge' due to 'lines of force being trapped in the topology of space' which formed the central point of Misner and Wheeler's 'Geometrodynamics'. In the context of string theory and Polchinski's non-perturbative Dirichlet-branes, some of the vector fields have their origin in the Ramond⊗Ramond sector and we have 'Ramond⊗Ramond charge without charge'.

- Because the fields are abelian and also because of the classical scale-invariance there is no possibility of a classical quantization of the charges. Quantization can only be achieved by going outside the framework of classical supergravity theory, for example by coupling to so-called 'fundamental' branes and applying a Dirac type argument or by applying Saha's well known argument for the angular momentum about the line of centre's joining an electric and magnetic charge. One then discovers that if the electric charges belong to some lattice $q_A \in \Lambda$ then the magnetic charges belong to the reciprocal lattice $p^A \in \Lambda^\star$. That is

$$\frac{2}{\hbar} p^A q_A \in \mathbb{Z}. \qquad (2.34)$$

2.6 Black holes

In the case of $N = 1$ one might be tempted to think that the only static solution with a regular event horizon, the Schwarzschild black hole, should be considered as some sort of soliton. However there are a number of reasons why this is not correct.

- Although the solution is regular outside its event horizon, inside it contains a spacetime singularity. This may not be fatal if Cosmic Censorship holds. In that case the spacetime outside the event horizon would be regular and predictable. In fact it is widely believed to be classically stable.

- However, the mass M is arbitrary and is not fixed by any quantization condition. Moreover, classically the black hole can absorb gravitons and gravitini leading to a mass increase. In fact classically the area A of the event horizon can never decrease. This irreversible behaviour is quite unlike what one expects of a classical soliton.

• Quantum mechanically, the Hawking effect means the Schwarzschild black hole is definitely unstable. The same is true of the Kerr solution. It does not seem reasonable therefore to expect that in the full quantum gravity theory one may associate with them a stable non-perturbative state in the quantum mechanical Hilbert space. In fact because of their thermal nature it is much more likely that the classical solutions should be associated with a density matrices representing a black hole in thermal equilibrium with its evaporation products.

• From the point of view of SUSY it is clear that the Schwarzschild and Kerr solutions do not correspond to BPS states since they do not admit any Killing spinors. For $N = 1$ the endomorphism E_μ vanishes and such Killing spinors would have to be covariantly constant, as would the Killing vector constructed from them This is impossible if the solution is not to be flat.

Later we shall see that SUSY is incompatible with a non-extreme Killing horizon.

2.6.1 Reissner–Nordström

If $N = 2$ the candidate static solitons would be Reissner–Nordström black holes. If the singularity is not to be naked we must have

$$M \geq |Z|, \tag{2.35}$$

where

$$|Z|^2 = \frac{q^2 + p^2}{G}. \tag{2.36}$$

Note that in this section I am reinstating Newton's constant G. The Hawking temperature is

$$T = \frac{1}{4\pi G} \frac{\sqrt{M^2 - |Z|^2}}{(M + \sqrt{M^2 - |Z|^2})^2}. \tag{2.37}$$

If $M > |Z|$ the temperature is non-zero and the solution is unstable against Hawking evaporation of gravitons, photons and gravitini. Moreover it cannot be a BPS state since the Killing vector which is timelike near infinity becomes spacelike inside the horizon.

Only in the extreme case $M = |Z|$ for which $T = 0$ is $\frac{\partial}{\partial t}$ never space-like. Moreover in that case there are multi-black hole solutions, the so-called Majumdar–Papapetrou solutions:

$$ds^2 = -H^{-2}dt^2 + H^2 d\mathbf{x}^2, \tag{2.38}$$

with, up to an electric-magnetic duality rotation,

$$F = dt \wedge d\left(\frac{1}{H}\right),\tag{2.39}$$

where H is an arbitrary harmonic function on $\mathbb{E}^3$.

It is an easy exercise to verify that the entire Majumdar–Papapetrou family of solutions admits a Killing spinor whose associated Killing vector is $\frac{\partial}{\partial t}$. In fact they are the only static solutions of Einstein–Maxwell theory admitting a Killing spinor. Thus the Majumdar–Papapetrou solutions correspond to BPS states.

One further property of the extreme holes, called *vacuum interpolation*, should be noted. This is while near infinity the solution tends to the flat maximally supersymmetric ground state, Minkowski spacetime $\mathbb{E}^{3,1}$, near the horizon the metric tends to the other maximally supersymmetric ground state, Bertotti–Robinson spacetime $AdS_2 \times S^2$. Thus, as is the case for many solitons, the solution spatially interpolates between different vacua or ground states of the theory.

The Bekenstein–Hawking entropy S of a general charged black hole is given by

$$S = \frac{A}{4G} = \pi G(M + \sqrt{M^2 - |Z|^2})^2.\tag{2.40}$$

For fixed $|Z|$ this is least in the extreme case when it attains

$$\pi(q^2 + p^2),\tag{2.41}$$

which is *independent* of Newton's constant G and depends *only* on the quantized charges (p, q).

2.6.2 Black holes and frozen moduli

If we now pass to the case when more than one vector and some scalars are present we find that in general that the scalars will vary with position. There is thus in a sense 'scalar hair'. However, spatial dependence of the scalar fields and non-vanishing scalar charges Σ^a are only present by virtue of the fact that the source term $(\partial V/\partial\bar{\phi})(\bar{\phi}, p, q)$ in the scalar equations of motion. If it happens that $\bar{\phi}_\infty = \bar{\phi}^{\text{frozen}}$ however, then the charges vanish and the scalars are constant. The moduli are then said to be frozen. The geometry of the black holes is then the same as the Reissner–Nordström case with

$$Z^2 = V(\bar{\phi}^{\text{frozen}}, p, q) = |Z|^2(p, q).\tag{2.42}$$

In general the moduli will not be frozen and for instance the value of the scalars on the horizon $\bar{\phi}^{\text{horizon}}$ will be different from its value $\bar{\phi}_\infty$ at infinity.

For extreme static black holes however, regularity of the horizon demands that

$$\overline{\phi}^{\text{horizon}} = \overline{\phi}^{\text{frozen}}. \tag{2.43}$$

As a consequence we have the important general fact that the Bekenstein–Hawking entropy of extreme holes is always independent of the moduli at infinity and depends only on the quantized charges (p, q). That is

$$S_{\text{extreme}} = \pi V(\overline{\phi}, p, q) = \pi |Z|^2 (p, q). \tag{2.44}$$

As explained in other lectures the Bekenstein–Hawking entropy of extreme holes can be obtained by D-brane calculations at weak coupling. It is vital for the consistency of this picture that S_{extreme} really is independent of the moduli $\overline{\phi}_\infty$ which label the vacua. A striking consequence of this is that the entropy of any initial data set with given (p, q) should never be less than $\pi |Z(p, q)|^2$ and its mass M should never be less than $|Z(p, q)|$.

2.6.3 The first law of thermodynamics and the Smarr virial theorem

For time-stationary fields we may define electrostatic potentials ψ^A and magnetostatic potentials χ_A by

$$F_{0i}^A = \partial_i \psi^A \tag{2.45}$$

and

$$G_{A0i} = \partial_i \chi_A. \tag{2.46}$$

The first law of classical black hole mechanics needs a modification if we consider variations of the moduli $\overline{\phi}_\infty$. It becomes

$$dM = TdS + \psi^A dq_A + \chi_A dp^A - \Sigma^a \overline{G}_{ab}(\overline{\phi}_\infty) d\overline{\phi}^b. \tag{2.47}$$

The last term is the new one. Note that, at the risk of causing confusion but in the interest of leaving the formulae comparatively uncluttered, I have not explicitly distinguished between the potential functions χ_A, ψ^A and their values at the horizon.

It follows using the scaling invariance that the mass is given by the Smarr formula

$$M = 2TS + \psi^A q_A + \chi_A p^A. \tag{2.48}$$

Thus, in the present circumstances, the Smarr formula is equivalent to the first law as a consequence of scaling symmetry. In other words we may regard the Smarr relation as a virial type theorem. This formula allows a simple derivation of the 'No solitons without horizons' result. If there is no horizon then $S = 0$ and by Gauss's theorem $q_A = 0 = p^A$. It follows that the mass $M = 0$ and hence by the positive mass theorem the solution must be flat.

2.7   Bogomol'nyi bounds

We shall now give a brief indication of how one identifies the central charges
and establishes Bogomol'nyi bounds. Let ϵ_∞^i be constant spinors at infinity. In what follows we shall sometimes omit writing out a summation of i
explicitly. The supercharges Q^i are defined by

$$\bar{\epsilon}_\infty Q = \frac{1}{4\pi G} \int_{S_\infty^2} \frac{1}{2} \bar{\epsilon}_\infty \gamma^{\mu\nu\lambda} \psi_\lambda d\Sigma_{\mu\nu}. \tag{2.49}$$

The Nester two-form $N^{\mu\nu}$ associated to a spinor field ϵ is defined by

$$N^{\mu\nu} = \bar{\epsilon}\gamma^{\mu\nu\lambda}\widehat{\nabla}_\lambda\epsilon. \tag{2.50}$$

Under a SUSY variation we have

$$\bar{\epsilon}_\infty \delta_\epsilon Q = \frac{1}{4\pi G} \int_{S_\infty^2} \frac{1}{2} N^{\mu\nu} d\Sigma_{\mu\nu}. \tag{2.51}$$

Stokes's theorem gives

$$\bar{\epsilon}_\infty \delta_\epsilon Q = \frac{1}{4\pi G} \int_\Sigma \nabla_\mu N^{\mu\nu} d\Sigma_\nu, \tag{2.52}$$

where Σ is a suitable spacelike surface whose boundary at infinity is S_∞^2 and
whose inner boundary either vanishes or is such that by virtue of suitable
boundary conditions one may ignore its contribution.

Now by the supergravity equations of motion one finds

$$\nabla_\mu N^{\mu\nu} = \widehat{\nabla}_\mu \widehat{\epsilon} \gamma^{\mu\nu\lambda} \widehat{\nabla}\epsilon + \overline{M\epsilon}\gamma^\nu M\epsilon \tag{2.53}$$

Now restricted to Σ, $\overline{M\epsilon}\gamma^0 M\epsilon \geq 0$ and

$$\widehat{\nabla}_\mu \bar{\epsilon} \gamma^{\mu 0 \lambda} \widehat{\nabla}\epsilon = |\widehat{\nabla}_a \epsilon|^2 - |\gamma^a \widehat{\nabla}_a \epsilon|^2, \tag{2.54}$$

where the derivative $\widehat{\nabla}_a$ is tangent to Σ. So far ϵ has been arbitrary. We pick
it such that

- $\gamma^a \widehat{\nabla}_a \epsilon = 0$

- $\epsilon \to \epsilon_\infty$ at infinity.

We also choose ϵ such that the inner boundary terms, such as might arise
at an horizon, vanish. It is not obvious but it is in fact true that this can
be done. It follows that the right-hand side of (2.52) is non-negative and
vanishes if and only if everywhere on Σ

$$M\epsilon = 0 \tag{2.55}$$

and

$$\hat{\nabla}_a \epsilon = 0. \tag{2.56}$$

Since Σ is arbitrary we may deduce that in fact the right-hand side of (2.52) is non-negative and vanishes if and only if everywhere in spacetime

$$M\epsilon = 0 \tag{2.57}$$

and

$$\hat{\nabla}_\mu \epsilon = 0. \tag{2.58}$$

This means that ϵ must be a Killing spinor.

Now the left-hand side of (2.52) may be shown to be

$$\bar{\epsilon}_\infty \gamma^\mu P_\mu \bar{\epsilon}_\infty + \bar{\epsilon}_\infty (U_{ij} + \gamma_5 V_{ij}) \epsilon_\infty. \tag{2.59}$$

Here P_μ may be identified with the ADM 4-momentum and U_{ij} and V_{ij} are central charges which depend on the magnetic and electric charges (p, q) and the moduli, i.e. of the values $\bar{\phi}_\infty$ of the scalar fields at infinity. In the case of $N = 2$ there are just two central charges which may be combined into a single complex central charge $Z(p, q, \bar{\phi}_\infty)$ and the Bogomol'nyi bound becomes

$$M \geq |Z(p, q, \bar{\phi}_\infty)|. \tag{2.60}$$

3 Finding solutions

We now look for local solutions on $\mathcal{M} = \Sigma \times \mathbb{R}$ which are independent of the time coordinate $t \in \mathbb{R}$. Globally of course the geometry is much more subtle because of the presence of horizons but that will not affect the local equations of motion. The basic idea used here is that in 3 dimensions we may use duality transformations to replace vectors by scalars. The resulting equations may be derived form an action describing three-dimensional gravity on Σ coupled to a sigma model.

3.1 Reduction from 4 to 3 dimensions

The metric is expressed as

$$ds^2 = -e^{2U}(dt + \omega_i dx^i)^2 + e^{-2U} \gamma_{ij} dx^i dx^j. \tag{3.1}$$

The effective Lagrangian is

$$\frac{1}{16\pi} R[\gamma] - \frac{1}{8\pi} G_{ab}(\bar{\phi}^a) \partial_i \phi^a \partial_j \phi^b \gamma^{ij}, \tag{3.2}$$

where ϕ^a is the collection of fields $(U, \psi, \overline{\phi}^a, \psi^A, \chi_A)$ taking values in an augmented target space M_ϕ where ψ is the twist potential, ψ^A the electrostatic potential and χ_A the magnetostatic potential. The twist potential arises by dualizing ω_i. Note that (U, ψ) are the gravitational analogues of electric and magnetic potentials respectively. Indeed for pure gravity,

$$\operatorname{curl} \omega = e^{-2U} \operatorname{grad} \psi, \qquad (3.3)$$

the internal space is $H^2 \equiv SO(2,1)/SO(2)$ and the metric is

$$dU^2 + e^{-2U} d\psi^2. \qquad (3.4)$$

The formula for the twist potential becomes more complicated in the presence of vectors.

Three important general features to note are

- If the signature of spacetime is $(3,1)$ then the signature of the σ-model metric G_{ab} is $(2+n_s, 2n_v)$. Physically this is because the n_s scalar fields ϕ^a and the gravitational scalars (U, ψ) give rise to *attractive* forces while the n_v vector fields give rise to *repulsive* forces.

- The metric $\overline{G}_{ab}$ admits $2n_v + 1$ commuting Killing fields $\frac{\partial}{\partial \psi}, \frac{\partial}{\partial \psi^A}, \frac{\partial}{\partial \chi_A}$ which give rise to $2n_v + 1$ charges, the last of which, the so-called NUT charge, vanishes for asymptotically flat solutions.

- Typically M_ϕ is also a symmetric space with indefinite metric of the form G/H, where G is an example U-duality group which includes both S and T duality groups. Of course H is no longer *compact*.

- If we were to include fermions we would get 3-dimensional SUGRA theory but with Euclidean signature.

3.1.1   Examples

Let us look at some examples of G/H.

- $N = 1$. This is pure gravity, there are no vectors and the signature is positive. The the coset is 2-dimensional hyperbolic space $H^2 = SO(2,1)/SO(2)$.

- $N = 2$, $SU(2,1)/S(SU(1,1) \times U(1))$, Einstein–Maxwell theory. In fact $\{M_\phi, G_{ab}\}$ is an analytic continuation of the Fubini–Study metric on $\mathbb{CP}^2$ to another real section with Kleinian signature $(2,2)$.

- $N = 4$ SUGRA, $SO(8,2)/SO(2) \times SO(6,2)$.

- $N = 4$ SUGRA plus supermatter, $SO(8,8)/SO(2,6) \times SO(6,2)$. This is what you get if you dimensionally reduce $N = 1$ supergravity from 10 dimensions.

- The reduction of the heterotic theory in 10 dimensions gives $SO(24,8)/SO(22,2) \times SO(2,6)$

- $N = 8$ SUGRA, $E_{8(+8)}/SO^*(16)$.

It is clear that the group G may be used as a solution generating group. Of course some elements of G may not take physically interesting solutions to physically distinct or physically interesting solutions. nevertheless one may anticipate that black hole solutions will fall into some sort of multiplets of a suitable subgroup of G and indeed this turns out to be the case.

3.1.2 Static truncations

In what follows we shall mainly be concerned with non-rotating holes and so we drop the twist potential and consider the *static truncation* with effective Lagrangian

$$+ \frac{1}{16}R[\gamma] + \frac{1}{8\pi}(\partial U)^2 + \frac{1}{8\pi}\overline{G}_{ab}\partial\overline{\phi}^a\partial\overline{\phi}^b - \frac{1}{8\pi}e^{-2U}(\partial\psi^A,\partial\chi_A)\mathcal{M}^{-1}(\partial\psi^A,\partial\chi_A)^t \tag{3.5}$$

3.1.3 Gravitational instantons

The methods we have just described may also be used to obtain solutions of the Einstein equations with positive-definite signature admitting a circle action. All that is required to get the equations is a suitable analytic continuation of the previous formulae. This entails a chage in the groups and the symmetric spaces. Thus in the case of pure gravity case the metric is

$$ds^2 = e^{2U}(d\tau + \omega_i dx^i)^2 + e^{-2U}\gamma_{ij}dx^i dx^j. \tag{3.6}$$

The twist potential still satisfies (3.3) but the internal space becomes $dS_2 = SO(2,1)/SO(1,1)$ with metric

$$ds^2 = dU^2 - e^{-2U}d\psi^2. \tag{3.7}$$

3.1.4 The equations of motion

The scalar equation of motion requires that ϕ gives a harmonic map from Σ to M_ϕ.

$$\nabla^2\phi = 0 \tag{3.8}$$

where the covariant derivative ∇ contains a piece corresponding to the pull-back under ϕ of the connection $\Gamma^a_{bc}(\phi)$ on M_ϕ; thus

$$\nabla_i \partial_j \phi^a = \partial_i \partial_j \phi^a - \Gamma_{ij}(x)^k \partial_k \phi^a + \partial_i \phi^c \partial_j \phi^b \Gamma^a_{bc}(\phi). \tag{3.9}$$

Variation with respect to the metric γ_{ij} gives an Einstein-type equation:

$$R_{ij} = 2\partial_i \phi^a \partial_j \phi^b G_{ab}. \tag{3.10}$$

If we pretend that we are thinking of Einstein's equations in three dimensions then the the left-hand side may be thought of as $T_{ij} - \gamma_{ij} \gamma^{mn} T_{mn}$ where T_{ij} is the stress tensor.

There are essentially three easy types of solutions of this system of equations which may be described using simple geometrical techniques.

- Spherically symmetric solutions

- Multi-centre (i.e. BPS) solutions

- cosmic string solutions.

3.2 Spherically symmetric solutions

The idea is to reduce the problem to one involving geodesics in M_ϕ. A consistent ansatz for the metric is γ_{ij} is

$$\gamma_{ij} dx^i dx^j = \frac{c^4 d\tau^2}{\sinh^4 c\tau} + \frac{c^2}{\sinh^2 c\tau} (d\theta^2 + \sin^2 \theta d\phi^2). \tag{3.11}$$

The radial coordinate τ is in fact a harmonic function on Σ with respect to the metric γ_{ij}. The range of τ is from the horizon at $-\infty$ to spatial infinity at $\tau = 0$. In these coordinates the only non-vanishing component of the Ricci tensor of γ_{ij} is the radial component and this has the constant value $2c^2$. For a regular solution the constant c is related to the temperature and entropy by

$$c = 2ST. \tag{3.12}$$

Now it is a fact about harmonic maps that the composition of a harmonic map with a geodesic map is harmonic. Thus to satisfy the scalar equations of motion, $\phi^a(\tau)$ must execute geodesic motion in $\{G_{ab}, M_\phi\}$ with τ serving as affine parameter along the geodesic. The Einstein equation then fixes the value of the constraint

$$G_{ab} \frac{d\phi^a}{d\tau} \frac{d\phi^a}{d\tau} = c^2. \tag{3.13}$$

Because of electromagnetic gauge-invariance there are $2n_v$ Noether constants of the motion, i,e, the electric and magnetic charges:

$$\begin{pmatrix} p^A \\ q_A \end{pmatrix} = \mathcal{M}^{-1} \begin{pmatrix} d\chi_A/d\tau \\ d\psi^A/d\tau \end{pmatrix}. \tag{3.14}$$

The remaining equations follow from the effective action

$$\left(\frac{dU}{d\tau}\right)^2 + \overline{G}_{ab}\frac{d\overline{\phi}^a}{d\tau}\frac{d\overline{\phi}^b}{d\tau} + e^{2U}V(\overline{\phi}, p, q) \tag{3.15}$$

and the constraint becomes

$$\left(\frac{dU}{d\tau}\right)^2 + \overline{G}_{ab}\frac{d\overline{\phi}^a}{d\tau}\frac{d\overline{\phi}^b}{d\tau} - e^{2U}V(\overline{\phi}, p, q) = (2ST)^2. \tag{3.16}$$

Evaluating this at infinity we get

$$M^2 + \overline{G}_{ab}\Sigma^a\Sigma^b - V(\overline{\phi}_\infty, p, q) = (2ST)^2. \tag{3.17}$$

The extreme case corresponds to $c = 0$. The metric γ_{ij} is now

$$\gamma_{ij}dx^i dx^j = \frac{d\tau^2}{\tau^4} + \frac{1}{\tau^2}(d\theta^2 + \sin\theta d\phi^2). \tag{3.18}$$

3.3 Toda and Liouville systems

The method just outlined has the advantage that it makes clear how the duality group acts on the solutions.

The procedure also makes it clear why the radial equations frequently give rise to ordinary differential equations of Toda type which are in principle exactly integrable. As we have seen, the problem of finding spherically symmetric solutions reduces to finding geodesics in the symmetric space G/H. We may think of this as solving a a Hamiltonian system on the cotangent space $T^*(M_\phi) = T^*(G/H)$. This symmetric space admits at at least $2n_v + 1$ and typically more commuting Killing vectors. These come from the fact that one may add an arbitrary constant to the twist potential ψ and the magnetic and electric potentials (χ_a, ψ^A). In addition there may be further symmetries arising from axion-like fields. Let us suppose that *in toto* there are r such commuting Killing vectors. Ignoring any possible identifications they will generate the group $\mathbb{R}^r$. One may eliminate the r commuting constants of the motion to obtain a dynamical system on the quotient configuration space $M_\phi/\mathbb{R}^r = \mathbb{R}^r\backslash G/H$. From a Hamiltonian point of view one is of course just performing a Marsden–Weinstein symplectic reduction.

Now it is known from the work of Perelomov and Olshanetsky that Toda systems arise precisely in this way. Thus it is no surprise that one encounters them in finding solutions of the Einstein equations depending on a single variable. The simplest example is when $M_\phi = G/H = SO(2,1)/SO(1,1) = AdS_2$ or its Riemannian version $SO(2,1)/SO(2) = H^2$. The internal metric is given by (3.7) or (3.4) respectively. If the constant of the motion $q = e^{-2U}(d\psi/d\tau)$ then the dynamical system has effective Lagrangian

$$\frac{1}{2}\left(\frac{dU}{d\tau}\right)^2 \pm \frac{1}{2}q^2 e^{2U} \tag{3.19}$$

with constant of the motion

$$\frac{1}{2}\left(\frac{dU}{d\tau}\right)^2 \mp \frac{1}{2}q^2 e^{2U} = \text{constant}, \tag{3.20}$$

where the upper sign refers the AdS_2 case and the minus to the H^2 case. The resulting dynamical system is of course a rather trivial Liouville system and may be integrated using elementary methods.

If the internal space decomposes into a product of such models, then the integration is equally easy. In practice most of the exact solutions in the literature may be obtained in this way. As I mentioned above, in principle, the general Toda system is exactly integrable but in practice it seems to be rather cumbersome to carry out the integration explicitly.

3.4   Multi-centre solutions

We make the ansatz that the metric γ_{ij} is flat and we may take Σ to be Euclidean 3-space $\mathbb{E}^3$.

$$\gamma_{ij} = \delta_{ij}. \tag{3.21}$$

This means that the coordinates $(t, \mathbf{x})$ are harmonic coordinates, i.e. we are using a gauge in which $\partial_\mu \mathfrak{g}^{\mu\nu} = 0$. The vanishing of the Ricci tensor then requires the vanishing stress tensor or local force balance condition

$$\partial_i \phi^a \partial_j \phi^b G_{ab} = 0. \tag{3.22}$$

We must also satisfy the harmonic condition. The following construction will do the job. We start with k ordinary harmonic functions $H^r(x^i)$, $r = 1, 2, \ldots, k$ on Euclidean space $\mathbb{E}^3$. These give a harmonic map from $H : \mathbb{E}^3 \to \mathbb{E}^k$.

We next find a k-dimensional *totally geodesic totally lightlike* submanifold of the target space M_ϕ. This is a map $f : \mathbb{E}^k \to M_\phi$ whose image N is

- Totally null, i.e. the induced metric $G_{rs} = G_{ab}\frac{\partial f^a}{\partial y^r}\frac{\partial f^b}{\partial y^s}$ vanishes.

- totally geodesic, which means that a (necessarily lightlike) geodesic which is initially tangent to N remains tangent to N.

The simplest example would be a null geodesic for which $k = 1$. The parameters y^r are affine parameters. In the case of Einstein–Maxwell theory we may take the so-called 'α' or 'β' 2-planes which play a role in twistor theory. The important point about totally geodesic maps is that they are harmonic.

Given our maps $H^r(x^i)$ and $f^a(y^r)$ we compose them, i.e. we set

$$\phi^a(x^i) = f^a(H^r(x^i)). \tag{3.23}$$

The result is a harmonic map and we are done.

This simple and elegant technique is in principle all that is required to construct all BPS solutions of relevance to four dimensions. As we shall see, it frequently works in higher dimensions. Of course, to check that they are BPS one has to check for the existence of Killing spinors. The technique also makes transparently clear the action of the U-duality group. Moreover, since we are dealing with symmetric spaces, everything can, in principle be reduced to calculations in the Lie algebra of G.

Consider the simplest case, the static truncation of Einstein–Maxwell theory. The internal space is $SO(2,1)/SO(1,1) = AdS_2$ with metric

$$dU^2 - e^{-2U}d\psi^2 \tag{3.24}$$

where ψ is the electrostatic potential. The null geodesics are given by

$$\psi = \frac{1}{H} \tag{3.25}$$

$$e^U = \frac{1}{H}. \tag{3.26}$$

We have recovered the Majumdar–Papapetrou solutions. For dilaton gravity with dimensionless coupling constant a which has the matter action

$$-\frac{1}{8p}\partial\sigma^2 - \frac{1}{16\pi}e^{-2a\sigma}F_{\mu\nu}^2 \tag{3.27}$$

the internal space carries the metric (in the electrostatic case)

$$dU^2 + d\sigma^2 - e^{-2(a\sigma+U)}d\psi^2. \tag{3.28}$$

The null geodesics are

$$\psi = \frac{1}{\sqrt{1+a^2}}\frac{1}{H} \tag{3.29}$$

$$e^U = \frac{1}{H^{\frac{1}{1+a^2}}} \tag{3.30}$$

$$e^{-a\sigma} = \frac{a^2}{H^{\frac{1}{1+a^2}}} \tag{3.31}$$

There are four interesting cases:

- $a = 0$: this is Einstein–Maxwell theory;

- $a = \frac{1}{\sqrt{3}}$: this is what you get if you reduce Einstein–Maxwell theory from five to four spacetime dimensions;

- $a = 1$: this corresponds to the reduction of string theory from 10 to 4 spacetime dimensions;

- $a = \sqrt{3}$: Kaluza–Klein theory. These solutions are S-dual to the Taub-NUT solutions, i.e. to Kaluza–Klein monopoles

Modulo duality transformations, these solutions are all special cases of the solutions with four $U(1)$ fields and three-scalar fields. In the case that two $U(1)$ fields are electrostatic and two are magnetostatic the internal space decomposes as the metric product of four copies of $SO(2,1)/SO(1,2)$. The spacetime metrics are given by:

$$ds^2 = -(H_1 H_2 H_3 H_4)^{-\frac{1}{2}} dt^2 + (H_1 H_2 H_3 H_4)^{\frac{1}{2}} d\mathbf{x}^2. \tag{3.32}$$

The totally null, totally geodesic submanifolds are such that

$$(\psi^1, \psi^3, \chi_2, \chi_4) = \left(\frac{1}{H_1}, \frac{1}{H_3}, \frac{1}{H_2}, \frac{1}{H_4} \right). \tag{3.33}$$

The entropy is given by

$$S = \pi \sqrt{q_1 q_3 p^2 p^4}. \tag{3.34}$$

Of course these solutions may be lifted to 11 dimensions, for example, where they may be thought of as intersecting 5-branes.

3.5   Other applications

The harmonic function technique works in other than 4 spacetime dimensions. We now give a few examples.

3.5.1   The D-instanton

Perhaps the simplest application of the technique described above yields the D-instantons of ten-dimensional type IIB theory. These are Riemannian solutions. If $\tau = a + ie^{-\Phi}$ where a is the pseudoscalar and Φ the dilaton are the only excited bosonic fields other than the metric then the Lorentzian equations come from the $SL(2, \mathbb{R})$-invariant action

$$R - \frac{1}{2}(\partial\Phi)^2 - e^{2\Phi}(\partial a)^2. \tag{3.35}$$

Note that we are using Einstein conformal gauge.

For the instantons $a = i\alpha$ with α real and the equations come from the action

$$R - \frac{1}{2}(\partial\Phi)^2 + e^{2\Phi}(\partial\alpha)^2. \tag{3.36}$$

This is effectively the same as the previous cases. One takes the Einstein metric to be flat and

$$\alpha + \text{constant} = e^{\Phi} = H \tag{3.37}$$

where H is a harmonic function on $\mathbb{E}^{10}$.

Weyl rescaling the metric to string gauge

$$ds^2 = e^{\frac{1}{2}\Phi}dx^2 = H^{\frac{1}{2}}dx^2 \tag{3.38}$$

gives an Einstein–Rosen bridge down which global Ramond⊗Ramond charge can be carried away. Note that large distances correspond to moderate string coupling $g = e^{\Phi}$ while being near the origin of $\mathbb{E}^{10}$ corresponds to strong coupling and the supergravity approximation cannot be trusted.

3.5.2 NS⊗NS 5-brane in 10 dimensions

The 10-dimensional metric in string conformal frame is

$$ds_S^2 = -dt^2 + (dx_9)^2 + (dx_8)^2 + (dx_7)^2 + (dx_6)^2 + (dx_5)^2 + e^{2\Phi}g_{\mu\nu}dx^\mu dx^\nu \tag{3.39}$$

where $g_{\mu\nu}$ is the 4-dimensional metric in Einstein gauge. If a is the 4-dimensional dual of the NS⊗NS three-form field strength then the equations follow from the Lagrangian

$$R - 2(\partial\Phi)^2 + \frac{1}{2}e^{4\Phi}(\partial a)^2. \tag{3.40}$$

Again one picks the metric $g_{\mu\nu}$ to be flat and

$$a + \text{constant} = e^{2\Phi} = H \tag{3.41}$$

where H is now a harmonic function on $\mathbb{E}^4$.

One may now apply a duality transformation taking one to the Ramond⊗Ramond five-brane. This leaves the Einstein metric invariant but takes $\Phi \to -\Phi$. The resulting metric in string conformal gauge is

$$ds_S^2 = H^{-\frac{1}{2}}(-dt^2 + (dx_9)^2 + (dx_8)^2 + (dx_7)^2 + (dx_6)^2 + (dx_5)^2) + H^{\frac{1}{2}}dx^2. \tag{3.42}$$

3.6   Cosmic string solutions

The 7-brane of Type IIB theory is an example of of how to construct cosmic string like solutions. The main difference in technique with the former case is that since the internal metric is positive definite the the spatial metric can no longer be flat and so we need to solve for it explicitly. This is simple if the spatial metric is 2-dimensional.

To get the 7-brane, we write the 10-dimensional metric in Einstein gauge as

$$ds^2 = -dt^2 + +(dx_9)^2 + (dx_8)^2 + (dx_7)^2 + (dx_6)^2 + (dx_5)^2$$
$$+(dx_4)^2 + (dx_4)^2 + e^\phi dz d\bar{z}. \qquad (3.43)$$

The static equations arise from the 2-dimensional Euclidean action

$$R - \frac{1}{2}\frac{(\partial \tau_1)^2 + (\partial \tau_2)^2}{\tau_2^2} \qquad (3.44)$$

where $\tau = \tau_1 + i\tau_2 = a + ie^{-\phi}$ gives a map into the fundamental domain of the modular group $SL(2,\mathbb{Z})\backslash SO(2,\mathbb{R})/SO(2)$. We may regard the 2-dimensional space sections as a Kahler manifold and the harmonic map equations are thus satisfied by a *holomorphic ansatz* $\tau = \tau(z)$. We must also satisfy the Einstein condition. Using the formula for the Ricci scalar of the 2-dimensional metric and the holomorphicity condition this reduces to the *linear* Poisson equation

$$\partial\bar{\partial}(\phi - \log\tau_2) = 0. \qquad (3.45)$$

To get the *fundamental string* one chooses

$$\phi = \Phi \qquad (3.46)$$

$$\tau \propto \log z. \qquad (3.47)$$

In four spacetime time dimensions the fundamental string is 'super-heavy', it is not asymptotically conical at infinity.

To get the *7-brane*, which does correspond to a more conventional cosmic string, one picks

$$j(\tau(z)) = f(z) = \frac{p(z)}{q(z)} \qquad (3.48)$$

where $j(\tau)$ is the elliptic modular function and $f(z) = (p(z)/q(z))$ is a rational function of degree k.

The appropriate solution for the metric is

$$e^\phi = \tau_2 \eta^2 \bar{\eta}^2 \left| \prod_{i=1}^{k}(z - z_i)^{-\frac{1}{12}} \right|^2. \qquad (3.49)$$

where $\eta(\tau)$ is the Dedekind eta-function. Asymptotically

$$e^\phi \sim (z\bar{z})^{-\frac{k}{12}}. \tag{3.50}$$

Therefore the spatial metric is that of a cone with deficit angle

$$\delta = \frac{4k\pi}{24}. \tag{3.51}$$

This may also be verified using the equations of motion and the Gauss–Bonnet theorem. As a result one can have up to 12 seven-branes in an open universe. To close the universe one needs 24 seven-branes.

The solution has the following 'F-theory' interpretation. One considers the metric

$$ds^2 = g_{ij}dy^i dy^j + e^\phi dz d\bar{z}, \tag{3.52}$$

where g_{ij} is the following unimodular metric on the torus T^2 with coordinates y^i

$$\begin{pmatrix} \tau_2^{-1} & \tau_1\tau_2^{-1} \\ \tau_1\tau_2^{-1} & \tau_1^2\tau_2^{-1} + \tau_2 \end{pmatrix}. \tag{3.53}$$

The metric (3.52)is self-dual or hyper-Kähler. If one takes 24 seven-branes one gets an approximation to a $K3$ surface elliptically fibered over $\mathbb{CP}^1$.

Another interesting special case arises as an orbifold. Consider $T^2 \times \mathbb{C}$ with coordinates (y^1, y^2, z). Quotient by the involution $(y^1, y^2, z) \to (-y^1, -y^2, -z)$. There are four fixed points which may be blown up to obtain a regular simply connected manifold on which there exists a twelve real-dimensional family of family of smooth hyper-Kähler metrics. The second Betti number is five and the intersection form of the five non-trivial cycles is given by the Cartan matrix of the extended Dynkin diagram $\overline{D_4}$. These metrics have been obtained as hyper-Kähler quotients by Kronheimer and by Nakajima. The smooth metrics are rotationally symmetric but $\frac{\partial}{\partial y^1}$ and $\frac{\partial}{\partial y^2}$ are only approximate Killing vectors. physically they are interesting as examples of 'Alice Strings' because, thinking of the two-torus as Kaluza–Klein type internal space with two approximate $U(1)$'s and two approximate charge conjugation operators $C_2 : (y^1, y^2, z) \to (-y^1, y^2, z)$ and $C_2 : (y^1, y^2, z) \to (y^1, -y^2, z)$, one finds that if Alice circumambulates the string, but staying very far away, she returns charge conjugated. Of course if she ventures into the core region she will find that the two electric charges are not strictly conserved. Because the solutions admit a non-triholomorphic circle action, they are given in terms of a solution of the $su(\infty)$ Toda equation. From this it is easy to check that the solutions approach the orbifold limit with exponential accuracy.

To see this explicitly note that the metric

$$ds^2 = \frac{1}{\nu'}(2d\theta + \nu_1 dy^2 - \nu_2 dy^2)^2 + \nu'\left\{d\rho^2 + e^\nu((dy^1)^2 + (dy^2)^2)\right\} \tag{3.54}$$

is hyper-Kähler if $\nu(\rho, y^1, y^2)$ satisfies

$$(e^\nu)'' + \nu_{11} + \nu_{22} = 0, \tag{3.55}$$

where the superscript $'$ denotes differentiation with respect to ρ and the subscripts 1 and 2 denote differentiation with respect to y^1 and y^2 respectively. The Killing vector $\partial/\partial\theta$ leaves invariant the privileged Kähler form

$$(2d\theta + \nu_1 dy^2 - \nu_2 dy^1) \wedge d\rho + +(e^\nu)' dy^1 \wedge dy^2. \tag{3.56}$$

whose closedness requires that (3.55) holds. Note that the geometrical significance of the coordinate ρ is that it is the moment map for the circle action. The simplest solution of (3.55) is $e^\nu = \rho$. This gives the flat metric

$$ds^2 = dr^2 + r^2 d\theta^2 + (dy^1)^2 + (dy^2)^2, \tag{3.57}$$

with $r = 2\sqrt{\rho}$. This is independent of y^1 and y^2. For solutions admitting an elliptic fibration we require a solution of (3.55) which is periodic in y^1 and y^2. To check the typical behaviour near infinity, one linearizes (3.55) about the solution $e^\nu = \rho$. The resulting equation admits solutions by separation of variables. It is then a routine exercise to convince oneself that the general solution must decay exponentially at infinity.

Before leaving these metrics it is perhaps worth pointing out that their relation to the much better known class of Ricci-flat Riemannnian metrics admitting a triholomorphic circle action and which depend on an arbitrary harmonic function H on $\mathbb{E}^3$. They are easily obtained using the technique described above. The metrics are

$$ds^2 = H^{-1}(dt + \omega_i dx^i)^2 + H\mathbf{dx}^2, \tag{3.58}$$

with

$$\text{curl}\,\omega = \text{grad}\,H. \tag{3.59}$$

with Kahler forms

$$Hdx^1 \wedge dx^2 + dx^3 \wedge (dt + \omega_i dx^i). \tag{3.60}$$

$$Hdx^2 \wedge dx^3 + dx^1 \wedge (dt + \omega_i dx^i). \tag{3.61}$$

$$Hdx^3 \wedge dx^1 + dx^2 \wedge (dt + \omega_i dx^i). \tag{3.62}$$

The closedness of these Kähler forms is equivalent to the condition (3.59).

If H is independent of $\arctan(x^2/y^2)$ there will be an additional circle action which preserves the first Kähler form but rotates the second into the third. This means that by taking an arbitrary axisymmetric harmonic function we can, in principle, obtain a solution of the the $su(\infty)$ Toda equation (3.55).

To get a complete metric one must choose H to be a finite sum of k poles with identical positive residues. The coordinate singularities at the poles may then be removed by periodically identifying the imaginary time coordinate t. If $H \to 1$ at infinity the metrics are asymptotically locally flat, 'ALF'. and represent k Kaluza–Klein monopoles. If $H \to 0$ at infinity the metrics are asymptotically locally Euclidean, 'ALE'. Thus if $0 \le t \le 2\pi$ and $h = 1/2r$ we get the flat metric on $\mathbb{R}^3$ while $H = 1 + (1/2r)$ we get the Taub–NUT metric on $\mathbb{R}^3$.

On the other hand if we take $\phi = \log \tau_2$ and identify $z = x^1 + ix^2$ and $t = y^1$ and $y^2 = x^3$ the metrics (3.52) amd (3.58) coincide. In fact $H = \tau_2$ and $\omega_3 = \tau_1$.

We see that $\partial/\partial y^1$ generates a triholomorphic circle action. The three Kähler forms are

$$\Omega_1 = \tau_2 dx^1 \wedge dx^2 + dy^2 \wedge dy^1, \qquad (3.63)$$

$$\Omega_2 = \tau_2 dx^2 \wedge dy^2 + dx^1 \wedge (dy^1 + \tau_1 dy^2) \qquad (3.64)$$

and

$$\Omega_3 = \tau_2 dy^2 \wedge dx^1 + dx^2 \wedge (dy^1 + \tau_1 dy^2). \qquad (3.65)$$

and they are closed by virtue of the Cauchy–Riemann equations for $\tau(x^1 + ix^2)$.

4 Conclusion

In these lectures I have tried to give some idea of what qualifies as a soliton in classical supergravity theories and how one finds the solutions. I have concentrated on general principles and largely restricted myself to four spacetime dimensions. The lectures were emphatically *not* intended as a comprehensive review. For recent applications the reader is referred to other articles in this volume or to the voluminous current literature. Appended below is a rather restricted list of references largely confined to papers that I have written, either alone or with collaborators, where the reader may find more details of the claims made above or from which the reader may trace back to the original sources. As I stated above, it was not my intention to provide a comprehensive review and no slight is intended against those not explicitly cited.

References

[1] G.W. Gibbons, 'Soliton states and central charges in extended supergravity theories'. In *Proceedings of the Heisenberg Memorial Symposium*, P.

Breitenlohner & H.P. Durr (eds.), *Springer Lecture Notes in Physics* **160** 145–151 (1982).

[2] G.W. Gibbons, 'The multiplet structure of solitons in the $N = 2$ supergravity theory'. In *Quantum Structure of Space and Time*, M.J. Duff & C.J. Isham (eds.), (Cambridge University Press) 317–321 (1983).

[3] G.W. Gibbons and C.M. Hull, 'A Bogomol'nyi bound for general relativity and solitons in $N = 2$ supergravity'. *Phys. Lett.* **109** 190–194 (1982).

[4] G.W. Gibbons 'The Bogomol'nyi inequality for Einstein–Maxwell theory'. In *Monopoles in Quantum Field Theory*, N. Craigie, P. Goddard & W. Nahm (eds.), (World Scientific, Singapore) 137–138 (1982).

[5] G.W. Gibbons, 'Antigravitating black hole solitons with scalar hair in $N = 4$ supergravity'. *Nucl. Phys.* **B207** 337–349 (1982).

[6] G.W. Gibbons, G.T. Horowitz, S.W. Hawking and M.J. Perry, 'Positive mass theorems for black holes'. *Comm. Math. Phys.* **88** 295–308 (1983).

[7] G.W. Gibbons, C.M. Hull and N.P. Warner, 'The stability of gauged supergravity'. *Nucl. Phys.* **B218** 173–190 (1983).

[8] G.W. Gibbons, 'Electrovac ground state in gauged $SU(2) \times SU(2)$ supergravity'. *Nucl. Phys.* **B233** 24–49 (1984).

[9] G.W. Gibbons, 'Vacua and solitons in gauged supergravity'. In *Relativity Cosmology Topological Mass and Supergravity*, C. Aragone (ed.), (World Scientific, Singapore) 163–177 (1984).

[10] G.W. Gibbons and M.J. Perry, 'Soliton-supermultiplets and Kaluza–Klein theory'. *Nucl. Phys.* **B248** 629–646 (1984).

[11] G.W. Gibbons, 'Solitons in general relativity and supergravity'. In *Non-Linear Phenomena in Physics*, F. Claro (ed.), *Springer Proceedings in Physics* **3** 255–290 (1985).

[12] G.W. Gibbons, 'Aspects of supergravity theories'. In *Supersymmetry, Supergravity and Related Topics*, F. del Aguila, A. Azcárraga & L.E. Ibáñez (eds.), (World Scientific, Singapore) (1985).

[13] G.W. Gibbons, 'Solitons and black holes in 4 and 5 dimensions'. In *Field Theory, Quantum Gravity and Strings*, H.J. de Vega & N. Sanchez (eds.), *Springer Lecture Notes in Physics* **246** 46–59 (1986).

[14] G.W. Gibbons and P.J. Ruback, 'Motion of extreme Reissner–Nordström black holes in the low-velocity limit'. *Phys. Rev. Lett.* **57** 1492–1495 (1986).

[15] G.W. Gibbons, 'Quantized flux-tubes in Einstein–Maxwell theory and non-compact internal spaces'. In *Fields and Geometry*, A. Jadczyk (ed.), (World Scientific, Singapore) 627–638 (1986).

[16] P.B. Breitenlohner, G.W. Gibbons and D. Maison, '4-dimensional black holes from Kaluza–Klein theories'. *Comm. Math. Phys.* **120** 295–334 (1988).

[17] G.W. Gibbons and K. Maeda, Black holes and membranes in higher-dimensional theories with dilaton fields'. *Nucl. Phys.* **B298** 741–775 (1988).

[18] G.W. Gibbons, M.E. Ortiz and F. Ruiz Ruiz, 'Stringy cosmic strings with horizons'. *Phys. Lett.* **240B** 50–54 (1990).

[19] A. Dabholkar, G.W. Gibbons J.A. Harvey and F. Ruiz Ruiz, 'Superstrings and solitons'. *Nucl. Phys.* **B340** 33–55 (1990).

[20] G.W. Gibbons and P.K. Townsend, 'Vacuum interpolation in supergravity via super *p*-branes'. *Phys. Rev. Lett.* **71** 3754–3757 (1994).

[21] G.W. Gibbons, D. Kastor, L.A.J. London, J. Traschen and P.K. Townsend *Nucl. Phys.* **B 416** 850–880 (1994).

[22] M.J. Duff, G.W. Gibbons and P.K. Townsend, 'Macroscopic superstrings as interpolating solitons'. *Phys. Lett.* **B 332** 321–328 (1994).

[23] G.W. Gibbons and R. Kallosh, 'The topology, entropy and Witten index of extreme black holes'. *Phys. Rev.* **D51** 2839–2862 (1995).

[24] G.W. Gibbons, G.T. Horowitz and P.K. Townsend, 'Higher-dimensional resolution of dilatonic black hole singularities'. *Class. Quant. Grav.* **12** 297–317 (1995).

[25] G.W. Gibbons and P.K. Townsend, 'Antigravitating BPS monopoles and dyons'. *Phys. Lett.* **B 356** 472–478 (1995) hep-th/9506131.

[26] F. Dowker, G.P. Gauntlett, G.W. Gibbons and G.T. Horowitz, 'The decay of magnetic fields in Kaluza–Klein theory'. *Phys. Rev.* **D 52** 6929–6940 (1995) hep-th/9507143.

[27] G.W. Gibbons, M.B. Green and M.J. Perry, 'Instantons and 7-branes in Type IIB superstring theory'. *Phys. Lett.* **B 370** 37–44 (1996) hep-th 9511080.

[28] F. Dowker, J.P. Gauntlett, G.W. Gibbons and G.T. Horowitz, 'Nucleation of p-Branes and fundamental strings'. *Phys. Rev.* **D53** 7115-7128 (1996) hep-th/9512154.

[29] G.W. Gibbons, R. Kallosh and B. Kol, 'Moduli, scalar charges, and the first law of thermodynamics'. *Phys. Rev. Lett.* **77** 4992-4995 (1996) hep-th/9607108.

[30] G.W. Gibbons, R. Goto and P. Rychenkova, 'Hyper-Kähler quotient construction of BPS monopole moduli spaces'. *Comm. Math. Phys.* **186** 581-600 (1997) hep-th/9608085.

[31] S.R. Das, G.W. Gibbons and S.D. Mathur, 'Universality of low energy absorption cross-sections for black holes'. *Phys. Rev. Lett.* **78** 417-419 (1997) hep-th/9609052.

[32] A. Chamseddine, S. Ferrara, G.W. Gibbons and R. Kallosh, 'Enhancement of supersymmetry near 5d black hole horizon'. *Phys. Rev.* **D55** 3647-3653 (1997) hep-th/960155.

[33] S. Ferrara, G.W. Gibbons and R. Kallosh, 'Black holes and critical points in moduli space'. *Nucl. Phys.* **B500** 75-93 (1997) hep-th/9702103.

[34] J.P. Gauntlett, G.W. Gibbons, G. Papadopoulos and P.K. Townsend, 'Hyper-Kähler manifolds and multiply intersecting branes'. *Nucl. Phys.* **B500** 133-162 (1997) hep-th/9702202.

[35] G.W. Gibbons, G. Papadopoulos and K. Stelle, 'HKT and OKT geometries on soliton black hole moduli spaces'. *Nucl. Phys.* **B**, in press, hep-th/9706207.

[36] B.R. Greene, A. Shapere, C. Vafa and S.-T. Yau, 'Stringy cosmics strings and noncompact Calabi-Yau manifolds'. *Nucl. Phys.* **B337** 1-36 (1990).

[37] M.A. Olshanetsky and A.M. Perelomov, 'Classical integrable finite-dimensional systems related to Lie algebras'. *Phys. Rep.* **71** 313-400 (1981).

An Introduction to Non-Perturbative String Theory

Ashoke Sen

Abstract

In this review I discuss some basic aspects of non-perturbative string theory. The topics include tests of duality symmetries based on the analysis of the low energy effective action and the spectrum of BPS states, relationship between different duality symmetries, an introduction to M- and F-theories, black hole entropy in string theory, and Matrix theory.

0   Introduction

During the last few years, our understanding of string theory has undergone a dramatic change. The key to this development is the discovery of duality symmetries, which relate the strong and weak coupling limits of apparently different string theories. These symmetries not only relate apparently different string theories, but give us a way to compute certain strong coupling results in one string theory by mapping it to a weak coupling result in a dual string theory. In this review I shall try to give an introduction to this exciting subject. However, instead of surveying all the important developments in this subject I shall try to explain the basic ideas with the help of a few simple examples. I apologise for the inherent bias in the choice of examples and the topics; this is solely due to the varied degree of familiarity that I have with this vast subject. I have also not attempted to give a complete list of references. Instead I have only included those references whose results have been directly used or mentioned in this article. A complete list of references may be obtained by looking at the citations to some of the original papers in the bibliography. There are also many other reviews in this subject where more references can be found (see [1]–[24]). I hope that this review will serve the limited purpose of initiating a person with a knowledge of perturbative string theory into this area. (For an introduction to perturbative string theory see [25]).

The review will be divided into ten main sections as described below.

1. A brief review of perturbative string theory: In this section I shall very briefly recollect some of the results of perturbative string theory

which will be useful to us in the rest of this article. This will in no way constitute an introduction to this subject; at best it will serve as a reminder to a reader who is already familiar with this subject.

2. Notion of duality symmetry: In this section I shall describe the notion of duality symmetry in string theory, a few examples of duality conjectures in string theory, and the general procedure for testing these duality conjectures.

3. Analysis of the low energy effective action: In this section I shall describe how one arrives at various duality conjectures by analyzing the low energy effective action of string theory.

4. Precision test of duality based on the spectrum of BPS states: In this section I shall discuss how one can device precision tests of various duality conjectures based on the analysis of the spectrum of a certain class of supersymmetric states in string theory.

5. Interrelation between various dualities: In this section I shall try to relate the various duality conjectures introduced in the Sections 2–4 by 'deriving' them from a basic set of duality conjectures. I shall also discuss what we mean by relating different dualities and try to formulate the rules that must be followed during such a derivation.

6. Duality in theories with < 16 supersymmetries: The discussion in Sections 3–5 is focussed on string theories with at least 16 supersymmetry generators. In this section I consider theories with fewer supersymmetries. Specifically we shall focus our attention on theories with 8 supercharges, which correspond to $N = 2$ supersymmetry in 4 dimensions.

7. M-theory: In this section I discuss the emergence of a new theory in 11 dimensions – now known as M-theory – from the strong coupling limit of type IIA string theory. I also discuss how compactification of M-theory gives rise to new theories that cannot be regarded as perturbative compactifications of a string theory.

8. F-theory: In this section I shall discuss yet another novel way of generating non-perturbative compactification of string theory based on a construction known as F-theory. This class of compactifications is non-perturbative in the sense that the string coupling necessarily becomes strong in some regions of the internal compact manifold, unlike conventional compactification where the string coupling can be kept small everywhere on the internal manifold.

9. Microscopic derivation of the black hole entropy: In this section I shall discuss how many of the techniques and ideas that were used to test various duality conjectures in string theory can be used to give a microscopic derivation of the Bekenstein–Hawking entropy and Hawking radiation from black holes.

10. Matrix theory: In this final section I shall discuss a proposal for a nonperturbative definition of M-theory and various other string theories in terms of quantum mechanics of $N \times N$ matrices in the large N limit.

Throughout this article I shall work in units where $\hbar = 1$ and $c = 1$.

1   A Brief Review of Perturbative String Theory

String theory is based on the simple idea that elementary particles, which appear as point-like objects to the present day experimentalists, are actually different vibrational modes of strings. The energy per unit length of the string, known as the string tension, is parametrized as $(2\pi\alpha')^{-1}$, where α' has the dimension of (length)^2. As we shall describe later, this theory automatically contains gravitational interaction between elementary particles, but in order to correctly reproduce the strength of this interaction, we need to choose $\sqrt{\alpha'}$ to be of the order of $10^{-33}cm$. Since $\sqrt{\alpha'}$ is the only length parameter in the theory, the typical size of a string is of the order of $\sqrt{\alpha'} \sim 10^{-33}cm$ – a distance that cannot be resolved by present day experiments. Thus there is no direct way of testing string theory, and its appeal lies in its theoretical consistency.

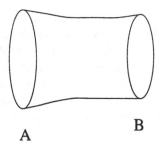

A   **B**

Figure 1: Propagation of a closed string.

The basic principle behind constructing a quantum theory of relativistic string is quite simple. Consider propagation of a string from a space-time configuration A to a space-time configuration B. During this motion the string

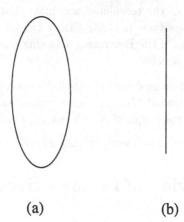

(a)   (b)

Figure 2: (a) A closed string, and (b) an open string.

sweeps out a 2-dimensional surface in space-time, known as the string world-sheet (see Figure 1). The amplitude for the propagation of the string from the space-time position A to space-time position B is given by the weighted sum over all world-sheets bounded by the initial and the final locations of the string. The weight factor is given by e^{-S} where S is the product of the string tension and the area of the world-sheet. It turns out that this procedure by itself does not give rise to a fully consistent string theory. In order to get a fully consistent string theory we need to add some internal fermionic degrees of freedom to the string and generalize the notion of area by adding new terms involving these fermionic degrees of freedom. The leads to 5 (apparently) different consistent string theories in $(9 + 1)$-dimensional space-time, as we shall describe.

In the first quantized formalism, the dynamics of a point particle is described by quantum mechanics. Generalizing this we see that the first quantized description of a string will involve a $(1 + 1)$-dimensional quantum field theory. However, unlike a conventional quantum field theory where the spatial directions have infinite extent, here the spatial direction, which labels the coordinate on the string, has finite extent. It represents a compact circle if the string is closed (Figure 2(a)) and a finite line interval if the string is open (Figure 2(b)). This $(1 + 1)$-dimensional field theory is known as the world-sheet theory. The fields in this $(1 + 1)$-dimensional quantum field theory and the boundary conditions on these fields vary in different string theories. Since the spatial direction of the world-sheet theory has finite extent, each world-sheet field can be regarded as a collection of an infinite number of harmonic oscillators labelled by the quantized momentum along this spatial direction. Different states of the string are obtained by acting on the Fock vacuum by these oscillators. This gives an infinite tower of states. Typically each string theory contains a set of massless states and an infinite tower of massive states. The massive string states typically have mass of the order of $(10^{-33}cm)^{-1} \sim 10^{19}GeV$ and are far beyond the reach of present day accelerators. Thus the interesting part of the theory is the one involving the massless states. We shall now briefly describe the spectrum and interaction in various string theories and their compactifications.

1.1 The spectrum

There are 5 known fully consistent string theories in 10 dimensions. They are known as type IIA, type IIB, type I, $E_8 \times E_8$ heterotic and $SO(32)$ heterotic string theories respectively. Here we give a brief description of the degrees of freedom and the spectrum of massless states in each of these theories. We shall give the description in the so-called light-cone gauge, which has the advantage that all states in the spectrum are physical states.

1. Type II string theories: In this case the world-sheet theory is a free field theory containing 8 scalar fields and 8 Majorana fermions. These 8 scalar fields are in fact common to all 5 string theories, and represent the 8 transverse coordinates of a string moving in a 9-dimensional space. It is useful to regard the 8 Majorana fermions as 16 Majorana–Weyl fermions, 8 of them having left-handed chirality and the other 8 having right-handed chirality. We shall refer to these as left- and right-moving fermions respectively. Both the type II string theories contain only closed strings; hence the spatial component of the world-sheet is a circle. The 8 scalar fields satisfy periodic boundary conditions as we go around the circle. The fermions have a choice of having periodic or anti-periodic boundary conditions. It is customary to refer to periodic boundary conditions as Ramond (R) boundary conditions (see [181]) and anti-periodic boundary conditions as Neveu–Schwarz (NS) boundary conditions (see [182]). It turns out that in order to get a consistent string theory we need to include in our theory different classes of string states, some of which have periodic and some of which have anti-periodic boundary conditions on the fermions. In all there are 4 classes of states which need to be included in the spectrum:

   - NS–NS where we put anti-periodic boundary conditions on both the left- and the right-moving fermions,

   - NS–R where we put anti-periodic boundary conditions on the left-moving fermions and periodic boundary conditions on the right-moving fermions,

- R–NS where we put periodic boundary conditions on the left-moving fermions and anti-periodic boundary conditions on the right-moving fermions,

- R–R where we put anti-periodic boundary conditions on both the left- and the right-moving fermions.

Finally, we keep only about 1/4 of the states in each sector by keeping only those states in the spectrum which have in them only an even number of left-moving fermions and an even number of right-moving fermions. This is known as the GSO projection (see [183]). The procedure has some ambiguity since in each of the 4 sectors we have the choice of assigning to the ground state either even or odd fermion number. Consistency of string theory rules out most of these possibilities, but at the end two possibilities remain. These differ from each other in the following way. In one possibility, the assignment of the left- and the right-moving fermion number to the left- and the right-moving Ramond ground states are carried out in an identical manner. This gives type IIB string theory. In the second possibility the GSO projections in the left- and the right-moving sector differ from each other. This theory is known as type IIA string theory.

Typically, states from the Ramond sector are in the spinor representation of the $SO(9,1)$ Lorentz algebra, whereas those from the NS sector are in the tensor representation. Since the product of two spinor representations gives us back a tensor representation, the states from the NS–NS and the R–R sectors are bosonic, and those from the NS–R and R–NS sectors are fermionic. It will be useful to list the massless bosonic states in these two string theories. Since the two theories differ only in their R-sector, the NS sector bosonic states are the same in the two theories. They constitute a symmetric rank-2 tensor field, an antisymmetric rank-2 tensor field, and a scalar field known as the dilaton.[1] The RR sector massless states of type IIA string theory consist of a vector and a rank-3 anti-symmetric tensor. On the other hand, the massless states from the RR sector of type IIB string theory consist of a scalar, a rank-2 anti-symmetric tensor field, and a rank-4 anti-symmetric tensor gauge field satisfying the constraint that its field strength is self-dual.

The spectra of both these theories are invariant under space-time supersymmetry transformations which transform fermionic states to bosonic states and vice-versa. The supersymmetry algebra for type IIB theory is known as the chiral $N = 2$ superalgebra and that of type IIA theory

---

[1]Although from string theory we get the spectrum of states, it is useful to organise the spectrum in terms of fields. In other words the spectrum of massless fields in string theory is identical to that of a free field theory with these fields.

is known as the non-chiral $N = 2$ superalgebra. Both superalgebras consist of 32 supersymmetry generators.

Often it is convenient to organise the infinite tower of states in string theory by their oscillator level, defined as follows. As has already been pointed out before, the world-sheet degrees of freedom of the string can be regarded as a collection of an infinite number of harmonic oscillators. For the creation operator associated with each oscillator we define the level as the absolute value of the number of units of world-sheet momentum that it creates while acting on the vacuum. The total oscillator level of a state is then the sum of the levels of all the oscillators that act on the Fock vacuum to create this state. (The Fock vacuum, in turn, is characterized by several quantum numbers, which are the momenta conjugate to the zero modes of various fields – modes carrying zero world-sheet momentum.) We can also separately define the left-(right-) moving oscillator level as the contribution to the oscillator level from the left- (right-) moving bosonic and fermionic fields. Finally, if E and P denote respectively the world-sheet energy and momentum[2] then we define $L_0 = (E + P)/2$ and $\bar{L}_0 = (E - P)/2$. Both L_0 and $\bar{L}_0$ include contributions from the oscillators as well as from the Fock vacuum. Thus, for example, the total contribution to L_0 will be given by the sum of the right-moving oscillator level and the contribution to L_0 from the Fock vacuum.

2. Heterotic string theories: The world-sheet theory of the heterotic string theories consists of 8 scalar fields, 8 right-moving Majorana–Weyl fermions and 32 left-moving Majorana–Weyl fermions. We have as before NS and R boundary conditions as well as GSO projection involving the right-moving fermions. Also, as in the case of type II string theories, the NS sector states transform in the tensor representation and the R sector states transform in the spinor representation of the $SO(9,1)$ Lorentz algebra. However, unlike in the case of type II string theories, in this case the boundary condition on the left-moving fermions do not affect the Lorentz transformation properties of the state. Thus bosonic states come from states with NS boundary condition on the right-moving fermions and fermionic states come from states with R boundary condition on the right-moving fermions.

There are two possible boundary conditions on the left-moving fermions which give rise to fully consistent string theories. They are:

---

[2]We should distinguish between world-sheet momentum and the momenta of the $(9+1)$-dimensional theory. The latter are the the momenta conjugate to the zero modes of various bosonic fields in the world-sheet theory.

- $SO(32)$ heterotic string theory: In this case we have two possible boundary conditions on the left-moving fermions: either all of them have periodic boundary conditions, or all of them have anti-periodic boundary conditions. In each sector we also have a GSO projection that keeps only those states in the spectrum which contain an even number of left-moving fermions. The massless bosonic states in this theory consist of a symmetric rank-2 field, an anti-symmetric rank-2 field, a scalar field known as the dilaton and a set of 496 gauge fields filling up the adjoint representation of the gauge group $SO(32)$.

- $E_8 \times E_8$ heterotic string theory: In this case we divide the 32 left-moving fermions into two groups of 16 each and use 4 possible boundary conditions: (1) all the left-moving fermions have periodic boundary condition; (2) all the left-moving fermions have anti-periodic boundary condition; (3) all the left-moving fermions in group 1 have periodic boundary conditions and all the left-moving fermions in group 2 have anti-periodic boundary conditions; (4) all the left-moving fermions in group 1 have anti-periodic boundary conditions and all the left-moving fermions from group 2 have periodic boundary conditions. In each sector we also have a GSO projection that keeps only those states in the spectrum which contain an even number of left-moving fermions from the first group, and alsoan even number of left-moving fermions from the second group. The massless bosonic states in this theory consist of a symmetric rank-2 field, an anti-symmetric rank-2 field, a scalar field known as the dilaton and a set of 496 gauge fields filling up the adjoint representation of the gauge group $E_8 \times E_8$.

The spectra of states in both the heterotic string theories are invariant under a set of space-time supersymmetry transformations. The relevant superalgebra is known as the chiral $N = 1$ supersymmetry algebra, and has 16 real generators.

Using the Bose-Fermi equivalence in (1+1) dimensions, we can reformulate both the heterotic string theories by replacing the 32 left-moving fermions by 16 left-moving bosons. In order to get a consistent string theory the momenta conjugate to these bosons must take discrete values. It turns out that there are only two consistent ways of quantizing the momenta, giving us back the two heterotic string theories.

3. Type I string theory: The world-sheet theory of type I theory is identical to that of type IIB string theory, with the following two crucial difference.

- Type IIB string theory has a symmetry that exchanges the left- and the right-moving sectors in the world-sheet theory. This transformation is known as the world-sheet parity transformation. (This symmetry is not present in type IIA theory since the GSO projection in the two sectors are different). In constructing type I string theory we keep only those states in the spectrum which are invariant under this world-sheet parity transformation.

- In type I string theory we also include open string states in the spectrum. The world-sheet degrees of freedom are identical to those in the closed string sector. Specifying the theory requires us to specify the boundary conditions on the various fields. We put Neumann boundary condition on the 8 scalars, and appropriate boundary conditions on the fermions.

The spectrum of massless bosonic states in this theory consists of a symmetric rank-2 tensor and a scalar dilaton from the closed string NS sector, an anti-symmetric rank-2 tensor from the closed string RR sector, and 496 gauge fields in the adjoint representation of $SO(32)$ from the open string sector. This spectrum is also invariant under the chiral $N = 1$ supersymmetry algebra with 16 real supersymmetry generators.

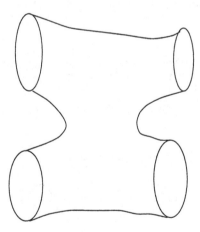

Figure 3: A string world-sheet bounded by 4 external strings.

1.2 Interactions

So far we have discussed the spectrum of string theory, but in order to fully describe the theory we must also describe the interaction between various particles in the spectrum. In particular, we would like to know how to compute a scattering amplitude involving various string states. It turns out that there is a unique way of introducing interaction in string theory. Consider for example a scattering involving 4 external strings, situated along some specific curves in space-time. The prescription for computing the scattering amplitude is to compute the weighted sum over all possible string world-sheets bounded by the 4 strings with weight factor e^{-S}, S being the string tension multiplied by the generalized area of this surface (taking into account the fermionic degrees of freedom of the world-sheet). One such surface is shown in Figure 3. If we imagine the time axis running from left to right, then this diagram represents two strings joining into one string and then splitting into two strings, – the analog of a tree diagram in field theory. A more complicated surface is shown in Figure 4. This represents two strings joining into one string, which then splits into two and joins again, and finally splits into two strings. This is the analog of a one-loop diagram in field theory. The relative normalization between the contributions from these two diagrams is not determined by any consistency requirement. This introduces an arbitrary parameter in string theory, known as the string coupling constant. However, once the relative normalization between these two diagrams is fixed, the relative normalization between all other diagrams is fixed due to various consistency requirements. Thus, besides the dimensionful parameter α', string theory has a single dimensionless coupling constant. As we shall see later, both these parameters can be absorbed into definitions of various fields in the theory.

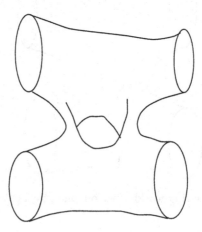

Figure 4: A more complicated string world-sheet.

What we have described so far is the computation of the scattering amplitude with fixed locations of the external strings in space-time. The more relevant quantity is the scattering amplitude where the external strings are in eigenstates of the energy and momenta operators conjugate to the coordinates of the (9 + 1)-dimensional space-time. This is done by simply taking the convolution of the above scattering amplitude with the wave-functions of the strings corresponding to the external states. In practice there is an extremely efficient method of doing this computation using the so-called vertex operators. It turns out that unlike in quantum field theory, all of these scattering amplitudes in string theory are ultraviolet finite. This is one of the major achievements of string theory.

Our main interest will be in the scattering involving the external massless states. The most convenient way to summarize the result of this computation in any string theory is to specify the effective action. By definition this effective action is such that if we compute the *tree level* scattering amplitude using this action, we should reproduce the S-matrix elements involving the massless states of string theory. In general such an action will have to contain infinite number of terms, but we can organise these terms by examining the number of space-time derivatives that appear in a given term in the action. Terms with the lowest number of derivatives constitute the *low energy effective action* − so-called because this gives the dominant contribution if we want to evaluate the scattering amplitude when all the external particles have small energy and momenta.

The low energy effective actions for all 5 string theories have been found. The actions for the type IIA and type IIB string theories correspond to those of two well known supergravity theories in 10 space-time dimensions, called type IIA and type IIB supergravity theories respectively. On the other hand the actions for the two heterotic and type I string theories correspond to another set of well-known supersymmetric theories in 10 dimensions, − $N = 1$ supergravity coupled to $N = 1$ super Yang–Mills theory. For type I and the $SO(32)$ heterotic string theories the Yang–Mills gauge group is $SO(32)$ whereas for the $E_8 \times E_8$ heterotic string theory the gauge group is $E_8 \times E_8$. The emergence of gravity in all the 5 string theories is the most striking result in string theory. Its origin can be traced to the existence of the symmetric rank-2 tensor state (the graviton) in all these theories. This, combined with the result on finiteness of scattering amplitudes, shows that string theory gives us a finite quantum theory of gravity. We shall explicitly write down the low energy effective action of some of the string theories in Section 3.

The effective actions of all 5 string theories are invariant under the transformation

$$\Phi \to \Phi - 2C, \qquad g_S \to e^C g_S, \qquad (1.1)$$

together with possible rescaling of other fields. Here Φ denotes the dilaton field, g_S denotes the string coupling, and C is an arbitrary constant. Using this scaling property, g_S can be absorbed in Φ. Put another way, the dimensionless coupling constant in string theory is related to the vacuum expectation value $\langle\Phi\rangle$ of Φ. The perturbative effective action does not have any potential for Φ, and hence $\langle\Phi\rangle$ can take arbitrary values. One expects that in a realistic string theory where supersymmetry is spontaneouly broken, there will be a potential for Φ, and hence $\langle\Phi\rangle$ will be determined uniquely.

In a similar vein one can argue that in string theory even the string tension, or equivalently the parameter α', has no physical significance. Since α' has the dimension of (length)^2 and is the only dimensionful parameter in the theory, the effective action will have an invariance under the simultaneous rescaling of α' and the metric $g_{\mu\nu}$:

$$\alpha' \to \lambda\alpha', \qquad g_{\mu\nu} \to \lambda g_{\mu\nu}, \tag{1.2}$$

together with possible rescaling of other fields. Using this scaling symmetry α' can be absorbed into the definition of $g_{\mu\nu}$. We shall discuss these two rescalings in detail in Section 3.1.

1.3 Compactification

So far we have described 5 different string theories, but they all live in 10 space-time dimensions. Since our world is $(3+1)$-dimensional, these are not realistic string theories. However one can construct string theories in lower dimensions using the idea of compactification. The idea is to take the $(9+1)$-dimensional space-time as the product of a $(9-d)$-dimensional compact manifold $\mathcal{M}$ with Euclidean signature and a $(d+1)$-dimensional Minkowski space $R^{d,1}$. Then, in the limit when the size of the compact manifold is sufficiently small so that the present day experiments cannot resolve this distance, the world will effectively appear to be $(d+1)$-dimensional. Choosing $d = 3$ will give us a $(3+1)$-dimensional theory. Of course we cannot choose any arbitrary manifold $\mathcal{M}$ for this purpose; it must satisfy the equations of motion of the effective field theory that comes out of string theory. One also normally considers only those manifolds which preserve part of the space-time supersymmetry of the original 10-dimensional theory, since this guarantees vanishing of the cosmological constant, and hence consistency of the corresponding string theory order-by-order in perturbation theory. There are many known examples of manifolds satisfying these restrictions, e.g. tori of different dimensions, $K3$, Calabi–Yau manifolds, etc. Instead of going via the effective action, one can also directly describe these compactified theories as string theories. For this one needs to modify the string world-sheet action in such a way that it describes string propagation in the new manifold $\mathcal{M} \times R^{d,1}$, instead of in flat 10-dimensional space-time. This modifies the world-sheet theory to an interacting non-linear σ-model instead of a free field theory. Consistency of string theory puts restriction on the kind of manifold on which the string can propagate. At the end both approaches yield identical results.

The simplest class of compact manifolds, on which we shall focus much of our attention in the rest of this article, are tori − product of circles. The effect of this compactification is to periodically identify some of the bosonic fields in the string world-sheet field theory − the fields which represent coordinates tangential to the compact circles. One effect of this is that the momentum carried by any string state along any of these circles is quantized in units of $1/R$, where R is the radius of the circle. But that is another novel effect: we now have new states that correspond to strings wrapped around a compact circle. For such a state, as we go once around the string, we also go once around the compact circle. These states are known as winding states and play a crucial role in the analysis of duality symmetries.

2 Notion of Duality Symmetries in String Theory

In this section I shall elaborate the notion of duality symmetries, the difficulties in testing them, and the way of avoiding these difficulties. We begin by introducing the notion of duality in string theory.

2.1 Duality symmetries: Definition and examples

As was described in the last section, there are 5 consistent string theories in 10 space-time dimensions. We also saw that we can get many different string theories in lower dimensions by compactifying these 5 theories on appropriate manifolds $\mathcal{M}$. Each of these theories is parametrized by a set of parameters known as moduli[3], e.g.

- String coupling constant (related to the vacuum expectation value of the dilaton field),

- Shape and size of $\mathcal{M}$ (information contained in the metric),

- various other background fields.

Inside the moduli space of the theory there is a certain region where the string coupling is weak and perturbation theory is valid. Elsewhere the theory is strongly coupled. This situation has been illustrated in Figure 5.

---

[3]In string theory these moduli are related to vacuum expectation values of various dynamical fields and are expected to take definite values when supersymmetry is broken.

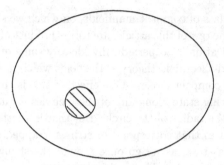

Figure 5: A schematic representation of the moduli space of a string theory. The shaded region denotes the weak coupling region, whereas the white region denotes the strong coupling region.

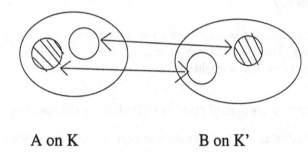

A on K   **B on K'**

Figure 6: A schematic representation of the duality map between the moduli spaces of two different string theories, A on K and B on K', where A and B are two of the 5 string theories in 10 dimensions, and K, K' are two compact manifolds. Under this duality the weak coupling region of the first theory (denoted by the shaded region) gets mapped to the strong coupling region of the second theory and vice versa.

String duality provides us with an equivalence map between two different string theories. In general this equivalence relation maps the weak coupling region of one theory to the strong coupling region of the second theory and vice versa. This situation is illustrated in Figure 6.

Before we proceed, let us give a few examples of dual pairs:

- Type I and $SO(32)$ heterotic string theories in $D = 10$ are conjectured to be dual to each other (see [26, 27, 28, 29]).

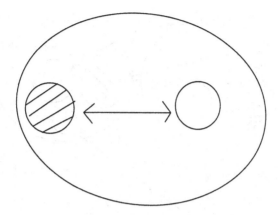

Figure 7: Schematic representation of the moduli space of a self-dual theory. Duality relates weak and strong coupling regions of the same theory.

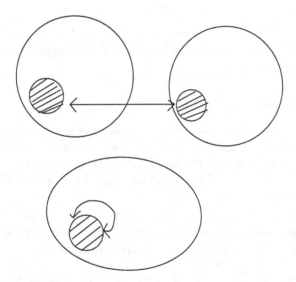

Figure 8: Examples of T-duality relating a weakly coupled theory to a different or the same weakly coupled theory.

- Type IIA string theory compactified on $K3$ and heterotic string theory compactified on a 4-dimensional torus T^4 are conjectured to be dual to

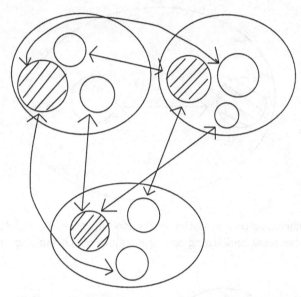

Figure 9: A schematic representation of the moduli spaces of a chain of theories related by duality. In each case the shaded region denotes the weak coupling region as usual.

each other (see [30, 31, 26, 32, 33]).[4]

Under duality, typically perturbation expansions get mixed up. Thus for example, tree level results in one theory might include perturbative and non-perturbative corrections in the dual theory. Also, under duality, many of the elementary string states in one theory get mapped to solitons and their bound states in the dual theory.

Although duality in general relates the weak coupling limit of one theory to the strong coupling limit of another theory, there are special cases where the situation is a bit different. For example, we can have:

- Self-duality: Here duality gives an equivalence relation between different regions of the moduli space of the same theory, as illustrated in Figure 7. In this case, duality transformations form a symmetry group that acts on the moduli space of the theory. For example, type IIB string theory in $D = 10$ is conjectured to have an $SL(2, Z)$ self-duality group (see [30]).

---

[4]Throughout this article a string theory on $\mathcal{M}$ will mean string theory in the background $\mathcal{M} \times R^{9-n,1}$ where n is the real dimension of $\mathcal{M}$, and $R^{9-n,1}$ denotes $(10-n)$-dimensional Minkowski space.

- T-duality: In this case, duality transformation maps the weak coupling region of one theory to the weak coupling region of another theory or the same theory as illustrated in Figure 8. For example, type IIA string theory compactified on a circle of radius R is dual to IIB string theory compactified on a circle of radius R^{-1} at the same value of the string coupling. Also, either of the two heterotic string theories compactified on a circle of radius R is dual to the same theory compactified on a circle of radius R^{-1} at the same value of the coupling constant. As a result the duality map does not mix up the perturbation expansions in the two theories. (For a review of this subject see [34].)

In a generic situation duality can relate not just two theories, but a whole chain of theories, as illustrated in Figure 9. Thus for example, type IIA string theory compactified on $K3$ is related to heterotic string theory compactified on T^4. On the other hand, due to the equivalence of the $SO(32)$ heterotic and type I string theory in 10 dimensions, $SO(32)$ heterotic string theory compactified on T^4 is related to type I string theory compactified on T^4. Thus these three theories are related by a chain of duality transformations.

From this discussion we see that the presence of duality in string theory has two important consequences. First of all, it reduces the degree of non-uniqueness of string theory, by relating various apparently unrelated (compactified) string theories. Furthermore, it allows us to study a strongly coupled string theory by mapping it to a weakly coupled dual theory whenever such a dual theory exists.

2.2   Testing duality conjectures

Let us now turn to the question of testing duality. As we have already emphasized, duality typically relates a weakly coupled string theory to a strongly coupled string theory. Thus in order to prove / test duality we must be able to analyze at least one of the theories at strong coupling. But in string theory we only know how to define the theory perturbatively at weak coupling. Thus it would seem impossible to prove or test any duality conjecture in string theory.[5] This is where supersymmetry comes to our rescue. Supersymmetry gives rise to certain non-renormalization theorems in string theory, due to which some of the weak coupling calculations can be trusted even at strong coupling. Thus we can focus our attention on such 'non-renormalized' quantities and ask if they are invariant under the proposed duality transformations.

---

[5]Note that this problem is absent for T-duality transformations which relate two weakly coupled string theories, and hence can be tested using string perturbation theory. All T-duality symmetries in string theory can be 'proved' in this way, at least to all orders in perturbation theory.

Testing duality invariance of these quantities provides us with various tests of various duality conjectures, and is in fact the basis of all duality conjectures.

The precise content of these non-renormalization theorems depends on the number of supersymmetries present in the theory. The maximum number of supersymmetry generators that can be present in a string theory is 32. This gives $N = 2$ supersymmetry in 10 dimensions, and N=8 supersymmetry in 4 dimensions. Examples of such theories are type IIA or type IIB string theories compactified on n-dimensional tori T^n. The next interesting class of theories are those with 16 supersymmetry generators. This corresponds to $N = 1$ supersymmetry in 10 dimensions and $N = 4$ supersymmetry in 4 dimensions. Examples of such theories are type IIA or type IIB string theories compactified on $K3 \times T^n$, heterotic string theory compactified on T^n, etc. Another class of theories that we shall discuss are those with 8 supersymmetry generators, e.g. heterotic string theory on $K3 \times T^n$, type IIA or IIB string theory on 6-dimensional Calabi–Yau manifolds, etc. For theories with 16 or more SUSY generators the non-renormalization theorems are particularly powerful. In particular,

• The form of the low energy effective action involving the massless states of the theory is completely fixed by the requirement of supersymmetry (and the spectrum); see [35]. Thus this effective action cannot get renormalized by string loop corrections. As a result, any valid symmetry of the theory must be a symmetry of this effective field theory.

• These theories contain special class of states which are invariant under part of the supersymmetry transformations. They are known as BPS states, named after Bogomol'nyi, Prasad and Sommerfield. The mass of a BPS state is completely determined in terms of its charge as a consequence of the supersymmetry algebra. Since this relation is derived purely from an analysis of the supersymmetry algebra, it is not modified by quantum corrections. Furthermore it can be argued that the degeneracy of BPS states of a given charge does not change as we move in the moduli space even from a weak to strong coupling region; see [36]. Thus the spectrum of BPS states can be calculated from weak coupling analysis and the result can be continued to the strong coupling region. Since any valid symmetry of the theory must be a symmetry of the spectrum of BPS states, we can use this to design non-trivial tests of duality; see [1].

For theories with 8 supersymmetries the non-renormalization theorems are less powerful. However, even in this case one can design non-trivial tests of various duality conjectures. We shall discuss these in Section 6.

3 Analysis of Low Energy Effective Field Theory

In this section I shall discuss tests of various dualities in string theories with ≥ 16 supersymmetries based on the analysis of their low energy effective action. As has been emphasized in the previous section, the form of this low energy effective action is determined completely by the requirement of supersymmetry and the spectrum of massless states in the theory. Thus it does not receive any quantum corrections, and if a given duality transformation is to be a symmetry of a string theory, it must be a symmetry of the corresponding low energy effective action. Actually, since the low energy *effective action* is to be used only for deriving the equations of motion from this action, and/or computing the tree level S-matrix elements using this action, but not to perform a full-fledged path integral, it is enough that only the equations of motion derived from this action are invariant under duality transformations. (This also guarantees that the tree level S-matrix elements computed from this effective action are invariant under the duality transformations.) It is not necessary for the action itself to be invariant.

Throughout this article we shall denote by $G_{\mu\nu}$ the string metric – the metric that is used in computing the area of the string world-sheet embedded in space time for calculating string scattering amplitudes. For a string theory compactified on a $(9-d)$-dimensional manifold $\mathcal{M}$, we shall denote by Φ the shifted dilaton, related to the dilaton $\Phi^{(10)}$ of the 10-dimensional string theory as

$$\Phi = \Phi^{(10)} - \ln V, \qquad (3.1)$$

where $(2\pi)^{9-d}V$ is the volume of $\mathcal{M}$ measured in the 10-dimensional string metric. The dilaton is normalized in such a way that $e^{\langle\Phi^{(10)}\rangle}$ corresponds to the square of the closed string coupling constant in 10 dimensions.[6] $g_{\mu\nu}$ will denote the canonical Einstein metric, which is related to the string metric by an appropriate conformal rescaling involving the dilaton field,

$$g_{\mu\nu} = e^{-\frac{2}{d-1}\Phi}G_{\mu\nu}. \qquad (3.2)$$

We shall always use this metric to raise and lower indices. The signature of space-time will be taken as $(-,+,\cdots,+)$. Finally, all fields will be made dimensionless by absorbing appropriate powers of α' in them.

We shall now consider several examples. The discussion will closely follow references [1, 30, 26]. For a detailed review of the material covered in this section, see reference [15].

---

[6]Φ is related to the more commonly normalized dilaton ϕ by a factor of two: $\Phi = 2\phi$.

3.1   Type I – $SO(32)$ heterotic duality in $D = 10$

In $SO(32)$ heterotic string theory, the massless bosonic states come from the
NS sector of the closed heterotic string, and contain the metric $g_{\mu\nu}^{(H)}$, the
dilaton $\Phi^{(H)}$, the rank-2 anti-symmetric tensor field $B_{\mu\nu}^{(H)}$, and gauge fields
$A_\mu^{(H)a}$ ($1 \leq a \leq 496$) in the adjoint representation of $SO(32)$. The low energy
dynamics involving these massless bosonic fields is described by the $N = 1$
supergravity coupled to $SO(32)$ super Yang–Mills theory in 10 dimensions;
see [147]. The action is given by (see [115]):

$$
\begin{aligned}
S^{(H)} \;=\; & \frac{1}{(2\pi)^7(\alpha'_H)^4 g_H^2} \int d^{10}x \sqrt{-g^{(H)}} \Big[R^{(H)} - \frac{1}{8} g^{(H)\mu\nu} \partial_\mu \Phi^{(H)} \partial_\nu \Phi^{(H)} \\
& - \frac{1}{4} g^{(H)\mu\mu'} g^{(H)\nu\nu'} e^{-\Phi^{(H)}/4} \mathrm{Tr}(F_{\mu\nu}^{(H)} F_{\mu'\nu'}^{(H)}) \\
& - \frac{1}{12} g^{(H)\mu\mu'} g^{(H)\nu\nu'} g^{(H)\rho\rho'} e^{-\Phi^{(H)}/2} H_{\mu\nu\rho}^{(H)} H_{\mu'\nu'\rho'}^{(H)} \Big] ,
\end{aligned} \tag{3.3}
$$

where $R^{(H)}$ is the Ricci scalar, $F_{\mu\nu}^{(H)}$ denotes the non-abelian gauge field
strength,

$$
F_{\mu\nu}^{(H)} = \partial_\mu A_\nu^{(H)} - \partial_\nu A_\mu^{(H)} + \sqrt{\frac{2}{\alpha'_H}} [A_\mu^{(H)}, A_\nu^{(H)}] , \tag{3.4}
$$

Here Tr denotes trace in the vector representation of $SO(32)$, and $H_{\mu\nu\rho}^{(H)}$ is
the field strength associated with the $B_{\mu\nu}^{(H)}$ field:

$$
H_{\mu\nu\rho}^{(H)} \;=\; \partial_\mu B_{\nu\rho}^{(H)} - \frac{1}{2} \mathrm{Tr}(A_\mu^{(H)} F_{\nu\rho}^{(H)} - \frac{1}{3}\sqrt{\frac{2}{\alpha'_H}} A_\mu^{(H)} [A_\nu^{(H)}, A_\rho^{(H)}])
$$

$$
+ \text{cyclic permutations of } \mu, \nu, \rho . \tag{3.5}
$$

$2\pi\alpha'_H$ and g_H are respectively the inverse string tension and the coupling
constant of the heterotic string theory. The rescalings (1.1), (1.2) take the
following form acting on the complete set of fields:

$$
g_H \to e^C g_H, \qquad \Phi^{(H)} \to \Phi^{(H)} - 2C, \qquad g_{\mu\nu}^{(H)} \to e^{C/2} g_{\mu\nu}^{(H)}
$$

$$
B_{\mu\nu}^{(H)} \to B_{\mu\nu}^{(H)}, \qquad A_\mu^{(H)a} \to A_\mu^{(H)a} , \tag{3.6}
$$

$$
\alpha'_H \to \lambda \alpha'_H, \qquad \Phi^{(H)} \to \Phi^{(H)}, \qquad g_{\mu\nu}^{(H)} \to \lambda g_{\mu\nu}^{(H)}
$$

$$
B_{\mu\nu}^{(H)} \to \lambda B_{\mu\nu}^{(H)}, \qquad A_\mu^{(H)a} \to \lambda^{1/2} A_\mu^{(H)a} , \tag{3.7}
$$

Since g_H and α'_H can be changed by this rescaling, these parameters cannot
have a universal significance. In particular, we can absorb g_H and α'_H into
the various fields by setting $e^{-C} = g_H$ and $\lambda = (\alpha'_H)^{-1}$ in (3.6), (3.7). This

is equivalent to setting $g_H = 1$ and $\alpha'_H = 1$. In this notation the physical coupling constant is given by the vacuum expectation value of $e^{\Phi^{(H)}/2}$, and the ADM mass per unit length of an infinitely long straight string, measured in the metric $e^{\langle\Phi^{(H)}\rangle/4}g_{\mu\nu}^{(H)}$ that approaches the string metric $G_{\mu\nu}^{(H)}$ far away from the string, is equal to $1/2\pi$. By changing $\langle\Phi^{(H)}\rangle$ we can get all possible values of string coupling and, using a metric that differs from the one used here by a constant multiplicative factor, we can get all possible values of the string tension.

For $\alpha'_H = 1$ and $g_H = 1$ eqs.(3.3)-(3.5) take the form:

$$S^{(H)} = \frac{1}{(2\pi)^7}\int d^{10}x\sqrt{-g^{(H)}}\Big[R^{(H)} - \frac{1}{8}g^{(H)\mu\nu}\partial_\mu\Phi^{(H)}\partial_\nu\Phi^{(H)}$$

$$-\frac{1}{4}g^{(H)\mu\mu'}g^{(H)\nu\nu'}e^{-\Phi^{(H)}/4}\mathrm{Tr}(F_{\mu\nu}^{(H)}F_{\mu'\nu'}^{(H)})$$

$$-\frac{1}{12}g^{(H)\mu\mu'}g^{(H)\nu\nu'}g^{(H)\rho\rho'}e^{-\Phi^{(H)}/2}H_{\mu\nu\rho}^{(H)}H_{\mu'\nu'\rho'}^{(H)}\Big], \quad (3.8)$$

$$F_{\mu\nu}^{(H)} = \partial_\mu A_\nu^{(H)} - \partial_\nu A_\mu^{(H)} + \sqrt{2}[A_\mu^{(H)}, A_\nu^{(H)}], \quad (3.9)$$

$$H_{\mu\nu\rho}^{(H)} = \partial_\mu B_{\nu\rho}^{(H)} - \frac{1}{2}\mathrm{Tr}\Big(A_\mu^{(H)}F_{\nu\rho}^{(H)} - \frac{\sqrt{2}}{3}A_\mu^{(H)}[A_\nu^{(H)}, A_\rho^{(H)}]\Big)$$

+cyclic permutations of μ, ν, ρ. \quad (3.10)

Let us now turn to the type I string theory. The massless bosonic states in type I theory come from three different sectors. The closed string Neveu-Schwarz–Neveu-Schwarz (NS) sector gives the metric $g_{\mu\nu}^{(I)}$ and the dilaton $\Phi^{(I)}$. The closed string Ramond-Ramond (RR) sector gives an anti-symmetric tensor field $B_{\mu\nu}^{(I)}$. Besides these, there are bosonic fields coming from the NS sector of the open string. This sector gives rise to gauge fields $A_\mu^{(I)a}$ ($a = 1,\ldots,496$) in the adjoint representation of the group $SO(32)$. (The superscript (I) refers to the fact that these are the fields in the type I string theory.) The low energy dynamics is again described by the $N = 1$ supergravity theory coupled to $SO(32)$ super Yang–Mills theory; see [148]. But it is instructive to rewrite the effective action in terms of the type I variables. For suitable choice of the string tension and the coupling constant, this is given by (see [115])

$$S^{(I)} = \frac{1}{(2\pi)^7}\int d^{10}x\sqrt{-g^{(I)}}\Big[R^{(I)} - \frac{1}{8}g^{(I)\mu\nu}\partial_\mu\Phi^{(I)}\partial_\nu\Phi^{(I)}$$

$$-\frac{1}{4}g^{(I)\mu\mu'}g^{(I)\nu\nu'}e^{\Phi^{(I)}/4}\mathrm{Tr}(F_{\mu\nu}^{(I)}F_{\mu'\nu'}^{(I)})$$

$$-\frac{1}{12}g^{(I)\mu\mu'}g^{(I)\nu\nu'}g^{(I)\rho\rho'}e^{\Phi^{(I)}/2}H_{\mu\nu\rho}^{(I)}H_{\mu'\nu'\rho'}^{(I)}\Big], \quad (3.11)$$

where $R^{(I)}$ is the Ricci scalar, $F_{\mu\nu}^{(I)}$ denotes the non-abelian gauge field strength,

$$F_{\mu\nu}^{(I)} = \partial_\mu A_\nu^{(I)} - \partial_\nu A_\mu^{(I)} + \sqrt{2}[A_\mu^{(I)}, A_\nu^{(I)}], \qquad (3.12)$$

and $H_{\mu\nu\rho}^{(I)}$ is the field strength associated with the $B_{\mu\nu}^{(I)}$ field:

$$H_{\mu\nu\rho}^{(I)} = \partial_\mu B_{\nu\rho}^{(I)} - \frac{1}{2}\mathrm{Tr}\left(A_\mu^{(I)} F_{\nu\rho}^{(I)} - \frac{\sqrt{2}}{3}A_\mu^{(I)}[A_\nu^{(I)}, A_\rho^{(I)}]\right)$$

$$+\text{cyclic permutations of } \mu, \nu, \rho. \qquad (3.13)$$

For both the type I and the $SO(32)$ heterotic string theory, the low energy effective action is derived from the string tree level analysis. However, to this order in the derivatives, the form of the effective action is determined completely by the requirement of supersymmetry for a given gauge group. Thus neither action can receive any quantum corrections.

It is straightforward to see that the actions (3.8) and (3.11) are identical provided we make the identification:

$$\Phi^{(H)} = -\Phi^{(I)}, \qquad g_{\mu\nu}^{(H)} = g_{\mu\nu}^{(I)}$$

$$B_{\mu\nu}^{(H)} = B_{\mu\nu}^{(I)}, \qquad A_\mu^{(H)a} = A_\mu^{(I)a}. \qquad (3.14)$$

This led to the hypothesis that the type I and the $SO(32)$ heterotic string theories in 10 dimensions are equivalent; see [26]. One can find stronger evidence for this hypothesis by analysing the spectrum of supersymmetric states, but the equivalence of the two effective actions was the reason for proposing this duality in the first place.

Note the $-$ sign in the relation between $\Phi^{(H)}$ and $\Phi^{(I)}$ in eq.(3.14). Recalling that $e^{\langle\Phi\rangle/2}$ is the string coupling, we see that the strong coupling limit of one theory is related to the weak coupling limit of the other theory and vice versa.

From now on I shall use the unit $\alpha' = 1$ for writing down the effective action of all string theories. Physically this would mean that the ADM mass per unit length of a test string, measured in the metric $e^{2\langle\Phi\rangle/(d-1)}g_{\mu\nu}$ that agrees with the string metric $G_{\mu\nu}$ defined in (3.2) far away from the test string, is given by $1/2\pi$. In future we shall refer to the ADM mass of a particle measured in this metric as the mass measured in the string metric.

3.2   Self-duality of heterotic string theory on T^6

In the previous subsection we have described the massless bosonic field content of the 10-dimensional $SO(32)$ heterotic string theory. When we compactify it on a 6-dimensional torus, we can get many other massless scalar fields from the internal components of the metric, the anti-symmetric tensor field and the gauge fields in the Cartan subalgebra of the gauge group.[7] This gives a total of $(21 + 15 + 96 = 132)$ scalar fields. It turns out that these scalars can be represented by a 28×28 matrix-valued field M satisfying[8]

$$MLM^T = L, \qquad M^T = M, \tag{3.15}$$

where

$$L = \begin{pmatrix} & I_6 & \\ I_6 & & \\ & & -I_{16} \end{pmatrix}. \tag{3.16}$$

I_n denotes an $n \times n$ identity matrix. We shall choose a convention in which $M = I_{28}$ corresponds to a compactification on $(S^1)^6$ with each S^1 having radius $\sqrt{\alpha'} = 1$ measured in the string metric, and without any background gauge or antisymmetric tensor fields. We can get another scalar field a by dualizing the gauge invariant field strength H of the antisymmetrix tensor field through the relation:

$$H^{\mu\nu\rho} = -(\sqrt{-g})^{-1} e^{2\Phi} \epsilon^{\mu\nu\rho\sigma} \partial_\sigma a, \tag{3.17}$$

where Φ denotes the 4-dimensional dilaton and $g_{\mu\nu}$ denotes the $(3 + 1)$-dimensional canonical metric defined in eqs.(3.1), (3.2) respectively. It is convenient to combine the dilaton Φ and the axion field a into a single complex scalar λ:

$$\lambda = a + ie^{-\Phi} \equiv \lambda_1 + i\lambda_2. \tag{3.18}$$

At a generic point in the moduli space, where the scalars M take arbitrary vacuum expectation values, the non-abelian gauge symmetry of the 10-dimensional theory is broken to its abelian subgroup $U(1)^{16}$. Besides these 16 $U(1)$ gauge fields we get 12 other $U(1)$ gauge fields from components $G_{m\mu}$, $B_{m\mu}$ ($4 \le m \le 9$, $0 \le \mu \le 3$) of the metric and the anti-symmetric tensor field respectively. Let us denote these 28 $U(1)$ gauge fields (after suitable normalization) by A_μ^a ($1 \le a \le 28$). In terms of these fields, the low energy effective action of the theory is given by (see [37, 38, 39, 41, 1]),[9]

$$S = \frac{1}{2\pi} \int d^4x \sqrt{-g} \Big[R - g^{\mu\nu} \frac{\partial_\mu \lambda \partial_\nu \bar{\lambda}}{2(\lambda_2)^2} + \frac{1}{8} g^{\mu\nu} \text{Tr}(\partial_\mu M L \partial_\nu M L)$$

$$- \frac{1}{4} \lambda_2 g^{\mu\mu'} g^{\nu\nu'} F_{\mu\nu}^a (LML)_{ab} F_{\mu'\nu'}^b + \frac{1}{4} \lambda_1 g^{\mu\rho} g^{\nu\sigma} F_{\mu\nu}^a L_{ab} \tilde{F}_{\rho\sigma}^b \Big], \tag{3.19}$$

---

[7]Only the 16 gauge fields in the Cartan subalgebra of the gauge group can develop vacuum expectation value, since such vacuum expectation values do not generate any field strength, and hence do not generate energy density.

[8]For a review of this construction, see [1].

[9]The normalization of the gauge fields used here differ from that in reference (see [1]) by a factor of two. Also there we used $\alpha' = 16$ whereas here we are using $\alpha' = 1$.

where $F^a_{\mu\nu}$ is the field strength associated with A^a_μ, R is the Ricci scalar. and

$$\tilde{F}^{a\mu\nu} = \frac{1}{2}(\sqrt{-g})^{-1}\epsilon^{\mu\nu\rho\sigma}F^a_{\rho\sigma}. \tag{3.20}$$

This action is invariant under an $O(6,22)$ transformation:[10]

$$M \to \Omega M \Omega^T, \qquad A^a_\mu \to \Omega^a_b A^b_\mu, \qquad g_{\mu\nu} \to g_{\mu\nu}, \qquad \lambda \to \lambda, \tag{3.21}$$

where Ω satisfies:

$$\Omega L \Omega^T = L. \tag{3.22}$$

An $O(6,22;Z)$ subgroup of this can be shown to be a T-duality symmetry of the full string theory; see [34]. This $O(6,22;Z)$ subgroup can be described as follows. Let Λ_{28} denote a 28-dimensional lattice obtained by taking the direct sum of the 12-dimensional lattice of integers, and the 16-dimensional root lattice of $SO(32)$.[11] $O(6,22;Z)$ is defined to be the subset of $O(6,22)$ transformations which leave Λ_{28} invariant, i.e. acting on any vector in Λ_{28} produces another vector in Λ_{28}. It will be useful for our future reference to undertstand why only an $O(6,22;Z)$ subgroup of the full $O(6,22)$ group is a symmetry of the full string theory. Since $O(6,22;Z)$ is a T-duality symmetry, this question can be answered within the context of perturbative string theory. The point is that although at a generic point in the moduli space the massless string states do not carry any charge, there are massive charged states in the spectrum of the full string theory. Since there are 28 charges associated with the 28 $U(1)$ gauge fields, a state can be characterized by a 28-dimensional charge vector. With appropriate normalization, this charge vector can be shown to lie in the lattice Λ_{28}, i.e. the charge vector of any state in the spectrum can be shown to be an element of the lattice Λ_{28}. Since the $O(6,22)$ transformation acts linearly on the $U(1)$ gauge fields, it also acts linearly on the charge vectors. As a result only those $O(6,22)$ elements can be genuine symmetries of string theory which preserve the lattice Λ_{28}. Any other $O(6,22)$ element, acting on a physical state in the spectrum, will take it to a state with charge vector outside the lattice Λ_{28}. Since such a state does not exist in the spectrum, such an $O(6,22)$ transformation cannot be a symmetry of the full string theory.

In order to see a specific example of a T-duality transformation, let us consider heterotic string theory compactified on $(S^1)^6$ with one of the circles having radius R measured in the string metric, and the rest having unit radius.

---

[10]$O(p,q)$ denotes the group of Lorentz transformations in p space-like and q time-like dimensions. (These have nothing to do with physical space-time, which always has only one time-like direction.) $O(p,q;Z)$ denotes a discrete subgroup of $O(p,q)$.

[11]More precisely we have to take the root lattice of $Spin(32)/Z_2$ which is obtained by adding to the $SO(32)$ root lattice the weight vectors of the spinor representations of $SO(32)$ with a definite chirality.

Let us also assume that there is no background gauge or anti-symmetric tensor field. Using the convention of reference [1] one can show that for this background

$$M^{(H)} = \begin{pmatrix} R^{-2} & & & \\ & I_5 & & \\ & & R^2 & \\ & & & I_5 \\ & & & & I_{16} \end{pmatrix}. \tag{3.23}$$

Consider now the $O(6, 22; Z)$ transformation with the matrix:

$$\Omega = \begin{pmatrix} 0 & & 1 & \\ & I_5 & & \\ 1 & & 0 & \\ & & & I_{21} \end{pmatrix}. \tag{3.24}$$

Using eq.(3.21) we see that this transforms $M^{(H)}$ to

$$M^{(H)} = \begin{pmatrix} R^2 & & & \\ & I_5 & & \\ & & R^{-2} & \\ & & & I_5 \\ & & & & I_{16} \end{pmatrix}. \tag{3.25}$$

Thus the net effect of this transformation is $R \rightarrow R^{-1}$. It says that the heterotic string theory compactified on a circle of radius R is equivalent to the same theory compactified on a circle of radius R^{-1}. For this reason $R = 1$ (i.e. $R = \sqrt{\alpha'}$) is known as the self-dual radius. Other $O(6, 22; Z)$ transformations acting on (3.23) will give rise to more complicated $M^{(H)}$ corresponding to a configuration with background gauge and/or anti-symmetric tensor fields.

Besides this symmetry, the equations of motion derived from this action can be shown to be invariant under an $SL(2, R)$ transformation of the form (see [37, 42, 43])

$$F_{\mu\nu}^a \rightarrow (r\lambda_1 + s)F_{\mu\nu}^a + r\lambda_2(ML)_b^a \tilde{F}_{\mu\nu}^b, \qquad \lambda \rightarrow \frac{p\lambda + q}{r\lambda + s},$$

$$g_{\mu\nu} \rightarrow g_{\mu\nu}, \qquad M \rightarrow M, \tag{3.26}$$

where p, q, r, s are real numbers satisfying $ps - qr = 1$. The existence of such symmetries (known as hidden non-compact symmetries) in this and in other supergravity theories were discovered in early days of supergravity theories and in fact played a crucial role in the construction of these theories in the first place; see [146, 37]. Since this $SL(2, R)$ transformation mixes the gauge field strength with its Poincaré dual, it is an electric–magnetic duality transformation. This leads to the conjecture that a subgroup of this continuous symmetry group is an exact symmetry of string theory; see [44, 45, 43, 46, 47, 48, 49, 1]. One might wonder why the conjecture refers to only a discrete subgroup of $SL(2, R)$ instead of the full $SL(2, R)$ group as the genuine symmetry group. This follows from the same logic that was responsible for breaking $O(6, 22)$ to $O(6, 22; Z)$; however since the $SL(2, R)$ transformation mixes electric field with magnetic field, we now need to take into account the quantization of magnetic charges. We have already described the quantization condition on the electric charges. Using the usual Dirac–Schwinger–Zwanziger rules one can show that, in an appropriate normalization, the 28-dimensional magnetic charge vectors also lie in the same lattice Λ_{28}. Also, with this normalization convention the electric and magnetic charge vectors transform as a doublet under the $SL(2, R)$ transformation; thus it is clear that the subgroup of $SL(2, R)$ that respects the charge quantization condition is $SL(2, Z)$. An arbitrary $SL(2, R)$ transformation acting on the quantized electric and magnetic charges will not give rise to electric and magnetic charges consistent with the quantization law. This is the reason behind the conjectured $SL(2, Z)$ symmetry of heterotic string theory on T^6. Note that since this duality acts non-trivially on the dilaton and hence the string coupling, this is a non-perturbative symmetry, and cannot be verified order-by-order in perturbation theory. Historically, this is the first example of a concrete duality conjecture in string theory. Later we shall review other tests of this duality conjecture.

3.3 Duality between heterotic on T^4 and type IIA on $K3$

The massless bosonic field content of heterotic string theory compactified on T^4 can be found in a manner identical to that in heterotic string theory on T^6. Besides the dilaton $\Phi^{(H)}$, we get many other massless scalar fields from the internal components of the metric, the anti-symmetric tensor field and the gauge fields. In this case these scalars can be represented by a 24×24 matrix-valued field $M^{(H)}$ satisfying

$$M^{(H)} L M^{(H)T} = L, \qquad M^{(H)T} = M^{(H)}, \qquad (3.27)$$

where

$$L = \begin{pmatrix} & I_4 & \\ I_4 & & \\ & & -I_{16} \end{pmatrix}. \qquad (3.28)$$

We again use the convention that $M^{(H)} = I_{24}$ corresponds to compactification on $(S^1)^4$ with each S^1 having self-dual radius ($\sqrt{\alpha'} = 1$), without any background gauge field or anti-symmetric tensor field. At a generic point in the moduli space, where the scalars $M^{(H)}$ take arbitrary vacuum expectation values, we get a $U(1)^{24}$ gauge group, with 16 gauge fields coming from the Cartan subalgebra of the original gauge group in 10 dimensions, and 8 other gauge fields from components $G_{m\mu}$, $B_{m\mu}$ ($6 \leq m \leq 9$, $0 \leq \mu \leq 5$) of the metric and the anti-symmetric tensor field respectively. Here x^m denote the compact directions, and x^μ denote the non-compact directions. Let us denote these 24 $U(1)$ gauge fields by $A_\mu^{(H)a}$ ($1 \leq a \leq 24$). Finally, let $g_{\mu\nu}^{(H)}$ and $B_{\mu\nu}^{(H)}$ denote the canonical metric and the anti-symmetric tensor field respectively. In terms of these fields, the low energy effective action of the theory is given by,

$$
\begin{aligned}
S_H = \ & \frac{1}{(2\pi)^3} \int d^6x \sqrt{-g^{(H)}} \Big[R^{(H)} - \frac{1}{2} g^{(H)\mu\nu} \partial_\mu \Phi^{(H)} \partial_\nu \Phi^{(H)} \\
& + \frac{1}{8} g^{\mu\nu} \mathrm{Tr}(\partial_\mu M^{(H)} L \partial_\nu M^{(H)} L) \\
& - \frac{1}{4} e^{-\Phi^{(H)}/2} g^{(H)\mu\mu'} g^{(H)\nu\nu'} F_{\mu\nu}^{(H)a} (L M^{(H)} L)_{ab} F_{\mu'\nu'}^{(H)b} \\
& - \frac{1}{12} e^{-\Phi^{(H)}} g^{(H)\mu\mu'} g^{(H)\nu\nu'} g^{(H)\rho\rho'} H_{\mu\nu\rho}^{(H)} H_{\mu'\nu'\rho'}^{(H)} \Big],
\end{aligned}
\tag{3.29}
$$

where $F_{\mu\nu}^{(H)a}$ is the field strength associated with $A_\mu^{(H)a}$, $R^{(H)}$ is the Ricci scalar, and $H_{\mu\nu\rho}^{(H)}$ is the field strength associated with $B_{\mu\nu}^{(H)}$:

$$
H_{\mu\nu\rho}^{(H)} = (\partial_\mu B_{\nu\rho}^{(H)} + \frac{1}{2} A_\mu^{(H)a} L_{ab} F_{\nu\rho}^{(H)b}) + \text{(cyclic permutations of } \mu, \nu, \rho).
\tag{3.30}
$$

This action is invariant under an $O(4, 20)$ transformation:

$$
M^{(H)} \to \Omega M^{(H)} \Omega^T, \qquad A_\mu^{(H)a} \to \Omega_b^a A_\mu^{(H)b}, \qquad g_{\mu\nu}^{(H)} \to g_{\mu\nu}^{(H)},
$$

$$
B_{\mu\nu}^{(H)} \to B_{\mu\nu}^{(H)}, \qquad \Phi^{(H)} \to \Phi^{(H)},
\tag{3.31}
$$

where Ω satisfies:

$$
\Omega L \Omega^T = L.
\tag{3.32}
$$

Again as in the case of T^6 compactification, only an $O(4, 20; Z)$ subgroup of this which preserves the charge lattice Λ_{24} is an exact T-duality symmetry of this theory. The lattice Λ_{24} is obtained by taking the direct sum of the 8-dimensional lattice of integers and the root lattice of $\mathrm{Spin}(32)/Z_2$.

Let us now turn to the spectrum of massless bosonic fields in type IIA string theory on $K3$. In 10 dimensions the massless bosonic fields in type IIA string theory are the metric g_{MN}, the rank-2 anti-symmetric tensor B_{MN} and the scalar dilation Φ coming from the NS sector, and a gauge field A_M and a rank-3 anti-symmetric tensor field C_{MNP} coming from the RR sector. The low energy effective action of this theory involving the massless bosonic fields is given by (see [97])

$$S_{IIA} = \frac{1}{(2\pi)^7} \int d^{10}x \sqrt{-g} \Big[R - \frac{1}{8} g^{\mu\nu} \partial_\mu \Phi \partial_\nu \Phi$$

$$- \frac{1}{12} e^{-\Phi/2} g^{\mu\mu'} g^{\nu\nu'} g^{\rho\rho'} H_{\mu\nu\rho} H_{\mu'\nu'\rho'} - \frac{1}{4} e^{3\Phi/4} g^{\mu\mu'} g^{\nu\nu'} F_{\mu\nu} F_{\mu'\nu'}$$

$$- \frac{1}{48} e^{\Phi/4} g^{\mu\mu'} g^{\nu\nu'} g^{\rho\rho'} g^{\sigma\sigma'} G_{\mu\nu\rho\sigma} G_{\mu'\nu'\rho'\sigma'}$$

$$- \frac{1}{(48)^2} (\sqrt{-g})^{-1} \varepsilon^{\mu_0 \cdots \mu_9} B_{\mu_0\mu_1} G_{\mu_2 \cdots \mu_5} G_{\mu_6 \cdots \mu_9} \Big], \qquad (3.33)$$

where R is the Ricci scalar, and

$$F_{\mu\nu} = \partial_\mu A_\nu - \partial_\nu A_\mu,$$

$$H_{\mu\nu\rho} = \partial_\mu B_{\nu\rho} + \text{cyclic permutations of } \mu, \nu, \rho,$$

$$G_{\mu\nu\rho} = \partial_\mu C_{\nu\rho\sigma} + A_\mu H_{\nu\rho\sigma} + (-1)^P \cdot \text{cyclic permutations}, \quad (3.34)$$

are the field strengths associated with A_μ, $B_{\mu\nu}$ and $C_{\mu\nu\rho}$ respectively. Upon compactification on $K3$ we get a new set of scalar fields from the Kähler and complex structure moduli of $K3$. These can be regarded as deformations of the metric and give a total of 58 real scalar fields. We get 22 more scalar fields $\phi^{(p)}$ by decomposing the anti-symmetric tensor field B_{MN} along the 22 harmonic 2-forms $\omega_{mn}^{(p)}$ on $K3$:

$$B_{mn}(x, y) \sim \sum_{p=1}^{22} \phi_p(x) \omega_{mn}^{(p)}(y) + \cdots . \qquad (3.35)$$

Here $\{x^\mu\}$ and $\{y^m\}$ denote coordinates along the non-compact and $K3$ directions respectively. These 80 scalar fields together parametrize a coset $O(4, 20)/O(4) \times O(20)$ and can be described by a matrix $M^{(A)}$ satisfying properties identical to those of $M^{(H)}$ described in (3.27). This theory also has 24 $U(1)$ gauge fields. 22 of the gauge fields arise from the components of the 3-form field C_{MNP}:

$$C_{mn\mu}(x, y) = \sum_{p=1}^{22} \omega_{mn}^{(p)}(y) \mathcal{A}_\mu^{(p)}(x) + \cdots . \qquad (3.36)$$

The $\mathcal{A}_\mu^{(p)}$ defined in (3.36) behave as gauge fields in 6 dimensions. One more gauge field comes from the original RR gauge field A_μ. The last one $\mathcal{A}_\mu$ comes from dualizing $C_{\mu\nu\rho}$:

$$G \sim {}^*(d\mathcal{A}), \qquad (3.37)$$

where * denotes Poincaré dual in 6 dimensions. Together we shall denote these gauge fields by $A_\mu^{(A)a}$ for $1 \leq a \leq 24$. Besides these fields, the theory contains the canonical metric and the anti-symmetric tensor field, which we shall denote by $g^{(A)}_{\mu\nu}$ and $B^{(A)}_{\mu\nu}$ respectively. The action involving these fields is given by

$$
\begin{aligned}
S_A = \frac{1}{(2\pi)^3} \int d^6x \sqrt{-g^{(A)}} \Big[& R^{(A)} - \frac{1}{2} g^{(A)\mu\nu} \partial_\mu \Phi^{(A)} \partial_\nu \Phi^{(A)} \\
& + \frac{1}{8} g^{\mu\nu} \mathrm{Tr}(\partial_\mu M^{(A)} L \partial_\nu M^{(A)} L) \\
& - \frac{1}{4} e^{\Phi^{(A)}/2} g^{(A)\mu\mu'} g^{(A)\nu\nu'} F^{(A)a}_{\mu\nu} (L M^{(A)} L)_{ab} F^{(A)b}_{\mu'\nu'} \\
& - \frac{1}{12} e^{-\Phi^{(A)}} g^{(A)\mu\mu'} g^{(A)\nu\nu'} g^{(A)\rho\rho'} H^{(A)}_{\mu\nu\rho} H^{(A)}_{\mu'\nu'\rho'} \\
& - \frac{1}{16} \varepsilon^{\mu\nu\rho\delta\epsilon\eta} (\sqrt{-g^{(A)}})^{-1} B^{(A)}_{\mu\nu} F^{(A)a}_{\rho\delta} L_{ab} F^{(A)b}_{\epsilon\eta} \Big] , \quad (3.38)
\end{aligned}
$$

where $F^{(A)a}_{\mu\nu}$ is the field strength associated with $A^{(A)a}_\mu$, $R^{(A)}$ is the Ricci scalar, and $H^{(A)}_{\mu\nu\rho}$ is the field strength associated with $B^{(A)}_{\mu\nu}$:

$$
H^{(A)}_{\mu\nu\rho} = \partial_\mu B^{(A)}_{\nu\rho} + (\text{cyclic permutations of } \mu, \nu, \rho) . \quad (3.39)
$$

In writing down the above action we have used the convention that $M^{(A)} = I_{24}$ corresponds to compactification on a specific reference $K3$, possibly with specific background B_{mn} fields. This action has an $O(4, 20)$ symmetry of the form:

$$
M^{(A)} \to \Omega M^{(A)} \Omega^T, \qquad A^{(A)a}_\mu \to \Omega^a_b A^{(A)b}_\mu, \qquad g^{(A)}_{\mu\nu} \to g^{(A)}_{\mu\nu},
$$

$$
B^{(A)}_{\mu\nu} \to B^{(A)}_{\mu\nu}, \qquad \Phi^{(A)} \to \Phi^{(A)} , \quad (3.40)
$$

where Ω satisfies:

$$
\Omega L \Omega^T = L . \quad (3.41)
$$

An $O(4, 20; Z)$ subgroup of this can be shown to be an exact T-duality symmetry of string theory; see [184]. The lattice Λ'_{24} which is preserved by this $O(4, 20; Z)$ subgroup of $O(4, 20)$ is not the lattice Λ_{24} defined earlier, but is in general an $O(4, 20)$ rotation of that lattice:

$$
\Lambda'_{24} = \Omega_0 \Lambda_{24} . \quad (3.42)
$$

Ω_0 depends on the choice of the special reference $K3$ mentioned earlier.

It is now a straightforward exercise to show that the equations of motion and the Bianchi identities derived from (3.29) and (3.38) are identical if we use the following map between the heterotic and the type II variables (see [50, 30]):

$$
g^{(H)}_{\mu\nu} = g^{(A)}_{\mu\nu}, \qquad M^{(H)} = \tilde{\Omega} M^{(A)} \tilde{\Omega}^T,
$$

$$\Phi^{(H)} = -\Phi^{(A)}, \qquad A_\mu^{(H)a} = \tilde{\Omega}_b^a A_\mu^{(A)a},$$

$$\sqrt{-g^{(H)}}\exp(-\Phi^{(H)})H^{(H)\mu\nu\rho} = \frac{1}{6}\varepsilon^{\mu\nu\rho\delta\epsilon\eta}H_{\delta\epsilon\eta}^{(A)}. \qquad (3.43)$$

where $\tilde{\Omega}$ is an arbitrary $O(4,20)$ matrix. This leads to the conjectured equivalence between heterotic string theory compactified on T^4 and type IIA string theory compactified on $K3$; see [30]. But clearly the two theories cannot be equivalent for all $\tilde{\Omega}$ since in the individual theories the $O(4,20)$ symmetry is broken down to $O(4,20;Z)$. $\tilde{\Omega}$ can be found (up to an $O(4,20;Z)$ transformation) by comparing the T-duality symmetry transformations in the two theories. To do this let us note that according to eq.(3.43) a transformation $M^{(H)} \to \Omega M^{(H)}\Omega^T$ will induce a transformation

$$M^{(A)} \to (\tilde{\Omega}^{-1}\Omega\tilde{\Omega})M^{(A)}(\tilde{\Omega}^{-1}\Omega\tilde{\Omega})^T. \qquad (3.44)$$

Thus if Ω preserves the lattice Λ_{24}, $\tilde{\Omega}^{-1}\Omega\tilde{\Omega}$ should preserve the lattice $\Lambda'_{24} = \Omega_0\Lambda_{24}$. This happens if we choose

$$\tilde{\Omega} = \Omega_0^{-1}. \qquad (3.45)$$

Note again that there is a relative minus sign that relates $\Phi^{(H)}$ and $\Phi^{(A)}$, showing that the strong coupling limit of one theory corresponds to the weak coupling limit of the other theory.

3.4 $SL(2,Z)$ self-duality of Type IIB in D=10

As described in Section 1.1, the massless bosonic fields in type IIB string theory come from two sectors – Neveu-Schwarz–Neveu-Schwarz (NS) and Ramond–Ramond (RR). The NS sector gives the graviton described by the metric $g_{\mu\nu}$, an anti-symmetric tensor field $B_{\mu\nu}$, and a scalar field Φ known as the dilaton. The RR sector contributes a scalar field a, sometimes called the axion, another rank-2 anti-symmetric tensor field $B'_{\mu\nu}$, and a rank-4 anti-symmetric tensor field $D_{\mu\nu\rho\sigma}$ whose field strength is self-dual.

It is often convenient to combine the axion and the dilaton into a complex scalar field λ as follows:[12]

$$\lambda = a + ie^{-\Phi/2} \equiv \lambda_1 + i\lambda_2. \qquad (3.46)$$

The low energy effective action in this theory can be determined either from the requirement of supersymmetry, or by explicit computation in string theory. Actually it turns out that there is no simple covariant action for this

---

[12]Note that this field λ has no relation to the field λ defined in Section 3.2 for heterotic string theory on T^6, although both transform as moduli under the respective $SL(2,Z)$ duality transformations in the two theories.

low energy theory, but there are covariant field equations (see [51]), which are in fact just the equations of motion of type IIB supergravity. Although in string theory this low energy theory is derived from the tree level analysis, non-renormalization theorems tell us that this is exact to this order in the space-time derivatives. Basically, supersymmetry determines the form of the equations of motion to this order in the derivatives completely, and so there is no scope for the quantum corrections to change the form of the action.

For the sake of brevity, we shall not explicitly write down the equations of motion. The main point is that these equations of motion are covariant (in the sense that they transform into each other) under an $SL(2, R)$ transformation (see [51]):

$$\lambda \to \frac{p\lambda + q}{r\lambda + s}, \quad \begin{pmatrix} B_{\mu\nu} \\ B'_{\mu\nu} \end{pmatrix} \to \begin{pmatrix} p & q \\ r & s \end{pmatrix} \begin{pmatrix} B_{\mu\nu} \\ B'_{\mu\nu} \end{pmatrix},$$

$$g_{\mu\nu} \to g_{\mu\nu}, \quad D_{\mu\nu\rho\sigma} \to D_{\mu\nu\rho\sigma}, \tag{3.47}$$

where p, q, r, s are real numbers satisfying,

$$ps - qr = 1. \tag{3.48}$$

The existence of this $SL(2, R)$ symmetry in the type IIB supergravity theory led to the conjecture that an $SL(2, Z)$ subgroup of this $SL(2, R)$, obtained by restricting p, q, r, s to be integers instead of arbitrary real numbers, is a symmetry of the full string theory; see [30]. The breaking of $SL(2, R)$ to $SL(2, Z)$ can be seen as follows. An elementary string is known to carry $B_{\mu\nu}$ charge. In a suitable normalization convention, it carries exactly one unit of $B_{\mu\nu}$ charge. This means that the $B_{\mu\nu}$ charge must be quantized in integer units, as the spectrum of string theory does not contain fractional strings carrying a fraction of the charge carried by the elementary string. From (3.47) we see that acting on an elementary string state carrying one unit of $B_{\mu\nu}$ charge, the $SL(2, R)$ transformation gives a state with p units of $B_{\mu\nu}$ charge and r units of $B'_{\mu\nu}$ charge. Thus p must be an integer. It is easy to see that the maximal subgroup of $SL(2, R)$ for which p is always an integer consists of matrices of the form

$$\begin{pmatrix} p & \alpha q \\ \alpha^{-1} r & s \end{pmatrix}, \tag{3.49}$$

with p, q, r, s integers satisfying $ps - qr = 1$, and α a fixed constant. Absorbing α into a redefinition of $B'_{\mu\nu}$ we see that the subgroup of $SL(2, R)$ matrices consistent with charge quantization are the $SL(2, Z)$ matrices $\begin{pmatrix} p & q \\ r & s \end{pmatrix}$ with p, q, r, s integers satisfying $ps - qr = 1$.

Note that this argument only shows that $SL(2, Z)$ is the maximal possible subgroup of $SL(2, R)$ that *can be a symmetry of the full string theory*, but does not prove that $SL(2, Z)$ is a symmetry of string theory. In particular, since $SL(2, Z)$ acts non-trivially on the dilaton, whose vacuum expectation value represents the string coupling constant, it cannot be verified order-by-order in string perturbation theory. We shall see later how one can find non-trivial evidence for this symmetry.

Besides this non-perturbative $SL(2, Z)$ transformation, type IIB theory has two perturbatively verifiable discrete Z_2 symmetries. They are as follows:

- $(-1)^{F_L}$: It changes the sign of all the Ramond sector states on the left-moving sector of the world-sheet. In particular, acting on the massless bosonic sector fields, it changes the sign of a, $B'_{\mu\nu}$ and $D_{\mu\nu\rho\sigma}$, but leaves $g_{\mu\nu}$, $B_{\mu\nu}$ and Φ invariant.

- Ω: This is the world-sheet parity transformation mentioned in Section 1.1 that exchanges the left- and the right-moving sectors of the world-sheet. Acting on the massless bosonic sector fields, it changes the sign of $B_{\mu\nu}$, a and $D_{\mu\nu\rho\sigma}$, leaving the other fields invariant.

From this description, we see that the effect of $(-1)^{F_L} \cdot \Omega$ is to change the sign of $B_{\mu\nu}$ and $B'_{\mu\nu}$, leaving the other massless bosonic fields invariant. Comparing this with the action of the $SL(2, Z)$ transformation laws of the massless bosonic sector fields, we see that $(-1)^{F_L} \cdot \Omega$ can be identified with the $SL(2, Z)$ transformation:

$$\begin{pmatrix} -1 & \\ & -1 \end{pmatrix}. \tag{3.50}$$

This information will be useful to us later.

Theories obtained by modding out (compactified) type IIB string theory by a discrete symmetry group, where some of the elements of the group involve Ω, are known as orientifolds; see [113, 114]. The simplest example of an orientifold is type IIB string theory modded out by Ω. This corresponds to type I string theory. The closed string sector of type I theory consists of the Ω-invariant states of type IIB string theory. The open string states of type I string theory are the analogs of twisted sector states in an orbifold, which must be added to the theory in order to maintain finiteness.

3.5 Other examples

Following the same procedure, namely, studying symmetries of the effective action together with charge quantization rules, we are led to many other duality conjectures in theories with 16 or more supersymmetry generators. Here we shall list the main series of such duality conjectures. We begin with the self-duality groups of type II string theories compactified on tori of different dimensions. As mentioned earlier, there is a T-duality that relates type IIA on a circle to type IIB on a circle of inverse radius. Thus, for $n \geq 1$, the self-duality groups of type IIA and type IIB theories compactified on an n-dimensional torus T^n will be identical. We now list the conjectured self-duality groups of type IIA/IIB string theory compactified on T^n for different values of n (see [30]):

$D = 10 - n$	Full Duality Group	T-duality Group
9	$SL(2, Z)$	–
8	$SL(2, Z) \times SL(3, Z)$	$SL(2, Z) \times SL(2, Z)$
7	$SL(5, Z)$	$SO(3, 3; Z)$
6	$SO(5, 5; Z)$	$SO(4, 4; Z)$
5	$E_{6(6)}(Z)$	$SO(5, 5; Z)$
4	$E_{7(7)}(Z)$	$SO(6, 6; Z)$
3	$E_{8(8)}(Z)$	$SO(7, 7; Z)$
2	$\widehat{E_{8(8)}}(Z)$	$SO(8, 8; Z)$

Note that besides the full duality group, we have also displayed the T-duality group of each theory which can be verified order-by-order in string perturbation theory. We denote by $E_{n(n)}$ a non-compact version of the exceptional group E_n for $n = 6, 7, 8$, and by $E_{n(n)}(Z)$ a discrete subgroup of $E_{n(n)}$. For any group G $\widehat{G}$ denotes the loop group of G based on the corresponding affine algebra and $\widehat{G}(Z)$ denotes a discrete subgroup of this loop group. Note that we have stopped at $D = 2$. We could in principle continue this all the way to $D = 1$ where all space-like directions are compactified. In this case one expects a very large duality symmetry group based on a hyperbolic Lie algebra (see [116]), which is not well understood to this date.

In each of the cases mentioned, the low energy effective field theory is invariant under the full continuous group (see [52]), but charge quantization breaks this symmetry to its discrete subgroup. As noted before, these symmetries were discovered in the early days of supergravity theories, and were known as hidden non-compact symmetries.

Next we turn to the self-duality conjectures involving compactified heterotic string theories. Although there are two distinct heterotic string theories in 10 dimensions, upon compactification on a circle, the two heterotic string theories can be shown to be related by a T-duality transformation. As a result, upon compactification on T^n, both of them will have the same self-duality group. We now display this self-duality group in various dimensions:

$D = 10 - n$	Full Duality Group	T-duality Group
9	$O(1, 17, Z)$	$O(1, 17; Z)$
8	$O(2, 18, Z)$	$O(2, 18; Z)$
7	$O(3, 19, Z)$	$O(3, 19; Z)$
6	$O(4, 20, Z)$	$O(4, 20; Z)$
5	$O(5, 21, Z)$	$O(5, 21; Z)$
4	$O(6, 22, Z) \times SL(2, Z)$	$O(6, 22; Z)$
3	$O(8, 24, Z)$	$O(7, 23; Z)$
2	$O(8, \widehat{24}, Z)$	$O(8, 24; Z)$

Since the type I and $SO(32)$ heterotic string theories are conjectured to be dual to each other in 10 dimensions, the second column of the above table also represents the duality symmetry group of type I string theory on T^n. However, in the case of type I string theory, there is no perturbatively realised self-duality group (except trivial transformations which are part of the $SO(32)$ gauge group and the group of global diffeomorphisms of T^n).

The effective action of type IIB string theory compactified on $K3$ has an $SO(5, 21)$ symmetry (see [50]), which leads to the conjecture that an $SO(5, 21; Z)$ subgroup of this is an exact self-duality symmetry of the type IIB string theory on $K3$. The conjectured duality between type IIA string theory compactified on $K3$ and heterotic string theory compactified on T^4 has already been discussed. Due to the equivalence of type IIB on S^1 and type IIA on S^1, type IIA on $K3 \times T^n$ is equivalent to type IIB on $K3 \times T^n$. Finally, due to the conjectured duality between type IIA on $K3$ and heterotic on T^4, type IIA/IIB on $K3 \times T^n$ are dual to heterotic string theory on T^{n+4} for $n \geq 1$. Thus the self-duality symmetry groups in these theories can be read off from the second column of the previous table displaying the self-duality groups of heterotic string theory on T^n.

Besides the theories discussed here, there are other theories with 16 or more supercharges obtained from non-geometric compactification of heterotic/type II string theories; see [53, 54, 55]. The duality symmetry groups of these theories can again be guessed from an analysis of the low energy effective field theory and the charge quantization conditions. Later we shall also describe a more systematic way of 'deriving' various duality conjectures from some basic set of dualities.

Although in this section I have focussed on duality symmetries of the low energy effective action which satisfies a non-renormalization theorem as a consequence of space-time supersymmetry, this is not the only part of the full effective action which satisfy such a non-renormalization theorem. Quite often the effective action contains another set of terms satisfying non-renormalization theorems. They are required for anomaly cancellation, and are known as Green–Schwarz terms. The Adler–Bardeen theorem guarantees that they are not renormalized beyond one loop. These terms have also been used effectively for testing various duality conjectures (see [185]), but I shall not discuss them in this article.

4 Precision Test of Duality: Spectrum of BPS States

Analysis of the low energy effective action, as discussed in the last section, provides us with only a crude test of duality. Its value lies in its simplicity. Indeed, most of the duality conjectures in string theory were arrived at by analysing the symmetries of the low energy effective action.

But once we have arrived at a duality conjecture based on the analysis of the low energy effective action, we can perform a much more precise test by analysing the spectrum of BPS states in the theories. BPS states are states which are invariant under part of the supersymmetry transformation, and are characterized by two important properties:

- They belong to a supermultiplet which has typically lower dimension than a non-BPS state. This has an analog in the theory of representations of the Lorentz group, where massless states form a shorter representation of the algebra than massive states. Thus, for example, a photon has only two polarizations but a massive vector particle has three polarizations.

- The mass of a BPS state is completely determined by its charge as a consequence of the supersymmetry algebra. This relation between the mass and the charge is known as the BPS mass formula. This statement also has an analog in the theory of representations of the Lorentz algebra, e.g. a spin 1 representation of the Lorentz algebra containing only two states must necessarily be massless.

We shall now explain the origin of these two properties; see [36]. Suppose the theory has N real supersymmetry generators Q_α $(1 \le \alpha \le N)$. Acting on a single particle state *at rest*, the supersymmetry algebra takes the form:

$$\{Q_\alpha, Q_\beta\} = f_{\alpha\beta}(m, \vec{Q}, \{y\}),$$
(4.1)

where $f_{\alpha\beta}$ is a real symmetric matrix which is a function of its arguments m, $\vec{Q}$ and $\{y\}$. Here m denotes the rest mass of the particle, $\vec{Q}$ denotes various gauge charges carried by the particle, and $\{y\}$ denotes the coordinates labelling the moduli space of the theory.[13] We shall now consider the following distinct cases:

---

[13]Only specific combinations of $\vec{Q}$ and $\{y\}$, known as central charges, appear in the algebra.

1. $f_{\alpha\beta}$ has no zero eigenvalue. In this case by taking appropriate linear combinations of Q_α we can diagonalize f. By a further appropriate rescaling of Q_α, we can bring f into the identity matrix. Thus in this basis the supersymmetry algebra has the form:

$$\{Q_\alpha, Q_\beta\} = \delta_{\alpha\beta}. \qquad (4.2)$$

This is the N-dimensional Clifford algebra. Thus the single-particle states under consideration form a representation of this Clifford algebra, which is $2^{N/2}$-dimensional. (We are considering the case where N is even.) Such states would correspond to non-BPS states.

2. f has $N - M$ zero eigenvalues for some $M < N$. In this case, by taking linear combinations of the Q_α we can bring the algebra into the form:

$$\{Q_\alpha, Q_\beta\} = \delta_{\alpha\beta}, \quad \text{for} \quad 1 \leq \alpha, \beta \leq M,$$

$$= 0 \quad \text{for} \quad \alpha \text{ or } \beta > M. \qquad (4.3)$$

We can form an irreducible representation of this algebra by taking all states to be annihilated by Q_α for $\alpha > M$. In that case the states will form a representation of an M-dimensional Clifford algebra generated by Q_α for $1 \leq \alpha \leq M$. This representation is $2^{M/2}$-dimensional for M even. Since $M < N$, we see that these are lower-dimensional representations compared to that of a generic non-BPS state. Furthermore, these states are invariant under part of the supersymmetry algebra generated by Q_α for $\alpha > M$. These are known as BPS states. We can get different kinds of BPS states depending on the value of M, i.e. depending on the number of supersymmetry generators that leave the state invariant.

From this discussion it is clear that in order to get a BPS state, the matrix f must have some zero eigenvalues. This in turn, gives a constraint involving mass m, charges $\vec{Q}$ and the moduli $\{y\}$, and is the origin of the BPS formula relating the mass and the charge of the particle.

Before we proceed, let us illustrate the preceeding discussion in the context of a string theory. Consider Type IIB string theory compactified on a circle S^1. The total number of supersymmetry generators in this theory is 32. Thus a generic non-BPS supermultiplet is $2^{16} = (256)^2$-dimensional. These are known as long multiplets. This theory also has BPS states breaking half the space-time supersymmetry. For these states $M = 16$ and hence we have a $2^8 = 256$-dimensional representation of the supersymmetry algebra. These states are known as ultra-short multiplets. We can also have BPS states breaking $3/4$ of the space-time supersymmetry ($M = 24$). These will form a $2^{12} = 256 \times 16$-dimensional representation, and are known as short multiplets. In each case there is a specific relation between the mass and the various charges carried by the state. We shall discuss this relation as well as the origin of these BPS states in more detail later.

As another example, consider heterotic string theory compactified on an n-dimensional torus T^n. The original theory has 16 supercharges. Thus a generic non-BPS state will belong to a $2^8 = 256$-dimensional representation of the supersymmetry algebra. But if we consider states that are invariant under half of the supercharges, then they belong to a $2^4 = 16$-dimensional representation of the supersymmetry algebra. This is known as the short representation of this superalgebra. We can also have states that break 3/4 of the supersymmetries.[14] These belong to a 64-dimensional representation of the supersymmetry algebra known as intermediate states.

BPS states are further characterized by the property that the degeneracy of BPS states with a given set of charge quantum numbers is independent of the value of the moduli fields $\{y\}$. Since string coupling is also one of the moduli of the theory, this implies that the degeneracy at any value of the string coupling is the same as that at weak coupling. This is the key property of the BPS states that makes them so useful in testing duality, so let us review the argument leading to this property; see [36]. We shall discuss this in the context of the specific example of type IIB string theory compactified on S^1, but it can be applied to any other theory. Suppose the theory has an ultra-short multiplet at some point in the moduli space. Now let us change the moduli. The question that we shall be asking is: can the ultra-short multiplet become a long (or any other) multiplet as we change the moduli? If we assume that the total number of states does not change discontinuously, then this is clearly not possible since other multiplets have different numbers of states. Thus as long as the spectrum varies smoothly with the moduli (which we shall assume), an ultra-short multiplet stays ultra-short as we move in the moduli space; see [95]. Furthermore, as long as it stays ultra-short, its mass is determined by the BPS formula. Thus we see that the degeneracy of ultra-short multiplets cannot change as we change the moduli of the theory. A similar argument can be given for other multiplets as well. Note that for this argument to be strictly valid, we require that the mass of the BPS state should stay away from the continuum, since otherwise the counting of states is not a well defined procedure. This requires that the mass of a BPS state should be strictly less than the total mass of any set of two or more particles carrying the same total charge as the BPS state.

Given this result, we can now adapt the following strategy to carry out tests of various duality conjectures using the spectrum of BPS states in the theory:

---

[14]It turns out that these states can exist only for $n \geq 5$. This constraint arises due to the fact that the unbroken supersymmetry generators must form a representation of the little group $SO(9 - n)$ of a massive particle in $(10 - n)$-dimensional space-time.

1. Identify BPS states in the spectrum of elementary string states. The spectrum of these BPS states can be trusted at all values of the coupling even though it is calculated at weak coupling.

2. Make a conjectured duality transformation. This typically takes a BPS state in the spectrum of elementary string states to another BPS state, but with quantum numbers that are not present in the spectrum of elementary string states. Thus these states must arise as solitons /composite states.

3. Try to explicitly verify the existence of these solitonic states with degeneracy as predicted by duality. This will provide a non-trivial test of the corresponding duality conjecture.

We shall now illustrate this procedure with the help of specific examples. We shall mainly follow ; see [58, 72, 69].

4.1 $SL(2, Z)$ S-duality in heterotic on T^6 and multi-monopole moduli spaces

As discussed in Section 3.2, heterotic string theory compactified on T^6 is conjectured to have an $SL(2, Z)$ duality symmetry. In this subsection we shall see how one can test this conjecture by examining the spectrum of BPS states.

Since the BPS spectrum does not change as we change the moduli, we can analyse the spectrum near some particular point in the moduli space. As discussed in Section 3.2, at a generic point in the moduli space the unbroken gauge group is $U(1)^{28}$. But there are special points in this moduli space where we get enhanced non-abelian gauge group; see [120]. Thus for example, if we set the internal components of the original 10-dimensional gauge fields to zero, we get unbroken $E_8 \times E_8$ or $SO(32)$ gauge symmetry. Let us consider a special point in the moduli space where an $SU(2)$ gauge symmetry is restored. This can be done for example by taking a particular S^1 in T^6 to be orthogonal to all other circles, taking the components of the gauge fields along this S^1 to be zero, and taking the radius of this S^1 to be the self-dual radius. In that case the effective field theory at energies much below the string scale will be described by an $N = 4$ supersymmetric $SU(2)$ gauge theory, together with a set of decoupled $N = 4$ supersymmetric $U(1)$ gauge theories and $N = 4$ supergravity. The conjectured $SL(2, Z)$ duality of the heterotic string theory will require the $N = 4$ supersymmetric $SU(2)$ gauge theory to have this $SL(2, Z)$ symmetry.[15] Thus by testing the duality invariance of the spectrum

---

[15]Independently of string theory, the existence of a strong-weak coupling duality in this theory was conjectured earlier; see [56, 57].

of this $N = 4$ supersymmetric $SU(2)$ gauge theory we can test the conjectured $SL(2, Z)$ symmetry of heterotic string theory.

The $N = 4$ supersymmetric $SU(2)$ gauge theory has a vector, 6 massless scalars and 4 massless Majorana fermions in the adjoint representation of $SU(2)$; see [57]. The form of the Lagrangian is fixed completely by the requirement of $N = 4$ supersymmetry up to two independent parameters – the coupling constant g that determines the strength of all interactions (gauge, Yukawa, scalar self-interaction etc.), and the vacuum angle θ that multiplies the topological term $\mathrm{Tr}(F\tilde{F})$ involving the gauge field. With the choice of suitable normalization conventions, g and θ are related to the vacuum expectation value of the field λ defined in (3.18) through the relation:

$$\langle \lambda \rangle = \frac{\theta}{2\pi} + i \frac{4\pi}{g^2} . \tag{4.4}$$

The potential involving the 6 adjoint representation scalar fields ϕ_m^α ($1 \leq \alpha \leq 3, 1 \leq m \leq 6$) is proportional to

$$\sum_{m<n} \sum_\alpha (\epsilon^{\alpha\beta\gamma} \phi_m^\beta \phi_n^\gamma)^2 . \tag{4.5}$$

This vanishes for

$$\phi_m^\alpha = a_m \delta_{\alpha 3} . \tag{4.6}$$

Vacuum expectation values of ϕ_m^α of the form (4.6) do not break supersymmetry, but break the gauge group $SU(2)$ to $U(1)$. The parameters $\{a_m\}$ correspond to the vacuum expectation values of a subset of the scalar moduli fields M in the full string theory. We shall work in a region in the moduli space where $a_m \neq 0$ for some m, but the scale of breaking of $SU(2)$ is small compared to the string scale ($|a_m| << (\sqrt{\alpha'})^{-1}$ for all m), so that gravity is still decoupled from this gauge theory. The BPS states in the spectrum of elementary particles in this theory are the heavy charged bosons $W^\pm$ and their superpartners. These break half of the 16 space-time supersymmetry generators and hence form a $2^{8/2} = 16$-dimensional representation of the supersymmetry algebra. These states can be found explicitly in the spectrum of elementary string states from the sector containing strings with one unit of winding and one unit of momentum along the special S^1 that is responsible for the enhanced $SU(2)$ gauge symmetry. As we approach the point in the moduli space where this special S^1 has self-dual radius, these states become massless and form part of the $SU(2)$ gauge multiplet.

When $SU(2)$ is broken to $U(1)$ by the vacuum expectation value of ϕ_m, the spectrum of solitons in this theory is characterized by two quantum numbers, the electric charge quantum number n_e and the magnetic charge quantum number n_m, normalized so that n_e and n_m are both integers. We shall denote such a state by $\begin{pmatrix} n_e \\ n_m \end{pmatrix}$. In this notation the elementary W^+ boson corresponds to a $\begin{pmatrix} 1 \\ 0 \end{pmatrix}$ state. By studying the action of the $SL(2,Z)$ transformation (3.26) on the gauge fields, we can easily work out its action on the charge quantum numbers $\begin{pmatrix} n_e \\ n_m \end{pmatrix}$; see [1]. The answer is

$$\begin{pmatrix} n_e \\ n_m \end{pmatrix} \rightarrow \begin{pmatrix} p & q \\ r & s \end{pmatrix} \begin{pmatrix} n_e \\ n_m \end{pmatrix}, \tag{4.7}$$

for appropriate choice of sign convention for n_e and n_m. Thus, acting on a $\begin{pmatrix} 1 \\ 0 \end{pmatrix}$ state it produces a $\begin{pmatrix} p \\ r \end{pmatrix}$ state. From the relation $ps - qr = 1$ satisfied by an $SL(2,Z)$ matrix, we can easily see that p and r are relatively prime. Furthermore, for every p and r relatively prime, we can find integers q and s satisfying $ps - qr = 1$. Thus $SL(2,Z)$ duality predicts that *for every p and r relatively prime, the theory must contain a unique short multiplet with charge quantum numbers $\begin{pmatrix} p \\ r \end{pmatrix}$; see [58].*

We can now directly examine the solitonic sector of the theory to check this prediction. The theory contains classical monopole solutions which break half of the supersymmetries of the original theory. These solutions are non-singular everywhere, and in fact, for a given r, there is a $4r$-parameter non-singular solution with r units of total magnetic charge; see [117, 59]. These $4r$ parameters correspond to the bosonic collective excitations of this system; see [118]. In order to study the spectrum of BPS solitons, we need to quantize these collective excitations and look for supersymmetric ground states of the corresponding quantum mechanical system. Each solution also has an infinite number of vibrational modes with non-zero frequency, but excitations of these modes are not relevant for finding supersymmetric ground states.

States with $r = 1$ come from 1-monopole solutions. This has 4 bosonic collective coordinates, 3 of which correspond to the physical position of the monopole in the 3-dimensional space, while the 4th is an angular variable describing the $U(1)$ phase of the monopole. The momenta conjugate to the first 3 coordinates correspond to the components of the physical momentum of the particle. These can be set to zero by working in the rest frame of the monopole. The 4th coordinate is periodically identified and hence its conjugate momentum is quantized in integer units. This integer p corresponds to the electric charge quantum number n_e. Thus the states obtained by quantizing the bosonic sector of the theory have charge quantum numbers $\begin{pmatrix} p \\ 1 \end{pmatrix}$ for all integer p.

The degeneracy comes from quantizing the fermionic sector. There are 8 fermionic zero modes, which describe the result of applying the 8 broken supersymmetry generators on the monopole solution. These form an

8-dimensional Clifford algebra. Thus the ground state has $2^4 = 16$-fold degeneracy, exactly as predicted by $SL(2, Z)$; see [57].

Let us now turn to the analysis of states with $r > 1$; see [58]. As has already been said, this system has $4r$ bosonic collective coordinates, which, when the monopoles are far away from each other, correspond to the spatial location and the $U(1)$ phase of each of the r monopoles. The total number of fermionic collective coordinates can be computed from an index theorem and is equal to $8r$; see [119]. We can divide this set into the 'center of mass' coordinates containing 4 bosonic and 8 fermionic coordinates, and the 'relative coordinates' containing $4(r-1)$ bosonic and $8(r-1)$ fermionic coordinates. The quantization of the center of mass system gives states carrying charge quantum numbers $\begin{pmatrix} p \\ r \end{pmatrix}$ with 16-fold degeneracy, p being the momentum conjugate to the overall $U(1)$ phase. This shows that the degeneracy is always a multiple of 16, consistent with the fact that a short multiplet is 16-fold degenerate. At this stage p can be any integer, not necessarily relatively prime to r. However, since the total wave-function is a product of the wave-function of the center of mass system and the relative system, in order to determine the number of short multiplets for a given value of p, we need to turn to the quantum mechanics of the relative coordinates.

It turns out that the bosonic coordinates in the relative coordinate system describe a non-trivial $4(r-1)$-dimensional manifold, known as the relative moduli space of r monopoles; see [118, 59, 60]. The quantum mechanics of the bosonic and fermionic relative coordinates can be regarded as that of a supersymmetric particle moving in this moduli space. There are several subtleties with this system. They are listed below:

• First of all, the center of mass and the relative coordinates do not completely decouple, although they decouple locally. The full moduli space has the structure (see [59]):

$$(R^3 \times S^1 \times \mathcal{M}_r)/Z_r \,, \qquad (4.8)$$

where R^3 is parametrized by the center of mass location, S^1 by the overall $U(1)$ phase, and $\mathcal{M}_r$ by the relative coordinates. There is an identification of points in the product space $R^3 \times S^1 \times \mathcal{M}_r$ by a Z_r transformation that acts as a shift by $2\pi/r$ on S^1 and as a diffeomorphism on $\mathcal{M}_r$ without any fixed point; see [59, 60]. Due to this identification, the total wave-function must be invariant under this Z_r transformation. Since the part of the wave-function involving the coordinate of S^1 picks up a phase $\exp(2\pi i p/r)$ under this Z_r, we see that the wave-function involving the relative coordinates must pick up a phase of $\exp(-2\pi i p/r)$ under this Z_r transformation.

- Normally the part of the wave-function involving the relative coordinates will be a function on $\mathcal{M}_r$. But it turns out that the effect of the $8(r-1)$ fermionic degrees of freedom in the quantum mechanical system makes the wave-function a differential form of arbitrary rank on $\mathcal{M}_r$; see [61, 62].

- Finally, among all the possible states, the ones saturating the Bogomol'nyi bound correspond to harmonic differential forms on $\mathcal{M}_r$. This can be understood as follows. It can be shown that the Hamiltonian of the relative coordinates correspond to the Laplacian on $\mathcal{M}$. Also it turns out that the BPS mass formula is saturated by contributions from the center of mass coordinates. Hence in order to get a BPS state, the part of the wave-function involving the relative coordinates must be an eigenstate of the corresponding Hamiltonian with zero eigenvalue, i.e. it must be a harmonic form on $\mathcal{M}_r$. Thus for every harmonic differential form we get a short multiplet, since the fermionic degrees of freedom associated with the center of mass coordinates supply the necessary 16-fold degeneracy.

Thus the existence of a short multiplet of charge quantum numbers $\begin{pmatrix} p \\ r \end{pmatrix}$ would require the existence of a harmonic form on $\mathcal{M}_r$ that picks up a phase of $\exp(2\pi i p/r)$ under the action of Z_r. According to the prediction of $SL(2, Z)$ *such a harmonic form should exist only for p and r relatively prime, and not for other values of p; see [58].*

For $r = 2$ the relevant harmonic form can be constructed explicitly (see [63, 64, 58]), therby verifying the existence of the states predicted by $SL(2, Z)$ duality. For $r > 2$ the analysis is more complicated since the metric in the multimonopole moduli space is not known. However, general arguments showing the existence of the necessary harmonic forms has been given; see [65, 66].

Besides the BPS states discussed here, the spectrum of elementary string states in the heterotic string theory on T^6 contains many other BPS states. In the world-sheet theory, a generic state is created by applying oscillators from the left- and the right-moving sector on the Fock vacuum. The Fock vacuum, in turn, is characterized by a pair of vectors $(\vec{k}_L, \vec{k}_R)$ specifying the charges (momenta) associated with the 6 right-handed and 22 left-handed currents on the world-sheet. From the viewpoint of the space-time theory, these 28 components of $(\vec{k}_L, \vec{k}_R)$ are just appropriate linear combinations of the charges carried by the state under the 28 $U(1)$ gauge fields. The tree level mass formula for an elementary string state in the NS sector is given by[16]

$$m^2 = \frac{4}{\lambda_2}\Big[\frac{\vec{k}_R^2}{2} + N_R - \frac{1}{2}\Big] = \frac{4}{\lambda_2}\Big[\frac{\vec{k}_L^2}{2} + N_L - 1\Big], \qquad (4.9)$$

where N_R and N_L denote respectively the oscillator levels of the state in the right- and the left-moving sectors of the world-sheet. In the above equation the terms in the square bracket denote the total contribution to L_0 and $\bar{L}_0$ from the oscillators, the internal momenta, and the vacua in the right- and the left-moving sectors respectively. Normally we do not have the factor of λ_2^{-1} in the mass formula since the formula refers to the ADM mass measured in the string metric $G_{\mu\nu} = \lambda_2^{-1} g_{\mu\nu}$. But here (and in the rest of the article) we quote the ADM mass measured in the canonical metric $g_{\mu\nu}$. This is more convenient for discussing duality invariance of the spectrum, since it is $g_{\mu\nu}$ and not $G_{\mu\nu}$ that remains invariant under a duality transformation. The additive factors of $-1/2$ and -1 can be interpreted as the contributions to L_0 and $\bar{L}_0$ from the vacuum. (In the covariant formulation these can be traced to the contributions from the world-sheet ghost fields).

It turns out that of the full set of elementary string states, only those states which satisfy the constraint (see [71])

$$N_R = \frac{1}{2}, \qquad (4.10)$$

correspond to BPS states (short multiplets). From eqs.(4.9) we see that for these states

$$N_L = \frac{1}{2}(\vec{k}_R^2 - \vec{k}_L^2) + 1. \qquad (4.11)$$

The degeneracy $d(N_L)$ of short multiplets for a given set of $\vec{k}_L$, $\vec{k}_R$ is determined by the number of ways a level N_L state can be created out of the Fock vacuum by the 24 left-moving bosonic oscillators (in the light-cone gauge) -8 from the transverse bosonic coordinates of the string and 16 from the bosonization of the 32 left-moving fermions on the world-sheet — and is given by the formula:

$$\sum_{N_L=0}^{\infty} d(N_L) q^{N_L} = \prod_{n=1}^{\infty} \frac{1}{(1-q^n)^{24}}. \qquad (4.12)$$

The BPS states discussed earlier — the ones which can be regarded as the massive gauge bosons of a spontaneously broken non-abelian gauge theory — correspond to the $N_L = 0$ states in this classification. From eq.(4.12) we see that we have only one short multiplet for states with this quantum number; this is consistent with their description as heavy gauge bosons in an $N = 4$ supersymmetric gauge theory. The next interesting class of states are

---

[16]In this and all subsequent mass formulas λ_2 should really be interpreted as the vacuum expectation value of λ_2.

the ones with $N_L = 1$. From (4.12) we see that they have degeneracy 24.[17] An $SL(2, Z)$ transformation relates these states to appropriate magnetically charged states with r units of magnetic charge and p units of electric charge for p and r relatively prime. Thus the $SL(2, Z)$ self-duality symmetry of the heterotic string theory predicts the existence of 24-fold degenerate solitonic states with these charge quantum numbers.

Verifying the existence of these solitonic states turns out to be quite difficult; see [67]. The main problem is that, unlike the $N_L = 0$ states, the solitonic states (known as H-monopoles) which are related to the $N_L = 1$ states by $SL(2, Z)$ duality turn out to be singular objects, and hence we cannot unambiguously determine the dynamics of collective coordinates of these solitons just from the low energy effective field theory. Nevertheless, the problem has now been solved for $r = 1$ (see [79, 80, 77, 81]), and one finds that these solitons have exactly the correct degeneracy 24.

Similar analysis based on soliton solutions of low energy supergravity theory has been used to test many other duality conjectures; see [30, 32, 33, 27, 28, 68]. One of the main problems with this approach has been that, unlike the example discussed in this section, most of these other solutions are either singular, or have strong curvature at the core where the low energy approximation breaks down. As a result, analysis based on these solutions has been of limited use. The situation changed after the advent of D-branes, to which we now turn.

4.2 $SL(2, Z)$ duality in type IIB on S^1 and D-branes

As discussed earlier, type IIB string theory in 10 dimensions has a conjectured $SL(2, Z)$ duality symmetry group. In this section I shall discuss the consequence of this conjectured symmetry for the spectrum of BPS states in type IIB string theory compactified on a circle S^1. For details, see [69, 70].

The spectrum of elementary string states in this theory is characterized by two charges k_L and k_R defined as

$$k_L = (k\lambda_2^{1/4}/R - wR/\lambda_2^{1/4})/\sqrt{2}, \qquad k_R = (k\lambda_2^{1/4}/R + wR/\lambda_2^{1/4})/\sqrt{2}, \quad (4.13)$$

where R denotes the radius of S^1 measured in the 10-dimensional canonical metric, k/R denotes the momentum along S^1 with k being an integer, and w, also an integer, denotes the number of times the elementary string is wound along S^1. As usual we have set $\alpha' = 1$. In the world-sheet theory describing first quantized string theory, k_L and k_R denote the left- and the right-moving momenta respectively. There is infinite tower of states with this quantum

---

[17]In counting degeneracy we are only counting the number of short multiplets, and ignoring the trivial factor of 16 that represents the degeneracy within each short multiplet.

number, obtained by applied appropriate oscillators, both from the left- and the right-moving sector of the world-sheet, on the Fock vacuum of the world-sheet theory carrying these quantum numbers. The mass formula for any state in this tower, measured in the 10-dimensional canonical metric, is given by

$$m^2 = \frac{2}{\sqrt{\lambda_2}}(k_L^2 + 2N_L) = \frac{2}{\sqrt{\lambda_2}}(k_R^2 + 2N_R), \qquad (4.14)$$

where N_L, N_R denote oscillator levels on the left- and the right-moving sectors of the world-sheet respectively.[18] In the normal convention, one does not have the factors of (λ_2) in the mass formula, but here it comes due to the fact that we are using the *10-dimensional* canonical metric instead of the string metric to define the mass of a state. (Note that if we had used the 9-dimensional canonical metric as defined in eqs.(3.1), (3.2), there will be an additional multiplicative factor of $R^{-2/9}$ in the expression for m^2.)

Most of these states are not BPS states as they are not invariant under any part of the supersymmetry transformation. It turns out that in order to be invariant under half of the space-time supersymmetry coming from the left- (right-) moving sector of the world-sheet, N_L (N_R) must vanish; see [71]. Thus a state with $N_L = N_R = 0$ will preserve half of the total number of supersymmetries and will correspond to ultra-short multiplets. From eq.(4.14) we see that the mass formula for these states takes the form

$$m^2 = \frac{2k_L^2}{\sqrt{\lambda_2}} = \frac{2k_R^2}{\sqrt{\lambda_2}}. \qquad (4.15)$$

This is the BPS mass formula for these ultra-short multiplets. This requires $k_L = \pm k_R$ or, equivalently, $k = 0$ or $w = 0$. On the other hand, a state with either $N_L = 0$ or $N_R = 0$ will break (3/4)th of the total number of supersymmetries in the theory, and will correspond to short multiplets. If, for definiteness, we consider states with $N_R = 0$, then the BPS mass formula takes the form

$$m^2 = \frac{2k_R^2}{\sqrt{\lambda_2}}. \qquad (4.16)$$

We determine N_L in terms of k_L and k_R through the relation

$$N_L = \frac{1}{2}(k_R^2 - k_L^2) = wk. \qquad (4.17)$$

There is no further constraint on w and k. Although we have derived these mass formulae by directly analysing the spectrum of elementary string states, they can also be derived by analyzing the supersymmetry algebra, as indicated earlier.

---

[18]We have stated the formula in the RR sector, but due to space-time supersymmetry we get the identical spectrum from the NS and the R sectors.

One can easily calculate the degeneracy of these states by analyzing the spectrum of elementary string states in detail. For example, for the states with $N_L = N_R = 0$, there is a 16-fold degeneracy of states in each (left- and right-) sector of the world-sheet: 8 from the NS sector and 8 from the R sector. Thus the net degeneracy of such a state is $16 \times 16 = 256$, showing that there is a unique ultra-short multiplet carrying given charges (k_L, k_R). The degeneracy of short multiplets can be found in a similar manner. Consider for example states with $N_R = 0$, $N_L = 1$. In this case there is a 16-fold degeneracy coming from the right-moving sector of the world-sheet. There is an 8-fold degeneracy from the Ramond sector Fock vacuum of the left-moving sector. There is also an extra degeneracy factor in the left-moving Ramond sector due to the fact that there are many oscillators that can act on the Fock vacuum of the world-sheet theory to give a state at oscillator level $N_L = 1$. For example we get 8 states by acting with the transverse bosonic oscillators α^i_{-1} ($1 \leq i \leq 8$), and 8 states by acting with the transverse fermionic oscillators ψ^i_{-1}.[19] This gives a total degeneracy factor of 8×16 in the left-moving Ramond sector. Due to supersymmetry, we get an identical factor from the left-moving NS sector as well. Thus we get a state with total degeneracy $16 \times 16 \times 16$: 16 from the right-moving sector, and 16×16 from the left-moving sector — which is the correct degeneracy of a single short multiplet. Similar counting can be done for higher values of N_L as well. It turns out that the total number of short multiplets $d(N_L)$ with $N_R = 0$ for some given value of $N_L \geq 1$ is given by the formula

$$\sum_{N_L} d(N_L) q^{N_L} = \frac{1}{16} \prod_{n=1}^{\infty} \left(\frac{1+q^n}{1-q^n} \right)^8 . \tag{4.18}$$

The $(1+q^n)^8$ and $(1-q^n)^8$ factors in the numerator and the denominator are related respectively to the fact that in the light-cone gauge there are 8 left-moving fermionic fields and 8 left-moving bosonic fields on the world-sheet. The overall factor of $1/16$ is due to the fact that the lowest level state is only 256-fold degenerate but a single short multiplet requires 16×256 states.

Let us first consider the ultra-short multiplet with $k = 0$, $w = 1$. These states have mass

$$m^2 = \frac{R^2}{\lambda_2} . \tag{4.19}$$

It is well known that an elementary string acts as a source of the $B_{\mu\nu}$ field (see e.g. reference [71]). Thus in the $(8 + 1)$-dimensional theory obtained by compactifying type IIB on S^1, the $w = 1$ state will carry one unit of $B_{9\mu}$

---

[19]Since ψ^i_{-1} has fermion number one, it has to act on the Fock vacua with odd fermion number in order that the states obtained after acting with ψ^i_{-1} on the vacua satisfy the GSO projection.

gauge field charge. Now, under $SL(2, Z)$

$$\begin{pmatrix} B_{9\mu} \\ B'_{9\mu} \end{pmatrix} \rightarrow \begin{pmatrix} p & q \\ r & s \end{pmatrix} \begin{pmatrix} B_{9\mu} \\ B'_{9\mu} \end{pmatrix} . \tag{4.20}$$

This converts the $w = 1$ state, which we shall denote by $\begin{pmatrix} 1 \\ 0 \end{pmatrix}$ reflecting the $\begin{pmatrix} B_{9\mu} \\ B'_{9\mu} \end{pmatrix}$ charge carried by the state, to a $\begin{pmatrix} p \\ r \end{pmatrix}$ state, i.e. a state carrying p units of $B_{9\mu}$ charge and r units of $B'_{9\mu}$ charge. The condition $ps - qr = 1$ implies that the pair of integers (p, r) are relatively prime. Thus $SL(2, Z)$ duality of type IIB string theory predicts that for all (p, r) relatively prime, the theory must have a unique ultra-short multiplet with p units of $B_{9\mu}$ charge and r units of $B'_{9\mu}$ charge; see [68]. The BPS mass formula for these states can be derived by analysing the supersymmetry algebra, as indicated earlier, and is given by

$$m^2 = \frac{R^2}{\lambda_2} |r\lambda - p|^2 . \tag{4.21}$$

Note that this formula is invariant under the $SL(2, Z)$ transformation

$$\lambda \rightarrow \frac{a\lambda + b}{c\lambda + d}, \quad \begin{pmatrix} p \\ r \end{pmatrix} \rightarrow \begin{pmatrix} a & b \\ c & d \end{pmatrix} \begin{pmatrix} p \\ r \end{pmatrix} , \tag{4.22}$$

where $\begin{pmatrix} a & b \\ c & d \end{pmatrix}$ is an $SL(2, Z)$ matrix.

A similar prediction for the spectrum of BPS states can be made for short multiplets as well. In this case the state is characterized by three integers p, r and k reflecting the $B_{9\mu}$, $B'_{9\mu}$ and $G_{9\mu}$ charge (momentum along S^1) respectively. Let us denote by $d(k, p, r)$ the degeneracy of such short multiplets. For (p, r) relatively prime, an $SL(2, Z)$ transformation relates these to elementary string states with one unit of winding and k units of momentum along S^1. Such states have degeneracy $d(k)$ given in eq.(4.18). Then by following the same logic as before, we see that the $SL(2, Z)$ duality predicts that for (p, r) relatively prime, $d(k, p, r)$ is independent of p and r and depends on k according to the relation:

$$\sum_k d(k, p, r) q^k = \frac{1}{16} \prod_{n=1}^{\infty} \left(\frac{1 + q^n}{1 - q^n} \right)^8 . \tag{4.23}$$

In other words, there should be a Hagedorn spectrum of short multiplets with charge $\begin{pmatrix} p \\ r \end{pmatrix}$.

A test of $SL(2, Z)$ symmetry involves explicitly verifying the existence of these states. To see what such a test involves, recall that $B'_{\mu\nu}$ arises in the

RR sector of string theory. In type II theory, all elementary string states are neutral under RR gauge fields, as can be seen by computing a 3-point function involving any two elementary string states and an RR sector gauge field. Thus a state carrying $B'_{9\mu}$ charge must arise as a soliton. The naive approach will involve constructing such a soliton solution as a solution to the low energy supergravity equations of motion, quantizing its zero modes, and seeing if we recover the correct spectrum of BPS states. However, in actual practice, when one constructs the solution carrying $B'_{\mu\nu}$ charge, it turns out to be singular. Due to this fact it is difficult to proceed further along this line, as identifying the zero modes of a singular solution is not a well defined procedure. In particular we need to determine what boundary condition the modes must satisfy at the singularity. Fortunately, in this theory, there is a novel way of constructing a soliton solution that avoids this problem. This construction uses Dirichlet ($D-$) branes; see [72, 73]. In order to compute the degeneracy of these solitonic states, we must understand the definition and some of the the properties of these D-branes. This is the subject to which we now turn.

Normally type IIA/IIB string theory contains closed string states only. But we can postulate the existence of solitonic extended objects in these theories such that in the presence of these solitons, there can be open string states whose ends lie on these extended objects (see Figure 10). This can in fact be taken to be the defining relation for these solitons, with the open string states with ends lying on the soliton corresponding to the (infinite number of) vibrational modes of the soliton. Of course, one needs to ensure that the soliton defined in this way satisfies all the properties expected of a soliton solution in this theory, e.g. partially unbroken supersymmetry, existence of static multi-soliton solutions etc. Since open strings satisfy Dirichlet boundary condition in directions transverse to these solitons, these solitons are called D-branes. In particular, we shall call a D-brane with Neumann boundary condition in $p + 1$ directions (including time) and Dirichlet boundary condition in $9 - p$ directions a Dirichlet p-brane, since it can be regarded as a soliton extending along p space-like directions on which we have put Neumann boundary condition. (Thus a 0-brane represents a particle-like object, a 1-brane a string-like object, and a 2-brane a membrane-like object.) To be more explicit, let us consider the following boundary condition on the open string:

$$X^m(\sigma = 0, \pi) \;=\; x_0^m \quad \text{for} \quad p + 1 \leq m \leq 9,$$
$$\partial_\sigma X^\mu(\sigma = 0, \pi) \;=\; 0 \quad \text{for} \quad 0 \leq \mu \leq p, \qquad (4.24)$$

where σ denotes the spatial direction on the string world-sheet. The boundary conditions on the world-sheet fermion fields are determined from (4.24) using various consistency requirements including world-sheet supersymmetry that relates the world-sheet bosons and fermions. Note that these boundary

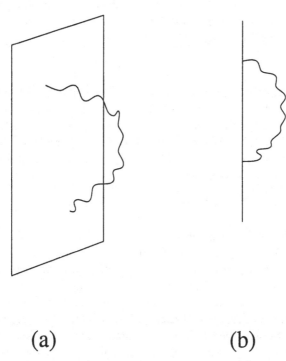

(a)   (b)

Figure 10: Open string states with ends attached to a (a) Dirichlet membrane, (b) Dirichlet string.

conditions break translational invariance along x^m. Since we want the full theory to be translationally invariant, the only possible interpretation of such a boundary condition is that there is a p-dimensional extended object situated at $x^m = x_0^m$ that is responsible for breaking this translational invariance. We call this a Dirichlet p-brane located at $x^m = x_0^m$ ($p + 1 \leq m \leq 9$), and extended along $x^1, \ldots, x^p$.

Let us now summarize some of the important properties of D-branes that will be relevant for understanding the test of $SL(2, Z)$ duality in type IIB string theory:

- The Dirichlet p-brane in IIB is invariant under half of the space-time supersymmetry transformations for odd p. To see how this property arises, let us denote by ϵ_L and ϵ_R the space-time supersymmetry transformation parameters in type IIB string theory, originating in the left- and the right-moving sector of the world-sheet theory respectively. ϵ_L and ϵ_R satisfy the chirality constraint

$$\Gamma^0 \cdots \Gamma^9 \epsilon_L = \epsilon_L, \qquad \Gamma^0 \cdots \Gamma^9 \epsilon_R = \epsilon_R, \qquad (4.25)$$

where Γ^μ are the 10-dimensional gamma matrices. The open string boundary conditions (4.24) together with the corresponding boundary conditions on the world-sheet fermions give further restriction on ϵ_L and ϵ_R of the form (see [72])

$$\epsilon_L = \Gamma^{p+1} \dots \Gamma^9 \epsilon_R. \qquad (4.26)$$

It is easy to see that the two equations (4.25) and (4.26) are compatible only for odd p. Thus in type IIB string theory Dirichlet p-branes are invariant under half of the space-time supersymmetry transformations for odd p. An identical argument shows that in type IIA string theory we have supersymmetric Dirichlet p-branes only for even p since in this theory eq.(4.25) is replaced by

$$\Gamma^0 \dots \Gamma^9 \epsilon_L = \epsilon_L, \qquad \Gamma^0 \dots \Gamma^9 \epsilon_R = -\epsilon_R. \qquad (4.27)$$

- Type IIB (IIA) string theory contains a p-form gauge field for even (odd) p. For example, in type IIB string theory these p-form gauge fields correspond to the scalar a, the rank-2 anti-symmetric tensor field $B'_{\mu\nu}$ and the rank-4 anti-symmetric tensor field $D_{\mu\nu\rho\sigma}$. It can be shown that a Dirichlet p-brane carries one unit of charge under the RR $(p+1)$-form gauge field; see [72]. More precisely, if we denote by $C_{\mu_1 \dots \mu_q}$ the q-form gauge potential, then a Dirichlet p-brane extending along the $1 \cdots p$ direction acts as a source of $C_{01 \dots p}$. (For $p = 5$ and 7 these correspond to magnetic dual potentials of $B'_{\mu\nu}$ and a respectively.) This result can be obtained by computing the 1-point function of the vertex operator for the field C in the presence of a D-brane. The relevant string world-sheet diagram has been indicated in Figure 11. We shall not discuss the details of this computation here.

From this discussion it follows that a Dirichlet 1-brane (D-string) in type IIB theory carries one unit of charge under the RR 2-form field $B'_{\mu\nu}$. This means that in type IIB on S^1 (labelled by the coordinate x^9) a D-string wrapped around the S^1 describes a particle charged under $B'_{9\mu}$. This then is a candidate soliton carrying charge quantum numbers $\begin{pmatrix} 0 \\ 1 \end{pmatrix}$ that is related to the $\begin{pmatrix} 1 \\ 0 \end{pmatrix}$ state via $SL(2, Z)$ duality. As we had seen earlier, $SL(2, Z)$ duality predicts that there should be a unique ultra-short multiplet carrying charge quantum numbers $\begin{pmatrix} 0 \\ 1 \end{pmatrix}$. Thus our task now is as follows:

- Quantize the collective coordinates of this soliton.

- Verify if we get an ultra-short multiplet in this quantum theory.

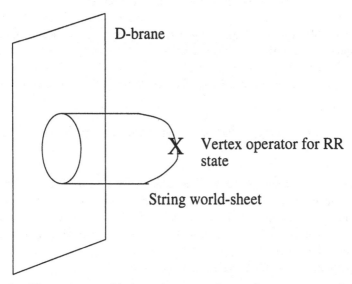

D-brane

X Vertex operator for RR state

String world-sheet

Figure 11: The string world-sheet diagram relevant for computing the coupling of the RR gauge field to the D-brane. It corresponds to a surface of the topology of a hemisphere with its boundary glued to the D-brane. The vertex operator of the RR-field is inserted at a point on the hemisphere.

Since the D-string is a 1-dimensional object, the dynamics of its collective coordinates should be described by a $(1 + 1)$-dimensional field theory. As we had discussed earlier, all the vibrational modes of the D-string are given by the open string states with ends attached to the D-string. In particular, the zero frequency modes (collective modes) of the D-string that are relevant for analyzing the spectrum of BPS states correspond to *massless* open string states propagating on the D-string. By analyzing the spectrum of these open string states one finds that the collective coordinates in this case correspond to

- 8 bosonic fields y^m denoting the location of this string in 8 transverse directions.

- A $U(1)$ gauge field.

- 8 Majorana fermions.

It can be shown that the dynamics of these collective coordinates is described by a $(1 + 1)$-dimensional supersymmetric quantum field theory which is the dimensional reduction of the $N = 1$ supersymmetric $U(1)$ gauge theory from

(9+1) to (1+1) dimensions. Normally in (1+1) dimensions, gauge fields have no dynamics. But here, since the space direction is compact, $y \equiv \oint A_1 dl$ is a physical variable. Furthermore, the compactness of $U(1)$ forces y to be periodically identified ($y \equiv y + a$ for some a). Thus the momentum p_y conjugate to y is quantized ($p_y = 2\pi k/a$ with k an integer.) It can be shown (see [69]) that this momentum, which represents electric flux along the D-string, is actually a source of $B_{9\mu}$ charge! Thus if we restrict to the $p_y = 0$ sector then these states carry $\begin{pmatrix} 0 \\ 1 \end{pmatrix}$ charge quantum numbers as discussed earlier, but by taking $p_y = 2\pi k/a$, we can get states carrying charge quantum numbers $\begin{pmatrix} k \\ 1 \end{pmatrix}$ as well.

Due to the compactness of the space direction, we can actually regard this as a quantum mechanical system instead of a $(1+1)$-dimensional quantum field theory. It turns out that in looking for ultra-short multiplets, we can ignore all modes carrying momentum along S^1. This corresponds to dimensionally reducing the theory to $(0+1)$ dimensions. The degrees of freedom of this quantum mechanical system are:

- 8 bosonic coordinates y^m,

- 1 compact bosonic coordinate y,

- 16 fermionic coordinates.

A quantum state is labelled by the momenta conjugate to y^m (ordinary momenta) and an integer labelling momentum conjugate to y which can be identified with the quantum number p labelling $B_{9\mu}$ charge. The fermionic coordinates satisfy the 16-dimensional Clifford algebra. Thus quantization of the fermionic coordinates gives $2^8 = 256$-fold degeneracy, which is precisely the correct degeneracy for an ultra-short multiplet. This establishes the existence of all the required states of charge $\begin{pmatrix} p \\ 1 \end{pmatrix}$ predicted by $SL(2, Z)$ symmetry.

What about $\begin{pmatrix} p \\ r \end{pmatrix}$ states with $r > 1$? These carry r units of $B'_{9\mu}$ charge and hence must arise as a bound state of r D-strings wrapped along S^1. Thus the first question we need to ask is: what is the $(1+1)$-dimensional quantum field theory governing the dynamics of this system? In order to answer this question we need to study the dynamics of r D-strings. This system can be described as easily as a single D-string: instead of allowing open strings to end on a single D-string, we allow it to end on any of the r D-strings situated at

$$x^m = x^m_{(i)}, \qquad 2 \leq m \leq 9, \qquad 1 \leq i \leq r, \qquad (4.28)$$

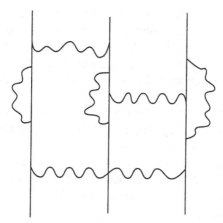

Figure 12: Possible open string states in the presence of three parallel D-strings.

where $\vec{x}_{(i)}$ denotes the location of the ith D-string. The situation is illustrated in Figure 12. Thus the dynamics of this system will now be described not only by the open strings starting and ending on the same D-string, but also by open strings whose two ends lie on two different D-strings.

For studying the spectrum of BPS states we need to focus our attention on the massless open string states. First of all, for each of the r D-strings we get a $U(1)$ gauge field, 8 scalar fields and 8 Majorana fermions from open strings with both ends lying on that D-string. But we can get extra massless states from open strings whose two ends lie on two different D-strings when these two D-strings coincide. It turns out that for r coincident D-strings the dynamics of massless strings on the D-string world-sheet is given by the dimensional reduction to $(1+1)$ dimensions of $N = 1$ supersymmetric $U(r)$ gauge theory in 10 dimensions, or equivalently, $N = 4$ supersymmetric $U(r)$ gauge theory in 4 dimensions; see [69]. Following a logic similar to that in the case of a single D-string, one can show that the problem of computing the degeneracy of $\binom{p}{r}$ states reduces to the computation of certain Witten index in this quantum theory. We shall not go through the details of this analysis, but just state the final result. It turns out that *there is a unique ultra-short multiplet for every pair of integers (p, r) which are relatively prime, precisely as predicted by $SL(2, Z)$;* see [69]!

A similar analysis can be carried out for the short multiplets that carry momentum k along S^1 besides carrying the B and B' charges p and r; see [69, 70]. In order to get these states from the D-brane spectrum, we can no longer dimensionally reduce the $(1 + 1)$-dimensional theory to $(0 + 1)$ dimensions. Instead we need to take into account the modes of the various fields of the $(1 + 1)$-dimensional field theory carrying momentum along the internal S^1. The BPS states come from configurations where only the left- (or right-) moving modes on S^1 are excited. The calculation of the degeneracy $d(k, p, r)$ of BPS states carrying given charge quantum numbers (p, r, k) is done by determining in how many ways the total momentum k can be divided among the various left-moving bosonic and fermionic modes. This counting problem turns out to be identical to the one used to get the Hagedorn spectrum of BPS states in the elementary string spectrum, except that the elementary string is replaced here by the solitonic D-string. Naturally, we get back the Hagedorn spectrum for $d(k, p, r)$ as well. Thus the answer agrees exactly with that predicted by $SL(2, Z)$ duality. This provides us with a test of the conjectured $SL(2, Z)$ symmetry of type IIB on S^1.

The method of using D-branes to derive the dynamics of collective coordinates has been used to verify the predictions of other duality conjectures involving various string compactifications. Among them are self-duality of type II string theory on T^4 (see [74, 75, 76, 77]), the duality between heterotic on T^4 and type IIA on $K3$ (see [78]), the duality between type I and $SO(32)$ heterotic string theory (see [29]), etc.

4.3   Massless solitons and tensionless strings

An interesting aspect of the conjectured duality between the heterotic string theory on T^4 and type IIA string theory on $K3$ is that at special points in the moduli space the heterotic string theory has enhanced non-abelian gauge symmetry, e.g. $E_8 \times E_8$ or $SO(32)$ in the absence of vacuum expectation values of the internal components of the gauge fields, $SU(2)$ at the self-dual radius etc. Perturbative type IIA string theory on $K3$ does not have any such gauge symmetry enhancement, since the spectrum of elementary string states does not contain any state charged under the $U(1)$ gauge fields arising in the RR sector. Thus, for example, we do not have the $W^{\pm}$ bosons that are required for enhancing a $U(1)$ gauge group to $SU(2)$. At first sight this seems to lead to a contradiction. However upon closer examination one realises that this cannot really be a problem; see [150]. To see this let us consider a point in the moduli space of heterotic string theory on T^4 where the non-abelian gauge symmetry is broken. At this point the would-be massless gauge bosons of the non-abelian gauge theory acquire mass by the Higgs mechanism, and appear as BPS states in the abelian theory. As we approach the point of enhanced gauge symmetry, the masses of these states vanish. Since the masses of BPS states are determined by the BPS formula, the vanishing of the masses must be a consequence of the BPS formula. Thus if we are able to find the images of these BPS states on the type IIA side as appropriate D-brane states, then the masses of these D-brane states must also vanish as we approach the point in the moduli space where the heterotic theory has enhanced gauge symmetry. These massless D-brane solitons will then provide the states necessary for enhancing the gauge symmetry.

To see this more explicitly, let us examine the BPS formula. It can be shown that in the variables defined in Section 3.3 the BPS formula is given by,

$$m^2 = e^{-\Phi^{(A)}/2}\alpha^T(LM^{(A)}L + L)\alpha, \qquad (4.29)$$

where α is a 24-dimensional vector belonging to the lattice Λ'_{24}, and represents the $U(1)$ charges carried by this particular state. For each $\vec{\alpha}$ we can assign an occupation number $n(\vec{\alpha})$ which gives the number of BPS multiplets carrying this specific set of charges. Since $M^{(A)}$ is a symmetric $O(4,20)$ matrix, we can express this as $\Omega^{(A)T}\Omega^{(A)}$ for some $O(4,20)$ matrix $\Omega^{(A)}$, and rewrite eq.(4.29) as

$$m^2 = e^{-\Phi^{(A)}/2}\alpha^T L\Omega^{(A)T}(I_{24} + L)\Omega^{(A)}L\alpha. \qquad (4.30)$$

As can be seen from eq.(3.28), $(I_{24} + L)$ has 20 zero eigenvalues. As we vary $M^{(A)}$ and hence $\Omega^{(A)}$, the vector $\Omega^{(A)}L\alpha$ rotates in the 24-dimensional space. If for some $\Omega^{(A)}$ it is aligned along one of the eigenvectors of $(I_{24} + L)$ with zero eigenvalue, we shall get massless solitons provided the occupation number $n(\vec{\alpha})$ for this specific $\vec{\alpha}$ is non-zero.

Although this argument resolves the problem at an abstract level, one would like to understand this mechanism directly by analysing the type IIA string theory, since, after all, we do not encounter massless solitons very often in physics. This has been possible through the work of [26, 151, 78]. For simplicity let us focus on the case of enhanced $SU(2)$ gauge symmetry. First of all, one finds that at a generic point in the moduli space where $SU(2)$ is broken, the images of the $W^{\pm}$ bosons in the type IIA theory are given by a $(D-2)$-brane wrapped around a certain 2-cycle (topologically non-trivial 2-dimensional surface) inside $K3$, the $+$ and the $-$ sign of the charge being obtained from two different orientations of the $(D-2)$-brane. Since the two tangential directions on the $(D-2)$-brane are directed along the two internal directions of $K3$ tangential to the 2-cycle, this object has no extension in any of the 5 non-compact spatial directions, and hence behaves like a particle.[20]. It turns out that as we approach the point in the moduli space where the theory on the heterotic side develops enhanced $SU(2)$ gauge symmetry, the $K3$ on which type IIA theory is compactified becomes singular. At this singularity the area of the topologically non-trivial 2-cycle mentioned above goes to zero. As a result, the mass of the wrapped $(D-2)$-brane, obtained by multiplying the tension of the $(D-2)$-brane by the area of the 2-cycle, vanishes. This gives us the massless solitons that are required for the gauge

---

[20]These states were analyzed in detail in [152]

symmetry enhancement. A similar mechanism works for getting other gauge groups as well. In fact it turns out that there is a one to one correspondence between the enhanced gauge groups, which are classified by ADE Dynkin diagram, and the singularity type of $K3$, which are also classified by the ADE Dynkin diagram; see [26]. This establishes an explicit physical relationship between ADE singularities and ADE Lie algebras.

The appearance of enhanced gauge symmetry in type IIA on $K3$ poses another puzzle. Let us compactify this theory on one more circle. Since such a compactification does not destroy gauge symmetry, this theory also has enhanced gauge symmetry when the $K3$ becomes singular. But type IIA on $K3 \times S^1$ is T-dual to type IIB on $K3 \times S^1$; thus type IIB on $K3 \times S^1$ must also develop enhanced gauge symmetry when $K3$ develops singularities. Does this imply that type IIB on $K3$ also develops enhanced gauge symmetry at these special points in the $K3$ moduli space? This does not seem possible, since type IIB string theory does not have any $(D-2)$-brane solitons which can be wrapped around the collapsed two cycles of $K3$. It turns out that instead of acquiring enhanced gauge symmetry, type IIB string theory acquires tension-less strings at these special points in the $K3$ moduli space; see [153]. These arise from taking a $(D-3)$-brane of type IIB string theory, and wrapping it on a 2-cycle of $K3$. Thus two of the tangential directions of the 3-brane are directed along the internal directions of $K3$, and the third direction of the 3-brane is along one of the non-compact spatial directions. Thus from the point of view of the $(5+1)$-dimensional theory such a configuration will appear as a string. The tension of this string is given by the product of the tension (energy per unit 3-volume) of the 3-brane and the area of the 2-cycle on which the 3-brane is wrapped. Thus as we approach the singular point on the $K3$ moduli space where the area of the 2-cycle vanishes, the tension of the string goes to zero. In other words, we get tensionless strings. Upon further compactification on a circle we get massless particles from configurations where this tensionless string is wound around the circle. These are precisely the massless gauge bosons required for the gauge symmetry enhancement in type IIB on $K3 \times S^1$.

5   Interrelation Between Different Duality Conjectures

In the last three sections we have seen many different duality conjectures and have learned how to test these conjectures. We shall now see that many of these conjectures are not independent, but can be 'derived' from each other. There are several different ways in which dualities can be related to each other. We shall discuss them one by one. The material covered in this section is taken mainly from [84, 85, 82].

5.1   Combining non-perturbative and T-dualities

Suppose a string theory A compactified on a manifold K_A has a conjectured duality symmetry group G. Now further compactify this theory on some manifold $\mathcal{M}$. Then the theory A on $K_A \times \mathcal{M}$ is expected to have the following set of duality symmetries:

- It inherits the original duality symmetry group G of A on K_A.

- It also has a perturbatively verifiable T-duality group. Let us call it H.

Quite often G and H do not commute and together generate a much bigger group; see [83, 30]. In that case, the existence of this bigger group of symmetries can be regarded as a consequence of the duality symmetry of A on K_A and T-dualities.

We shall illustrate this with a specific example; see [30]. We have seen that in 10 dimensions type IIB string theory has a conjectured duality group $SL(2, Z)$ that acts non-trivially on the coupling constant. From the table given in Section 3.5 we see that type IIB on T^n also has a T-duality group $SO(n, n; Z)$, whose existence can be verified order-by-order in string perturbation theory. It turns out that typically these two duality groups do not commute, and in fact generate the full duality symmetry group of type IIB on T^n as given in the table of Section 3.5. Thus we see that the existence of the full duality symmetry group of type IIB on T^n can be infered from the $SL(2, Z)$ duality symmetry of the 10-dimensional type IIB string theory, and the perturbatively verifiable T-duality symmetries of type IIB on T^n.

5.2   Duality of dualities

Suppose two theories are conjectured to be dual to each other, and each theory in turn has a conjectured self-duality group. Typically part of this self-duality group is T-duality, and the rest involves non-trivial transformation of the coupling constant. But quite often the non-perturbative duality transformations in one theory correspond to T-duality in the dual theory and vice versa. As a result, the full self-duality group in both theories follows from the conjectured duality between the two theories.

Again we shall illustrate this with an example; see [31, 26]. Let us start with the conjectured duality between heterotic on T^4 and type IIA on $K3$. Now let us compactify both theories further on a 2-dimensional torus T^2. This produces a dual pair of theories: type IIA on $K3 \times T^2$ and heterotic on T^6. Now, heterotic on T^6 has a T-duality group $O(6, 22; Z)$ that can be verified using heterotic perturbation theory. On the other hand , type IIA on $K3 \times T^2$ has a T-duality group $O(4, 20; Z) \times SL(2, Z) \times SL(2, Z)'$ that can be verified using type II perturbation theory. The full conjectured duality group in both theories is $O(6, 22, Z) \times SL(2, Z)$.

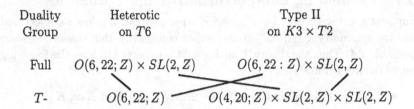

Figure 13: The embedding of the T-duality groups in the full duality group in heterotic on T^6 and type IIA on $K3 \times T^2$.

Now the question we would like to address is, how are the T-duality symmetry groups in the two theories embedded in the full conjectured $O(6, 22; Z) \times SL(2, Z)$ duality group? This has been illustrated in Figure 13. In particular we find that the $SL(2, Z)$ factor of the full duality group is a subgroup of the T-duality group in type IIA on $K3 \times T^2$, and hence can be verified in this theory order-by-order in perturbation theory. On the other hand, the $O(6, 22; Z)$ factor of the duality group appears as a T-duality symmetry of the heterotic string theory, and hence can be verified order-by-order in perturbation theory in this theory. Thus assuming that T-duality in either theory is a valid symmetry, and the duality between the heterotic on T^4 and type IIA on $K3$, we can establish the existence of the self-duality group $O(6, 22; Z) \times SL(2, Z)$ in heterotic on T^6 and type IIA on $K3 \times T^2$.

Using the results of this and the previous subsection, we see that so far among all the conjectured non-perturbative duality symmetries, the independent ones are:

1. $SL(2, Z)$ of type IIB in D=10,

2. type I $\leftrightarrow$ $SO(32)$ heterotic in D=10, and

3. IIA on $K3$ $\leftrightarrow$ heterotic on T^4.

We shall now show how to 'derive' (3) from (1) and (2).

5.3 Fiberwise duality transformation

In this subsection we shall describe the idea of constructing dual pairs of theories using fiberwise duality transformation; see [84]. Suppose (Theory A on K_A) has been conjectured to be dual to (Theory B on K_B). Here A and B are two of the 5 different string theories in D=10, and K_A, K_B are two different manifolds (in general). This duality involves a precise map between the moduli spaces of the two theories. Now construct a pair of new manifolds $\mathcal{M}_A$, $\mathcal{M}_B$ by starting from some other manifold $\mathcal{B}$, and erecting at every point on $\mathcal{B}$ a copy of $\mathcal{K}_A$, $\mathcal{K}_B$. The moduli of $\mathcal{K}_A$, $\mathcal{K}_B$ vary slowly over $\mathcal{B}$ and are related to each other via the duality map that relates (A on $\mathcal{K}_A$) to (B on $\mathcal{K}_B$). Then we would expect a duality

$$\text{Theory } A \text{ on } \mathcal{M}_A \leftrightarrow \text{Theory } B \text{ on } \mathcal{M}_B$$

by applying the duality transformation fiberwise. This then gives rise to a new duality conjecture. This situation has been illustrated in Figure 14.

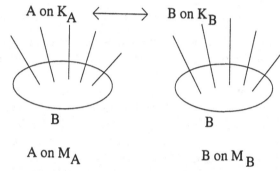

Figure 14: Application of fiberwise duality transformation. In each local neighbourhood of the base manifold $\mathcal{B}$, the two theories are equivalent due to the equivalence of the theories living on the fiber ($\times$ any manifold). Thus we would expect the theories A on $\mathcal{M}_A$ and B on $\mathcal{M}_B$ to be equivalent.

Now suppose that at some isolated points (or subspaces of codimension ≥ 1) on $\mathcal{B}$ the fibers $\mathcal{K}_A$ and $\mathcal{K}_B$ degenerate. (We shall see some explicit examples of this later.) Is the duality between (A on $\mathcal{M}_A$) and (B on $\mathcal{M}_B$) still valid? We might expect that even in this case the duality between the two theories holds since the singularities occur on subspaces of 'measure zero'. Although there is no rigorous argument as to why this should be so, this appears to be the case in all known examples. Conversely, assuming that this is the case, we can derive the existence of many new duality symmetries from a given duality symmetry.

A special case of this construction involves Z_2 orbifolds. Suppose we have a dual pair (A on $\mathcal{K}_A$) $\leftrightarrow$ (B on $\mathcal{K}_B$). Further suppose that (A on $\mathcal{K}_A$) has a Z_2 symmetry generated by h_A. Then the dual theory must also have a Z_2 symmetry generated by h_B. h_A and h_B are mapped to each other under duality. Now compactify both theories on another manifold $\mathcal{D}$ with a Z_2 isometry generated by g, and compare the two quotient theories

$$(A \text{ on } \mathcal{K}_A \times \mathcal{D}/h_A \cdot g) \text{ and } (B \text{ on } \mathcal{K}_B \times \mathcal{D}/h_B \cdot g)$$

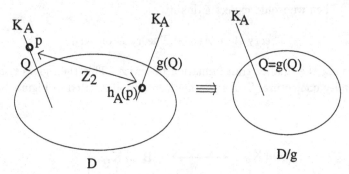

Figure 15: Representation of a Z_2 orbifold as a fibered space. The Z_2 transformation relates the point (Q, p) on $\mathcal{D} \times \mathcal{K}_A$ to the point $(g(Q), h_A(p))$.

$(\mathcal{K}_A \times \mathcal{D}/h_A \cdot g)$ is obtained from the product manifold $\mathcal{K}_A \times \mathcal{D}$ by identifying points that are related by the Z_2 transformation $h_A \cdot g$. This situation is illustrated in Figure 15. As shown in this figure, $(\mathcal{K}_A \times \mathcal{D}/h_A \cdot g)$ admits a fibration with base $\mathcal{D}/g$ and fiber $\mathcal{K}_A$. In particular, note that since $h_A \cdot g$ takes a point $(p \in \mathcal{K}_A, Q \in \mathcal{D})$ to $(h_A(p), g(Q))$, if we focus our attention on a definite point Q on $\mathcal{D}$, then there is no identification of the points in the copy of $\mathcal{K}_A$ that is sitting at Q. This shows that the fiber is $\mathcal{K}_A$ and *not* $\mathcal{K}_A/h_A$. As we go from Q to $g(Q)$, which is a closed cycle on $\mathcal{D}/g$, the fiber gets twisted by the transformation h_A.

The second theory, B, on $(\mathcal{K}_B \times \mathcal{D})/(h_B \cdot g)$, has an identical structure. Thus we can now apply fiberwise duality transformation to derive a new duality:

$$(A \text{ on } \mathcal{K}_A \times \mathcal{D}/h_A \cdot g) \leftrightarrow (B \text{ on } \mathcal{K}_B \times \mathcal{D}/h_B \cdot g)$$

Note that if $P_0 \in \mathcal{D}$ is a fixed point of g (i.e. if $g(P_0) = P_0$) then in $\mathcal{K}_A \times \mathcal{D}/h_A \cdot g$ there is an identification of points (p, P_0) and $(h_A(p), P_0)$. Similarly in $\mathcal{K}_B \times \mathcal{D}/h_B \cdot g$ there is an identification of points (p', P_0) and $(h_B(p'), P_0)$. Thus at P_0 the fibers degenerate to $\mathcal{K}_A/h_A$ and $\mathcal{K}_B/h_B$ respectively. At these points the argument in support of duality between the two theories breaks down. However, as we have discuused earlier, since these are points of 'measure zero' on $\mathcal{D}$, we would expect that the two quotient theories are still dual to each other; see [85]. We shall now illustrate this construction in the context of a specific example.

We start with type IIB string theory in 10 dimensions. This has a conjectured $SL(2, Z)$ symmetry. Let S denote the $SL(2, Z)$ element

$$\begin{pmatrix} 0 & 1 \\ -1 & 0 \end{pmatrix}. \tag{5.1}$$

Recall that this theory also has two global discrete symmetries $(-1)^{F_L}$ and Ω. The actions of S, $(-1)^{F_L}$ and Ω on the massless bosonic fields in this theory were described in Section 3.4. From this one can explicitly compute the action of $S(-1)^{F_L}S^{-1}$ on these massless fields. This action turns out to be identical to that of Ω. A similar result holds for their action on the massless fermionic fields as well. Finally, since the action of S on the massive fields is not known, one can define this action in such a way that the actions of $S(-1)^{F_L}S^{-1}$ and Ω are identical on all states. This gives:

$$S(-1)^{F_L}S^{-1} = \Omega. \qquad (5.2)$$

We are now ready to apply our formalism. We take $(A$ on $\mathcal{K}_A)$ to be type IIB in $D = 10$, $(B$ on $\mathcal{K}_B)$ to be type IIB in $D = 10$ transformed by S, h_A to be $(-1)^{F_L}$, h_B to be Ω, $\mathcal{D}$ to be T^4, and g to be the transformation $\mathcal{I}_4$ that changes the sign of all the coordinates on T^4. This gives the duality:

$$(\text{IIB on } T^4/(-1)^{F_L} \cdot \mathcal{I}_4) \leftrightarrow (\text{IIB on } T^4/\Omega \cdot \mathcal{I}_4)$$

Note that in this case the fibers $\mathcal{K}_A$ and $\mathcal{K}_B$ are points, but this does not prevent us from applying our method of constructing dual pairs. Also there are 16 fixed points on T^4 under $\mathcal{I}_4$ where the application of fiberwise duality transformation breaks down, but as has been argued before, we still expect the duality to hold since these are points of measure zero on T^4.

We shall now bring this duality into a more familiar form via T-duality transformation. Let us make the $R \to 1/R$ duality transformation on one of the circles of T^4 in the theory on the left-hand side. This converts type IIB theory to type IIA. This also transforms $(-1)^{F_L} \cdot \mathcal{I}_4$ to $\mathcal{I}_4$, which can be checked by explicitly studying the action of these transformations on the various massless fields. Thus the theory on the left-hand side is T-dual to type IIA on $T^4/\mathcal{I}_4$. This of course is just a special case of type IIA on $K3$.

Let us now take the theory on the right-hand side and make the $R \to 1/R$ duality transformations on all 4 circles. This takes type IIB theory to type IIB theory. But this transforms $\Omega \cdot \mathcal{I}_4$ to Ω, which can again be seen by studying the action of these transformations on the massless fields. Thus the theory on the right is T-dual to type IIB on T^4/Ω. Since type I string theory can be regarded as type IIB string theory modded out by Ω, we see that the theory on the right-hand side is type I on T^4. But by (heterotic–type I) duality in 10 dimensions this is dual to heterotic on T^4. Thus we have 'derived' the duality

$$(\text{Type IIA on } K3) \leftrightarrow (\text{Heterotic on } T^4)$$

from other conjectured dualities in $D = 10$. Although in this way the duality has been established only at a particular point in the moduli space (the orbifold limit of $K3$), the argument can be generalized to establish this duality at a generic point in the moduli space as well; see [85].

There are many other applications of fiberwise duality transformation. Some of them will be discussed later in this review.

5.4 Recovering higher-dimensional dualities from lower-dimensional ones

So far we have discussed methods of deriving dualities involving compactified string theories by starting with the duality symmetries of string theories in higher dimensions. But we can also proceed in the reverse direction. Suppose a string theory compactified on a manifold $\mathcal{M}_1 \times \mathcal{M}_2$ has a self-duality symmetry group G. Now consider the limit when the size of $\mathcal{M}_2$ goes to infinity. A generic element of G, acting on this configuration, will convert this configuration to one where $\mathcal{M}_2$ has small or finite size. However, there may be a subgroup H of G that commutes with this limit, i.e. any element of this subgroup, acting on a configuration where $\mathcal{M}_2$ is big, gives us back a configuration where $\mathcal{M}_2$ is big. Thus we would expect that H is the duality symmetry group of the theory in the decompactification limit, i.e. of the original string theory compactified on $\mathcal{M}_1$. The same argument can be extended to the case of a pair of dual theories.

A priori this procedure does not appear to be very useful, since one normally likes to derive more complicated duality transformations of lower-dimensional theories from the simpler ones in the higher-dimensional theory. But we shall now show how this procedure can be used to derive the $SL(2, Z)$ duality symmetry of type IIB string theory from the conjectured duality between type I and $SO(32)$ heterotic string theories, and T-duality symmetries of the heterotic string theory. We shall describe the main steps in this argument; for details, see [82]. We start with the duality between type I on T^2 and heterotic on T^2 that follows from the duality between these theories in 10 dimensions. Now heterotic string theory on T^2 has a T-duality group $O(2,18;Z)$. We shall focus our attention on an $SL(2, Z) \times SL(2, Z)$ subgroup of this T-duality group. One of these two $SL(2, Z)$ factors is associated with the global diffeomorphisms of T^2, and the other one is associated with the $R \rightarrow 1/R$ duality symmetries on the two circles. By the 'duality of dualities' argument, this must also be a symmetry of type I on T^2. Since type I string theory can be regarded as type IIB string theory modded out by the world-sheet parity transformation Ω discussed in Section 3.4, we conclude that $SL(2, Z) \times SL(2, Z)$ is a subgroup of the self-duality group of type IIB on T^2/Ω. Let us now make an $R \rightarrow 1/R$ duality transformation on both the circles of this T^2. This converts type IIB on T^2 to type IIB theory compactified on a dual T^2, and Ω to $(-1)^{F_L} \cdot \Omega \cdot \mathcal{I}_2$, where $\mathcal{I}_2$ denotes the reversal of orientation of both the circles of T^2. (This can be seen by studying the action of various transformations on the massless fields.) Geometrically, this model describes type IIB string theory compactified on the surface of a tetrahedron (which is geometrically $T^2/\mathcal{I}_2$), with an added twist of $(-1)^{F_L} \cdot \Omega$ as we go around any of the 4 vertices of the tetrahedron (the fixed points of $\mathcal{I}_2$). (This theory will be discussed in more detail in Section 8). Thus we conclude that $SL(2, Z) \times SL(2, Z)$ is a subgroup of the self-duality group of type IIB on a tetrahedron. Now take the limit where the size of the tetrahedron goes to infinity. It turns out that both the $SL(2, Z)$ factors commute with this limit. One of these $SL(2, Z)$ groups becomes part of the diffeomorphism group of type IIB string theory and does not correspond to anything new, but the other $SL(2, Z)$ factor represents the S-duality transformation discussed in Section 3.4. Since this limit gives us back the decompactified type IIB string theory, we conclude that type IIB string theory in 10 dimensions has a self-duality group $SL(2, Z)$.

Thus we see that all the dualities discussed so far can be 'derived' from a single duality conjecture: the one between type I and $SO(32)$ heterotic string theories in 10 dimensions. In the next section we shall see more examples of dualities which can be derived from the ones that we have already discussed.

6 Duality in Theories with Less than 16 Supersymmetry Generators

So far our discussion has been focussed on theories with 16 or more supersymmetry charges. As was pointed out in Section 3, for these theories the non-renormalization theorems for the low energy effective action and the spectrum of BPS states are particularly powerful. This makes it easy to test duality conjectures involving these theories. In this section we shall extend our discussion to theories with 8 supercharges. Examples of such theories are provided by $N = 2$ supersymmetric theories in 4 dimensions. We shall see that these theories have a very rich structure, and although the non-renormalization theorems are less powerful here, they are still powerful enough to provide us with some of the most striking tests of duality conjectures involving these theories. The material covered in this section is based mainly on references [86, 87, 84].

6.1 Construction of dual pair of theories with 8 supercharges

For definiteness we shall focus our attention on $N = 2$ supersymmetric theories in 4 dimensions. There are several ways to get theories with $N = 2$

supersymmetry in 4 dimensions. Two of them are:

1. Type IIA/IIB on Calabi–Yau 3-folds: In our convention an n-fold describes an n complex or $2n$ real-dimensional manifold. In 10 dimensions type II theories have 32 supersymmetry generators. Compactification on a Calabi–Yau 3-fold breaks 3/4 of the supersymmetry. Thus we are left with 8 supersymmetry generators in D=4, giving rise to $N = 2$ supersymmetry.

2. Heterotic string theory on $K3 \times T^2$: In 10 dimensions heterotic string theory has 16 supersymmetry generators. Compactification on $K3 \times T^2$ breaks half of the supersymmetry. Thus we have a theory with 8 supersymmetry generators, again giving $N = 2$ supersymmetry in 4 dimensions. It is also possible to construct a more general class of 4-dimensional heterotic string theories with the same number of supersymmetries where the background does not have the product structure $K3 \times T^2$; see [122, 86].

The question we would like to ask is: is it possible to construct pairs of $N = 2$ supersymmetric type II and heterotic string compactifications in 4 dimensions which will be non-perturbatively dual to each other? Historically such dual pairs were first constructed by trial and error (see [86]) and then a more systematic approach was developed; see [87, 84, 88, 89, 90]. However we shall begin by describing the systematic approach, and then describe how one tests these dualities. The systematic construction of such dual pairs can be carried out by application of fiberwise duality transformations as described in the last section. The steps involved in this construction are as follows:

- Start from the conjectured duality (Type IIA on $K3$) $\leftrightarrow$ (Heterotic on T^4).

- Choose a CP^1 base.

- Construct a Calabi–Yau 3-fold by fibering $K3$ over the base CP^1. One can construct a whole class of Calabi–Yau manifolds in this way by choosing different ways of varying $K3$ over CP^1.

- For type IIA on each such Calabi–Yau 3-fold we can get a dual heterotic compactification by replacing the type IIA on $K3$ by heterotic on T^4 on each fiber according to the duality map. This gives heterotic string theory on a manifold obtained by varying T^4 on CP^1 according to the duality map. Typically this manifold turns out to be $K3 \times T^2$ or some variant of this. This model is expected to be dual to the type IIA string theory on the Calabi–Yau manifold that we started with. Thus we get a duality map

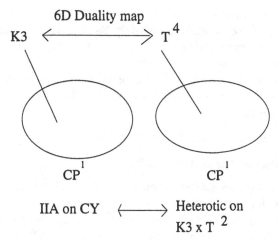

Figure 16: Construction of a dual pair of $N = 2$ supersymmetric string theories in 4 dimensions from the dual pair of theories in 6 dimensions.

$$(\text{Type IIA on CY}) \leftrightarrow (\text{Heterotic on } K3 \times T^2)$$

This construction has been illustrated in Figure 16. Note that the original duality map gives a precise relationship between the moduli of type IIA on $K3$ and heterotic on T^4. On the heterotic side the moduli involve background gauge fields on T^4 besides the shape and size of T^4. Thus for a specific Calabi–Yau, knowing how $K3$ varies over CP^1, we can find out how on the heterotic side the background gauge fields on T^4 vary as we move along CP^1. This gives the gauge field configuration on $K3 \times T^2$. Different Calabi–Yau manifolds will give rise to different gauge fields on $K3 \times T^2$.

We shall illustrate this procedure with the example of a pair of Z_2 orbifolds of the form (see [87]):

$$(\text{IIA on } K3 \times T^2/h_A \cdot g) \leftrightarrow (\text{Heterotic on } T^4 \times T^2/h_B \cdot g)$$

where g acts on T^2 by changing the sign of both its coordinates, h_A is a specific involution of $K3$ known as the Enriques involution, and h_B is the image of this transformation on the heterotic side. By our previous argument relating orbifolds to fibered spaces, these two theories are expected to be dual to each other via fiberwise duality transformation. $K3 \times T^2/(h_A \cdot g)$ can be shown to describe a Calabi–Yau manifold. Thus the theory on the left-hand side corresponds to type IIA string theory compactified on this Calabi–Yau manifold. In order to determine the theory on the heterotic side, we need to determine h_B. We shall now describe this procedure in some detail.

In order to determine h_B, we need to study the relationship between the fields appearing in type IIA on $K3$ and heterotic on T^4. The low energy effective action of both the theories and the origin of the various massless fields in these theories were discussed in Section 3.3. We shall focus our attention on the gauge fields. As discussed there, in the type IIA on $K3$, 22 of the gauge fields come from decomposing the 3-form field along the harmonic 2-forms on $K3$. Now, h_A, being a geometric transformation on $K3$, has known action on the harmonic forms $\omega^{(p)}$. For this particular example, h_A corresponds to

- exchanging 10 of the $\omega^{(p)}$ with 10 others and

- changing the sign of 2 more $\omega^{(p)}$.

This translates into a similar action on the fields $\mathcal{A}_\mu^{(p)}$ defined in eq.(3.36). Furthermore, h_A leaves invariant the other two gauge fields, coming from the 10-dimensional gauge field A_μ and the dual of $C_{\mu\nu\rho}$. We can now translate this into an action on the gauge fields in heterotic on T^4. It turns out that the action on the heterotic side is given by

- exchanging the gauge fields in the two E_8 factors ,

- exchanging $(G_{9\mu}, B_{9\mu})$ with $(G_{8\mu}, B_{8\mu})$, and,

- changing the sign of $(G_{7\mu}$ and $B_{7\mu})$.

This translates into the following geometric action in heterotic string theory on T^4.[21]

- exchanging the two E_8 factors in the gauge group,

- $x^8 \leftrightarrow x^9$,

- $x^7 \to -x^7$.

This is h_B.[22] It turns out that modding out heterotic string theory on T^6 by the transformation $h_B \cdot g$ produces an $N = 2$ supersymmetric theory. Thus this construction gives a type II–heterotic dual pair with $N = 2$ supersymmetry.

Using the idea of fiberwise duality transformation we can construct many more examples of heterotic–type IIA dual pairs in 4 dimensions with $N = 2$ supersymmetry. Quite often, using mirror symmetry (see [123]), we can also relate this to IIB string theory on a mirror Calabi–Yau manifold.

---

[21]Here we are regarding this theory as the $E_8 \times E_8$ heterotic string theory compactified on T^4. By the duality between the two heterotic string theories upon compactification on a circle, this is equivalent to $SO(32)$ heterotic string theory compactified on T^4.

[22]We need to add to this a non-geometrical shift involving half of a lattice vector in Λ_{24} in order to get a modular invariant theory on the heterotic side. This transformation is not visible in perturbative type IIA theory.

6.2 Test of duality conjectures involving theories with 8 supercharges

Given such a dual pair of theories constructed by application of fiberwise duality transformation, the next question will be: how do we test if these theories are really dual to each other? After all, as we have seen, there is no rigorous proof that fiberwise duality transformation always produces a correct dual pair of theories, particularly when the fiber degenerates at some points / regions in the base. Unlike in the case of theories with 16 supercharges, one cannot directly compare the tree level low energy effective action in the two theories, as they undergo quantum corrections in general. Furthermore, in this theory the spectrum of BPS saturated states can change discontinuously as we move in the moduli space; see [95]. Hence the spectrum computed at weak coupling cannot always be trusted at strong coupling. Nevertheless there are some non-renormalization theorems which allow us to test these proposed dualities, as we shall now describe.

Matter multiplets in $N = 2$ supersymmetric theories in 4 dimensions are of two types. (For a review, see [95].) They are

- vector multiplet containing 1 vector, 1 complex scalar, and 2 Majorana fermions, and

- hypermultiplet containing 2 complex scalars and 2 Majorana fermions.

Let us consider a theory at a generic point in the moduli space where the massless matter fields include only abelian gauge fields and neutral hyper-multiplets. Let $\vec{\phi}$ denote the complex scalars in the vector multiplet, and $\vec{\psi}$ denote the complex scalars in the hypermultiplet. The $N = 2$ supersymmetry requires that there is no coupling between the vector and the hypermultiplets in those terms in the low energy effective action S_{eff} which contain at most two space-time derivatives; see [91]. Thus the scalar kinetic terms appearing in the Lagrangian density associated with S_{eff} must be of the form

$$G_{m\bar{n}}^{V}(\vec{\phi})\partial_{\mu}\phi^{m}\partial^{\mu}\bar{\phi}^{n} + G_{\alpha\bar{\beta}}^{H}(\vec{\psi})\partial_{\mu}\psi^{\alpha}\partial^{\mu}\bar{\psi}^{\beta}, \tag{6.1}$$

where G^{V} and G^{H} are appropriate metrics in the vector and the hypermul-tiplet moduli spaces. The kinetic terms of the vectors and the fermionic fields are related to these scalar kinetic terms by the requirement of $N = 2$ supersymmetry.

This decoupling between the hyper- and the vector- multiplet moduli spaces by itself is not of much help, since each term may be independently modified by quantum corrections.[23] But in string theory we have some extra

---

[23]There are however strong restrictions on what kind of metric G_V and G_H should describe. In particular G_V must describe a special Kahler geometry (see [124, 91]), whereas G_H must describe a quaternionic geometry; see [125]. However, these restrictions do not fix G_V and G_H completely.

ingredient; see [86, 84]. Recall that the coupling constant in string theory involves the dilaton. Thus quantum corrections to a given term must involve a coupling to the dilaton. Now consider the following two special cases:

1. The dilaton belongs to a hypermultiplet. Then there can be no correction to the vector multiplet kinetic term since such corrections will give a coupling between the dilaton and the vector multiplet.

2. The dilaton belongs to a vector multiplet. In this case the same argument shows that there can be no correction to the hypermultiplet kinetic term.

In type IIA/IIB string theory on Calabi–Yau manifold the dilaton belongs to a hypermultiplet. Thus in these theories the vector multiplet kinetic term, calculated at the tree level, is exact. On the other hand, in heterotic on $K3 \times T^2$, the dilaton is in the vector multiplet. Thus the hypermultiplet kinetic term, calculated at the tree level, is exact. Using this information we can adopt the following strategy for testing duality.[24]

1. Take a type II–heterotic dual pair and calculate the vector multiplet kinetic term exactly from the tree level analysis on the type II side.

2. Using the map between the fields in the type II and the heterotic theory, we can rewrite the exact vector multiplet kinetic term in terms of the heterotic variables.

3. In particular the heterotic variables include the heterotic dilaton Φ_H which is in the vector multiplet. So we can now expand the exact answer in powers of e^{Φ_H} and compare this answer with the explicit calculations in heterotic string perturbation theory. Typically the expansion involves tree, one-loop, and non-perturbative terms. (There is no perturbative contribution in the heterotic theory beyond one loop due to some Adler-Bardeen type non-renormalization theorems.) Thus one can compare the expected tree and one-loop terms, calculated explicitly in the heterotic string theory, with the expansion of the exact answer.

The results of the above calculation in heterotic and type II string theories agree in all the cases tested; see [86, 92, 93]! This agreement is quite remarkable, since the one-loop calculation is highly non-trivial on the heterotic side, and involves integrals over the moduli space of the torus. Indeed, the agreement between the two answers is a consequence of highly non-trivial mathematical identities.

---

[24]Here we describe the test using the vector multiplet kinetic term, but a similar analysis should be possible with the hypermultiplet kinetic term as well.

Given that the tree and one-loop results in the heterotic string theory agree with the expansion of the exact result on the type II side, one might ask if a similar agreement can be found for the non-perturbative contribution from the heterotic string theory as well. From the exact answer calculated from the type II side we know what this contribution should be. But we cannot calculate it directly on the heterotic side, since there is no non-perturbative formulation of string theory. However, one can take an appropriate limit in which the stringy effects on the heterotic side disappear and the theory reduces to some appropriate $N = 2$ supersymmetric quantum field theory.[25] Thus now the calculation of these non-perturbative effects on the heterotic side reduces to a calculation in the $N = 2$ supersymmetric field theory. This can be carried out using the method developed by Seiberg and Witten; see [95]. Again there is perfect agreement with the results from the type II side; see [94]. Besides providing a non-trivial test of string duality, this also shows that the complete Seiberg–Witten (see [95]) results (and more) are contained in the classical geometry of Calabi–Yau spaces!

7 M-theory

So far we have discussed dualities that relate known string theories. However, sometimes analysis similar to those that lead to various duality conjectures can also lead to the discovery of new theories. One such theory is a conjectured theory living in 11 dimensions. This theory is now known as M-theory. In this section we shall give a brief description of this theory following references [26, 100, 101, 103, 104, 102, 105].

7.1 M-theory in 11 dimensions

The arguments leading to the existence of M-theory go as follows; see [96, 26]. Take type IIA string theory in 10 dimensions. The low energy effective action of this theory is non-chiral $N = 2$ supergravity in 10 dimensions. It is well known that this can be obtained from the dimensional reduction of $N = 1$ supergravity in 11 dimensions; see [97]. More specifically, the relationship between the two theories is as follows. The bosonic fields in $N = 1$ supergravity theory in 11 dimensions consist of the metric $g_{MN}^{(S)}$ and a rank-3 anti-symmetric tensor field $C_{MNP}^{(S)}$ ($0 \leq M, N \leq 10$). The bosonic part of the action of this theory is given by (see [126])

$$
S_{SG} = \frac{1}{(2\pi)^8} \int d^{11}x \left[\sqrt{-g^{(S)}} \left(R^{(S)} - \frac{1}{48} G^{(S)2} \right) \right.
$$
$$
\left. - \frac{1}{(12)^4} \varepsilon^{\mu_0 \cdots \mu_{10}} C_{\mu_0 \mu_1 \mu_2}^{(S)} G_{\mu_3 \cdots \mu_6}^{(S)} G_{\mu_7 \cdots \mu_{10}}^{(S)} \right], \tag{7.1}
$$

---

[25]This is in the same spirit as in the case of toroidal compactification of heterotic string theory, where, by going near a special point in the moduli space, we can effectively get an $N = 4$ supersymmetric Yang–Mills theory.

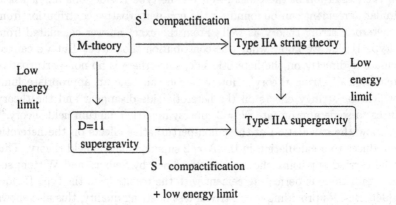

Figure 17: The relationship between M-theory and various other supergravity / string theories.

where $G^{(S)} \sim dC^{(S)}$ is the 4-form field strength associated with the 3-form field $C^{(S)}$. In writing down the above equation we have set the 11-dimensional Planck mass to unity (or equivalently we can say that we have absorbed it into a redefinition of the metric.) Let us now compactify this supergravity theory on a circle of radius $R(\sim \sqrt{g_{10,10}^{(S)}})$ measured in the supergravity metric $g_{MN}^{(S)}$ and ignore (for the time being) the Kaluza–Klein modes carrying momentum in the internal direction. Then the effective action in the dimensionally reduced theory agrees with that of type IIA string theory given in (3.33) under the identification (see [97])

$$\sqrt{g_{10,10}^{(S)}} = e^{\Phi/3}, \qquad g_{\mu\nu}^{(S)} \simeq e^{-\Phi/12} g_{\mu\nu} \qquad g_{10\mu}^{(S)} \simeq e^{2\Phi/3} A_\mu,$$
$$C_{\mu\nu\rho}^{(S)} \simeq C_{\mu\nu\rho}, \qquad C_{10\mu\nu}^{(S)} \simeq B_{\mu\nu}, \qquad (0 \le \mu, \nu \le 9). \qquad (7.2)$$

Here $\simeq$ denotes equality up to additive terms involving second and higher powers in fields. We are using the convention that $\Phi = 0$ corresponds to compactification on a circle of unit radius. Note that as the radius $R(\sim \sqrt{g_{10,10}^{(S)}})$ approaches ∞, $\Phi \to \infty$. This corresponds to strong coupling limit of the type IIA string theory. This leads one to the conjecture (see [96, 26]) that *in the strong coupling limit type IIA string theory approaches an 11-dimensional Lorentz invariant theory, whose low energy limit is 11-dimensional $N = 1$ supergravity*. This theory has been called M-theory. The situation is illustrated in Figure 17. Part of the conjecture is just the definition of M-theory as the strong coupling limit of type IIA string theory. The non-trivial part of the conjecture is that it describes a Lorentz invariant theory in 11 dimensions.

The evidence for the existence of an 11-dimensional theory, as discussed so far, has been analogous to the evidence for various duality conjectures based on the comparison of their low energy effective action. One might ask if there are more precise tests involving the spectrum of BPS states. There are indeed such tests. M theory on S^1 will have Kaluza–Klein modes representing states in the 11-dimensional $N = 1$ supergravity multiplet carrying momentum along the compact x^{10} direction. These are BPS states, and can be shown to belong to the 256-dimensional ultra-short representation of the supersymmetry algebra. The charge quantum number characterizing such a state is the momentum k/R along S^1. Thus for every integer k we should find such BPS states in type IIA string theory in 10 dimensions. In M-theory these states carry k units of $g_{10\mu}^{(S)}$ charge. Since $g_{10\mu}^{(S)}$ gets mapped to A_μ under the M-theory - IIA duality, these states must carry k units of A_μ charge in type IIA string theory. If we now recall that in type IIA string theory A_μ arises in the RR sector, we see that these states cannot come from elementary string states, as elementary string excitations are neutral under RR sector gauge fields. However, Dirichlet 0-branes in this theory do carry A_μ charge. In particular the state with $k = 1$ corresponds to a single Dirichlet 0-brane. As usual, the collective coordinate dynamics of the 0-branes is determined from the dynamics of massless open string states with ends lying on the $D0$-brane, and in this case is described by the dimensional reduction of $N = 1$ super-Maxwell theory from $(9 + 1)$ to $(0 + 1)$ dimensions. This theory has 16 fermion zero modes whose quantization leads to a $2^8 = 256$-fold degenerate state. Thus we see that we indeed have an ultra-short multiplet with unit A_μ charge, as predicted by the M-theory–IIA duality conjecture.

What about states with $k > 1$? In type IIA string theory these must arise as bound states of k $D0$-branes. The dynamics of collective coordinates of k $D0$-branes is given by the dimensional reduction of $N = 1$ supersymmetric $U(k)$ gauge theory from $(9 + 1)$ to $(0 + 1)$ dimensions. Thus the number of ultra-short multiplets with k units of A_μ charge is determined in terms of the number of normalizable supersymmetric ground states of this quantum mechanical system. Finding these bound states is much more difficult than the bound state problems discussed earlier. The main obstacle to this analysis is that a charge k state has the same energy as k charge 1 states at rest. Thus the bound states we are looking for sit at the bottom of a continuum. Such states are difficult to study. For $k = 2$ such a bound state with the correct degeneracy has been found; see [98]. The analysis for higher k still remains to be done.

The analysis can be simplified by compactifying M-theory on T^2 and considering the Kaluza–Klein modes carrying (k_1, k_2) units of momenta along the two S^1's. Assuming that the two S^1's are orthogonal, and have radii R_1 and R_2 respectively, the mass of such a state, up to a proportionality factor, is

$$\sqrt{\left(\frac{k_1}{R_1}\right)^2 + \left(\frac{k_2}{R_2}\right)^2}. \tag{7.3}$$

For (k_1, k_2) relatively prime, such a state has strictly less energy than the sum of the masses of any other set of states with the same total charge; see [99]. Thus one should be able to find these states in type IIA string theory on S^1 (which, according to the conjecture, is equivalent to M-theory on T^2) without encountering the difficulties mentioned earlier. By following the same kind of argument, these states can be shown to be in one to one correspondence to a class of supersymmetric vacua in a $(1 + 1)$-dimensional supersymmetric gauge theory compactified on a circle.[26] All such states have been found with degeneracy as predicted by the M-theory–IIA duality.

There are also other consistency checks on the proposed M-theory–IIA duality. Consider M-theory on T^2. According to M-theory–type IIA duality, it is dual to IIA on S^1. But we know that IIA on S^1 is related by T-duality to IIB on S^1. Thus we have a duality between M-theory on T^2 and IIB on S^1. Now IIB on S^1 has an $SL(2, Z)$ strong-weak coupling duality inherited from 10-dimensional type IIB string theory. Thus one might ask, what does it correspond to in M-theory on T^2? One can find the answer by using the known map between the massless fields in the two theories, and the action of $SL(2, Z)$ in type IIB string theory. It turns out that this $SL(2, Z)$ symmetry in M-theory is simply the group of global diffeomorphisms of T^2; see [68, 101, 100]. Thus we again have an example of 'duality of dualities'. The $SL(2, Z)$ of IIB is a non-perturbative symmetry. But in M-theory on T^2 it is simply a consequence of the diffeomorphism invariance of the 11-dimensional theory.

Turning this analysis around we see that this also supports the ansatz that M-theory, defined as the strong coupling limit of IIA, is a fully Lorentz invariant theory in 11 dimensions. The argument goes as follows:

- First of all, from Lorentz invariance of type IIA string theory we know that we have Lorentz invariance in coordinates $x^0, \ldots x^9$ when all the coordinates $x^0, \ldots, x^9$ are non-compact.

- Then from the conjectured $SL(2, Z)$ duality symmetry of type IIB string theory we know that we have an exchange symmetry between the 9th and the 10th coordinate of M-theory when these coordinates are compact. In the limit when the radius of both the compact circles are taken to be large, this would mean that we should have Lorentz invariance in coordinates $x^0, \ldots x^{10}$.

---

[26]In fact, these states are related via an $R \to 1/R$ duality transformation to the ultra-short multiplets in type IIB on S^1 discussed in Section 4.2.

7.2 Compactification of M-theory

Given the existence of M-theory, we can now construct new vacua of the theory by compactifying M-theory on various manifolds. (For a review of compactification of 11-dimensional supergravity, see [198]. For example, we can consider M-theory compactified on $K3$, Calabi–Yau, and various orbifolds. These can all be regarded as appropriate strong coupling limits of type IIA compactification on the same manifold. But in general these cannot be regarded as perturbative string vacua. The essential feature of this strong coupling limit is the emergence of Lorentz invariance in one higher dimension. For example, M-theory on a Calabi–Yau manifold gives a 5-dimensional theory with $N = 1$ supersymmetry; see [197]. Such a theory cannot be constructed by conventional compactification of type IIA string theory at weak coupling.

Of course in many cases these non-perturbative vacua are related to perturbative string vacua by conjectured duality relations. These duality conjectures can be arrived at by using arguments very similar to those used in arriving at string duality conjectures. Some examples of such conjectured dualities are given below (see [102, 103, 104, 105]):

M-theory on

S^1/Z_2	$\leftrightarrow$	$(E_8 \times E_8)$ heterotic in D=10
$K3$	$\leftrightarrow$	Heterotic/Type I on T^3
T^5/Z_2	$\leftrightarrow$	IIB on $K3$
T^8/Z_2	$\leftrightarrow$	Type I/Heterotic on T^7
T^9/Z_2	$\leftrightarrow$	Type IIB on T^8/Z_2

In each case Z_2 acts by reversing the sign of all the coordinates of T^n; for odd n this is also accompanied by a reversal of sign of $C_{MNP}^{(S)}$. Each of these duality conjectures satisfies the consistency condition that the theory on the right-hand side, upon further compactification on a circle, is dual to type IIA string theory compactified on the manifold on the left hand side.

The duality between M-theory on S^1/Z_2 and the $E_8 \times E_8$ heterotic string theory is particularly amusing. Here the Z_2 transformation acts by reversing the orientation of S^1, together with a change of sign of the 3-form field $C_{MNP}^{(S)}$. S^1/Z_2 denotes a real line segment bounded by the two fixed points on S^1. It turns out that the two E_8 gauge multiplets arise from a 'twisted sector' of the theory and sit at the two ends of this line segment. The supergravity sector, on the other hand, sits in the bulk. Now in the conventional heterotic string compactification on Calabi–Yau spaces, all the observed gauge bosons and charged particles come from one E_8 and are neutral under the second E_8; see [187]. The second E_8, known as the hidden sector or the shadow world, is expected to be responsible for supersymmetry breaking. In the M-theory picture these two sectors are physically separated in space. In other words, the real world and the shadow world live at two ends of the line and interact only via the exchange of supergravity multiplets propagating in the bulk; see [196]! It has been suggested that this physical separation could be as large as a millimeter; see [106]! This limit comes from the analysis of the fifth force experiment, since if this dimension is too large, we should have an inverse cube law for the gravitational force instead of an inverse square law. No such direct limit comes from the inverse square law of gauge interaction, since gauge fields live on the boundary of S^1/Z_2 and hence do not get affected by the existence of this extra dimension.

Many of the listed duality conjectures involving M-theory (in fact all except the first one) can be derived by fiberwise duality transformation; see [105].[27] Let us for example consider the duality

$$(M\text{-theory on } T^5/Z_2) \leftrightarrow (\text{type IIB on } K3)$$

The Z_2 generator is $\mathcal{I}_5 \cdot \sigma$, where $\mathcal{I}_5$ changes the sign of all 5 coordinates $(x^6, \ldots, x^{10})$ on T^5, and σ denotes the transformation $C^{(S)}_{MNP} \rightarrow -C^{(S)}_{MNP}$. Let us express this as $(\mathcal{I}_1 \cdot \sigma) \cdot \mathcal{I}_4$ where $\mathcal{I}_1$ changes the sign of x^{10}, and $\mathcal{I}_4$ changes the sign of $(x^6, \ldots, x^9)$. We now use the result of fiberwise duality transformation

$$(A \text{ on } K_A \times \mathcal{D}/(h_A \cdot g)) \equiv (B \text{ on } K_B \times \mathcal{D}/(h_B \cdot g))$$

by choosing A on K_A to be M-theory on S^1, B on K_B to be type IIA string theory, h_A to be $\mathcal{I}_1 \cdot \sigma$, h_B to be $(-1)^{F_L}$ (this can be shown to be the image of h_A in the type IIA string theory), $\mathcal{D}$ to be T^4 spanned by $x^6, \ldots, x^9$, and g to be $\mathcal{I}_4$. Thus we get the duality

$$M\text{-theory on } (S^1 \times T^4/\mathcal{I}_1 \cdot \sigma \cdot \mathcal{I}_4) \leftrightarrow \text{IIA on } T^4/(-1)^{F_L} \cdot \mathcal{I}_4$$

The theory on the left hand side is M-theory on T^5/Z_2. On the other hand, if we take the theory on the right hand side and make an $R \rightarrow 1/R$ duality transformation on one of the circles, it converts

- type IIA theory to type IIB theory, and

- $(-1)^{F_L} \cdot \mathcal{I}_4$ into $\mathcal{I}_4$.

Thus the theory on the right is dual to type IIB on $T^4/\mathcal{I}_4$, which is a special case of type IIB on $K3$. Thus we get the duality:

---

[27]The duality between $E_8 \times E_8$ heterotic string theory and M-theory on S^1 can be 'derived' from other known duality conjectures by taking the infinite radius limit of a lower-dimensional duality relation; see [82].

$$(M\text{-theory on } T^5/Z_2) \leftrightarrow (\text{IIB on } K3)$$

This duality was first conjectured in [103, 104].

As in the case of type IIA string theory, we can get non-perturbative enhancement of gauge symmetries in M-theory when the compact manifold develops singularities; see [26, 193, 194]. M-theory contains classical membrane and 5-brane soliton solutions carrying electric and magnetic charges of $C^{(S)}_{MNP}$ respectively; see [8]. The extra massless states required for this symmetry enhancement come from membranes wrapped around the collapsed 2-cycles of the singular manifold.

8 F-Theory

Just as M-theory can be used to describe non-perturbative compactification of type IIA string theory, F-theory describes non-perturbative compactification of type IIB string theory; see [107, 108]. However, unlike M-theory, it does not correspond to a Lorentz invariant higher-dimensional theory, although, as we shall see, an auxiliary manifold with two extra dimensions plays a crucial role in the construction of F-theory compactification. This section will be based mainly on references [107, 108, 112].

8.1 Definition of F-theory

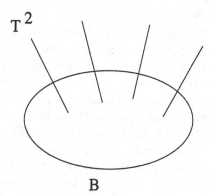

Figure 18: An elliptically fibered manifold $\mathcal{M}$ with base $\mathcal{B}$.

In conventional perturbative type IIB compactification one takes the dilaton–axion field λ (defined in Section 3.4) to be constant. F-theory is a novel way of compactifying type IIB theory that avoids this restriction and allows the string coupling to vary over the compact manifold. The starting

point in this construction is an elliptically fibered Calabi–Yau manifold defined as follows. Let $\mathcal{B}$ be a manifold (which we shall call the base manifold)
of real dimension d and $\mathcal{M}$ be another manifold of real dimension $d + 2$, obtained by erecting at every point of $\mathcal{B}$ a copy of a torus T^2, with the moduli
of T^2 varying over $\mathcal{B}$. This situation is illustrated in Figure 18. $\mathcal{M}$ is called
an elliptically fibered manifold. We shall choose the base $\mathcal{B}$ and the fibration
in such a way that $\mathcal{M}$ describes a Calabi–Yau manifold. Let $\vec{z}$ denote the
complex coordinate on $\mathcal{B}$, and $\tau(\vec{z})$ denote the complex structure of T^2 as a
function of $\vec{z}$. Then F-theory on $\mathcal{M}$ is defined to be type IIB string theory
compactified on $\mathcal{B}$ with

$$\lambda(\vec{z}) = \tau(\vec{z}). \tag{8.1}$$

Note that the size of the fiber torus does not appear in eq.(8.1). Thus F-
theory on a manifold $\mathcal{M}$ is insensitive to a subset of the moduli of $\mathcal{M}$ which
describe how the size of the fiber torus varies over the base.

In order that $\mathcal{M}$ be well defined, $\lambda = \tau(\vec{z})$ must come back to its original
value only up to an $SL(2, Z)$ transformation as we move along a closed cycle
on $\mathcal{B}$. Due to the presence of this non-perturbative duality transformation in
the monodromy group, conventional type IIB perturbation theory cannot be
used to describe this system. In particular, there are points in $\mathcal{B}$ where $\Im(\lambda)$
is of order unity, and hence type IIB theory is strongly coupled. For example,
suppose that $\Im(\lambda)$ is large in one region of the manifold, and also that as
we go around a closed curve starting from this region, there is an $SL(2, Z)$
transformation by the matrix $\begin{pmatrix} p & q \\ r & s \end{pmatrix}$ so that λ comes back to a value near
p/r. Then at some point on the curve $\Im(\lambda)$ must be finite, and hence the
string theory is strongly coupled.

From this note it would seem that although F-theory describes a novel
way of compactifying type IIB string theory, we cannot extract any information about such a theory, since string perturbation theory cannot be used
to analyse this system. However, it turns out that we can learn quite a lot
about these theories by using various known duality relations. For this con-
sider F-theory on $\mathcal{M} \times S^1$, i.e. type IIB theory on $\mathcal{B} \times S^1$ with $\lambda(\vec{z}) = \tau(\vec{z})$.
Now, as we have discussed in the last section, type IIB theory on S^1 is dual
to M-theory on T^2, with λ being the modular parameter of T^2. Thus we can
now apply fiberwise duality transformation illustrated in Figure 19 to relate
the F-theory compactification on $\mathcal{M} \times S^1$ to an M-theory compactification.
Under this duality the modulus of T^2 on the right hand side must be set equal
to $\lambda(\vec{z})$ of the theory on the left, which, according to eq.(8.1), is just $\tau(\vec{z})$.
This means that the manifold on the right hand side is the original manifold
$\mathcal{M}$ that we started with. This gives the duality

$$F\text{-theory on } \mathcal{M} \times S^1 \leftrightarrow M\text{-theory on } \mathcal{M}$$

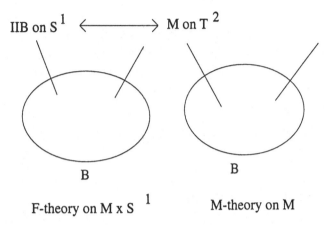

F-theory on M x S^1 M-theory on M

Figure 19: Fiberwise application of duality to relate F-theory on $\mathcal{M} \times S^1$ to M-theory on $\mathcal{M}$.

Thus many of the properties of F-theory on $\mathcal{M}$ (e.g. the number of super-symmetries, spectrum of massless states, etc.) can be studied from that of M-theory on $\mathcal{M}$ and then taking an appropriate limit in which the size of S^1 on the F-theory side goes to ∞. Using this one can show for example that if $\mathcal{M}$ is $K3$, then we preserve half of the space-time supersymmetries of the type IIB theory, whereas if $\mathcal{M}$ is a Calabi–Yau manifold then we preserve $1/4$ of the supersymmetry.

In order to gain more insight into various F-theory compactifications, we need to develop a convenient formalism for describing elliptically fibered manifolds. Our starting point will be the equation describing a torus:

$$y^2 = x^3 + fx + g \,, \tag{8.2}$$

where x and y are complex variables, and f and g are complex parameters. For every pair of constants f and g the above equation describes a 1-complex-dimensional surface, which can be shown to be a torus. The modular parameter τ of the torus is given by

$$j(\tau) = \frac{4 \cdot (24f)^3}{27g^2 + 4f^3} \,, \tag{8.3}$$

where $j(\tau)$ is a known function of τ. In fact it is the unique modular invariant function with a single pole at $\tau = i\infty$ and zeroes at $\tau = e^{i\pi/3}$. The overall normalization of j is chosen in such a way that $j(i) = (24)^3$.

In order to describe an elliptically fibered manifold on some base $\mathcal{B}$ we simply need to make f and g depend on the coordinates of $\mathcal{B}$. We shall illustrate this with the help of an elliptically fibered $K3$. Here we choose the base $\mathcal{B}$ to be CP^1. Thus the elliptically fibered manifold $\mathcal{M}$ is described by the equation

$$y^2 = x^3 + f(z)x + g(z),\qquad(8.4)$$

where z is the coordinate on CP^1. It can be shown that $\mathcal{M}$ describes a $K3$ manifold provided $f(z)$ is a polynomial of degree 8, and $g(z)$ is a polynomial of degree 12 in z. The modular parameter of the torus varies over the base CP^1 according to the relation

$$j(\tau(z)) = \frac{4 \cdot (24f(z))^3}{27g(z)^2 + 4f(z)^3}.\qquad(8.5)$$

By definition F-theory on this elliptically fibered $K3$ is type IIB string theory compactified on CP^1 with

$$\lambda = \tau(z).\qquad(8.6)$$

In order to specify the background completely we also need to specify the metric on the base CP^1. This can be calculated from the low energy effective field theory when the size of $\mathcal{B}$ is sufficiently large, and the answer is (see [109])

$$ds^2 = F(\tau(z), \bar\tau(\bar z))\, dzd\bar z \Big(\prod_{i=1}^{24}(z - z_i)^{-1/12}(\bar z - \bar z_i)^{-1/12}\Big),\qquad(8.7)$$

where z_i are the zeroes of $\Delta \equiv 4f^3 + 27g^2$, and

$$F(\tau, \bar\tau) = (\tau_2)\eta(\tau)^2\bar\eta(\bar\tau)^2.\qquad(8.8)$$

$\eta(\tau)$ denotes the Dedekind function and τ_2 is the imaginary part of τ.

Similarly we can describe more complicated F-theory compactifications by choosing more complicated bases $\mathcal{B}$. For example, consider the base $CP^1 \times CP^1$ labelled by a pair of complex coordinates (z, w); see [108]. We can get an elliptically fibered manifold on this base by the equation

$$y^2 = x^3 + f(z, w)x + g(z, w).\qquad(8.9)$$

In order that this manifold be Calabi–Yau, we need $f(z, w)$ to be a polynomial of degree $(8,8)$ in (z, w), and $g(z, w)$ to be a polynomial of degree $(12,12)$ in (z, w). F-theory on such a manifold is by definition a configuration of $\lambda(z, w)$ described by the equation

$$j(\lambda(z, w)) = \frac{4.(24f)^3}{27g^2 + 4f^3}.\qquad(8.10)$$

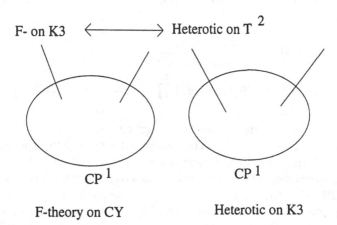

F-theory on CY   Heterotic on K3

Figure 20: Fiberwise application of duality to relate F-theory on Calabi–Yau to heterotic string theory on $K3$.

8.2 Dualities involving F-theory

There are many conjectured dualities involving F-theory compactifications. Some examples are given below (see [108]):

$$F\text{-theory on } K3 \quad \leftrightarrow \quad \text{Heterotic on } T^2$$
$$F\text{-theory on CY 3-fold} \quad \leftrightarrow \quad \text{Heterotic on } K3$$

All the duality conjectures involving F-theory have the following property: *If F-theory on $\mathcal{M}$ is dual to some string theory S compactified on a manifold $\mathcal{K}$, then M-theory on $\mathcal{M}$ must be dual to the same string theory compactified on $\mathcal{K} \times S^1$, and type IIA on $\mathcal{M}$ must be dual to the same string theory compactified on $\mathcal{K} \times T^2$. These results follow from the duality between type IIB on S^1 and M-theory on T^2, and that between M-theory on S^1 and type IIA string theory.*

All conjectured dualities involving F-theory on Calabi–Yau 3-folds and more complicated manifolds can be derived from the fiberwise duality transformation; see [108, 110, 111]. For example, for a Calabi–Yau 3-fold, this is done by representing the Calabi–Yau 3-fold as $K3$ fibered over CP^1, and replacing F-theory on $K3$ by heterotic on T^2 fiberwise. This has been illustrated in Figure 20. The theory on the right hand side of this figure represents heterotic string theory on $K3$ with apropriate gauge field background.

We shall now show how to derive the parent duality

$$(F\text{-theory on } K3) \leftrightarrow (\text{Heterotic on } T^2)$$

from other known duality conjectures; see [112]. Recall that for this background

$$j(\lambda(z)) = \frac{4 \cdot (24f(z))^3}{27g(z)^2 + 4f(z)^3}$$

$$ds^2 = F(\lambda, \bar{\lambda}) \, dz d\bar{z} \Big(\prod_{i=1}^{24} (z - z_i)^{-1/12} (\bar{z} - \bar{z}_i)^{-1/12} \Big), \quad (8.11)$$

where $f(z)$ and $g(z)$ are polynomials of degree 8 and 12 respectively, and z_i are the zeroes of $\Delta \equiv 4f^3 + 27g^2$. The strategy is to try to go to a special point in the moduli space where λ, instead of varying over CP^1, becomes a constant. At this special point the theory reduces to a conventional compactification of type IIB string theory. Examining eq.(8.11) we see that this requires f^3/g^2 to be a constant. If we now recall that f is a polynomial of degree 8 in z and g is a polynomial of degree 12 in z, we see that for f^3/g^2 to be constant, we need

$$f = \alpha\phi^2, \qquad g = \phi^3, \quad (8.12)$$

where ϕ is a polynomial of degree 4 in z, and α is a constant. Using the freedom of an overall rescaling of ϕ which does not change the value of λ, we can take

$$\phi = \prod_{m=1}^{4} (z - z_m). \quad (8.13)$$

This gives

$$\Delta \equiv 4f^3 + 27g^2 = (4\alpha^3 + 27) \prod_{m=1}^{4} (z - z_m)^6, \quad (8.14)$$

$$j(\lambda) = 4 \cdot (24\alpha)^3 / (4\alpha^3 + 27), \quad (8.15)$$

and

$$ds^2 = F(\lambda, \bar{\lambda}) \, dz d\bar{z} \prod_{m=1}^{4} (z - z_m)^{-1/2} (\bar{z} - \bar{z}_m)^{-1/2}. \quad (8.16)$$

Since λ is now a constant, the metric can be simplified by going to a new coordinate system w defined by

$$dw = \prod_{m=1}^{4} (z - z_m)^{-1/2} dz. \quad (8.17)$$

Then

$$ds^2 = C \, dw \, d\bar{w}, \quad (8.18)$$

where $C = F(\lambda, \bar{\lambda})$ is a constant. Thus the metric is flat! But this poses a puzzle, since we know that the base is CP^1, and that we cannot put a flat metric on CP^1 since it has non-zero Euler number. The resolution to this

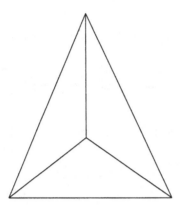

Figure 21: CP^1 with a flat metric except at 4 points.

puzzle comes from noting that if w_m are the images of $z = z_m$ in the w plane, then, near $z = z_m$,

$$(w - w_m) \sim (z - z_m)^{1/2} . \qquad (8.19)$$

This gives rise to a deficit angle of π at each of these 4 points. Thus the base has flat metric everywhere except for conical singularities at these 4 points. This represents a regular tetrahedron as shown in Figure 21. This can be also be identified as the orbifold $T^2/\mathcal{I}_2$. Here w is the complex coordinate on T^2, and $\mathcal{I}_2$ denotes the transformation $w \rightarrow -w$. z on the other hand, is the coordinate on $T^2/\mathcal{I}_2$. The points z_m are the fixed points of $\mathcal{I}_2$.

This analysis would suggest that at this special point in the moduli space F-theory on $K3$ has reduced to type IIB on $T^2/\mathcal{I}_2$ with constant λ given in eq.(8.15). However, there is a further subtlety. Recall that going once around a fixed point in the z-plane corresponds to going from w to $-w$ as illustrated in Figure 22. The relevant question to ask would be: is there any twist by some internal symmetry transformation g of type IIB theory as we go around a fixed point of z? If there is such a twist, then the Z_2 orbifold group will be generated by $w \rightarrow -w$, accompanied by the transformation g. In order to find g we need to study the effect of going around the point $z = z_m$ once. To do this, recall the equation describing this particular $K3$:

$$y^2 = x^3 + \alpha x \prod_{m=1}^{4} (z - z_m)^2 + \prod_{m=1}^{4} (z - z_m)^3 . \qquad (8.20)$$

Let us take z around z_1 once through the parametrization

$$\prod_{m=1}^{4} (z - z_m) = e^{2\pi i t} \prod_{m=1}^{4} (z_{\text{initial}} - z_m) , \qquad (8.21)$$

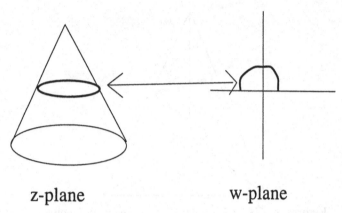

z-plane   w-plane

Figure 22: A closed curve around $z = z_m$ in the z-plane and its image in the w-plane.

and continuously changing t from 0 to 1. Also, during this change, focus on a point on the fiber torus and follow its trajectory. This can be achieved by choosing:

$$x = x_{\text{initial}}e^{2\pi it}, \qquad y = y_{\text{initial}}e^{3\pi it}. \qquad (8.22)$$

This point lies on the surface (8.20) for all values of t if the initial point $(x_{\text{initial}}, y_{\text{initial}}, z_{\text{initial}})$ lies on this surface. At the end of this process when $t = 1$, we do not return to the original point but to $(x_{\text{initial}}, -y_{\text{initial}}, z_{\text{initial}})$. To see what this transformation correponds to if we choose a more conventional coordinate system to the fiber torus, let us denote by u the conventional flat coordinate on the fiber torus, so that the torus is described through the identification $u \equiv u + 1 \equiv u + \tau$. It can be shown that u is related to (x, y) through the relation $du = Kdx/y$, where K is a constant. Thus $(x, y) \to (x, -y)$ corresponds to $u \to -u$, i.e. reversing the orientation of both the circles on the fiber torus. This is nothing but an $SL(2, Z)$ transformation with matrix $\begin{pmatrix} -1 & \\ & -1 \end{pmatrix}$. Thus we see that as we move around any of the points $z = z_m$ on a the base once, we make an $SL(2, Z)$ transformation with this matrix. But now recall that the $SL(2, Z)$ transformation with the matrix $\begin{pmatrix} -1 & \\ & -1 \end{pmatrix}$ can be identified with the transformation $(-1)^{F_L} \cdot \Omega$ as discussed in Section 3.4. Thus as we go around the point $z = z_m$ we make a global symmetry transformation by $(-1)^{F_L} \cdot \Omega$. This shows that the F-theory on $K3$ at this particular point in the moduli space can be identified to

$$\text{Type IIB on } T^2/\mathcal{I}_2 \cdot (-1)^{F_L} \cdot \Omega$$

We have encountered this theory earlier in Section 5.4. As discussed there, by making an $R \to 1/R$ duality transformation on both coordinates of T^2, we can relate this theory type IIB on T^2/Ω. But since type IIB modded out by Ω is type I theory, we see that this model can be identified to type I on T^2. Finally, using type I–$SO(32)$ heterotic duality in 10 dimensions, we can relate this model to heterotic string theory on T^2. This finally establishes the equivalence between

$$(F\text{-theory on } K3) \text{ and } (\text{Heterotic on } T^2)$$

Similar strategy has been used to establish many other dualities involving F-theory compactification; see [188].

As in the case of type IIA string theory and M-theory, F-theory on a singular manifold can also develop enhanced non-abelian gauge symmetry; see [108, 189, 190]. In this case the extra massless states required for the symmetry enhancement come from open string states lying on the base; see [191, 192].

The F-theory type compactification can be generalized in the following way. One starts with type II string theory compactified on T^n with a duality group G, and compactifies it further on a base B with the monodromy on the base being a subgroup of G. Such compactifications have been discussed in [195].

9 Microscopic Derivation of Black Hole Entropy

One of the major stumbling blocks to our understanding of nature has been the apparent incompatibility between quantum mechanics and general relativity. Three of the 4 known forces of nature – strong, weak and electromagnetic interactions – are very well explained by quantum field fheory. The current model of elementary particle physics – the standard model – has explained most of the observed phenomena involving these three interactions. However, this is not so for gravity.

There are many problems in quantizing gravity using quantum field theory. First of all, gravity is not perturbatively renormalizable, i.e. the usual rules in quantum field theory for extracting finite answers for all physical quantities from infinite answers at the intermediate stages of calculation are not valid for gravity. As mentioned in Section 1, this problem is automatically solved in string theory. However, during the last 25 years a more serious objection to the compatibility of general relativity and quantum mechanics has been raised; see [127, 128]. This is the problem to which we now turn. The discussion in this section will follow closely references [133, 136, 137, 138, 121, 142].

9.1 Problem with black holes in quantum mechanics

The starting point is the existence of a class of classical solutions in general relativity (possibly coupled to other fields) known as black holes. Classically, black holes are completely black. In other words, objects can fall into a black hole, but nothing can ever come out of one. (In more technical terms, one says that black holes have event horizons.) Black holes in general relativity also satisfy a classical no-hair theorem which states that a black hole solution is completely characterized by its mass, angular momentum, and gauge charges. Thus all other information (quantum numbers) carried by an object falling into the black hole is lost forever.

However, due to the work of Bekenstein, Hawking and others during the last 25 years it has become clear that once quantum effects are taken into account, this picture of black holes undergoes dramatic modification. In particular two things happen.

- Black holes emit thermal radiation at temperature (see [127])

$$T = \frac{\kappa}{2\pi}, \tag{9.1}$$

  where κ is the surface gravity of the black hole (the acceleration due to gravity felt by a static observer at the event horizon).

- Black holes carry entropy (see [128])

$$S = \frac{1}{4G_N} A. \tag{9.2}$$

  where A is the area of the event horizon and G_N is Newton's constant.

Black holes satisfy the usual laws of thermodynamics in terms of these variables. In particular the first law of black hole thermodynamics states that

$$dM = TdS. \tag{9.3}$$

This relates the change in the mass M of a black hole to its change in entropy S and the Hawking temperature T. The second law of black hole thermodynamics states that

$$dS \geq 0, \tag{9.4}$$

i.e. the sum of the entropy of the black hole and the usual thermodynamic entropy of its surroundings increases with time. In the presence of $U(1)$ gauge charges we can also define a chemical potential associated with each gauge charge. In the presence of these charges the laws of thermodynamics are modified in the usual manner.

It is this thermodynamic description of the black hole that causes an apparent conflict with quantum mechanics. This is best illustrated by considering the following thought experiment.

An Introduction to Non-Perturbative String Theory   381

- Consider a black hole formed out of the collapse of a pure state. We can imagine a spherical shell of matter described by an s-wave state collapsing to form a black hole.

- It will then emit thermal radiation and at the end evaporate completely. If the outgoing radiation is really thermal, then the final state is a mixed state.

Thus the net result of this two-step process is the evolution of a pure state to a mixed state, in conflict with the rules of quantum mechanics.

At this stage, it is useful to compare this with the phenomenon of thermal radiation from a star (or any other hot object). If an object in a pure quantum state is thrown into a star, it comes out as thermal radiation, so why doesn't this contradict quantum mechanics? The answer to this is that the thermal description of the radiation from a star is a result of averaging over the microstates of the star. We could, in principle, start from a pure quantum state of the star, and give a microscopic description of the radiation coming out of the star. In this description a pure state evolves to a pure state. In other words, although on average the star emits thermal radiation, the radiation coming out of the star has subtle dependence on what goes into the star, and a detailed analysis of this radiation can be used to completely reconstruct the initial state.

Let us now come back to black holes. Why can't the same reasoning be used for black holes to resolve the apparent conflict with quantum mechanics? The reason is that there is no similar microscopic description of the radiation from a black hole in conventional semiclassical gravity: the approximation that is used in demonstrating that black holes emit thermal radiation. This is related to the problem of there being no understanding of the black hole entropy in terms of counting microstates. In other words, the entropy formula (9.2) is derived purely in analogy with thermodynamics, but not as the logarithm of the density of states as in statistical mechanics. Thus there is no possibility of giving a quantum mechanical description of Hawking radiation by studying the evolution of individual microstates of the black hole.

Since string theory claims to be a consistent quantum theory of gravity, it should be able to explain black hole entropy and Hawking radiation in terms of conventional quantum mechanics. We shall now discuss some of the attempts to explain black hole thermodynamics in string theory.

9.2   Black holes as elementary string states

As we have discussed earlier (see, for example, eq.(4.9)) the spectrum of an elementary string contains an infinite tower of states. Since the Schwarzschild radius (the radius of the event horizon) of a black hole is proportional to its mass, for sufficiently large mass the Schwarzschild radius of a black hole will be larger than the string scale. In that case an elementary string state of the same mass will lie inside its Schwarzschild radius, and become a black hole; see [129]. This opens up the possibility of a statistical description of black hole entropy as follows; see [130]. For a given mass M, the microscopic entropy S_{micro} can be defined to be the logarithm of the number of elementary string states at that mass level. We can then compare this with the Bekenstein–Hawking entropy S_{BH} of the black hole with the same mass. If the two expressions agree, then we can give a statistical interpretation of black hole entropy by attributing it to the degeneracy of string states. Unfortunately this attempt fails, since one finds that

$$S_{\text{micro}} \propto M, \qquad S_{\text{BH}} \propto M^2 . \qquad (9.5)$$

The above formulae hold for chargeless black holes, but a similar discrepancy is present even for black holes carrying electric charge. This seems to be a severe blow to the attempt at giving a microscopic description of Bekenstein–Hawking entropy in string theory. But one should keep in mind that in the region of parameter space where the elementary string state becomes a black hole, there are strong coupling effects, and hence the mass of an elementary string state can get renormalized; see [130]. Thus the parameter M that appears in the computation of S_{BH} and the M that appears in the computation of S_{micro} may not be the same, but may be related by a renormalization factor.

Various attempts have been made to get out of this difficulty for Schwarzschild black holes (see [130, 131, 132, 186]), but we shall not discuss them here. Instead, we shall try to get out of this impasse by working with states for which there is no mass renormalization, namely the BPS states; see [133]. Since the degeneracy of a BPS state does not change as we change the coupling (at least for theories with ≥ 16 supersymmetry charges) we can proceed as follows. First we compute the degeneracy of BPS states at weak coupling where the microscopic description of the state is reliable. Then we increase the string coupling constant to a sufficiently large value where the state becomes a black hole. For this black hole, S_{micro} should be given by the logarithm of the degeneracy computed at weak coupling. This leads to the following strategy for comparing S_{micro} and S_{BH}:

1. Identify the BPS states among elementary string states and calculate their degeneracy. This gives S_{micro}.

2. Identify BPS black holes (also known as extremal black holes) with the same quantum numbers and find S_{BH} by computing the area of the event horizon.

3. Compare the two expressions.

We shall illustrate this with the help of a specific model − heterotic string theory compactified on T^6; see [133]. As discussed above eq.(4.9), this theory has two classes of $U(1)$ charges, $\vec{k}_L$ and $\vec{k}_R$. From eqs.(4.9), (4.10) we see that among the elementary string states, BPS states are those satisfying[28]

$$m^2 = (2\vec{k}_R^2/\langle\lambda_2\rangle), \qquad N_L = \frac{1}{4}\langle\lambda_2\rangle\Big(m^2 - \frac{2\vec{k}_L^2}{\langle\lambda_2\rangle}\Big) + 1. \qquad (9.6)$$

The degeneracy of these states can be calculated from eq.(4.12). For large N_L, this gives

$$d(m, k_L, \langle\lambda_2\rangle) \sim \exp(4\pi\sqrt{N_L}) \sim \exp\Big(2\pi\sqrt{\langle\lambda_2\rangle}\sqrt{m^2 - \frac{2\vec{k}_L^2}{\langle\lambda_2\rangle}}\Big). \qquad (9.7)$$

This gives

$$S_{\text{micro}} \simeq 2\pi\sqrt{\langle\lambda_2\rangle}\sqrt{m^2 - \frac{2\vec{k}_L^2}{\langle\lambda_2\rangle}}. \qquad (9.8)$$

Our next task is to calculate the Bekenstein–Hawking entropy S_{BH} for black holes carrying the same quantum numbers. It turns out that the black holes carrying these quantum numbers have vanishing area of the event horizon, and hence $S_{\text{BH}} = 0$. Thus again we seem to have run into a contradiction!

However, again there is a subtlety. In constructing the black hole solution, one uses the low energy effective field theory which is valid only when the curvature is much smaller than the string scale. But since this black hole has vanishing area of the event horizon, it follows that the solution actually has a curvature singularity at the horizon.[29] Thus we would expect that the solution near the horizon will be modified by the higher derivative terms in the string effective action, and hence might give a different answer for the area of the event horizon.

In order to estimate what the modified area will be, one needs to understand what kind of corrections we must include in the effective action. Typically in string theory there are two types of correction − the ones due to string world-sheet effects (the higher derivative terms), and the ones due to

---

[28]When we regard elementary strings as classical black hole solutions, the field λ_2 varies as a function of the radial distance from the origin. Due to this fact we are specifically using the notation $\langle\lambda_2\rangle$ to denote the asymptotic value of λ_2, i.e. the expectation value of λ_2 in the vacuum.

[29]One might wonder whether such a solution can be called a black hole at all, but the reason that they are called black holes is that they can be obtained as a limit of black hole solutions with non-singular horizons with finite area. As the mass of the black hole approaches the Bogomol'nyi bound (9.6), we get the singular black hole.

the string loop effects. In the present case, when one examines the classical black hole solution one finds that the field λ_2 approaches ∞ as we approach the event horizon. Thus the string coupling $\lambda_2^{-1/2}$ vanishes in this region and we do not expect the string loop corrections to be significant. This leaves us with the string world-sheet corrections. Although we cannot explicitly calculate the effect of these corrections, we can use a scaling argument to determine the form of these corrections up to an overall numerical factor. The argument goes as follows. If we use the string metric (as opposed to the canonical metric) to describe the black hole solution, then one finds that with a suitable choice of coordinate system the solution near the origin becomes completely independent of all parameters m, $\vec{k}_L$ and $\langle \lambda_2 \rangle$, except for an additive factor of

$$- \ln \left(\sqrt{\langle \lambda_2 \rangle} \sqrt{m^2 - \frac{2\vec{k}_L^2}{\langle \lambda_2 \rangle}} \right), \tag{9.9}$$

in the expression for the dilaton. This additive constant does not affect the string world-sheet Lagrangian. Thus, whatever be the effect of the corrections due to the string world-sheet effects, these corrections are universal, and do not depend on any parameters. Thus after taking into account the string world-sheet corrections, the area of the event horizon, as measured in the string metric, will be a universal numerical constant, independent of all external parameters. However, since the Bekenstein–Hawking entropy is to be identified with the area of the event horizon measured in the *canonical metric*, which differs from the string metric by a factor of $e^{-\Phi}$, and since Φ has an additive factor (9.9) that depends on the parameters, the modified Bekenstein–Hawking entropy will be given by,

$$S_{\text{BH}} = C \sqrt{\langle \lambda_2 \rangle} \sqrt{m^2 - \frac{2\vec{k}_L^2}{\langle \lambda_2 \rangle}}, \tag{9.10}$$

where C is an unknown numerical constant. This is in complete agreement with the answer for S_{micro} given in (9.8) if we choose

$$C = 2\pi. \tag{9.11}$$

Note, in particular, that S_{micro} and S_{BH} have the same functional dependence on m, $\vec{k}_L$ and $\langle \lambda_2 \rangle$. Considering the fact that these are all dimensionless parameters (we are working in units $\hbar = 1, c = 1, \alpha' = 1$) this agreement is impressive. This calculation can also be extended to other toroidal compactification of heterotic string theory (see [134]), and to non-toroidal compactification of heterotic and type II string theories; see [135].

9.3   Black holes and D-branes

Although the result described in the previous subsection is encouraging, we would like to do better and compare S_{BH} and S_{micro} without encountering any undetermined numerical factor. The strategy is to try to identify black hole solutions which are

- BPS states, and

- have non-vanishing area of the event horizon even without stringy corrections.

It turns out that there are indeed such black holes present in the theory, but they do not carry the same quantum numbers as elementary string states. Instead they carry the same quantum numbers as a configuration of D-branes. Thus in order to calculate the microscopic entropy we need to calculate the degeneracy of this D-brane configuration. As has already been discussed earlier, the dynamics of collective coordinates of D-branes is given by the massless open string states propagating on the D-branes. Thus we can explicitly determine the Hamiltonian describing this dynamics and calculate the degeneracy of states to calculate S_{micro}. This is precisely what is done.

Thus our strategy is as follows:

1. Identify a BPS black hole with non-vanishing area of the event horizon and calculate S_{BH} from this area.

2. Identify the D-brane configuration carrying the same quantum numbers as this black hole and calculate S_{micro} by computing the degeneracy of these states.

3. Compare the two answers.

The analysis is simplest in 5 dimensions, so we shall concentrate on this case; see [136, 18]. We focus on type IIB string theory on T^5. The D-brane configuration that we consider has

- Q_5 $(D-5)$-branes wrapped on T^5,

- Q_1 $(D-1)$-branes wrapped on one of the circles S^1 of T^5, and

- $-n$ units of momentum along S^1. If R denotes the radius of S^1, this corresponds to a momentum of $-n/R$.

The counting of states for this D-brane system can be done as follows. The world-volume theory of a system of Q_5 parallel $(D-5)$-branes is a supersymmetric $U(Q_5)$ gauge theory in $(5+1)$ dimensions, obtained by the dimensional reduction of $N = 1$ supersymmetric $U(Q_5)$ gauge theory in $(9+1)$ dimensions. It can be shown that a single $(D-1)$-brane inside Q_5 coincident $(D-5)$-branes can be identified to a single instanton in this $U(Q_5)$ gauge theory; see [154].[30]. Thus a system of Q_1 parallel $(D-1)$-branes inside Q_5 coincident $(D-5)$-branes can be described as a system of Q_1 instantons in the $U(Q_5)$ gauge theory. The moduli of this Q_1 instanton solution act as collective coordinates of this system. These moduli span a $4Q_1Q_5$-dimensional hyper-Kähler manifold. As a result the low energy dynamics of a system of Q_1 $(D-1)$-branes inside Q_5 $(D-5)$-branes is described by a $(1+1)$-dimensional supersymmetric σ-model with this instanton moduli space as the target space. Since this space is hyper-Kähler, the corresponding supersymmetric σ-model is conformally invariant; see [155]. Furthermore, since the moduli space has dimension $4Q_1Q_5$, the central charge is given by

$$c = \frac{3}{2} \cdot 4Q_1Q_5 = 6Q_1Q_5, \qquad (9.12)$$

taking into account the contribution of 1 from each scalar and $1/2$ from each fermion. The BPS states in this theory correspond to states with $L_0 = 0$. $L_0 - \bar{L}_0$ represents the total number of momentum units carried by the system along S^1; for this system this is equal to $-n$. For large n the degeneracy of such states can be computed; see [199]. The answer is

$$d(Q_1, Q_5, n) \sim \exp\left(2\pi\sqrt{\frac{cn}{6}}\right) = \exp(2\pi\sqrt{Q_1Q_5n}). \qquad (9.13)$$

Since this argument is somewhat abstract, we shall now give a simplified description of the counting of states of this system; see [137, 70, 138, 156]. First consider the case when there is one $(D-5)$-brane and Q_1 $(D-1)$-branes inside the $(D-5)$-brane (so that the $(D-1)$-branes are free to move inside the $D5$-brane but not free to leave the $(D-5)$-brane). Now, since Q_1 $(D-1)$-branes, each wrapped once around S^1, has the same charge as a single $(D-1)$-brane wrapped Q_1 times, we must also include this configuration in our counting of states. It turns out that the contribution to the total degeneracy is dominated by the latter configuration, so we can restrict our attention to this configuration. Now consider the case when there are Q_5 $(D-5)$-branes instead of just one. Again the dominant contribution comes from the configuration where instead of Q_5 $(D-5)$-branes each wrapped once around $S^1 \times T^4$, we have a single $(D-5)$-brane wrapped Q_5 times. Thus we have a configuration of a single $D1$-brane of length $2\pi R Q_1$ and a single $(D-5)$-brane of length $2\pi R Q_5$ along the direction of S^1. We shall call these branes long $D-1$ and long $(D-5)$-branes respectively. Now take Q_1 and Q_5 to be relatively prime. In that case, when we go around the long $(D-1)$-brane once by travelling Q_1 times around the circle S^1, we do not come back to the same point on the long $(D-5)$-brane. On the other hand, if we go around

---

[30]An instanton is a classical solution in Yang–Mills theory in 4 Euclidean dimensions. Thus in the $(5+1)$-dimensional gauge theory it represents a solution that is independent of time and one spatial direction, i.e. a static string

the long $(D-5)$-brane once by travelling Q_5 times around S^1, we do not come back to the same point on the long $(D-1)$-brane. In fact, we need to go around the long $(D-1)$-brane Q_5 times in order to come back to the same point on the long $(D-5)$-brane and the long $(D-1)$-brane. This amounts to going around S^1 $Q_1 Q_5$ times. Thus to an open string stretched between the $(D-1)$-brane and the $(D-5)$-brane the configuration will appear to be that of a single D-string of length $2\pi R Q_1 Q_5$, which is free to move in the 4 transverse directions inside a $(D-5)$-brane; see [138]. This gives 4 bosonic collective coordinates X^i. Due to supersymmetry, this system will also have 4 Majorana fermions λ^i moving on the D-string world-sheet.

For each of these 4 bosonic and fermionic coordinates there are left-moving modes as well as right-moving modes on the $(D-1)$-brane. A quantum of the mth left- (right-) moving mode carries $-m$ (m) units of momentum along S^1, with each unit of momentum now being equal to $1/(RQ_1Q_5)$. We need a state with a total of $-nQ_1Q_5$ units of momentum. It turns out that in order to saturate the BPS mass formula, we need to concentrate on states containing only quanta of left-moving modes and no quanta of right-moving modes. If there are N_m^i quanta of the mth left-moving mode of X^i, and n_m^i quanta of the mth left-moving mode of λ^i, then in order to get $-Q_1Q_5 n$ units of momentum along S^1 we need

$$Q_1 Q_5 n = \sum_{i=1}^{4} \sum_{m=1}^{\infty} m(N_m^i + n_m^i). \tag{9.14}$$

The degeneracy $d(Q_1, Q_5, n)$ is the number of ways we can choose integers N_m^i and n_m^i satisfying the above relation. This can be computed using standard procedure, and the answer is

$$d(Q_1, Q_5, n) \sim \exp(2\pi\sqrt{Q_1 Q_5 n}), \tag{9.15}$$

which is the same as (9.13). This gives,

$$S_{\text{micro}} = \ln d \simeq 2\pi\sqrt{Q_1 Q_5 n}. \tag{9.16}$$

We now need to compare this result with the Bekenstein–Hawking entropy of the black hole solution of the low energy effective field theory carrying the same set of charges and the same mass. In the normalization convention of eqs.(3.1), (3.2) and (3.33) the 5-dimensional Newton's constant is given by,

$$G_N^5 = \frac{\pi}{4}. \tag{9.17}$$

Instead of writing down the full black hole solution, we shall only write down the canonical metric for this black hole in the 5 non-compact directions, since this is what is required to compute the area of the event horizon, and hence $S_{\rm BH}$. The metric is (see [18, 19])[31]

$$ds^2 = -\lambda^{-2/3}dt^2 + \lambda^{1/3}[dr^2 + r^2 d\Omega_3^2] \,, \tag{9.18}$$

where,

$$\lambda = (1 + r_1^2/r^2)(1 + r_5^2/r^2)(1 + r_n^2/r^2) \,, \tag{9.19}$$

$$r_1^2 = (RV)^{2/3}g^{-1/2}Q_1/V, \qquad r_5^2 = (RV)^{2/3}g^{1/2}Q_5, \qquad r_n^2 = (RV)^{2/3}n/R^2V \,. \tag{9.20}$$

Here $d\Omega_3$ denotes line element on a unit 3-sphere, $(2\pi)^4 V$ is the volume of T^4 and R is the radius of S^1, both measured in the 10-dimensional canonical metric, and $g(\equiv e^{\langle \Phi^{(10)} \rangle /2})$ is the string coupling constant *in 10 dimensions*. Here T^4 denotes the subspace of the full T^5 that does not include the special S^1 on which the $(D-1)$-branes are wrapped. The event horizon is located at $r = 0$, and the area A of the event horizon can be easily computed from eqs.(9.18), (9.19) to be $2\pi^2 r_1 r_5 r_n$. This gives,

$$S_{\rm BH} = \frac{A}{4G_N^5} = 2\pi\sqrt{Q_1 Q_5 n} \,. \tag{9.21}$$

This is in exact agreement with $S_{\rm micro}$ computed in eq.(9.16).

Similar agreement between the Bekenstein–Hawking entropy and the microscopic entropy has been demonstrated for black holes carrying angular momentum (see [144]), and also for black holes in 4 dimensions; see [145]. The analysis can also be easily extended to compactification of type II string theory on $K3 \times S^1$ with 16 supersymmetry charges. A more non-trivial case is the extension to black holes in theories with 8 supercharges. This has been done in many cases, and in every case that has been studied, the Bekenstein–Hawking entropy of the BPS saturated black hole agrees with the microscopic entropy computed from analyzing the dynamics of the brane configuration; see [143]. An alternate approach to calculating $S_{\rm micro}$ has been advocated in [149] that also gives an answer in agreement with $S_{\rm BH}$.

Given the success of the D-brane dynamics in giving a microscopic description of the entropy of an extremal black hole, one might wonder if a similar analysis can be carried out for black holes which are not extremal, but are nearly extremal; see [137, 139]. For the D-brane configuration such states are obtained by relaxing the requirement that there is no quanta of right-moving modes. However, we restrict the number of such excitations so that the interaction between the left- and the right-moving modes can be neglected; this is known as the dilute gas approximation. Let the left-moving modes carry a

---

[31]Our 10-dimensional metric, in which R and V are measured, differs from that of [18, 19] by a factor of $g^{-1/2}$, and our 5-dimensional metric differs from that of [18, 19] by a factor of $(RV)^{2/3}g^{-1/2}$.

total momentum $-N_L/(Q_1Q_5R)$ along S^1 and the right-moving modes carry a total momentum $N_R/(Q_1Q_5R)$ along S^1. Then we require

$$N_L - N_R = Q_1Q_5n.$$ (9.22)

The degeneracy of states is obtained by computing the number of ways these momenta can be distributed between different left- and the right-moving modes. The microscopic entropy, computed this way, is given by

$$S_{\text{micro}} = 2\pi(\sqrt{N_L} + \sqrt{N_R}).$$ (9.23)

Since for this configuration the mass is no longer given by the BPS formula, the black hole solution also gets modified and gives a new S_{BH}. The answer turns out to be

$$S_{\text{BH}} = 2\pi(\sqrt{N_L} + \sqrt{N_R}),$$ (9.24)

again in perfect agreement with (9.23). In the expression for S_{BH} the combination $N_L - N_R$ enters through the dependence of the solution on the various charges, whereas the combination $N_L + N_R$ enters through the dependence of the solution on the mass of the black hole, which is now an independent parameter.

A priori we should not have expected such an agreement between S_{BH} and S_{micro} for these black holes, since for non-BPS states one did not expect any non-renormalization theorem to hold and there was no reason why the two answers computed in different domains of validity should agree. An explanation for this agreement was provided later with the help of a new non-renormalization theorem; see [140].

Having found agreement between S_{BH} and S_{micro}, one can now ask if we can also reproduce Hawking radiation from these black holes from the dynamics of D-branes; see [141, 121, 142]. It turns out that extremal black holes have zero temperature and hence do not Hawking radiate. This is consistent with the fact that in the microscopic description, BPS states are stable and hence cannot decay into other states. But non-extremal black holes do Hawking radiate and we can try to compare this radiation with the radiation due to the decay of a non-BPS D-brane. Computation of the Hawking radiation rate from the near extremal black hole can be done by standard techniques. One subtlety comes from the fact that although the black hole horizon gives out thermal radiation, there is a frequency dependent filtering of this radiation as it passes through the black hole background and reaches the asymptotic observer. This effect is known as the grey-body factor, and can be computed by knowing the background fields associated with the specific black hole solution under consideration. For the specific non-extremal black holes that we are considering, the net Hawking radiation of a specific class of scalar particles, as seen by an asymptotic observer, is given by (see [142]),[32]

$$\Gamma_H \;=\; 2\pi^2 (RV)^{4/3} \frac{Q_1 Q_5}{V} \frac{\pi k_0}{2} \frac{d^4 k}{(2\pi)^4} \frac{1}{e^{\frac{k_0}{2T_L}} - 1} \frac{1}{e^{\frac{k_0}{2T_R}} - 1} \qquad (9.25)$$

where

$$(T_R) \;=\; \frac{(RV)^{-1/3}}{\pi R Q_1 Q_5} \sqrt{N_R}$$

$$(T_L) \;=\; \frac{(RV)^{-1/3}}{\pi R Q_1 Q_5} \sqrt{N_L}, \qquad (9.26)$$

and $d^4 k$ denotes the 4-dimensional phase space in this $(4 + 1)$-dimensional theory.

In the D-brane description, the radiation of these specific scalar particles is due to the annihilation of the left- and right-moving modes on the $(D - 1)$-brane. This decay rate can be calculated by using standard string theoretic techniques, and the answer is (see [121]),

$$\Gamma_{micro} \;=\; 2\pi^2 (RV)^{4/3} \frac{Q_1 Q_5}{V} \frac{\pi k_0}{2} \frac{d^4 k}{(2\pi)^4} \frac{1}{e^{\frac{k_0}{2T_L}} - 1} \frac{1}{e^{\frac{k_0}{2T_R}} - 1}. \qquad (9.27)$$

Again we see that this is in exact agreement with Γ_H!

Thus we see that at least for a class of black holes we now have a concrete microscopic derivation of the Bekenstein–Hawking entropy and the phenomenon of Hawking radiation. The description is completely quantum mechanical. The only caveat is that these microscopic calculations are done in regions of parameter space where the system is not a black hole, and then, with the help of non-renormalization theorems, the answer is continued to the region where the system is a black hole. If we could understand in more detail how this continuation works, then perhaps we shall be able to show that there is really no conflict between Hawking radiation and quantum mechanics, in the same sense that thermal radiation from a star is not in conflict with quantum mechanics.

10   Matrix Theory

In Section 7 we postulated the existence of an 11-dimensional theory, known as M-theory, with the following two properties:

- The low energy limit of M-theory is 11-dimensional $N = 1$ supergravity.

---

[32] Again in comparing this expression to those in references [18, 19] we need to take into account the rescaling of the metric.

- M-theory compactified on a circle is dual to type IIA string theory. As the radius of this circle goes to infinity, the type IIA coupling constant also goes to infinity.

From this we can define M-theory as the strong coupling limit of type IIA string theory. However, this definition does not give us any clue as to how to systematically compute S-matrix elements in M-theory, since the type IIA string theory is defined only by the rules for its perturbation expansion in the coupling constant. Thus, finding a non-perturbative definition of M-theory is of importance. One might try various approaches:

1. M-theory is the $N = 1$ supergravity theory in $D = 11$: This proposal by itself does not make sense since this supergravity theory is not renormalizable and hence is plagued by the usual ultraviolet divergences. One might argue that there is some intrinsic regularization that accompanies this supergravity theory, but till we find such a regularisation, describing M-theory as the supergravity theory remains an empty statement.

2. M-theory is a string theory: One might imagine that M-theory can be made finite by regarding it as the low energy limit of some string theory in the same way that the various supergravity theories in 10 dimensions are made finite. However, nobody has been able to find any such string theory so far. There is also another compelling reason to believe that M-theory is not a string theory. With the sole exception of type I string theory, every other string theory has the property that the corresponding low energy supergravity theory contains the fundamental string as a supersymmetric soliton. $N = 1$ supergravity in 11 dimensions does not contain any such soliton solution.

3. M-theory is a theory of membranes or 5-branes: the 11-dimensional supergravity theory does contain membrane and 5-brane like soliton solutions; see [8]. Thus one might argue that M-theory should be formulated as a theory of membranes or 5-branes. The difficulty with this proposal is that unlike string theory, which is tractable due to the infinite-dimensional conformal symmetry on its world-sheet, the world-volume theory of membranes or 5-branes have no such infinite-dimensional symmetry and hence are extremely difficult to handle. However, the Matrix theory that we are about to discuss does in some sense regard M-theory as a theory of membranes; see [157].

These difficulties in formulating M-theory have led to a radically new way of thinking about M-theory, and in fact string theories in general. This proposal, known as Matrix theory (see [158, 159]), is based on describing the theory in terms of its Hamiltonian in the discrete light cone quantization (DLCQ). In this section we shall give a brief overview of this formulation. The contents of this section will form a small fraction of the material covered in [21, 22].

10.1   Discrete light-cone quantization (DLCQ)

We shall illustrate the procedure of DLCQ in the context of a scalar field theory. Let us begin with a free scalar field theory in $d + 1$ dimensions. The action of the system can be expressed as

$$S = \int dx^+ dx^- d^{d-1}\vec{x}_\perp (\partial_+\phi\partial_-\phi - \frac{1}{2}\sum_{i=1}^{d-1}\partial_i\phi\partial_i\phi) , \qquad (10.1)$$

where

$$x^\pm = \frac{1}{\sqrt{2}}(x^0 \pm x^d), \qquad \partial_\pm = \frac{\partial}{\partial x^\pm}, \qquad (10.2)$$

and $\vec{x}_\perp \equiv (x^1, \ldots, x^{d-1})$ denotes the transverse coordinates. We shall now canonically quantize this theory by regarding x^+ as time. Thus the momentum conjugate to ϕ is given by

$$\pi = \frac{\delta S}{\delta(\partial_+\phi(x))} = \partial_-\phi. \qquad (10.3)$$

Note that the relationship between ϕ and π given in eq.(10.3) does not involve a 'time' derivative, reflecting the fact that the original Lagrangian is linear in $\partial_+\phi$. As a result eq.(10.3) represents a constraint. The canonical equal time commutation relations can be found by using the standard formalism for quantization of constrained system. However, we can simplify this problem by going to the Fourier transformed variables defined through the decomposition

$$\phi(x^+, x^-, \vec{x}_\perp) = \int_0^\infty \frac{dk_-}{\sqrt{4\pi k_-}}[a(x^+, k_-, \vec{x}_\perp)e^{-ik_-x^-} + a^*(x^+, k_-, \vec{x}_\perp)e^{ik_-x^-}]$$
$$(10.4)$$

The action (10.1) now takes the form

$$S = i\int dx^+ \int_0^\infty dk_- \int d^{d-1}\vec{x}_\perp [a^*(x^+, k_-, \vec{x}_\perp)\partial_+a(x^+, k_-, \vec{x}_\perp) + \ldots], \quad (10.5)$$

where ... denote terms without any x^+ derivatives. Thus if we now regard a as the coordinate, its canonically conjugate momentum is given by

$$\frac{\delta S}{\delta(\partial_+a)} = ia^* \qquad (10.6)$$

Upon going to the quantum theory, a^* should be regarded as the Hermitian conjugate of the operator a. This gives the following equal time commutation rules:

$$[a(x^+, k_-, \vec{x}_\perp), a(x^+, l_-, \vec{y}_\perp)] = 0 = [a^\dagger(x^+, k_-, \vec{x}_\perp), a^\dagger(x^+, l_-, \vec{y}_\perp)]$$
$$[a(x^+, k_-, \vec{x}_\perp), a^\dagger(x^+, l_-, \vec{y}_\perp)] = \delta(k_- - l_-)\delta(\vec{x}_\perp - \vec{y}_\perp). \tag{10.7}$$

From this we see that $a^\dagger$ and a behave respectively as creation and annihilation operators for particles carrying momentum k_- located at the transverse location $\vec{x}_\perp$. Note that the argument k_- of a and $a^\dagger$ extends over positive values only. This can be traced to the fact that (10.3) is a constraint equation, hence only half of the degrees of freedom of ϕ are true coordinates, the other half being momenta.

We shall now compactify the light-like direction x^- on a circle of radius L. This of course is not a physically meaningful system since it has closed light-like curves and hence violates causality, but we simply use it as an infrared regulator and at the end take the $L \to \infty$ limit to recover physically meaningful answers for various processes. In this case the momentum k_- is quantized as

$$k_- = \frac{n}{L}. \tag{10.8}$$

Eqs.(10.4) and(10.5) are now modified as:

$$\phi(x^+, x^-, \vec{x}_\perp) = a_0(x^+, \vec{x}_\perp) + \frac{1}{\sqrt{4\pi}} \sum_{n=1}^{\infty} \frac{1}{\sqrt{n}} \Big[a_n(x^+, \vec{x}_\perp)e^{-inx^-/L}$$
$$+ a_n^*(x^+, \vec{x}_\perp)e^{inx^-/L}\Big] \tag{10.9}$$

$$S = i \int dx^+ \int d^{d-1}\vec{x}_\perp \sum_{n=1}^{\infty} (a_n^* \partial_+ a_n + \ldots), \tag{10.10}$$

This shows that ia_n^* is the momentum conjugate to a_n. In quantum theory a_n^* will be represented by the Hermitian conjugate $a_n^\dagger$ of the operator a_n. This gives the following commutation relations:

$$[a_n(x^+, \vec{x}_\perp), a_m(x^+, \vec{y}_\perp)] = 0 = [a_n^\dagger(x^+, \vec{x}_\perp), a_m^\dagger(x^+, \vec{y}_\perp)],$$
$$[a_n(x^+, \vec{x}_\perp), a_m^\dagger(x^+, \vec{y}_\perp)] = \delta_{mn}\delta(\vec{x}_\perp - \vec{y}_\perp). \tag{10.11}$$

Note that the action does not contain any term containing x^+ derivative of a_0. Thus we can interprete it as the Lagrange multiplier field and integrate it out. The Hamiltonian computed from the action (10.1) takes the form

$$H = \sum_{n=1}^{\infty} \frac{L}{2n} \int d^{d-1}\vec{x}_\perp \partial_i a_n^\dagger(x^+, k_-, \vec{x}_\perp) \partial_i a_n(x^+, k_-, \vec{x}_\perp). \tag{10.12}$$

We can define the ground state of this system as the state annihilated by all the a_n's. The p-particle state is created by acting with p of the $a^\dagger$'s on this ground state.

Let us now consider the effect of introducing interactions in the original action (10.1). This will add new terms to the Hamiltonian. In particular, since the interaction terms in the Lagrangian will involve a_0, integrating out a_0 will in general produce an infinite series of additional terms in H even if the Lagrangian itself contains only a finite number of terms.[33] But as long as the interactions are invariant under translation, all the terms in the Hamiltonian will conserve k_-. Let us now focus on a sector with $k_- = N/L$. This sector contains N-particle states with each particle carrying momentum $1/L$, but it may also contain states with less than N particles, with some particles carrying momentum larger than $1/L$. However, this sector does not cantain states with more than N particles, since all particles in this description carry strictly positive k_- in units of $1/L$. This suggests that the dynamics in this sector should be describable by an N-body *quantum mechanical Hamiltonian* $\mathcal{H}_N^{DLCQ}$. (One should distinguish this from the Hamiltonian H of the second quantized theory.) By definition, $\mathcal{H}_N^{DLCQ}$ describes the complete dynamics in the $k_- = N/L$ sector. Thus for example it should correctly reproduce the spectrum, as well as all scattering amplitudes in this sector. In particular, in this description a single particle state carrying N units of momentum should appear as a bound state of N particles in the spectrum of $\mathcal{H}_N^{DLCQ}$. Thus $\mathcal{H}_N^{DLCQ}$ must possess an appropriate bound state of this form. Similarly it must contain in its spectrum appropriate bound states representing p-particle states carrying total momentum N/L for all $p < N$, and for all possible distribution of the minus component of the total momentum N/L among these p particles.

Given the action of an interacting quantum field theory, it should in principle be possible to construct $\mathcal{H}_N^{DLCQ}$. Conversely, if we are given $\mathcal{H}_N^{DLCQ}$ for all N, we can, in principle, completely reconstruct the spectrum and the S-matrix elements of the theory. For example, calculation of a specific n-particle $\to$ m-particle process will involve the following steps:

1. Calculate the scattering amplitude for a set of n states carrying momenta $k_-^i = n_i/L$ going to a set of m states carrying momenta $l_-^i = m_i/L$ with

$$\sum_{i=1}^{n} n_i = \sum_{i=1}^{m} m_i = N. \qquad (10.13)$$

---

[33]If the theory contains moduli fields which have flat potential and hence can acquire arbitrary vacuum expectation values, then the process of integrating out the zero modes of these fields requires choosing a definite set of vacuum expectation values of these fields. This corresponds to choosing a specific point in the moduli space. At different points of the moduli space we shall get a different DLCQ theory. Thus the DLCQ formalism is not independent of the choice of background.

2. Now take the limit $N \to \infty$, $L \to \infty$, $n_i \to \infty$ and $m_i \to \infty$ keeping fixed

$$N/L, \qquad k^i_- = n_i/L, \qquad l^i_- = m_i/L. \qquad (10.14)$$

In this limit we shall reproduce the physical scattering amplitude involving $m + n$ external legs. We are of course implicitly assuming that there is an unambiguous procedure for taking the large N limit.

Given that $\mathcal{H}_N^{DLCQ}$ defines a theory, we can now try to define M-theory by specifying $\mathcal{H}_N^{DLCQ}$ for M-theory. This is what is done in the Matrix theory approach to M-theory.

10.2 DLCQ of M-theory

Although DLCQ is used for giving a non-perturbative definition of M-theory, it can in principle be used to describe any theory. We shall first give a general recipe for constructing $\mathcal{H}_N^{DLCQ}$ for any (compactified) string theory or M-theory, and then specialize to the case of M-theory; see [160, 161]. Let $\mathcal{T}$ be such a (compactified) string or M-theory. Typically $\mathcal{T}$ has one mass parameter M, which can be taken to be the Planck mass for M-theory, and $(\sqrt{\alpha'})^{-1}$ for string theory, and a set of dimensionless parameters, e.g. string coupling constant, the dimensions of the compact manifold measured in units of M^{-1}, etc. We shall label all these dimensionless parameters by $\{\vec{y}\}$. (From now on we shall display all factors of the mass parameter explicitly, and not work in the $\alpha' = 1$ unit as we have been doing till now.) Let us denote by $\mathcal{H}_N^{DLCQ}(M, L, \{\vec{y}\})$ the DLCQ Hamiltonian for this system. L as usual denotes the radius of the compact light-cone direction. We now propose the following recipe for constructing $\mathcal{H}_N^{DLCQ}$; see [160, 161].

1. Consider the same theory $\mathcal{T}$ with the same values of the dimensionless parameters $\{\vec{y}\}$, but a different value m of the mass parameter. Let us compactify this theory on a space-like circle S^1 of radius R. We shall call this theory the auxiliary theory.

2. When R is small, the Kaluza–Klein modes carrying momentum along S^1 are heavy, and one would expect that the dynamics of this system will be described by a non-relativistic quantum mechanical Hamiltonian. Let us focus on the sector with N particles each carrying momentum $1/R$ along S^1, and let us denote the N-body Hamiltonian describing the dynamics of this system by $\mathcal{H}_N^{KK}(m, R, \{\vec{y}\})$. (We subtract off the rest mass energy N/R of these particles in defining $\mathcal{H}_N^{KK}$.)

3. $\mathcal{H}_N^{DLCQ}$ is constructed from $\mathcal{H}_N^{KK}$ by taking the limit

$$\mathcal{H}_N^{DLCQ}(M, L, \{\vec{y}\}) = \lim_{R \to 0} \mathcal{H}_N^{KK}(m = M\sqrt{L/R}, R, \{\vec{y}\}). \qquad (10.15)$$

We shall first explore some of the consequences of this recipe. Later we shall discuss how this recipe might be 'derived'.

First of all, note that since the duality transformations in T (before compactification on S^1) leave the momenta along the non-compact directions unchanged, it leaves the sector with N units of momentum along S^1 invariant after we compactify T on S^1. As a result $\mathcal{H}_N^{KK}$, and hence $\mathcal{H}_N^{DLCQ}$ defined in (10.15), is expected to possess the full set of duality symmetries of T; see [162]–[171].

Second, note that although this recipe gives $\mathcal{H}_N^{DLCQ}$ for any theory T, it is particularly useful for (compactified) M-theory, since in the $R \to 0$ limit M-theory on a circle of radius R gets mapped to *weakly coupled* type IIA string theory as discussed in Section 7. As discussed there, states carrying momenta along S^1 correspond to $D0$-branes in the type IIA theory. Thus the recipe given above relates $\mathcal{H}_N^{DLCQ}$ to a specific weak coupling limit of the Hamiltonian describing N $D0$-branes in type IIA string theory. We shall later construct this Hamiltonian explicitly in some cases.

One might feel that the assertion made above is a bit too simplified, since $D0$-branes represent only a subset of particles in M-theory carrying momenta along S^1 – the 11-dimensional graviton and its supersymmetric partners. For example, if T corresponds to M-theory compactified on T^2, then T has solitonic states corresponding to the membrane of M-theory wrapped around T^2. These states are distinct from the supergravitons. Thus one would expect that in constructing $\mathcal{H}_N^{KK}$ (and hence $\mathcal{H}_N^{DLCQ}$) for this theory one needs to add to the $D0$-brane Hamiltonian new degrees of freedom which are capable of describing these wrapped membrane states carrying momentum along S^1. However, due to a truly marvellous property of the $D0$-brane system, this is not necessary. It turns out that the $D0$-brane Hamiltonian automatically contains the required degrees of freedom that gives rise to these new states. We shall explicitly see an example of this later.

Let us now apply this recipe to construct $\mathcal{H}_N^{DLCQ}$ for M-theory on T^n. Let M_p be the Planck mass of the M-theory, L_i be the radii of the n circles[34] which make up T^n and L be the radius of the light-like circle. Let m_p be the Planck mass of the auxiliary M-theory, R_i be the radii of the circles that make up the n-dimensional torus in this auxiliary M-theory, and R be the radius of the extra S^1 on which this auxiliary M-theory on T^n is further compactified. Then from (10.15) we get the following relation between the parameters of the two theories:

$$m_p = M_p\sqrt{L/R}, \qquad m_p R_i = M_p L_i, \tag{10.16}$$

---

[34]For simplicity we are assuming that the torus is made up of product of n circles, without any background field.

where the second equation reflects that the dimensionless parameters obtained by taking the product of the Planck mass and the radii of the compact directions must be the same in the two theories. We now map this to the Hamiltonian of a set of $D0$-branes in type IIA string theory by identifying M-theory on S^1 of radius R with type IIA string theory according to the rules given in Section 7. Let $g_S(\equiv e^{\Phi/2})$ and $m_S(\equiv (\alpha')^{-1/2})$ denote the coupling constant and the string mass in this type IIA theory. Then the first of eq.(7.2) gives

$$m_p R = g_S^{2/3} \, , \qquad (10.17)$$

where we have explicitly put in the 11-dimensional Planck mass that was set to unity in (7.2). Another relation between m_p, m_S and g_S comes from restoring the appropriate factors of m_p, m_S and g_S in (7.1) and (3.33). In particular (7.1) contains a multiplicative factor of $(m_p)^9$, whereas (3.33) contains a multiplicative factor of $(\alpha')^{-4} g_S^{-2} = m_S^8 g_S^{-2}$ (analog of eq.(3.3) for the heterotic theory). Thus the equality between the two actions upon compactification of the 11-dimensional theory on S^1 requires that

$$m_p^9 R = m_S^8 g_S^{-2} \, . \qquad (10.18)$$

Inverting the relations (10.17) and (10.18) we get,

$$m_S = m_p^{3/2} R^{1/2}, \qquad g_S = (m_p R)^{3/2} \, . \qquad (10.19)$$

If $\mathcal{H}_N^{D0}$ denotes the Hamiltonian for N $D0$-branes (with the rest mass subtracted) in this theory, then, using eqs.(10.15), (10.16) and (10.19) we get,

$$\mathcal{H}_N^{DLCQ}(M_p, L, \{R_i\})$$
$$= \lim_{R \to 0} \mathcal{H}_N^{D0}(m_S = M_p^{3/2} L^{3/4} R^{-1/4}, g_S = M_p^{3/2}(LR)^{3/4}, R_i = R^{1/2} L^{-1/2} L_i) \, . $$
$$(10.20)$$

This gives a sensible answer for all n up to 5 and does not give a sensible answer for $n \geq 6$; see [160, 161]. Here we shall only discuss two special cases, $n = 0$ and $n = 2$.

First we consider the case $n = 0$. This corresponds to 11-dimensional M-theory. As seen from (10.20), as $R \to 0$, $g_S \to 0$ and $m_S \to \infty$. Thus we can use the low energy and weak coupling approximation of type IIA string theory. This limit has been studied in detail in [172]. The action governing the dynamics of the $D0$-brane system in this limit is given by the dimensional reduction of $N = 1$ supersymmetric $U(N)$ gauge theory in 10 dimensions:

$$S \sim m_S^{-3} g_S^{-1} \int dt \Big[\sum_{m=1}^{9} \text{Tr}(\partial_t \Phi^m \partial_t \Phi^m) - \sum_{m<n=1}^{9} \text{Tr}([\Phi^m, \Phi^n]^2) \Big]$$
$$+ \text{fermionic terms}$$

$$= M_p^{-6} L^{-3} \int dt \Big[\sum_{m=1}^{9} \text{Tr}(\partial_t \Phi^m \partial_t \Phi^m) - \sum_{m<n=1}^{9} \text{Tr}([\Phi^m, \Phi^n]^2) \Big]$$
+fermionic terms

$$(10.21)$$

where Φ^m are $N \times N$ Hermitian matrices. In going from the first to the second line we have used eq.(10.20). This gives a well-defined Hamiltonian for DLCQ M-theory. The flat direction in the potential corresponds to a configuration where all the Φ^m's are simultaneously diagonalized. The N eigenvalues of Φ^m represent the mth coordinate of the N different $D0$-branes.

Let us now consider M-theory on T^2. From eq.(10.20) we see that in the $R \to 0$ limit, the radii R_i vanish. We can remedy this problem by giving a different description of the same system by making an $R_i \to 1/m_S^2 R_i$ duality transformation on both circles. This converts the original type IIA theory to type IIA theory on a dual torus $\tilde{T}^2$ and the system of N $D0$-branes to a system of N $(D-2)$-branes wrapped on $\tilde{T}^2$.[35] The new theory has parameters:

$$\tilde{g}_S = g_S/(m_S^2 R_1 R_2) = M_p^{-3/2} L^{1/4} R^{1/4} L_1^{-1} L_2^{-1},$$
$$\tilde{m}_S = m_S = M_p^{3/2} L^{3/4} R^{-1/4}, \qquad \tilde{R}_i = R_i^{-1} m_S^{-2} = M_p^{-3} L^{-1} L_i^{-1}.$$

$$(10.22)$$

In the $R \to 0$ limit this again gives a theory at weak coupling and large string mass. Furthermore the new radii $\tilde{R}_i$ are finite. The dynamics of wrapped $(D-2)$-branes in this limit is described by a $(2+1)$-dimensional $N = 8$ supersymmetric $U(N)$ Yang–Mills theory compactified on $\tilde{T}^2$. This theory has seven scalars Φ^m $(1 \le m \le 7)$ in the adjoint representation of the gauge group. The bosonic part of the Lagrangian is given by

$$S \sim \tilde{m}_S^{-1} \tilde{g}_S^{-1} \int dt \int_0^{\tilde{R}_1} dx^8 \int_0^{\tilde{R}_2} dx^9 \Big[\text{Tr}(F_{\mu\nu} F^{\mu\nu})$$
$$+ \text{Tr}(D_\mu \Phi^m D^\mu \Phi^m) - \sum_{m<n=1}^{7} \text{Tr}([\Phi^m, \Phi^n]^2) \Big]$$

$$= L^{-1} L_1 L_2 \int dt \int_0^{\tilde{R}_1} dx^8 \int_0^{\tilde{R}_2} dx^9 \Big[\text{Tr}(F_{\mu\nu} F^{\mu\nu})$$
$$+ \text{Tr}(D_\mu \Phi^m D^\mu \Phi^m) - \sum_{m<n=1}^{7} \text{Tr}([\Phi^m, \Phi^n]^2) \Big],$$

$$(10.23)$$

---

[35]Quite generally one can show that an $R_i \to 1/m_S^2 R_i$ duality transformation converts a Dirichlet boundary condition to a Neumann boundary condition and vice versa; see [5]. Thus this duality transformation converts a $D0$-brane with Dirichlet boundary conditions along T^2 into a $(D-2)$-brane wrapped on $\tilde{T}^2$ which has Neumann boundary condition along $\tilde{T}^2$.

where x^μ ($\mu = 0, 8, 9$) denote coordinates along the $(D-2)$-brane world-volume, and the N eigenvalues of Φ_m represent the mth transverse coordinate of the N different $(D-2)$-branes.

Let us now address the problem alluded to earlier, namely that M-theory on T^2 contains solitonic states in the form of wrapped membranes. Thus the complete $\mathcal{H}_N^{DLCQ}$ must contain these states as well. In the auxiliary type IIA theory, these wrapped membranes correspond to $(D-2)$-branes wrapped on T^2. Upon T-duality in both circles, these become $D0$-branes moving on the dual torus $\tilde{T}^2$. Thus the relevant question is, does the system described in (10.23) automatically contain these states, or do we need to add new degrees of freedom in this system so as to be able to describe these states? It turns out that the $D0$-brane charge in this dual theory simply corresponds to the flux of the $U(1)$ component of the magnetic field through $\tilde{T}^2$. Thus a state with k $D0$-branes (which corresponds to a membrane wrapped k times on the original torus) can be described by a specific excitation of the system (10.23) carrying k units of magnetic flux through $\tilde{T}^2$. There is no need to add new degrees of freedom.

Finally let us give a 'derivation' of the recipe described at the beginning of this section following [161]. First of all, we note that in the auxiliary theory $\mathcal{T}$ on S^1, if we multiply all the masses by some constant λ, and simultaneously multiply all the lengths by λ^{-1}, then the Hamiltonian gets multiplied by λ due to purely dimensional reasons. This gives the following identity:

$$\mathcal{H}_N^{KK}(m, R, \{\vec{y}\}) = \lambda^{-1} \mathcal{H}_N^{KK}(\lambda m, \lambda^{-1} R, \{\vec{y}\}). \qquad (10.24)$$

where the first and the second arguments denote respectively the overall mass scale and the radius of S^1, as usual. Let us now choose

$$r = \sqrt{RL}, \qquad m = M\sqrt{L/R}, \qquad \lambda = \sqrt{R/L} = r/L. \qquad (10.25)$$

Substituting this in (10.24) we get

$$\mathcal{H}_N^{KK}(M\sqrt{L/R}, R, \{\vec{y}\}) = \frac{L}{r}\mathcal{H}_N^{KK}(M, r, \{\vec{y}\}). \qquad (10.26)$$

Thus the recipe (10.15) can now be rewritten as

$$\mathcal{H}_N^{DLCQ}(M, L, \{\vec{y}\}) = \lim_{r \to 0} \frac{L}{r}\mathcal{H}_N^{KK}(M, r, \{\vec{y}\}). \qquad (10.27)$$

It is this form of the identity that we shall attempt to prove.

The basic idea behind this proof is to regard the light-like circle as an infinitely boosted space-like circle of zero radius; see [173, 174]. Let us start with theory $\mathcal{T}$ compactified on a space-like circle S^1 of radius r. If x and t

denote the coordinate along S^1 and the time coordinate respectively, then we have an identification:

$$\begin{pmatrix} x \\ t \end{pmatrix} \equiv \begin{pmatrix} x \\ t \end{pmatrix} + 2\pi \begin{pmatrix} r \\ 0 \end{pmatrix} . \tag{10.28}$$

Let us now define new coordinates (x', t') and $x'^{\pm}$ as follows:

$$\begin{pmatrix} x' \\ t' \end{pmatrix} = \begin{pmatrix} x \cosh \alpha - t \sinh \alpha \\ t \cosh \alpha - x \sinh \alpha \end{pmatrix} , \tag{10.29}$$

$$x'^{\pm} = \frac{1}{\sqrt{2}} (t' \pm x') . \tag{10.30}$$

In this coordinate system eq.(10.28) takes the form:

$$\begin{pmatrix} x'^+ \\ x'^- \end{pmatrix} \equiv \begin{pmatrix} x'^+ \\ x'^- \end{pmatrix} - \sqrt{2}\pi r \begin{pmatrix} -e^{-\alpha} \\ e^{\alpha} \end{pmatrix} . \tag{10.31}$$

Now consider the limit $r \to 0$, $\alpha \to \infty$ keeping fixed

$$L \equiv \frac{r}{\sqrt{2}} e^{\alpha}. \tag{10.32}$$

In this limit eq.(10.31) reduces to

$$\begin{pmatrix} x'^+ \\ x'^- \end{pmatrix} \equiv \begin{pmatrix} x'^+ \\ x'^- \end{pmatrix} - 2\pi \begin{pmatrix} 0 \\ L \end{pmatrix} . \tag{10.33}$$

This is equivalent to compactifying x'^- on a circle of radius L.

Under this map, a system carrying momentum N/R along S^1 gets mapped to a system carrying total momentum $k'^- = N/L$ along the x'^- direction. Thus it is not surprising that there is a relation between the Hamiltonians describing the two systems. To find the precise relation between these two Hamiltonians, we need to study the relation between the usual time coordinate t of the original theory and the light-cone time x'^+ of the boosted theory. From eqs.(10.29), (10.30) it follows that:

$$\frac{\partial}{\partial x'^+} = \frac{1}{\sqrt{2}} e^{\alpha} \left(\frac{\partial}{\partial t} + \frac{\partial}{\partial x} \right) . \tag{10.34}$$

Since the quantum operators which generate $i(\partial/\partial x'^+)$, $i(\partial/\partial t)$ and $i(\partial/\partial x)$ are $\mathcal{H}_N^{DLCQ}$, $\mathcal{H}_N^{KK} + M_N$ and $-N/r$ respectively, with M_N being the rest mass of the N Kaluza–Klein modes in $\mathcal{T}$ on S^1, we see from eq.(10.32), (10.34) that in the $r \to 0$ limit with L fixed,

$$\mathcal{H}_N^{DLCQ}(M, L, \{\vec{y}\}) = \frac{L}{r} (\mathcal{H}_N^{KK}(M, r, \{\vec{y}\}) + M_N - \frac{N}{r} = \frac{L}{r} \mathcal{H}_N^{KK}(M, r, \{\vec{y}\}) , \tag{10.35}$$

since $M_N = N/r$. This reproduces (10.27).

If we recall that $\mathcal{H}_N^{DLCQ}$ is supposed to describe the theory $\mathcal{T}$, whereas $\mathcal{H}_N^{KK}$ describes the theory $\mathcal{T}$ compactified on a small circle, then by the above argument, quite generally we can reconstruct a theory by knowing its behaviour when compactified on a small circle. This seems counterintuitive, so let us examine the steps leading to this conclusion. They may be summarized as follows:

1. We start with a small circle.

2. We convert this to an almost light-like circle of finite radius via a large boost.

3. We then take the limit where the radius of this light-like circle goes to infinity.

As we can see, the key point in this proof is the assumption that a light-like circle can be considered as a space-like circle of zero radius in the limit of infinite boost. Of course, this may be taken as a definition of the light-like circle. However, we are interested in a definition in which the radius of the light-like circle acts as an infrared regulator in the uncompactified theory, so that in the end by taking the $L \to \infty$ limit we recover the amplitudes in the uncompactified theory. Clearly there is a possibility that these two definitions do not match; see [174]. Indeed there are explicit computations which show that these two definitions do not always match for finite N (see [175, 176, 177, 178, 179]), although they do match for some specific terms in the supergraviton scattering amplitudes; see [158, 180, 173]. (Note that the 'proof' given above did not involve taking the $N \to \infty$ limit.) It has been suggested (see [21, 22]) that this problem might go away in the $N \to \infty$ limit, but there is no compelling argument as of now in favour of this. This of course does not mean that Matrix theory is wrong, it is just that we do not know for sure if it is right, and even if it is right, we do not quite know why it is right. Perhaps the arguments of reference [161] together with supersymmetry non-renormalization theorems and properties of the large N limit can be combined to constitute such a proof.

Acknowledgement   I wish to thank S. Panda for useful comments and suggestions.

References

[1] A. Sen, Int. J. Mod. Phys. **A9** (1994) 3707 [hep-th/9402002].

[2] M. Duff, R. Khuri and J. Lu, Phys. Rep. **259** (1995) 213 [hep-th/9412184].

[3] J. Schwarz, Nucl. Phys. Proc. Suppl. **55B** 1 [hep-th/9607201].

[4] S. Chaudhuri, C. Johnson and J. Polchinski, hep-th/9602052.

[5] J. Polchinski, Rev. Mod. Phys. **68** (1996) 1245 [hep-th/9607050].

[6] J. Polchinski, hep-th/9611050. In *Fields, Strings and Duality*, TASI 96, World Scientific, (1997), 293.

[7] P. Townsend, hep-th/9612121; gr-qc/9707012; hep-th/9712004.

[8] M. Duff, hep-th/9611203. In *Fields, Strings and Duality*, TASI 96, World Scientific, (1997), 219.

[9] M. Douglas, hep-th/9610041. In *Quantum Symmetries* Les Houches 1995, North-Holland (1998), 519.

[10] W. Lerche, hep-th/9611190, in *Nucl. Phys. Proc. Suppl.* **55b** 83; hep-th/9710246, in *Fortsch. Phys.* **45** (1997) 293.

[11] S. Förste and J. Louis, hep-th/9612192.

[12] C. Vafa, hep-th/9702201.

[13] A. Klemm, hep-th/9705131.

[14] E. Kiritsis, hep-th/9708130.

[15] B. de Wit and J. Louis, hep-th/9801132.

[16] M. Trigiante, hep-th/9801144.

[17] D. Youm, hep-th/9710046.

[18] J. Maldecena, hep-th/9705078.

[19] A. Peet, hep-th/9712253.

[20] A. Bilal, hep-th/9710136.

[21] T. Banks, hep-th/9710231.

[22] D. Bigatti and L. Susskind, hep-th/9712072.

[23] W. Taylor, hep-th/9801182.

[24] S. Mukhi, hep-ph/9710470.

[25] M. Green, J. Schwarz and E. Witten, Superstring Theory vol. 1 and 2, Cambridge University Press (1986); D. Lust and S. Theisen, Lectures on String Theory, Springer (1989); J. Polchinski, hep-th/9411028.

[26] E. Witten, Nucl. Phys. **B443** (1995) 85 [hep-th/9503124].

[27] A. Dabholkar, Phys. Lett. **B357** (1995) 307 [hep-th/9506160].

[28] C. Hull, Phys. Lett. **B357** (1995) 545 [hep-th/9506194].

[29] J. Polchinski and E. Witten, Nucl. Phys. **B460** (1996) 525 [hep-th/9510169].

[30] C. Hull and P. Townsend, Nucl. Phys. **B438** (1995) 109 [hep-th/9410167].

[31] M. Duff, Nucl. Phys. **B442** (1995) 47 [hep-th/9501030]; M. Duff and R. Khuri, Nucl. Phys. **B411** (1994) 473 [hep-th/9305142].

[32] A. Sen, Nucl. Phys. **B450** (1995) 103 [hep-th/9504027].

[33] J. Harvey and A. Strominger, Nucl. Phys. **B449** (1995) 535 [hep-th/9504047].

[34] A. Giveon, M. Porrati and E. Rabinovici, Phys. Rep. **244** (1994) 77 [hep-th/9401139].

[35] P. Fre, Nucl. Phys. **B [Proc. Sup.]** 45B,C (1996) 59 [hep-th/9512043] and references therein.

[36] E. Witten and D. Olive, Phys. Lett. **B78** (1978) 97.

[37] M. de Roo, Nucl. Phys. **B255** (1985) 515.

[38] S. Ferrara, C. Kounnas and M. Porrati, Phys. Lett. **B181** (1986) 263.

[39] M. Terentev, Sov. J. Nucl. Phys. **49** (1989) 713.

[40] S.F. Hassan and A. Sen, Nucl. Phys. **B375** (1992) 103 [hep-th/9109038].

[41] J. Maharana and J. Schwarz, Nucl. Phys. **B390** (1993) 3 [hep-th/9207016].

[42] A. Shapere, S. Trivedi and F. Wilczek, Mod. Phys. Lett. **A6** (1991) 2677.

[43] A. Sen, Nucl. Phys. **B404** (1993) 109 [hep-th/9207053].

[44] A. Font, L. Ibanez, D. Lust and F. Quevedo, Phys. Lett. **B249** (1990) 35.

[45] S.J. Rey, Phys. Rev. **D43** (1991) 526.

[46] J. Schwarz, hep-th/9209125.

[47] A. Sen, Phys. Lett. **B303** (1993) 22 [hep-th/9209016].

[48] A. Sen, Mod. Phys. Lett. **A8** (1993) 2023 [hep-th/9303057].

[49] J. Schwarz and A. Sen, Nucl. Phys. **B411** (1994) 35 [hep-th/9304154]; Phys. Lett. **B312** (1993) 105 [hep-th/9305185].

[50] N. Seiberg, Nucl. Phys. **B303** (1988) 286.

[51] M. Green and J. Schwarz, Phys. Lett. **122B** (1983) 143;
      J. Schwarz and P. West, Phys. Lett. **126B** (1983) 301;
      J. Schwarz, Nucl. Phys. **B226** (1983) 269;
      P. Howe and P. West, Nucl. Phys. **B238** (1984) 181.

[52] E. Cremmer, in 'Unification of Fundamental Particle Interactions', Plenum (1980);
      B. Julia, in 'Superspace and Supergravity', Cambridge Univ. Press (1981).

[53] S. Ferrara and C. Kounnas, Nucl. Phys. **B328** (1989) 406.

[54] S. Chaudhuri, G. Hockney and J. Lykken, Phys. Rev. Lett. **75** (1995) 2264 [hep-th/9505054].

[55] S. Chaudhuri and J. Polchinski, Phys. Rev. **D52** (1995) 7168 [hep-th/9506048].

[56] C. Montonen and D. Olive, Phys. Lett. **B72** (1977) 117.

[57] H. Osborn, Phys. Lett. **B83** (1979) 321.

[58] A. Sen, Phys. Lett. **B329** (1994) 217 [hep-th/9402032].

[59] M. Atiyah and N. Hitchin, Phys. Lett. **107A** (1985) 21; Phil. Trans. Roy. Soc. Lond. **A315** (1985) 459; Geometry and Dynamics of Magnetic Monopoles, Cambridge Univ. Press.

[60] G. Gibbons and N. Manton, Nucl. Phys. **B274** (1986) 183.

[61] J. Gauntlett, Nucl. Phys. **B400** (1993) 103 [hep-th/9205008]; Nucl. Phys. **B411** (1994) 443 [hep-th/9305068].

[62] J. Blum, Phys. Lett. **B333** (1994) 92 [hep-th/9401133].

[63] N. Manton and B. Schroers, Annals Phys. **225** (1993) 290.

[64] G. Gibbons and P. Ruback, Comm. Math. Phys. **115** (1988) 267.

[65] G. Segal and A. Selby, Comm. Math. Phys. **177** (1996) 775.

[66] M. Porrati, Phys. Lett. **B377** (1996) 67 [hep-th/9505187].

[67] J. Gauntlett and J. Harvey, [hep-th/9407111].

[68] J. Schwarz, Phys. Lett. **B360** (1995) 13 [hep-th/9508143].

[69] E. Witten, Nucl. Phys. **B460** (1996) 335 [hep-th/9510135].

[70] S. Das and S. Mathur, Phys. Lett. **B375** (1996) 103 [hep-th/9601152].

[71] A. Dabholkar and J. Harvey, Phys. Rev. Lett. **63** (1989) 478;
A. Dabholkar, G. Gibbons, J. Harvey and F. Ruiz, Nucl. Phys. **B340** (1990) 33.

[72] J. Polchinski, Phys. Rev. Lett. **75** (1995) 4724 [hep-th/9510017].

[73] J. Dai, R. Leigh and J. Polchinski, Mod. Phys. Lett. **A4** (1989) 2073;
R. Leigh, Mod. Phys. Lett. **A4** (1989) 2767;
J. Polchinski, Phys. Rev. **D50** (1994) 6041 [hep-th/9407031].

[74] M. Bershadsky, V. Sadov and C. Vafa, Nucl. Phys. **B463** (1996) 398 [hep-th/9510225].

[75] A. Sen, Phys. Rev. **D53** (1996) 2874 [hep-th/9711026].

[76] C. Vafa, Nucl. Phys. **B469** (1996) 415 [hep-th/9511088]; Nucl. Phys. **B463** (1996) 435 [hep-th/9512078].

[77] S. Sethi and M. Stern, Phys. Lett. **B398** (1997) 47 [hep-th/9607145].

[78] M. Bershadsky, V. Sadov and C. Vafa, Nucl. Phys. **B463** (1996) 420 [hep-th/9511222].

[79] E. Witten, Nucl. Phys. **B460** (1996) 541 [hep-th/9511030].

[80] M. Porrati, Phys. Lett. **B387** (1996) 492 [hep-th/9607082].

[81] J. Blum, Nucl. Phys. **B506** (1997) 223 [hep-th/9705030].

[82] A. Sen, hep-th/9609176.

[83] A. Sen, Nucl. Phys. **B434** (1995) 179 [hep-th/9408083]; Nucl. Phys. **B447** (1995) 62 [hep-th/9503057].

[84] C. Vafa and E. Witten, hep-th/9507050.

[85] A. Sen, Nucl. Phys. **B474** (1996) 361 [hep-th/9604070].

[86] S. Kachru and C. Vafa, Nucl. Phys. **B450** ((1995) 69 [hep-th/9605105].

[87] S. Ferrara, J. Harvey, A. Strominger and C. Vafa, Phys. Lett. **B361** (1995) 59 [hep-th/9505162].

[88] A. Klemm, W. Lerche and P. Mayr, Phys. Lett. **B357** (1995) 313 [hep-th/9506112].

[89] G. Aldazabal, A. Font, L. Ibanez and F. Quevedo, Nucl. Phys. **B461** (1996) 537 [hep-th/9510093].

[90] B. Hunt and R. Schimmrigk, Phys. Lett. **B381** (1996) 427 [hep-th/9512138];
B. Hunt, M. Lynker and R. Schimmtigk, hep-th/9609082.

[91] B. de Wit, P. Lauwers and A. van Proeyen, Nucl. Phys. **B255** (1985) 569.

[92] V. Kaplunovsky, J. Louis and S. Theisen, Phys. Lett. **B357** (1995) 71 [hep-th/9506110].

[93] I. Antoniadis, E. Gava, K. Narain and T. Taylor, Nucl. Phys. **B455** (1995) 109 [hep-th/9507115].

[94] S. Kachru, A. Klemm, W. Lerche, P. Mayr and C. Vafa, Nucl. Phys. **B459** (1996) 537 [hep-th/9508155];
A. Klemm, W. Lerche, P. Mayr, C. Vafa and N. Warner, Nucl. Phys. **B477** (1996) 746 [hep-th/9604034].

[95] N. Seiberg and E. Witten, Nucl. Phys. **B426** (1994) 19 [hep-th/9407087]; Nucl. Phys. **B431** (1994) 484 [hep-th/9408099].

[96] P. Townsend, Phys. Lett. **350B** (1995) 184 [hep-th/9501068].

[97] F. Giani and M. Pernici, Phys. Rev. **D30** (1984) 325;
I. Campbell and P. West, Nucl. Phys. **B243** (1984) 112;
M. Huq and M. Namazie, Class. Quant. Grav. **2** (1985) 293.

[98] S. Sethi and M. Stern, hep-th/9705046.

[99] A. Sen, Phys. Rev. **D54** (1996) 2964 [hep-th/9510229].

[100] J. Schwarz, Phys. Lett. **B367** (1996) 97 [hep-th/9510086].

[101] P. Aspinwall, Nucl. Phys. Proc. Suppl. **46** (1996) 30 [hep-th/9508154].

[102] P. Horava and E. Witten, Nucl. Phys. **B460** (1996) 506 [hep-th/9510209]; Nucl. Phys. **B475** (1996) 94 [hep-th/9603142].

[103] K. Dasgupta and S. Mukhi, Nucl. Phys. **B465** (1996) 399 [hep-th/9512196].

[104] E. Witten, Nucl. Phys. **B463** (1996) 383 [hep-th/9512219].

[105] A. Sen, Mod. Phys. Lett. **A11** (1996) 1339 [hep-th/9603113].

[106] E. Caceres, V. Kaplunovsky and M. Mandelberg, Nucl. Phys. **B493** (1997) 73 [hep-th/9606036].

[107] C. Vafa, Nucl. Phys. **B469** (1996) 403 [hep-th/9602022].

[108] D. Morrison and C. Vafa, Nucl. Phys. **B473** (1996) 74 [hep-th/9602114]; Nucl. Phys. **B476** (1996) 437 [hep-th/9603161].

[109] B. Greene, A. Shapere, C. Vafa and S.T. Yau, Nucl. Phys. **B337** (1990) 1.

[110] R. Friedman, J. Morgan and E. Witten, Comm. Math. Phys. **187** (1997) 679 [hep-th/9701162].

[111] M. Bershadsky, A. Johansen, T. Pantev and V. Sadov, Nucl. Phys. **B505** (1997) 165 [hep-th/9701165].

[112] A. Sen, Nucl. Phys. **B475** (1996) 562 [hep-th/9605150].

[113] A. Sagnotti, 'Open Strings and their Symmetry Groups', Talk at Cargese Summer Inst., 1987;
G. Pradisi and A. Sagnotti, Phys. Lett. **B216** (1989) 59;
M. Bianchi, G. Pradisi and A. Sagnotti, Nucl. Phys. **B376** (1992) 365;
P. Horava, Nucl. Phys. **B327** (1989) 461; Phys. Lett. **B231** (1989) 251.

[114] E. Gimon and J. Polchinski, Phys. Rev. **D54** (1996) 1667 [hep-th/9601038].

[115] A. Chamseddine, Phys. Rev. **D24** (1981) 3065;
E. Bergshoeff, M. de Roo, B. de Wit and P. van Niewenhuizen, Nucl. Phys. **B195** (1982) 97;
E. Bergshoeff, M. de Roo and B. de Wit, Nucl. Phys. **B217** (1983) 143;
G. Chapline and N. Manton, Phys. Lett. **120B** (1983) 105.

[116] H. Nicolai, Phys. Lett. **B276** (1992) 333.

[117] E. Weinberg, Phys. Rev. **D20** (1979) 936;
E. Corrigan and P. Goddard, Comm. Math. Phys. **80** (1981) 575;
C. Taubes, Comm. Math. Phys. **91** (1983) 235.

[118] N. Manton, Phys. Lett. **110B** (1982) 54.

[119] C. Callias, Comm. Math. Phys. **62** (1978) 213.

[120] K. Narain, Phys. Lett. **169B** (1986) 41;
K. Narain, H. Sarmadi and E. Witten, Nucl. Phys. **B279** (1987) 369.

[121] S. Das and S. Mathur, Nucl. Phys. **B478** (1996) 561 [hep-th/9606185].

[122] T. Banks, L. Dixon, D. Friedan and E. Martinec, Nucl. Phys. **B299** (1988) 613;
T. Banks and L. Dixon, Nucl. Phys. **B307** (1988) 93.

[123] B. Greene, hep-th/9702155 and references therein.

[124] B. de Wit and A. van Proeyen, Nucl. Phys. **B245** (1984) 89;
E. Cremmer, C. Kounnas, A. van Proeyen, J. Derendinger, S. Ferrara,
B. de Wit and L. Girardello, Nucl. Phys. **B250** (1985) 385;
S. Ferrara and A. Strominger, in 'Strings 89', World Scientific (1989);
A. Strominger, Comm. Math. Phys. **133** (1990) 163.

[125] J. Bagger and E. Witten, Nucl. Phys. **B222** (1983) 1;
K. Galicki, Comm. Math. Phys. **108** (1987) 117.

[126] E. Cremmer and B. Julia, Nucl. Phys. **B159** (1979) 141.

[127] S. Hawking, Nature **248** (1974) 30; Comm. Math. Phys. **43** (1975) 199.

[128] J. Bekenstein, Lett. Nuov. Cim. **4** (1972) 737; Phys. Rev. **D7** (1973) 2333; Phys. Rev. **D9** (1974) 3192;
G. Gibbons and S. Hawking, Phys. Rev. **D15** (1977) 2752..

[129] G. 't Hooft, Nucl. Phys. **B335** (1990) 138.

[130] L. Susskind, hep-th/9309145;
L. Susskind and J. Uglam, Phys. Rev. **D50** (1994) 2700 [hep-th/9401070];
J. Russo and L. Susskind, Nucl. Phys. **B437** (1995) 611 [hep-th/9405117].

[131] G. Horowitz and J. Polchinski, Phys. Rev. **D55** (1997) 6189 [hep-th/9612146].

[132] K. Sfetsos and K. Skenderis, hep-th/9711138.

[133] A. Sen, Mod. Phys. Lett. **A10** (1995) 2081 [hep-th/9504147].

[134] A. Peet, Nucl. Phys. **B456** (1995) 732 [hep-th/9506200].

[135] A. Sen, hep-th/9712150.

[136] A. Strominger and C. Vafa, Phys. Lett. **B379** (1996) 99 [hep-th/9601029].

[137] C. Callan and J. Maldacena, Nucl. Phys. **B472** (1996) 591 [hep-th/9602043].

[138] J. Maldacena and L. Susskind, Nucl. Phys. **B475** (1996) 679 [hep-th/9604042].

[139] G. Horowitz and A. Strominger, Phys. Rev. Lett. **77** (1996) 2368 [hep-th/9602051].

[140] J. Maldacena, Phys. Rev. **D55** (1997) 7645 [hep-th/9611125].

[141] A. Dhar, G. Mandal and S. Wadia, Phys. Lett. **B388** (1996) 51 [hep-th/9605234].

[142] J. Maldacena and A. Strominger, Phys. Rev. **D55** (1997) 861 [hep-th/9609026].

[143] S. Ferrara, R. Kallosh and A. Strominger, Phys. Rev. **D52** (1995) 5412 [hep-th/9508072];
A. Strominger, Phys. Lett. **B383** (1996) 39 [hep-th/9602111];
S. Ferrara and R. Kallosh, Phys. Rev. **D54** (1996) 1514 [hep-th/9602136]; Phys. Rev. **D54** (1996) 1525 [hep-th/9603090];
M. Shmakova, Phys. Rev. **D56** (1997) 540 [hep-th/9612076];
K. Behrndt, G. Lopez Cardoso, B. de Wit, R. Kallosh, D. Lust and T. Mohaupt, Nucl. Phys. **B488** (1997) 236 [hep-th/9610105];
S. Rey, Nucl. Phys. **B508** (1997) 569 [hep-th/9610157];
D. Kaplan, D. Lowe, J. Maldacena and A. Strominger, Phys. Rev. **D55** (1997) 4898 [hep-th/9609204];
K. Behrndt and T. Mohaupt, Phys. Rev. **D56** (1997) 2206 [hep-th/9611140];
J. Maldacena, Phys. Lett. **B403** (1997) 20 [hep-th/9611163];
A. Chou, R. Kallosh, J. Rahmfeld, S. Rey, M. Shmakova and W. Wong, Nucl. Phys. **B508** (1997) 147 [hep-th/9704142];
J. Maldacena, A. Strominger and E. Witten, hep-th/9711053;
C. Vafa, hep-th/9711067.

[144]   J. Bekenridge, R. Myers, A. Peet and C. Vafa, Phys. Lett. **B391** (1996)
        93 [hep-th/9602065];
        J. Bekenridge, D. Lowe, R. Myers, A. Peet, A. Strominger and C. Vafa,
        Phys. Lett. **B381** (1996) 423 [hep-th/9603078].

[145]   C. Johnson, R. Khuri and R. Myers, Phys. Lett. **B378** (1996) 78 [hep-
        th/9603061];
        J. Maldacena and A. Strominger, Phys. Rev. Lett. **77** (1996) 428 [hep-
        th/9603060];
        G. Horowitz, D. Lowe and J. Maldacena, Phys. Rev. Lett. **77** (1996)
        430 [hep-th/9603195].

[146]   M. Gaillard and B. Zumino, Nucl. Phys. **B193** (1981) 221;
        S. Ferrara, J. Scherk and B. Zumino, Nucl. Phys. **B121** (1977) 393;
        E. Cremmer, J. Scherk and S. Ferrara, Phys. Lett. **74B** (1978) 61;
        B. de Wit, Nucl. Phys. **B158** (1979) 189;
        B. de Wit and H. Nicolai, Nucl. Phys. **B208** (1982) 323;
        E. Cremmer and B. Julia, Phys. Lett. **80B** (1978) 48;
        B. de Wit and A. van Proeyen, Nucl. Phys. **B245** (1984) 89.

[147]   D. Gross, J. Harvey, E. Martinec and R. Rohm, Phys. Rev. Lett. **54**
        (1985) 502; Nucl. Phys. **B256** (1985) 253; Nucl. Phys. **B267** (1986) 75.

[148]   M. Green and J. Schwarz, Phys. Lett. **109B** (1982) 444; Phys. Lett.
        149B (1984) 117; Phys. Lett. **151B** (1985) 21; Nucl. Phys. **B255** (1985)
        93.

[149]   F. Larsen and F. Wilczek, Phys. Lett. **B375** (1996) 37 [hep-
        th/9511064]; Nucl. Phys. **B475** (1996) 627 [hep-th/9604134]; Nucl.
        Phys. **B488** (1997) 261 [hep-th/9609084];
        M. Cvetic and A. Tseytlin, Phys. Rev. **D53** (1996) 5619 [hep-
        th/9512031];
        A. Tseytlin, Mod. Phys. Lett. **A11** (1996) 689 [hep-th/9601177]; Nucl.
        Phys. **B477** (1996) 431 [hep-th/9605091].

[150]   C. Hull and P. Townsend, Nucl. Phys. **B451** (1995) 525 [hep-
        th/9505073].

[151]   A. Strominger, Nucl. Phys. **B451** (1995) 96 [hep-th/9504090];
        B. Greene, D. Morrison and A. Strominger, Nucl. Phys. **B451** (1995)
        109 [hep-th/9504145].

[152]   M. Douglas and G. Moore, hep-th/9603167;
        J. Polchinski, Phys. Rev. **B55** (1997) 6423 [hep-th/9606165];
        C. Johnson and R. Myers, Phys. Rev. **D55** (1997) 6382 [hep-
        th/9610140];

M. Douglas, [hep-th/9612126] published in JHEP electronics journal; D. Diaconescu, M. Douglas and J. Gomis, [hep-th/9712230].

[153] E. Witten, [hep-th/9507121].

[154] M. Douglas, [hep-th/9512077]; [hep-th/9604198].

[155] L. Alvarez-Gaume, D. Freedman and S. Mukhi, Ann. Phys. (NY) **134** (1981) 85;
L. Alvarez-Gaume and D. Freedman, Comm. Math. Phys. **80** (1981) 443.

[156] S. Hassan and S. Wadia, [hep-th/9712213].

[157] E. Bergshoeff, E. Sezgin and P. Townsend, Phys. Lett. **B189** (1987) 75;
Ann. Phys. (NY) **185** (1988) 330;
M. Duff, Class. Quant. Grav. **5** (1988) 189;
B. de Wit, M. Luscher and H. Nicolai, Nucl. Phys. **B320** (1989) 135;
B. de Wit, J. Hoppe and H. Nicolai, Nucl. Phys. **B305** (1988) 545.

[158] T. Banks, W. Fischler, S. Shenker and L. Susskind, Phys. Rev. **D55** (1997) 5112 [hep-th/9610043].

[159] L. Susskind, [hep-th/9704080].

[160] A. Sen, hep-th/9709220, to appear in Adv. Theor. Math. Phys.

[161] N. Seiberg, Phys. Rev. Lett. **79** (1997) 3577 [hep-th/9710009].

[162] O. Ganor, S. Ramgoolam and W. Taylor, Nucl. Phys. **B492** (1997) 191 [hep-th/9611202].

[163] L. Susskind, [hep-th/9611164].

[164] M. Rozali, Phys. Lett. **B400** (1997) 260 [hep-th/9702136].

[165] M. Berkooz, M. Rozali and N. Seiberg, Phys. Lett. **B408** (1997) 105 [hep-th/9704089].

[166] N. Seiberg, Phys. Lett. **B408** (1997) 98 [hep-th/9705221].

[167] S. Govindarajan, Phys. Rev. **D56** (1997) 5276 [hep-th/9705113].

[168] M. Berkooz and M. Rozali, [hep-th/9705175].

[169] M. Berkooz, [hep-th/9712012].

[170] M. Blau and M. O'Loughlin, [hep-th/9712047].

[171] N. Obers, B. Pioline and E. Rabinovici, [hep-th/9712084].

[172] M. Douglas, D. Kabat, S. Poulio and S. Shenker, Nucl. Phys. **B485** (1997) 85 [hep-th/9608024].

[173] K. Becker, M. Becker, J. Polchinski and A. Tseytlin, Phys. Rev. **D56** (1997) 3174 [hep-th/9706072].

[174] S. Hellerman and J. Polchinski, [hep-th/9711037].

[175] M. Douglas, H. Ooguri and S. Shenker, Phys. Lett. **B402** (1997) 36 [hep-th/9702203].

[176] O. Ganor, R. Gopakumar and S. Ramgoolam, hep-th/9705188.

[177] J. David, A. Dhar and G. Mandal, hep-th/9707132.

[178] M. Douglas and H. Ooguri, [hep-th/9710178].

[179] M. Dine and A. Rajaraman, [hep-th/9710174].

[180] K. Becker and M. Becker, Nucl. Phys. **B506** (1997) 48 [hep-th/9705091].

[181] P. Ramond, Phys. Rev. **D3** (1971) 2415.

[182] A. Neveu and J. Schwarz, Nucl. Phys. **B31** (1971) 86; Phys. Rev. **D4** (1971) 1109.

[183] F. Gliozzi, J. Scherk and D. Olive, Phys. Lett. **65B** (1976) 282; Nucl. Phys. **B122** (1977) 253.

[184] P. Aspinwall and D. Morrison, [hep-th/9404151];
P. Aspinwall, [hep-th/9611137] and references therein.

[185] M. Duff and R. Minasian, Nucl. Phys. **B436** (1995) 507 [hep-th/9406198];
C. Vafa and E. Witten, Nucl. Phys. **B447** (1995) 261 [hep-th/9505053];
M. Duff, J. Liu and R. Minasian, Nucl. Phys. **B452** (1995) 261 [hep-th/9506126];
M. Duff, R. Minasian and E. Witten, Nucl. Phys. **B463** (1996) 435 [hep-th/9601036];
M. Berkooz, R. Leigh, J. Polchinski, J. Schwarz, N. Seiberg and E. Witten, Nucl. Phys. **B475** (1996) 115 [hep-th/9605184].

[186] S. Das, S. Kalyanarama, P. Ramadevi and S. Mathur, [hep-th/9711003].

[187] P. Candelas, G. Horowitz, A. Strominger and E. Witten, Nucl. Phys. **B258** (1985) 46.

[188] E. Gimon and C. Johnson, Nucl. Phys. **B479** (1996) 285 [hep-th/9606176];
J. Blum and A. Zaffaroni, Phys. Lett. **B387** (1996) 71 [hep-th/9607019];
A. Dabholkar and J. Park, Phys. Lett. **B394** (1997) 302 [hep-th/9607041];
R. Gopakumar and S. Mukhi, Nucl. Phys. **B479** (1996) 260 [hep-th/9607057];
J. Park, [hep-th/9611119];
A. Sen, Nucl. Phys. **B489** (1997) 139 [hep-th/9611186]; Nucl. Phys. **B498** (1997) 135; [hep-th/9702061]; Phys. Rev. **D55** (1997) 7345; [hep-th/9702165];
O. Aharony, J. Sonneschein, S. Yankielowicz and S. Theisen, Nucl. Phys. **B493** (1997) 177 [hep-th/9611222].

[189] M. Bershadsky, K. Intrilligator, S. Kachru, D. Morrison, V. Sadov and C. Vafa, Nucl. Phys. **B481** (1996) 215 [hep-th/9605200];
P. Aspinwall and M. Gross, Phys. Lett. **B387** (1996) 735 [hep-th/9605131].

[190] K. Dasgupta and S. Mukhi, Phys. Lett. **B385** (1996) 125 [hep-th/9606044].

[191] A. Johansen, Phys. Lett. **B395** (1997) 36 [hep-th/9608186].

[192] M. Gaberdiel and B. Zwiebach, [hep-th/9709013];
M. Gaberdiel, T. Hauer and B. Zwiebach, [hep-th/9801205].

[193] E. Witten, Nucl. Phys. **B500** (1997) 3 [hep-th/9703166].

[194] A. Sen, [hep-th/9707123] published in JHEP electronic journal.

[195] A. Kumar and C. Vafa, Phys. Lett. **B396** (1997) 85 [hep-th/9611007].

[196] E. Witten, Nucl. Phys. **B471** (1996) 135 [hep-th/9602070].

[197] A. Cadavid, A. Ceresole, R. D'Auria and S. Ferrara, [hep-th/9506144];
I. Antoniadis, S. Ferrara and T. Taylor, Nucl. Phys. **B460** (1996) 489;
S. Ferrara, R. Khuri and R. Minasian, [hep-th/9602102];
E. Witten, Nucl. Phys. **B471** (1996) 195 [hep-th/9603150].

[198] M. Duff, B. Nilsson and C. Pope, Phys. Rep. **130** (1986) 1.

[199] J. Cardy, Nucl. Phys. **B270** (1986) 186.

Lectures on D-branes

Constantin Bachas

1 Foreword

Referring in his *Republic* to stereography – the study of solid forms – Plato was saying: ... *for even now, neglected and curtailed as it is, not only by the many but even by professed students, who can suggest no use for it, nevertheless in the face of all these obstacles it makes progress on account of its elegance, and it would not be astonishing if it were unravelled.*[1] Two and a half millennia later, much of this could have been said for string theory. The subject has progressed over the years by leaps and bounds, despite periods of neglect and (understandable) criticism for lack of direct experimental input. To be sure, the construction and key ingredients of the theory – gravity, gauge invariance, chirality – have a firm empirical basis, yet what has often catalyzed progress is the power and elegance of the underlying ideas, which look (at least a posteriori) inevitable. And whether the ultimate structure will be unravelled or not, there is already a name waiting for 'it': $\mathcal{M}$ theory.

Few of the features of the theory uncovered so far exemplify this power and elegance better than D-branes. Their definition as allowed endpoints for open strings generalizes the notion of quarks on which the QCD string can terminate. In contrast to the quarks of QCD, D-branes are however intrinsic excitations of the fundamental theory: their existence is required for consistency, and their properties – mass, charges, dynamics – are unambiguously determined in terms of the Regge slope α' and the asymptotic values of the dynamical moduli. They resemble in these respects conventional field-theory solitons, from which however they differ in important ways. D-particles, for instance, can probe distances much smaller than the size of the fundamental-string quanta, at weak coupling. In any case, D-branes, fundamental strings and smooth solitons together fill the multiplets of the various (conjectured) dualities, which connect all string theories to each other. D-branes have, in this sense, played a crucial role in delivering the important message of the 'second string revolution', that the way to reconcile quantum mechanics and Einstein gravity may be so constrained as to be 'unique'.

Besides filling duality multiplets, D-branes have however also opened a window into the microscopic structure of quantum gravity. The D-brane model of black holes may prove as important for understanding black-hole

---

[1]Translated by Ivor Thomas in *Greek Mathematical Works*, Loeb Classical Library, Harvard University Press 1939.

thermodynamics as has the Ising model proven in the past for understanding second-order phase transitions. Technically, the D-brane concept is so powerful because of the surprising relations it has revealed between supersymmetric gauge theories and geometry. These relations follow from the fact that Riemann surfaces with boundaries admit dual interpretations as field-theory diagrams along various open- or closed-string channels. Thus, in particular, the counting of microscopic BPS states of a black hole, an ultraviolet problem of quantum gravity, can be mapped to the more familiar problem of studying the moduli space of supersymmetric gauge theories. Conversely, 'brane engineering' has been a useful tool for discussing Seiberg dualities and other infrared properties of supersymmetric gauge theories, while low-energy supergravity, corrected by classical string effects, may offer a new line of attack on the old problem of solving gauge theories in the planar ('t Hooft) limit.

Most of these exciting developments will not be discussed in the present lectures. The material included here covers only some of the earlier papers on D-branes, and is a modest expansion of a previous 'half lecture' by the author (Bachas 1997a). The main difference from other existing reviews of the same subject (Polchinski *et al.* 1996, Polchinski 1996, Douglas 1996, Thorlacius 1998, Taylor 1998) is in the emphasis and presentation style. The aim is to provide the reader (i) with a basis, from which to move on to reviews of related and/or more advanced topics, and (ii) with an extensive (though far from complete) guide to the literature. I will be assuming a working knowledge of perturbative string theory at the level of Green, Schwarz and Witten (1987) (see also Ooguri and Yin 1996, Kiritsis 1997, Dijkgraaf 1997, and volume one of Polchinski 1998, for recent reviews), and some familiarity with the main ideas of string duality, for which there exist many nice and complementary lectures (Townsend 1996b and 1997, Aspinwall 1996, Schwarz 1997a and 1997b, Vafa 1997, Dijkgraaf 1997, Förste and Louis 1997, de Wit and Louis 1998, Lüst 1998, Julia 1998, West in this volume, Sen in this volume).

A list of pedagogical reviews for further reading includes: Bigatti and Susskind (1997), Bilal (1997), Banks (1998), Dijkgraaf *et al.* (1998), and de Wit (1998) for the Matrix-model conjecture, Giveon and Kutasov (1998) for brane engineering of gauge theories, Maldacena (1996) and Youm (1997) for the D-brane approach to black holes, Duff *et al.* (1995), Duff (1997), Stelle (1997,1998), Youm (1997) and Gauntlett (1997) for reviews of branes from the complementary, supergravity viewpoint. I am not aware of any extensive reviews of type-I compactifications, of D-branes in general curved backgrounds, and of semiclassical calculations using D-brane instantons. Some short lectures on these subjects, which the reader may consult for further references, include Sagnotti (1997), Bianchi (1997), Douglas (1997), Green (1997), Gutperle (1997), Bachas (1997b), Vanhove (1997) and Antoniadis *et*

416   *Bachas*

al. (1998). Last but not least, dualities in rigid supersymmetric field theories
– a subject intimately tied to D-branes – are reviewed by Intriligator and
Seiberg (1995), Harvey (1996), Olive (1996), Bilal (1996), Alvarez–Gaumé
and Zamora (1997), Lerche (1997), Peskin (1997), Di Vecchia (1998) and
West (1998).

2 Ramond–Ramond fields

With the exception of the heterotic string, all other consistent string theories contain in their spectrum antisymmetric tensor fields coming from the
Ramond–Ramond sector. This is the case for the type-IIA and type-IIB superstrings, as well as for the type-I theory whose closed-string states are a
subset of those of the type-IIB. One of the key properties of D-branes is that
they are the elementary charges of Ramond–Ramond fields, so let us begin
the discussion by recalling some basic facts about these fields.

2.1 Chiral bispinors

The states of a closed-string theory are given by the tensor product of a
left- and a right-moving worldsheet sector. For type-II theory in the covariant (NSR) formulation, each sector contains at the massless level a ten-
dimensional vector and a ten-dimensional Weyl-Majorana spinor. This is
depicted figuratively as follows:

$$\Big(|\mu\rangle \oplus |a\rangle\Big)_{\text{left}} \otimes \Big(|\nu\rangle \oplus |b\rangle\Big)_{\text{right}} \ ,$$

where $\mu, \nu = 0, \ldots, 9$ and $a, b = 1, dots, 16$ are, respectively, vector and spinor
indices. Bosonic fields thus include a two-index tensor, which can be decomposed into a symmetric traceless, a trace, and an antisymmetric part: these
are the usual fluctuations of the graviton ($G_{\mu\nu}$), dilaton (Φ) and Neveu–
Schwarz antisymmetric tensor ($B_{\mu\nu}$). In addition, massless bosonic fields
include a Ramond–Ramond bispinor H_{ab}, defined as the polarization in the
corresponding vertex operator

$$V_{\text{RR}} \sim \int d^2\xi \, e^{ip^\mu X_\mu} \, \overline{S}^{\text{T}} \Gamma^0 H(p) S \ . \tag{2.1}$$

In this expression S^a and $\overline{S}^b$ are the covariant left- and right-moving fermion
emission operators – a product of the corresponding spin-field and ghost operators (Friedan *et al.* 1986), p^μ is the ten-dimensional momentum, and Γ^0
the ten-dimensional gamma matrix.

The bispinor field H can be decomposed in a complete basis of all gamma-matrix antisymmetric products

$$H_{ab} = \sum_{n=0}^{10} \frac{i^n}{n!} H_{\mu_1\ldots\mu_n}(\Gamma^{\mu_1\ldots\mu_n})_{ab} . \tag{2.2}$$

Here $\Gamma^{\mu_1\ldots\mu_n} \equiv \frac{1}{n!}\Gamma^{[\mu_1}\ldots\Gamma^{\mu_n]}$, where square brackets denote the alternating sum over all permutations of the enclosed indices, and the $n = 0$ term stands by convention for the identity in spinor space. I use the following conventions: the ten-dimensional gamma matrices are purely imaginary and obey the algebra $\{\Gamma^\mu,\Gamma^\nu\} = -2\eta^{\mu\nu}$ with metric signature $(-+\cdots+)$. The chirality operator is $\Gamma_{11} = \Gamma^0\Gamma^1\ldots\Gamma^9$, Majorana spinors are real, and the Levi-Civita tensor $\epsilon^{01\ldots9} = 1$.

In view of the decomposition (2.2), the Ramond–Ramond massless fields are a collection of antisymmetric Lorentz tensors. These tensors are not independent because the bispinor field is subject to definite chirality projections,

$$H = \Gamma_{11} H = \pm H \Gamma_{11} . \tag{2.3}$$

The choice of sign distinguishes between the type-IIA and type-IIB models. For the type-IIA theory S and $\overline{S}$ have opposite chirality, so one should choose the sign plus. In the type-IIB case, on the other hand, the two spinors have the same chirality and one should choose the sign minus. To express the chirality constraints in terms of the antisymmetric tensor fields we use the gamma-matrix identities

$$\Gamma^{\mu_1\ldots\mu_n}\Gamma_{11} = (-)^n\Gamma_{11}\Gamma^{\mu_1\ldots\mu_n} = \frac{\epsilon^{\mu_1\ldots\mu_{10}}}{(10-n)!}\Gamma_{\mu_{10}\ldots\mu_{n+1}} \tag{2.4}$$

It follows easily that only even-n (odd-n) terms are allowed in the type-IIA (type-IIB) case. Furthermore the antisymmetric tensors obey the duality relations

$$H^{\mu_1\ldots\mu_n} = \frac{\epsilon^{\mu_1\ldots\mu_{10}}}{(10-n)!} H_{\mu_{10}\ldots\mu_{n+1}} , \quad \text{or equivalently} \quad H^{(n)} = {}^*H^{(10-n)} . \tag{2.5}$$

As a check note that the type-IIA theory has independent tensors with $n = 0$, 2 and 4 indices, while the type-IIB theory has $n = 1,3$ and a self-dual $n = 5$ tensor. The number of independent tensor components adds up in both cases to $16 \times 16 = 256$:

$$\text{IIA}: \quad 1 + \frac{10\times9}{2!} + \frac{10\times9\times8\times7}{4!} = 256 ,$$

$$\text{IIB}: \quad 10 + \frac{10\times9\times8}{3!} + \frac{10\times9\times8\times7\times6}{2\times5!} = 256 .$$

This is precisely the number of components of a bispinor.

Finally let us consider the type-I theory, which can be thought of as an orientifold projection of type-IIB (Sagnotti 1988, Hořava 1989a). This projection involves an interchange of left- and right-movers on the worldsheet. The surviving closed-string states must be symmetric in the Neveu–Schwarz sector and antisymmetric in the Ramond–Ramond sector, consistently with supersymmetry and with the fact that the graviton should survive. This implies the extra condition on the bispinor field

$$(\Gamma^0 H)^{\mathrm{T}} = -\Gamma^0 H . \tag{2.6}$$

Using $(\Gamma^\mu)^{\mathrm{T}} = -\Gamma^0 \Gamma^\mu \Gamma^0$ we conclude, after some straightforward algebra, that the only Ramond–Ramond fields surviving the extra projection are $H^{(3)}$ and its dual, $H^{(7)}$.

2.2   Supergraviton multiplets

The mass-shell or super-Virasoro conditions for the vertex operator V_{RR} imply that the bispinor field obeys two massless Dirac equations,

$$\not{p} H = H \not{p} = 0 . \tag{2.7}$$

To convert these to equations for the tensors we need the gamma identities

$$
\Gamma^\mu \Gamma^{\nu_1 \ldots \nu_n} = \Gamma^{\mu\nu_1 \ldots \nu_n} - \frac{1}{(n-1)!} \, \eta^{\mu[\nu_1} \, \Gamma^{\nu_2 \ldots \nu_n]}
$$
$$
\Gamma^{\nu_1 \ldots \nu_n} \Gamma^\mu = \Gamma^{\nu_1 \ldots \nu_n \mu} - \frac{1}{(n-1)!} \, \eta^{\mu[\nu_n} \, \Gamma^{\nu_1 \ldots \nu_{n-1}]}
\tag{2.8}
$$

and the decomposition (2.2) of a bispinor. After some straightforward algebra one finds

$$p^{[\mu} H^{\nu_1 \ldots \nu_n]} = p_\mu H^{\mu\nu_2 \ldots \nu_n} = 0 . \tag{2.9}$$

These are the Bianchi identity and free massless equation for an antisymmetric tensor field strength in momentum space, which we may write in more economic form as

$$dH^{(n)} = d \, {}^* H^{(n)} = 0 \tag{2.10}$$

The polarizations of covariant Ramond–Ramond emission vertices are therefore field-strength tensors rather than gauge potentials.

Solving the Bianchi identity locally allows us to express the n-form field strength as the exterior derivative of an $(n-1)$-form potential

$$H_{\mu_1 \ldots \mu_n} = \frac{1}{(n-1)!} \, \partial_{[\mu_1} C_{\mu_2 \ldots \mu_n]}, \quad \text{or} \quad H^{(n)} = dC^{(n-1)} . \tag{2.11}$$

Thus the type-IIA theory has a vector (C_μ) and a three-index tensor potential $(C_{\mu\nu\rho})$, in addition to a constant non-propagating zero-form field strength $(H^{(0)})$, while the type-IIB theory has a zero-form (C), a two-form $(C_{\mu\nu})$ and a four-form potential $(C_{\mu\nu\rho\sigma})$, the latter with self-dual field strength. Only the two-form potential survives the type-I orientifold projection. These facts are summarized in table 1. A $(p + 1)$-form 'electric' potential can of course be traded for a $(7 - p)$-form 'magnetic' potential, obtained by solving the Bianchi identity of the dual field strength.

	Neveu–Schwarz	Ramond–Ramond
type-IIA	$G_{\mu\nu}, \Phi, B_{\mu\nu}$	$C_\mu, C_{\mu\nu\rho}$; $H^{(0)}$
type-IIB	$G_{\mu\nu}, \Phi, B_{\mu\nu}$	$C, C_{\mu\nu}, C_{\mu\nu\rho\tau}$
type-I	$G_{\mu\nu}, \Phi$	$C_{\mu\nu}$
heterotic	$G_{\mu\nu}, \Phi, B_{\mu\nu}$	

Table 1: String origin of massless fields completing the $N = 1$ or $N = 2$ supergraviton multiplet of the various theories in ten dimensions.

From the point of view of low-energy supergravity all Ramond–Ramond fields belong to the ten-dimensional graviton multiplet. For $N = 2$ supersymmetry this contains 128 bosonic helicity states, while for $N = 1$ supersymmetry it only contains 64. For both the type-IIA and type-IIB theories, half of these states come from the Ramond–Ramond sector, as can be checked by counting the transverse physical components of the gauge potentials:

$$\text{IIA}: \quad 8 + \frac{8 \times 7 \times 6}{3!} = 64 \,,$$

$$\text{IIB}: \quad 1 + \frac{8 \times 7}{2!} + \frac{8 \times 7 \times 6 \times 5}{2 \times 4!} = 64 \,.$$

This counting is simpler in the light-cone Green–Schwarz formulation, where the Ramond–Ramond fields correspond to a chiral SO(8) bispinor.

2.3   Dualities and RR charges

A $(p + 1)$-form potential couples naturally to a p-brane, i.e. an excitation extending over p spatial dimensions. Let $Y^\mu(\zeta^\alpha)$ be the worldvolume of the brane $(\alpha = 0, \ldots, p)$, and let

$$\widehat{C}^{(p+1)} \equiv C_{\mu_1 \ldots \mu_{p+1}}(Y) \, \partial_0 Y^{\mu_1} \ldots \partial_p Y^{\mu_{p+1}} \qquad (2.12)$$

be the pull-back of the $(p + 1)$-form on this worldvolume. The natural ('electric') coupling is given by the integral

$$I_{\mathrm{WZ}} = \rho_{(p)} \int d^{p+1}\zeta \; \widehat{C}^{(p+1)} , \qquad (2.13)$$

with $\rho_{(p)}$ the charge-density of the brane. Familiar examples are the coupling of a point-particle ('0-brane') to a vector potential, and of a string ('1-brane') to a two-index antisymmetric tensor. Since the dual of a $(p+1)$-form potential in ten dimensions is a $(7 - p)$-form potential, there exists also a natural ('magnetic') coupling to a $(6 - p)$-brane. The sources for the field equation and Bianchi identity of a $(p + 1)$-form are thus p-branes and $(6 - p)$-branes.

Now within type-II perturbation theory there are no such elementary RR sources. Indeed, if a closed-string state were a source for a RR $(p + 1)$-form, then the trilinear coupling

$$\langle \mathrm{closed}| \; C^{(p+1)} \; |\mathrm{closed}\rangle$$

would not vanish. This is impossible because the coupling involves an odd number of left-moving (and of right-moving) fermion emission vertices, so that the corresponding correlator vanishes automatically on any closed Riemann surface. What this arguments shows in particular, is that fundamental closed strings do not couple 'electrically' to the Ramond–Ramond two-form. It is significant, as we will see, that in the presence of worldsheet boundaries this simple argument will fail.

Most non-perturbative dualities require, on the other hand, the existence of such elementary RR charges. The web of string dualities in nine or higher dimensions, discussed in more detail in this volume by Sen (see also the other reviews listed in the introduction), has been drawn in figure 1. The web holds together the five ten-dimensional superstring theories, and the eleven-dimensional $\mathcal{M}$ theory, whose low-energy limit is eleven-dimensional supergravity (Cremmer $et\ al.$ 1978), and which has a (fundamental?) supermembrane (Bergshoeff $et\ al.$ 1987). The black one-way arrows denote compactifications of $\mathcal{M}$ theory on the circle S^1, and on the interval S^1/Z_2. In the small-radius limit these are respectively described by type-IIA string theory (Townsend 1995, Witten 1995), and by the $E_8 \times E_8$ heterotic model (Hořava and Witten 1996a, 1996b). The two-way black arrows identify the

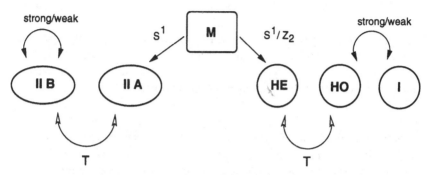

Figure 1: The web of dualities relating the ten-dimensional superstring theories and $\mathcal{M}$ theory, as described in the text.

strong-coupling limit of one theory with the weak-coupling limit of another. The type-I and heterotic SO(32) theories are related in this manner (Witten 1995, Polchinski and Witten 1996), while the type-IIB theory is self-dual (Hull and Townsend 1995). Finally, the two-way white arrows stand for perturbative T-dualities, after compactification on an extra circle (for a review of T-duality see Giveon *et al.* 1994).

Consider first the type-IIA theory, whose massless fields are given by dimensional reduction from eleven dimensions. The bosonic components of the eleven-dimensional multiplet are the graviton and a antisymmetric three-form, and they decompose in ten dimensions as follows:

$$G_{MN} \to G_{\mu\nu}, C_{\mu}, \Phi; \quad A_{MNR} \to C_{\mu\nu\rho}, B_{\mu\nu}, \qquad (2.14)$$

where M, N, $R = 0, \ldots, 10$. The eleven-dimensional supergravity has, however, also Kaluza–Klein excitations which couple to the off-diagonal metric components C_{μ}. Since this is a RR field in type-IIA theory, duality requires the existence of non-perturbative 0-brane charges. As concerns type-IIB string theory, its conjectured self-duality exchanges the two-forms ($B_{\mu\nu}$ and $C_{\mu\nu}$). Since fundamental strings are sources for the Neveu–Schwarz $B_{\mu\nu}$, this duality requires the existence of non-perturbative 1-branes coupling to the Ramond–Ramond $C_{\mu\nu}$ (Schwarz 1995).

Higher p-branes fit similarly in the conjectured web of dualities. This can be seen more easily after compactification to lower dimensions, where dualities typically mix the various fields coming from the Ramond–Ramond and Neveu–Schwarz sectors. For example, type-IIA theory compactified to six dimensions on a $K3$ surface is conjectured to be dual to the heterotic string compactified on a four-torus (Duff and Minasian 1995, Hull and Townsend

1995, Duff 1995, Witten 1995). The latter has extended gauge symmetry at special points of the Narain moduli space. On the type-IIA side the maximal abelian gauge symmetry has gauge fields that descend from the Ramond–Ramond three-index tensor. These can be enhanced to a non-abelian group only if there exist charged 2-branes wrapping around shrinking 2-cycles of the $K3$ surface (Bershadsky *et al.* 1996b). A similar phenomenon occurs for Calabi–Yau compactifications of type-IIB theory to four dimensions. The low-energy Lagrangian of Ramond–Ramond fields has a logarithmic singularity at special (conifold) points in the Calabi–Yau moduli space. This can be understood as due to nearly-massless 3-branes, wrapping around shrinking 3-cycles of the compact manifold, and which have been effectively integrated out (Strominger 1995). Strominger's observation was important for two reasons: (i) it provided the first example of a brane that becomes massless and can eventually condense (Ferrara *et al.* 1995, Kachru and Vafa 1995), and (ii) in this context the existence of RR-charged branes is not only a prediction of conjectured dualities – they *have to* exist because without them string theory would be singular and hence inconsistent.

3   D-brane tension and charge

The only fundamental quanta of string perturbation theory are elementary strings, so all other p-branes must arise as (non-perturbative) solitons. The effective low-energy supergravities exhibit, indeed, corresponding classical solutions (for reviews see Duff, Khuri and Lu 1995, Stelle 1997 and 1998, Youm 1997), but these are often singular and require the introduction of a source. One way to handle the corrections at the string scale is to look for (super)conformally-invariant σ-models, a lesson learned from the study of string compactifications. Callan *et al.* (1991a, 1991b) found such solitonic five-branes in both the type-II and the heterotic theories. Their branes involved only Neveu–Schwarz backgrounds – being ('magnetic') sources, in particular, for the two-index tensor $B_{\mu\nu}$. Branes with Ramond–Ramond backgrounds looked, however, hopelessly intractable: the corresponding σ-model would have to involve the vertex (2.1), which is made out of ghosts and spin fields and cannot, furthermore, be written in terms of two-dimensional superfields. Amazingly enough, these Ramond–Ramond charged p-branes turn out to admit a much simpler, exact and universal description as allowed endpoints for open strings, or D(irichlet)-branes (Polchinski 1995).

3.1   Open-string endpoints as defects

The bosonic part of the Polyakov action for a free fundamental string in flat space-time and in the conformal gauge reads[2]

$$I_{\rm F} = \int_\Sigma \frac{d^2\xi}{4\pi\alpha'}\, \partial_a X^\mu \partial^a X_\mu \,, \tag{3.1}$$

with Σ some generic surface with boundary. For its variation

$$\delta I_{\rm F} = -\int_\Sigma \frac{d^2\xi}{2\pi\alpha'}\, \delta X^\mu\, \partial_a\partial^a X_\mu + \int_{\partial\Sigma} \frac{d\xi^a}{2\pi\alpha'}\, \delta X^\mu \varepsilon_{ab}\partial^b X_\mu \tag{3.2}$$

to vanish, the X^μ must be harmonic functions on the worldsheet, and either of the following two conditions must hold on the boundary $\partial\Sigma$,

$$\begin{aligned} \partial_\perp X^\mu &=0 \quad \text{(Neumann)}, \\ \text{or} \quad \delta X^\mu &=0 \quad \text{(Dirichlet)} \,. \end{aligned} \tag{3.3}$$

Neumann conditions respect Poincaré invariance and are hence momentum-conserving. Dirichlet conditions, on the other hand, describe space-time defects. They have been studied in the past in various guises, for instance as sources for partonic behaviour in string theory (Green 1991 and references therein), as heavy-quark endpoints (Lüscher *et al.* 1980, Alvarez 1981), and as backgrounds for open-string compactification (Pradisi and Sagnotti 1989, Hořava 1989b, Dai *et al.* 1989). Their status of intrinsic non-perturbative excitations was not, however, fully appreciated in these earlier studies.

A static defect extending over p flat spatial dimensions is described by the boundary conditions

$$\partial_\perp X^{\alpha=0,\cdots,p} = X^{m=p+1,\cdots,9} = 0 \,, \tag{3.4}$$

which force open strings to move on a $(p+1)$-dimensional (worldvolume) hyperplane spanning the dimensions $\alpha=0,\cdots,p$. Since open strings do not propagate in the bulk in type-II theory, their presence is intimately tied to the existence of the defect, which we will refer to as a Dp-brane. Consider complex radial-time coordinates for the open string – these map a strip worldsheet onto the upper-half plane,

$$z = e^{\xi^0+i\xi^1} \quad (0<\xi^0<\infty,\ 0<\xi^1<\pi) \,. \tag{3.5}$$

The boundary conditions for the bosonic target-space coordinates then take the form

$$\partial X^\alpha = \bar\partial X^\alpha \Big|_{{\rm Im}z=0} \quad \text{and} \quad \partial X^m = -\bar\partial X^m \Big|_{{\rm Im}z=0} \,. \tag{3.6}$$

---

[2]I use the label $a,b\cdots$ both for space-time spinors and for the (Euclidean) worldsheet coordinates of a fundamental string – the context should, hopefully, help to avoid confusion.

Worldsheet supersymmetry imposes, on the other hand, the following boundary conditions on the worldsheet supercurrents (Green *et al.* 1987): $J_F = \epsilon \bar{J}_F$, where $\epsilon = +1$ in the Ramond sector, and $\epsilon = \text{sign}(\text{Im}z)$ in the Neveu–Schwarz sector. As a result the fermionic coordinates must obey

$$\psi^\alpha = \epsilon \, \bar{\psi}^\alpha \Big|_{\text{Im}z=0} \quad \text{and} \quad \psi^m = -\epsilon \, \bar{\psi}^m \Big|_{\text{Im}z=0} . \tag{3.7}$$

To determine the boundary conditions on spin fields, notice that their operator-product expansions with the fermions read (Friedan *et al.* 1986)

$$\psi^\mu(z)S(w) \sim (z-w)^{-1/2} \, \Gamma^\mu S(w) , \tag{3.8}$$

with a similar expression for right movers. Consistency with (3.7) imposes therefore the conditions,

$$S = \Pi_{(p)} \, \overline{S} \Big|_{\text{Im}z=0} , \tag{3.9}$$

where

$$\Pi_{(p)} = (i\Gamma_{11}\Gamma^{p+1})(i\Gamma_{11}\Gamma^{p+2}) \cdots (i\Gamma_{11}\Gamma^9) \tag{3.10}$$

is a real operator which commutes with all Γ^α and anticommutes with all Γ^m. Since $\Pi_{(p)}$ flips the spinor chirality for p even, only even-dimensional Dp-branes are allowed in type-IIA theory. For the same reason type-IIB and type-I theories allow only for odd-dimensional Dp-branes. In the type-I case we furthermore demand that (3.9) be symmetric under the interchange $S \leftrightarrow \overline{S}$. This implies $\Pi_{(p)}^2 = 1$, which is true only for $p = 1, 5$ and 9. All these facts are summarized in table 2.

The case $p = 9$ is degenerate, since it implies that open strings can propagate in the bulk of space-time. This is only consistent in type-I theory, i.e. when there are 32 D9-branes and an orientifold projection. The other Dp-branes listed in the table are in one-to-one correspondence with the 'electric' Ramond-Ramond potentials of table 1, and their 'magnetic' duals. We will indeed verify that they couple to these potentials as elementary sources. The effective action of a Dp-brane, with tension $T_{(p)}$ and charge density under the corresponding Ramond–Ramond $(p+1)$-form $\rho_{(p)}$, reads

$$I_{Dp} = \int d^{p+1}\zeta \left(T_{(p)} \, e^{-\Phi} \sqrt{-\det \widehat{G}_{\alpha\beta}} + \rho_{(p)} \, \widehat{C}^{(p+1)} \right), \tag{3.11}$$

where

$$\widehat{G}_{\alpha\beta} = G^{\mu\nu} \partial_\alpha Y_\mu \partial_\beta Y_\nu \tag{3.12}$$

is the induced worldvolume metric. The cases $p = -1, 7, 8$ are somewhat special. The D(-1)-brane sits at a particular space-time point and must be interpreted as a (Euclidean) instanton with action

$$I_{D(-1)} = T_{(-1)} \, e^{-\Phi} + i\rho_{(-1)} \, C^{(0)} \Big|_{\text{position}} . \tag{3.13}$$

type-IIA	$p = 0, 2, 4, 6, 8$
type-IIB	$p = -1, 1, 3, 5, 7, (9)$
type-I	$p = 1, 5, 9$

Table 2: The Dp-branes of the various string theories are (with the exception of the D9-brane) in one-to-one correspondence with the 'electric' Ramond–Ramond potentials of table 1, and their 'magnetic' duals. The two heterotic theories have no Ramond–Ramond fields and no Dp-branes.

Its 'magnetic' dual, in a sense to be made explicit below, is the D7-brane. Finally the D8-brane is a domain wall coupling to the non-propagating nine-form, i.e. separating regions with different values of $H^{(0)}$ (Polchinski and Witten 1996, Bergshoeff *et al.* 1996).

The values of $T_{(p)}$ and $\rho_{(p)}$ could be extracted in principle from one-point functions on the disk. Following Polchinski (1995) we will prefer to extract them from the interaction energy between two static identical D-branes. This approach will spare us the technicalities of normalizing vertex operators correctly, and will furthermore extend naturally to the study of dynamical D-brane interactions (Bachas 1996).

3.2 Static force: field-theory calculation

Viewed as solitons of ten-dimensional supergravity, two D-branes interact by exchanging gravitons, dilatons and antisymmetric tensors. This is a good approximation, provided their separation r is large compared to the fundamental string scale. The supergravity Lagrangian for the exchanged bosonic fields reads (see Green *et al.* 1987)

$$I_{\text{IIA,B}} = -\frac{1}{2\kappa_{(10)}^2} \int d^{10}x\sqrt{-G}\left[e^{-2\Phi}\left(R - 4(d\Phi)^2 + \frac{1}{12}(dB)^2\right) + \sum\frac{1}{2n!}H^{(n)\,2}\right]$$
(3.14)

where $n = 0, 2, 4$ for type-IIA theory, $n = 1, 3$ for type-IIB, while for the self-dual field-strength $H^{(5)}$ there is no covariant action we may write down. Since this is a tree-Lagrangian of closed-string modes, it is multiplied by the usual factor $e^{-2\Phi}$ corresponding to spherical worldsheet topology. The D-brane Lagrangian (3.11), on the other hand, is multiplied by a factor $e^{-\Phi}$, corresponding to the topology of the disk. The disk is indeed the lowest-genus diagram with a worldsheet boundary which can feel the presence of the D-brane. These dilaton pre-factors have been absorbed in the terms involving Ramond–Ramond fields through a rescaling

$$C^{(p+1)} \to e^{\Phi} C^{(p+1)} \; . \tag{3.15}$$

A careful analysis shows indeed that it is the field strengths of the rescaled potentials which satisfy the usual Bianchi identity and Maxwell equation when the dilaton varies (Callan et al. 1988, Li 1996b, Polyakov 1996).

To decouple the propagators of the graviton and dilaton, we pass to the Einstein metric

$$g_{\mu\nu} = e^{-\Phi/2} G_{\mu\nu} \; , \tag{3.16}$$

in terms of which the effective actions take the form

$$I_{\text{IIA,B}} \;=\; -\frac{1}{2\kappa_{(10)}^2} \int d^{10}x \sqrt{-g} \left[R + \frac{1}{2}(d\Phi)^2 + \frac{1}{12} e^{-\Phi}(dB)^2 \right.$$
$$\left. + \sum \frac{1}{2(p+2)!} e^{(3-p)\Phi/2} (dC^{(p+1)})^2 \right] \tag{3.17}$$

and

$$I_{\text{D}p} \;=\; \int d^{p+1}\zeta \left(T_{(p)} \, e^{(p-3)\Phi/4} \sqrt{-\det \widehat{g}_{\alpha\beta}} \,+\, \rho_{(p)} \, \widehat{C}^{(p+1)} \right). \tag{3.18}$$

To leading order in the gravitational coupling the interaction energy comes from the exchange of a single graviton, dilaton or Ramond–Ramond field, and reads

$$\mathcal{E}(r) \, \delta T = -2\kappa_{(10)}^2 \int d^{10}x \int d^{10}\tilde{x} \left[j_\Phi \Delta \tilde{j}_\Phi - j_C \Delta \tilde{j}_C + T_{\mu\nu} \Delta^{\mu\nu,\rho\tau} \tilde{T}_{\rho\tau} \right] \tag{3.19}$$

Here j_Φ, j_C and $T_{\mu\nu}$ are the sources for the dilaton, Ramond–Ramond form and graviton obtained by linearizing the worldvolume action for one of the branes, while the tilde quantities refer to the other brane. Δ and $\Delta^{\mu\nu,\rho\tau}$ are the scalar and the graviton propagators in ten dimensions, evaluated at the argument $(x - \tilde{x})$, and δT the total interaction time. To simplify notation, and since only one component of $C^{(p+1)}$ couples to a static planar Dp-brane, we have dropped the obvious tensor structure of the antisymmetric field.

The sources for a static planar defect take the form

$$j_\Phi \;=\; \frac{p-3}{4} \, T_{(p)} \, \delta(x^\perp)$$
$$j_C \;=\; \rho_{(p)} \, \delta(x^\perp) \tag{3.20}$$

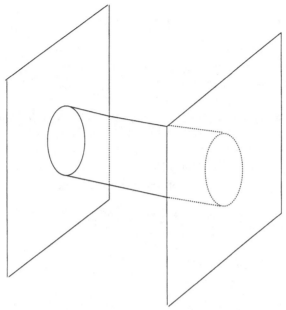

Figure 2: Two D-branes interacting through the exchange of a closed string. The diagram has a dual interpretation as Casimir force due to vacuum fluctuations of open strings.

$$T_{\mu\nu} = \frac{1}{2}T_{(p)}\, \delta(x^{\perp}) \times \left\{ \begin{array}{ll} \eta_{\mu\nu} & \text{if } \mu,\nu \leq p \\ 0 & \text{otherwise} \end{array} \right.$$

where the δ-function localizes the defect in transverse space. The tilde sources are taken identical, except that they are localized at distance r away in the transverse plane. The graviton propagator in the De Donder gauge and in d dimensions reads (Veltman 1975)

$$\Delta_{(d)}^{\mu\nu,\rho\tau} = \left(\eta^{\mu\rho}\eta^{\nu\tau} + \eta^{\mu\tau}\eta^{\nu\rho} - \frac{2}{d-2}\eta^{\mu\nu}\eta^{\rho\tau}\right)\Delta_{(d)}\,, \qquad (3.21)$$

where

$$\Delta_{(d)}(x) = \int \frac{d^d p}{(2\pi)^d}\frac{e^{ipx}}{p^2}\,. \qquad (3.22)$$

Putting all this together and doing some straightforward algebra we obtain

$$\mathcal{E}(r) = 2V_{(p)}\kappa_{(10)}^2\,[\rho_{(p)}^2 - T_{(p)}^2]\,\Delta_{(9-p)}^E(r)\,, \qquad (3.23)$$

where $V_{(p)}$ is the (regularized) p-brane volume and $\Delta_{(9-p)}^E(r)$ is the (Euclidean) scalar propagator in $(9-p)$ transverse dimensions. The net force is as should be expected the difference between Ramond–Ramond repulsion and gravitational plus dilaton attraction.

3.3   Static force: string calculation

The exchange of all closed-string modes, including the massless graviton, dilaton and $(p+1)$-form, is given by the cylinder diagram of figure 2. Viewed as an annulus, this same diagram also admits a dual and, from the field-theory point of view, surprising interpretation: the two D-branes interact by modifying the vacuum fluctuations of (stretched) open strings, in the same way that two superconducting plates attract by modifying the vacuum fluctuations of the photon field. It is this simple-minded duality which may, as we will see below, revolutionize our thinking about space-time.

The one-loop vacuum energy of oriented open strings reads

$$\mathcal{E}(r) = -\frac{V_{(p)}}{2} \int \frac{d^{p+1}k}{(2\pi)^{p+1}} \int_0^\infty \frac{dt}{t}\, \text{Str}\, e^{-\pi t(k^2+M^2)/2} =$$

$$= -2 \times \frac{V_{(p)}}{2} \int_0^\infty \frac{dt}{t}\, (2\pi^2 t)^{-(p+1)/2}\, e^{-r^2 t/2\pi}\, Z(t)\,,$$

(3.24)

where

$$Z(t) = -\frac{1}{2} \sum_{s=2,3,4} (-)^s\, \frac{\theta_s^4\left(0\mid\frac{it}{2}\right)}{\eta^{12}\left(\frac{it}{2}\right)}$$

(3.25)

is the usual spin structure sum obtained by supertracing over open-string oscillator states (see Green *et al.* 1987), and we have set $\alpha' = 1/2$. Strings stretching between the two D-branes have at the Nth oscillator level a mass $M^2 = (r/\pi)^2 + 2N$, so that their vacuum fluctuations are modified when we separate the D-branes. The vacuum energy of open strings with both endpoints on the same defect is, on the other hand, r-independent and has been omitted. Notice also the (important) factor of 2 in front of the second line: it accounts for the two possible orientations of the stretched string,

The first remark concerning the above expression is that it vanishes by the well-known θ-function identity. Comparing with eq. (3.23) we conclude that

$$T_{(p)} = \rho_{(p)}\,,$$

(3.26)

so that Ramond–Ramond repulsion cancels exactly the gravitational and dilaton attraction. As will be discussed in detail later on, this cancellation of the static force is a consequence of space-time supersymmetry. It is similar to the cancellation of Coulomb repulsion and Higgs-scalar attraction between 't Hooft–Polyakov monopoles in $N = 4$ supersymmetric Yang–Mills (see for example Harvey 1996) .

To extract the actual value of $T_{(p)}$ we must separate in the diagram the exchange of RR and NS–NS closed-string states. These are characterized by worldsheet fermions which are periodic, respectively antiperiodic around the cylinder, so that they correspond to the $s = 4$, respectively $s = 2, 3$ open-string spin structures. In the large-separation limit $(r \to \infty)$ we may furthermore expand the integrand near $t \sim 0$:

$$Z(t) \simeq (8 - 8) \times \left(\frac{t}{2}\right)^4 + o(e^{-1/t}) , \qquad (3.27)$$

where we have here used the standard θ-function asymptotics. Using also the integral representation

$$\Delta_{(d)}^E(r) = \frac{\pi}{2} \int_0^\infty dl \, (2\pi^2 l)^{-d/2} \, e^{-r^2/2\pi l} , \qquad (3.28)$$

and restoring correct mass units we obtain

$$\mathcal{E}(r) = V_{(p)} \, (1 - 1) \, 2\pi(4\pi^2\alpha')^{3-p} \, \Delta_{(9-p)}^E(r) \, + \, o(e^{-r/\sqrt{\alpha'}}) . \qquad (3.29)$$

Comparing with the field-theory calculation we can finally extract the tension and charge-density of type-II Dp-branes ,

$$T_{(p)}^2 = \rho_{(p)}^2 = \frac{\pi}{\kappa_{(10)}^2}(4\pi^2\alpha')^{3-p} . \qquad (3.30)$$

These are determined unambiguously, as should be expected for intrinsic excitations of a fundamental theory. Notice that in the type-I theory the above interaction energy should be multiplied by one half, because the stretched open strings are unoriented. The tensions and charge densities of type-I D-branes are, therefore, smaller than those of their type-IIB counterparts by a factor of $1/\sqrt{2}$.

4 Consistency and duality checks

String dualities and non-perturbative consistency impose a number of relations among the tensions and charge densities of D-branes, which we will now discuss. We will verify, in particular, that the values (3.30) are consistent with T-duality, with Dirac charge quantization, as well as with the existence of an eleventh dimension. From the string-theoretic point of view, the T-duality relations are the least surprising, since the symmetry is automatically built into the genus expansion. Verifying these relations is simply a check of the annulus calculation of the previous section. That the results obey also the Dirac conditions is more rewarding, since these test the non-perturbative consistency of the theory. What is, however, most astonishing is the fact that the annulus calculation 'knows' about the existence of the eleventh dimension.

4.1   Charge quantization

Dirac's quantization condition for electric and magnetic charge (Dirac 1931) has an analog for extended objects in higher dimensions (Nepomechie 1985, Teitelboim 1986a,b).[3] Consider a Dp-brane sitting at the origin, and integrate the field equation of the Ramond–Ramond form over the transverse space. Using Stokes' theorem one finds

$$\int_{S_{(8-p)}} {}^*H^{(p+2)} = 2\kappa_{(10)}^2 \, \rho_{(p)} \tag{4.1}$$

where $S_{(8-p)}$ is a (hyper)sphere, surrounding the defect, in transverse space. This equation is the analog of Gauss' law. Now Poincaré duality tells us that

$$ {}^*H^{(p+2)} = H^{(8-p)} \simeq dC^{(7-p)} \, , \tag{4.2}$$

where the potential $C^{(7-p)}$ is not globally defined because the Dp-brane is a source in the Bianchi identity for $H^{(8-p)}$. Following Dirac we may define a smooth potential everywhere, except along a singular (hyper)string which drills a hole in $S_{(8-p)}$. The hole is topologically equivalent to the interior of a hypersphere $S_{(7-p)}$. These facts are easier to visualize in three-dimensional space, where a point defect creates a string singularity which drills a disk out of a two-sphere, while a string defect creates a sheet singularity which drills a segment out of a circle, as in figure 3.

The Dirac singularity is dangerous because a Bohm–Aharonov experiment involving $(6-p)$-branes might detect it. Indeed, the wave-function of a $(6-p)$-brane transported around the singularity picks up a phase

$$\text{Phase} = \rho_{(6-p)} \int_{S_{(7-p)}} C^{(7-p)} = \rho_{(6-p)} \int_{S_{(8-p)}} H^{(8-p)} \, . \tag{4.3}$$

For the (hyper)string to be unphysical, this phase must be an integer multiple of 2π. Putting together equations (4.1-4.3) we thus find the condition

$$\text{Phase} = 2\kappa_{(10)}^2 \rho_{(p)} \rho_{(6-p)} = 2\pi n \, . \tag{4.4}$$

The charge densities (3.30) satisfy this condition with $n = 1$. D-branes are therefore the minimal Ramond–Ramond charges allowed in the theory, so one may conjecture that there are no others.

Dirac's argument is, strictly speaking, valid only for[4] $0 \le p \le 6$. In order to extend it to the pair $p = -1, 7$, note that a D7-brane creates a (hyper)-sheet

---

[3]Schwinger (1968) and Zwanzinger (1968) extended Dirac's argument to dyons. The generalization of their argument to higher dimensions involves a subtle sign discussed recently by Deser *et al.* (1997, 1998).

[4]The D3-brane is actually also special, since it couples to a self-dual four-form.

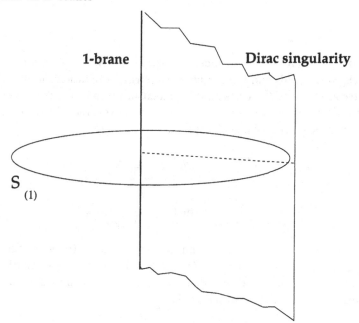

Figure 3: A 1-brane creates a 3-index 'electric' field $H^{(3)}$. Electric flux in $d = 4$ space-time dimensions is given by an integral of the dual vector over a circle $S_{(1)}$. The 'magnetic' potential is a scalar field, coupling to point-like (Euclidean) instantons, and jumping discontinuously across the depicted sheet singularity.

singularity across which the Ramond–Ramond scalar, $C^{(0)}$, jumps discontinuously by an amount $2\kappa^2_{(10)}\rho_{(7)}$. Dirac quantization ensures that the exponential of the (Euclidean) instanton action (3.13) has no discontinuity across the sheet, whose presence cannot therefore be detected by non-perturbative physics. It is the four-dimensional analog of this special case that is, as a matter of fact, illustrated in figure 3.

A final comment concerns the type-I theory, where the extra factor of $\frac{1}{\sqrt{2}}$ in the charge densities seems to violate the quantization condition. The puzzle is resolved by the observation (Witten 1996b) that the dynamical five-brane excitation consists of a *pair* of coincident D5-branes, so that

$$\rho^I_{(1)} = \sqrt{\frac{\pi}{2\kappa^2_{(10)}}}\,(4\pi^2\alpha') \quad \text{and} \quad \rho^I_{(5)} = 2 \times \sqrt{\frac{\pi}{2\kappa^2_{(10)}}}\,(4\pi^2\alpha')^{-1}\,. \qquad (4.5)$$

This is consistent with heterotic/type-I duality, as well as with the fact that the orientifold projection removes the collective coordinates of a single, isolated D5-brane (Gimon and Polchinski 1996).

4.2 *T*-duality

T-duality is a discrete gauge symmetry of string theory, that transforms both
the background fields and the perturbative (string) excitations around them
(see Giveon *et al.* 1994). The simplest context in which it occurs is compactification of type-II theory on a circle. The general expression for the compact
(ninth) coordinate of a closed string is

$$z\partial X^9 = \frac{i}{2}\left(\frac{n_9\alpha'}{R_9} + m_9 R_9\right) + i\sqrt{\frac{\alpha'}{2}}\sum_{k\neq 0} a_k^9\, z^{-k}$$

$$\bar{z}\bar{\partial} X^9 = \frac{i}{2}\left(\frac{n_9\alpha'}{R_9} - m_9 R_9\right) + i\sqrt{\frac{\alpha'}{2}}\sum_{k\neq 0} \tilde{a}_k^9\, \bar{z}^{-k} \tag{4.6}$$

Here n_9 and m_9 are the quantum numbers corresponding to momentum and
winding, and $z = e^{\xi^0+i\xi^1}$ with $0 \leq \xi^1 < 2\pi$. A *T*-duality transformation in-
verts the radius of the circle, interchanges winding with momentum numbers,
and flips the sign of right-moving oscillators:

$$R_9' = \frac{\alpha'}{R_9}, \quad (n_9', m_9') = (m_9, n_9) \text{ and } \tilde{a}_k^9{}' = -\tilde{a}_k^9. \tag{4.7}$$

It also shifts the expectation value of the dilaton, so as to leave the nine-
dimensional Planck scale unchanged,

$$\frac{R_9'}{\kappa_{(10)}'^2} = \frac{R_9}{\kappa_{(10)}^2}. \tag{4.8}$$

The transformation (4.7) can be thought of as a (hybrid) parity operation
restricted to the antiholomorphic worldsheet sector:

$$\bar{\partial} X^9{}' = -\bar{\partial} X^9. \tag{4.9}$$

Since the parity operator in spinor space is $i\Gamma^9\Gamma_{11}$, bispinor fields will transform accordingly as follows:

$$H' = iH\,\Gamma^9\Gamma_{11}. \tag{4.10}$$

Using the gamma-matrix identities of section 2, we may rewrite this relation
in component form,

$$H'_{\mu_1...\mu_n} = H_{9\,\mu_1...\mu_n} \text{ and } H'_{9\,\mu_1...\mu_n} = -H_{\mu_1...\mu_n}, \tag{4.11}$$

for any $\mu_i \neq 9$. *T*-duality exchanges therefore even-*n* with odd-*n* antisymmetric field strengths, and hence also type-IIA with type-IIB backgrounds.
Consistency requires that it also transform even-*p* to odd-*p* D-branes and vice
versa.

To see how this comes about let us consider a $D(p + 1)$-brane wrapping around the ninth dimension. We concentrate on the ninth coordinate of an open string living on this D-brane. It can be expressed as the sum of the holomorphic and anti-holomorphic pieces (4.6), with an extra factor two multiplying the zero modes because the open string is parametrized by $\xi^1 \in [0, \pi]$. The Neumann boundary condition $\partial X^9 = \bar{\partial} X^9$ at real z forces furthermore the identifications

$$a_k = \tilde{a}_k \ , \quad \text{and} \quad m_9 = 0 \ . \tag{4.12}$$

This is consistent with the fact that open strings can move freely along the ninth dimension on the D-brane, but cannot wind.

Now a T-duality transformation flips the sign of the antiholomorphic piece, changing the Neumann to a Dirichlet condition,[5]

$$a'_k = -\tilde{a}'_k \ , \quad \text{and} \quad n'_9 = 0 \ . \tag{4.13}$$

The wrapped $D(p + 1)$-brane is thus transformed, in the dual theory, to a Dp-brane localized in the ninth dimension (Hořava 1989b, Dai *et al.* 1989, Green 1991). Open strings cannot move along this dimension anymore, but since their endpoints are fixed on the defect they can now wind. The inverse transformation is also true: a Dp-brane, originally transverse to the ninth dimension, transforms to a wrapped $D(p + 1)$-brane in the dual theory. All this is compatible with the transformation (4.11) of Ramond–Ramond fields, to which the various D-branes couple. Furthermore, since a gauge transformation should not change the (nine-dimensional) tension of the defect, we must have

$$2\pi R_9 \, T_{(p+1)} = T'_{(p)} \ . \tag{4.14}$$

Using the formulae (4.7-4.8) one can check that the D-brane tensions indeed verify this T-duality constraint. Conversely, T-duality plus the minimal Dirac quantization condition fix unambiguously the expression (3.30) for the D-brane tensions.

4.3 Evidence for d=11

The third and most striking set of relations are those derived from the conjectured duality between type-IIA string theory and $\mathcal{M}$ theory compactified on a circle (Witten 1995, Townsend 1995). The eleven-dimensional supergravity couples consistently to a supermembrane (Bergshoeff *et al.* 1987), and has furthermore a ('magnetic') five-brane (Güven 1992) with a non-singular geometry (Gibbons *et al.* 1995). After compactification on the circle there exist also Kaluza-Klein modes, as well as a Kaluza-Klein monopole given by the

---

[5]For general curved backgrounds with abelian isometries this has been discussed by Alvarez *et al.* 1996, and by Dorn and Otto 1996.

Taub-NUT$\times R^7$ space (Sorkin 1983, Gross and Perry 1983). The correspondence between these excitations and the various branes on the type-IIA side is shown in table 3. The missing entry in this table is the eleven-dimensional counterpart of the D8-brane, which has not yet been identified (for a recent attempt see Bergshoeff et $al.$ 1997). The problem is that massive type-IIA supergravity (Romans 1986), which prevails on one side of the wall (Polchinski and Witten 1996, Bergshoeff et $al.$ 1996), seems to have no ancestor in eleven dimensions (Bautier et $al.$ 1997, Howe et $al.$ 1998).

Setting aside the D8-brane, let us consider the tensions of the remaining excitations listed in table 3. The tensions are expressed in terms of $\kappa_{(10)}$ and the Regge slope on the type-IIA side, and in terms of $\kappa_{(11)}$ and the compactification radius on the $\mathcal{M}$-theory side. To compare sides we must identify the ten-dimensional Planck scales,

$$\frac{1}{\kappa_{(10)}^2} = \frac{2\pi R_{11}}{\kappa_{(11)}^2} \ . \tag{4.15}$$

Equating the fundamental string tension (T_F) with the tension of a wrapped membrane fixes also α' in terms of eleven-dimensional parameters. This leaves us with five consistency checks of the conjectured duality, which are indeed explicitly verified.

How much of this truly tests the eleven-dimensional origin of string theory? To answer the question we must first understand how the entries on the $\mathcal{M}$-theory side of table 3 are obtained. Because of the scale invariance of the supergravity equations, the tensions of the classical membrane and five-brane solutions are a priori arbitrary. Assuming minimal Dirac quantization, and the BPS equality of mass and charge, fixes the product

$$2\kappa_{(11)}^2 T_2^M T_5^M = 2\pi \ . \tag{4.16}$$

An argument fixing each of the tensions separately was first given by Duff, Liu and Minasian (1995) and further developed by de Alwis (1996,1997) and Witten (1997a).[6] It uses the Chern–Simons term of the eleven-dimensional Lagrangian,

$$I_{11d} = -\frac{1}{2\kappa_{(11)}^2} \int d^{11}x \left[\sqrt{-G} \left(R + \frac{1}{48}(dA)^2 \right) + \frac{1}{6} A \wedge dA \wedge dA \right] \ , \tag{4.17}$$

where $A \equiv \frac{1}{3!} A_{MNR} \, dx^M \wedge dx^N \wedge dx^R$ is the three-index antisymmetric form encountered already in section 2. In a nutshell, the coefficient of this Chern–Simons term is fixed by supersymmetry (Cremmer et $al.$ 1978), but in the presence of electric and magnetic sources it is also subject to an independent quantization condition.[7]

---

[6]See also Lu (1997), Brax and Mourad (1997, 1998) and Conrad (1997).

[7]The quantization of the abelian Chern–Simons term in the presence of a magnetic source was first discussed in 2+1 dimensions (Henneaux and Teitelboim 1986, Polychronakos 1987).

tension	type-IIA	$\mathcal{M}$ on S^1	tension
$\dfrac{\sqrt{\pi}}{\kappa_{(10)}}(2\pi\sqrt{\alpha'})^3$	D0-brane	K-K excitation	$\dfrac{1}{R_{11}}$
$T_F = (2\pi\alpha')^{-1}$	string	wrapped membrane	$2\pi R_{11}\left(\dfrac{2\pi^2}{\kappa_{(11)}^2}\right)^{1/3}$
$\dfrac{\sqrt{\pi}}{\kappa_{(10)}}(2\pi\sqrt{\alpha'})$	D2-brane	membrane	$T_2^M = \left(\dfrac{2\pi^2}{\kappa_{(11)}^2}\right)^{1/3}$
$\dfrac{\sqrt{\pi}}{\kappa_{(10)}}(2\pi\sqrt{\alpha'})^{-1}$	D4-brane	wrapped five-brane	$R_{11}\left(\dfrac{2\pi^2}{\kappa_{(11)}^2}\right)^{2/3}$
$\dfrac{\pi}{\kappa_{(10)}^2}(2\pi\alpha')$	NS-five-brane	five-brane	$\dfrac{1}{2\pi}\left(\dfrac{2\pi^2}{\kappa_{(11)}^2}\right)^{2/3}$
$\dfrac{\sqrt{\pi}}{\kappa_{(10)}}(2\pi\sqrt{\alpha'})^{-3}$	D6-brane	K-K monopole	$\dfrac{2\pi^2 R_{11}^2}{\kappa_{(11)}^2}$
$\dfrac{\sqrt{\pi}}{\kappa_{(10)}}(2\pi\sqrt{\alpha'})^{-5}$	D8-brane	?	?

Table 3: Correspondence of BPS excitations of type-IIA string theory, and of $\mathcal{M}$ theory compactified on a circle. Equating tensions and the ten-dimensional Planck scale on both sides gives seven relations for two unknown parameters. Supersymmetry and consistency imply three Dirac quantization conditions, leaving us with two independent checks of the conjectured duality.

Let me describe the argument in the simpler context of five-dimensional Maxwell theory with a (abelian) Chern–Simons term,

$$I_{5d}^{MCS} = -\frac{1}{2\kappa_{(5)}^2} \int d^5x \left(\frac{1}{4}F^2 + \frac{k}{6} \, A \wedge F \wedge F \right) . \qquad (4.18)$$

Assume that the theory has both elementary electric charges q (coupling through $I_{WZ} = q \int A_\mu dx^\mu$), and dual minimally-charged magnetic strings. If we compactify the fourth spatial dimension on a circle of radius L, the effective four-dimensional action reads

$$I_{4d}^{MCS} = -\frac{1}{2\kappa_{(4)}^2} \int d^4x \left(\frac{1}{4}F^2 + \frac{1}{2}(da)^2 + \frac{k}{2} \, a \, F \wedge F \right) , \qquad (4.19)$$

where $\kappa_{(5)}^2 = 2\pi L \kappa_{(4)}^2$ and $a = A_4$. The scalar field a must be periodically identified, since under a large gauge transformation

$$a \to a + \frac{1}{qL} . \qquad (4.20)$$

Such a shift changes, however, the θ-term of the four-dimensional Lagrangian, and is potentially observable through the Witten effect, namely as a shift in the electric charge of a magnetic monopole (Witten 1979). This latter is a magnetic string wrapping around the compact fourth dimension. To avoid an immediate contradiction we must require that the induced charge be an integer multiple of q, so that it can be screened by elementary charges bound to the monopole.

In order to quantify this requirement, consider the θ-term resulting from the shift (4.20). In the background of a monopole field it will give rise to an interaction (Coleman 1981)

$$-\frac{k}{2\kappa_{(4)}^2 qL} \int d^4x \, F_{r0}^* {}^*F_{r0}^{(monopole)} = \frac{2\pi^2 k}{\kappa_{(5)}^2 q^2} \int dt \, A_0 , \qquad (4.21)$$

where we have here integrated by parts and used the monopole equation $\partial_r {}^*F_{r0}^{(monopole)} = (2\pi/q) \, \delta^{(3)}(\vec{r})$. The interaction (4.21) describes precisely the Witten effect, i.e. the fact that the magnetic monopole has acquired a non-vanishing electric charge. Demanding that the induced charge be an integer multiple of q leads, finally, to the quantization condition

$$q^3 = \frac{2\pi^2 k}{\kappa_{(5)}^2 n} . \qquad (4.22)$$

This is the sought-for relation between the coefficient of the (abelian) Chern–Simons term and the elementary electric charge of the theory.

Let us apply now the same reasoning to $\mathcal{M}$-theory. Compactifying to eight dimensions on a three-torus gives an effective eight-dimensional theory with both electric and magnetic membranes. The latter are the wrapped five-branes of $\mathcal{M}$-theory, which may acquire an electric charge through a generalized Witten effect. Demanding that a large gauge transformation induce a charge that can be screened by elementary membranes leads to the quantization condition

$$(T_2^M)^3 = \frac{2\pi^2}{\kappa_{(11)}^2 n} \ . \tag{4.23}$$

This relates the electric charge density or membrane tension, T_2^M, to the coefficient, $k = 1$, of the Chern–Simons term. The membrane tension predicted by duality corresponds to the maximal allowed case $n = 1$.

We can finally return to our original question: How much evidence for the existence of an eleventh dimension in string theory does the 'gedanken data' of table 3 contain? Note first that Dirac quantization relates the six tensions pairwise. Furthermore, since the maximal non-chiral 10d supergravity is unique, it must contain a $B \wedge H^{(4)} \wedge H^{(4)}$ term obtained from the 11d Chern–Simons term by dimensional reduction. An argument similar to the one described above can then be used to fix the product $T_{(2)}^2 T_F$ of D2-brane and fundamental-string tensions. Thus, supersymmetry and consistency determine (modulo integer ambiguities) all but two of the tensions of table 3, without any reference either to the ultraviolet definition of the theory or to the existence of an eleventh dimension. We are therefore left with a single truly independent check of the conjectured duality, which we can take to be the relation

$$T_{(0)} T_F = 2\pi\, T_{(2)} \ . \tag{4.24}$$

This is a trivial geometric identity in $\mathcal{M}$-theory, which had no a priori reason to be satisfied from the ten-dimensional viewpoint.

The sceptical reader may find that a single test constitutes little evidence for the duality conjecture. To be sure, the existence of threshold bound states of D-particles – the Kaluza–Klein modes of the supergraviton – constitutes further, a priori independent, evidence for the duality conjecture (Yi 1997, Sethi and Stern 1998, Porrati and Rozenberg 1998). The above discussion, however, underscores what might be the main lesson of the 'second string revolution': the ultimate theory may be unique *precisely* because reconciling quantum mechanics and gravity is such a constraining enterprise.

5   D-brane interactions

D-branes in supersymmetric configurations exert no net static force on each other, because (unbroken) supersymmetry ensures that the Casimir energy of open strings is zero. Setting the branes in relative motion (or rotating them) breaks generically all the supersymmetries, and leads to velocity- or orientation-dependent forces. We will now extend Polchinski's calculation to study such D-brane interactions. Some surprising new insights come from the close relationship between brane dynamics and supersymmetric gauge theory – a theme that will be recurrent in this and in the subsequent sections. Two results of particular importance, because they lie at the heart of the M(atrix)-model conjecture of Banks *et al.* (1997), are the dynamical appearance of the eleven-dimensional Planck length, and the simple scaling with distance of the leading low-velocity interaction of D-particles. Since space-time supersymmetry plays a key role in our discussion, we will first describe in some more detail the general BPS configurations of D-branes.

5.1 BPS configurations

A planar static D-brane is a BPS defect that leaves half of the space-time supersymmetries unbroken. This follows from the equality $T_{(p)} = \rho_{(p)}$, and the (rigid) supersymmetry algebra, appropriately extended to take into account p-brane charges (de Azcarraga *et al.* 1989, see Townsend 1997 for a detailed discussion). Alternatively, we can draw this conclusion from a worldsheet point of view. On a closed-string worldsheet the 32 space-time supercharges are given by contour integrals of the fermion-emission operators,

$$Q = \oint \frac{dz}{z} S \quad \text{and} \quad \overline{Q} = -\oint \frac{d\bar{z}}{\bar{z}} \overline{S} \ . \tag{5.1}$$

Holomorphicity allows us to deform the integration contours, picking (eventually) extra contributions only from points where vertex operators have been inserted. This leads to supersymmetric Ward identities for the perturbative closed-string S-matrix in flat ten-dimensional space-time.

Now in the background of a Dp-brane we must also define the action of the (unbroken) supercharges on the open strings. The corresponding integrals, at fixed radial time ξ^0, run over a semi-circle as in figure 4. Moving the integration to a later time is allowed only if the contributions of the worldsheet boundary vanish. This is the case for the sixteen linear combinations

$$Q + \Pi_{(p)}\overline{Q} = \int_0^\pi d\xi^1 \left(S + \Pi_{(p)}\overline{S} \right) \ , \tag{5.2}$$

for which the holomorphic and antiholomorphic pieces add up to zero on the real axis by virtue of the boundary conditions (3.9). The remaining sixteen supersymmetries are broken spontaneously by the Dp-brane, and cannot thus be realized linearly within the perturbative string expansion.

Consider next a background with two planar static D-branes, to which are associated two operators, $\Pi_{(p)}$ and $\tilde{\Pi}_{(\tilde{p})}$. These operators depend on the

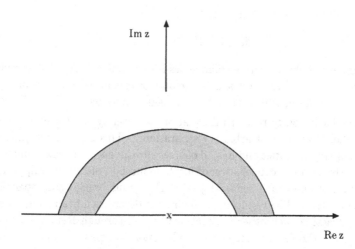

Figure 4: The semi-circles are two snapshots of an open string at fixed radial time $\xi^0 = \log|z|$. A charge is conserved when its time variation can be expressed as a holomorphic plus antiholomorphic contour integral around the shaded region in the upper complex plane. This means that the contributions of the linear segments on the worldsheet boundary should vanish.

orientation, but not on the position, of the branes. More explicitly, we can put equation (3.10) in covariant form

$$\Pi_{(p)} = -\frac{i^{p+1}}{(p+1)!}\,\omega^{(p)}_{\mu_0\cdots\mu_p}\,\Gamma^{\mu_0\cdots\mu_p} \tag{5.3}$$

where

$$\omega^{(p)} \equiv \frac{1}{(p+1)!}\,\omega^{(p)}_{\mu_0\cdots\mu_p}\,dY^{\mu_0}\wedge\cdots\wedge dY^{\mu_p} = \sqrt{-\hat{g}}\,d\zeta^0\wedge\cdots\wedge d\zeta^p \tag{5.4}$$

is the (oriented) volume form of the Dp-brane, and we have done some simple Γ-matrix rearrangements. There is of course a similar expression for the tilde brane. In the background of these two D-branes, the linearly realized supercharges are a subset of (5.2), namely

$$\mathcal{P}(Q + \Pi_{(p)}\overline{Q}) = \int_0^\pi d\xi^1\,\mathcal{P}(S + \Pi_{(p)}\overline{S}) \ , \tag{5.5}$$

with $\mathcal{P}$ an appropriate projection operator. Demanding that the corresponding contour integrals cancel out on a worldsheet boundary that is stuck on the tilde brane leads to the condition

$$\mathcal{P}\widetilde{\Pi}_{(\tilde{p})} = \mathcal{P}\Pi_{(p)} \ , \tag{5.6}$$

which admits a non-vanishing solution if and only if

$$\det\left(1 - \tilde{\Pi}_{(\tilde{p})}\Pi_{(p)}^{-1}\right) = 0 \ . \tag{5.7}$$

The number of unbroken supersymmetries is the number of zero eigenvalues of the above matrix. Every extra D-brane and/or orientifold imposes of course one extra condition, which has to be satisfied simultaneously.

A trivial solution to these BPS equations is given by two (or more) identical, parallel Dp-branes at arbitrary separation r. This background preserves sixteen supersymmetries and has, of course, an r-independent vacuum energy, consistently with the cancellation of forces found by Polchinski. Flipping the orientation of one brane sends $\Pi_{(p)} \rightarrow -\Pi_{(p)}$, thus breaking all space-time supersymmetries. The resulting configuration describes a brane and an anti-brane, attracting each other gravitationally, and through Ramond–Ramond exchange. In the force calculation of subsection 3.3, this amounts to reversing the sign of the $s = 4$ spin structure, i.e. of the GSO projection for the stretched open string. The surviving Neveu–Schwarz ground state becomes, in this case, tachyonic at a critical separation $r_{\mathrm{cr}} = \sqrt{2\pi^2\alpha'}$, beyond which the attractive force between the brane and the anti-brane diverges (Banks and Susskind 1995, Arvis 1983).

Other solutions to the BPS conditions can be found with two orthogonal D-branes. For such a configuration

$$\tilde{\Pi}_{(\tilde{p})}\Pi_{(p)}^{-1} = \pm \prod_{m \in \, p \sqcup \tilde{p}} \Gamma^m \ , \tag{5.8}$$

where $p \sqcup \tilde{p}$ denotes the set of dimensions spanned by one or other of the branes but not both, and the overall sign depends on the choice of orientations. The eigenvalues of the above operator depend only on the even number $(d_\perp)$ of dimensions in $p \sqcup \tilde{p}$. For $d_\perp = 4n + 2$ the eigenvalues are all purely imaginary, and supersymmetry is completely broken. For $d_\perp = 4$ or 8, on the other hand, half of the eigenvalues are $+1$, so eight of the supersymmetries are linearly realized in the background. Examples of $d_\perp = 4$ configurations (for early discussions see Bershadsky *et al.* 1996a, Sen, 1996, Douglas 1995) include a D4-brane and a D-particle, a D5-brane and a parallel D-string, or two completely transverse D2-branes. Examples of $d_\perp = 8$ configurations are a D8-brane and a D-particle, or two completely transverse D4-branes. As the reader can verify easily, all configurations with the same value of $d_\perp$ can be (at least formally) related by the T-duality transformations of subsection 4.2.

It will be useful later on to know the spectrum of an open string stretching between two orthogonal D-branes. Such a string has $d_\perp$ coordinates obeying mixed (DN) boundary conditions: Neumann at one endpoint and Dirichlet at the other. A bosonic DN coordinate has a half-integer mode expansion, while its fermionic partner is integer modded in the Neveu–Schwarz sector and half-integer modded in the Ramond sector. Using the standard expressions for the subtraction constants of integer or half-integer modded fields, we find the mass formula (in units $2\alpha' = 1$),

$$M^2 = \left(\frac{r}{\pi}\right)^2 + 2N_{\text{osc}} + \begin{cases} d_\perp/4 - 1 & \text{NS} \\ 0 & \text{R} \end{cases} , \qquad (5.9)$$

with N_{osc} the sum of the oscillator frequencies. Moreover, Neveu–Schwarz and Ramond states are spinors of $SO(d_\perp)$ and $SO(1, 9 - d_\perp)$ – the two maximal Lorentz subgroups that such a brane configuration could leave unbroken. Note in particular that for $d_\perp = 4$ the massless states have the content of a six-dimensional hypermultiplet, while for $d_\perp = 8$ the only massless state is a two-dimensional (anti)chiral fermion, which is a singlet of the unbroken chiral (8,0) supersymmetry (Banks, Seiberg and Silverstein 1997, Rey 1997).

There exist also BPS configurations with D-branes at arbitrary angles (Berkooz *et al.* 1996). A solution, for instance, of equation (5.6) with eight unbroken supersymmetries is given by $\mathcal{P} = \frac{1}{2}(1 - \Gamma^6\Gamma^7\Gamma^8\Gamma^9)$, and

$$\tilde{\Pi}_{(\tilde{p})}\Pi_{(p)}^{-1} = -(\cos\theta \; \Gamma^6 + \sin\theta \; \Gamma^8)(\cos\theta \; \Gamma^7 + \sin\theta \; \Gamma^9) \; \Gamma^6\Gamma^7 . \qquad (5.10)$$

It describes two identical D-branes, one of which spans the dimensions (67) and is transverse to the dimensions (89), whereas the second has undergone a (relative) unitary rotation in the $\mathbf{C}^2$-plane $(x^6 + ix^7, x^8 + ix^9)$. The case $d_\perp = 4$ discussed above corresponds to the special angle $\theta = \frac{\pi}{2}$. A more exotic example with six unbroken supersymmetries can be obtained by a rotation that preserves a quaternionic structure (Gauntlett *et al.* 1997). For a review of BPS configurations of intersecting and/or overlapping branes see Gauntlett (1997).

5.2   D-brane scattering

The velocity-dependent forces between D-branes can be analyzed by calculating the semi-classical phase shift for two moving external sources. I will here follow the original calculation (Bachas 1996) for two identical Dp-branes in the Neveu-Ramond-Schwarz formulation. The same results can be obtained in the light-cone boundary-state formalism (Callan and Klebanov 1996, Green and Gutperle 1996, Billo *et al.* 1997), and can be furthermore extended to different D-branes (Lifschytz 1996), non-vanishing worldvolume fields (Lifschytz 1997, Lifschytz and Mathur 1997, Matusis 1997), orbifold backgrounds (Hussain *et al.* 1997), type-I theory (Danielsson and Ferretti 1997), and to study spin-dependent interactions (Morales *et al.* 1997, 1998).

Consider two identical parallel Dp-branes, one of which is at rest, while the second is moving with velocity v and impact parameter b, as shown in

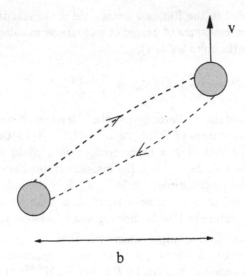

b

Figure 5: Two D-particles scattering with relative velocity v and impact parameter b. The broken lines depict a virtual pair of oriented strings being stretched by the relative motion. The imaginary part of the phase shift gives the probability that these virtual strings materialize.

figure 5. It is convenient to define the boost parameter corresponding to this relative motion,

$$v \equiv \tanh(\pi\epsilon) . \tag{5.11}$$

Thinking of the D-brane interaction as Casimir force leads us to study the spectrum of an open string stretched between these two external sources. If the motion is along the ninth dimension, the coordinates $X^{1,\cdots,8}$ retain their conventional mode expansions. The mode expansion of the light-cone coordinates $X^\pm = (X^0 \mp X^9)/\sqrt{2}$, on the other hand, is modified to

$$X^\pm = -i\sqrt{\frac{\alpha'}{2}} \sum_{k=-\infty}^{\infty} \left[\frac{a_k^\pm}{k \pm i\epsilon} z^{-k \mp i\epsilon} + \frac{a_k^\mp}{k \mp i\epsilon} \bar{z}^{-k \pm i\epsilon} \right] . \tag{5.12}$$

It is indeed easy to verify that X^0 and X^9 obey, respectively, Neumann and Dirichlet conditions at $\xi^1 = 0$, so that one end-point of the open string is fixed on the static Dp-brane. Furthermore, $X^\pm(\xi^0, \pi) = e^{\pm\pi\epsilon} X^\pm(\xi^0, 0)$, so that the other endpoint is boosted with velocity v, consistently with the fact that it is fixed on the moving Dp-brane. The mode expansions of the fermionic light-cone coordinates can be derived similarly with the result

$$\psi^\pm = \sqrt{\alpha'} \sum_k \psi_k^\pm z^{-k \mp i\epsilon} , \quad \bar{\psi}^\pm = \sqrt{\alpha'} \sum_k \psi_k^\mp \bar{z}^{-k \pm i\epsilon} , \tag{5.13}$$

where $k \in \mathbf{Z}$ in the Ramond sector of the open string, while $k \in \mathbf{Z} + \frac{1}{2}$ in the Neveu–Schwarz sector.

The relevant feature in the above expressions is the shift of all oscillator frequencies by an amount $\pm i\epsilon$. Similar expansions arise in the twisted sectors of an orbifold, with $i\epsilon$ replaced by a (real) rotation angle. Using the standard formulae for the partition functions of free massless fields with twisted boundary conditions we find (here $2\alpha' = 1$)

$$\delta(b,v) = -2 \times \frac{V_{(p)}}{2} \int_0^\infty \frac{dt}{t} \, (2\pi^2 t)^{-p/2} e^{-b^2 t/2\pi} \, Z(\epsilon,t) \,, \qquad (5.14)$$

where

$$Z(\epsilon,t) = -\frac{1}{2} \sum_{s=2,3,4} (-)^s \frac{\theta_s\left(\frac{\epsilon t}{2} \mid \frac{it}{2}\right)}{\theta_1\left(\frac{\epsilon t}{2} \mid \frac{it}{2}\right)} \frac{\theta_s^3\left(0 \mid \frac{it}{2}\right)}{\eta^9\left(\frac{it}{2}\right)} \,. \qquad (5.15)$$

Expressions (5.14–5.15) generalize Polchinski's calculation to the case of moving D-branes. As a check of normalizations notice that in the leading $v \to 0_+$ approximation the above result reduces correctly to the quasi-static phase shift

$$\delta(b,v) \simeq \int_{-\infty}^\infty d\tau \, \mathcal{E}(\sqrt{b^2 + v^2\tau^2}) + \cdots, \qquad (5.16)$$

with $\mathcal{E}(r)$ the static interaction energy, eq. (3.24). This follows from the fact that for small (first) argument the function θ_1 vanishes linearly, with $\theta_1'(0|\tau) = 2\pi\eta^3(\tau)$. Of course, since the D-branes feel no static force, this leading quasi-static phase shift is zero.

The supergravity, $b \to \infty$, limit of the phase shift can be obtained from the $t \to 0$ corner of the integration region. It is to this end convenient to first put the partition function, using Jacobi's identity, in the simpler form (Green and Gutperle 1996),

$$Z(\epsilon,t) = \frac{\theta_1^4\left(\frac{\epsilon t}{4} \mid \frac{it}{2}\right)}{\theta_1\left(\frac{\epsilon t}{2} \mid \frac{it}{2}\right) \eta^9\left(\frac{it}{2}\right)} \,. \qquad (5.17)$$

With the help of the modular transformations

$$\theta_1\left(-\frac{\nu}{\tau} \,\middle|\, -\frac{1}{\tau}\right) = \sqrt{i\tau} \, e^{i\pi\nu^2/\tau} \, \theta_1(\nu|\tau) \,, \quad \eta(-1/\tau) = \sqrt{-i\tau} \, \eta(\tau) \,, \quad (5.18)$$

as well as of the product representations (here $q = e^{2i\pi\tau}$)

$$\frac{\theta_1(\nu|\tau)}{\eta(\tau)} = 2q^{\frac{1}{12}} \sin(\pi\nu) \prod_{n=1}^\infty (1 - q^n e^{2\pi i\nu})(1 - q^n e^{-2\pi i\nu}) \qquad (5.19)$$

and

$$\eta(\tau) = q^{\frac{1}{24}} \prod_{n=1}^\infty (1 - q^n), \qquad (5.20)$$

we can easily extract the $t \to 0$ behaviour of the partition function. This leads to the following asymptotic behaviour for the phase shift, in the limit $b \to \infty$:

$$\delta \simeq -\frac{V_{(p)}}{(2\pi\sqrt{\alpha'})^p} \, \Gamma\left(3 - \frac{p}{2}\right) \left(\frac{4\pi\alpha'}{b^2}\right)^{3-p/2} \frac{\sinh^4(\pi\epsilon/2)}{\sinh(\pi\epsilon)} + \cdots , \qquad (5.21)$$

where the corrections come from the exchange of massive closed strings, and hence fall off exponentially with distance. It is a simple (but tedious) exercise to recover the above result by repeating the calculation of subsection 3.2, with one of the two external sources boosted to a moving frame. Alternatively, this result can be compared to the classical action for geodesic motion in the appropriate supergravity background (Balasubramanian and Larsen 1997).

5.3   The size of D-particles

That the string calculation should reproduce the supergravity result at sufficiently large impact parameter is reassuring, but hardly surprising. A more interesting question to address is what happens if we try to probe a D-brane at short, possibly substringy scales. In order to answer this question let us note that the partition function $Z(\epsilon, t)$ has poles along the integration axis, at $\epsilon t/2 = k\pi$ with k any odd positive integer. These correspond to zeroes of the trigonometric sine in the product representation of θ_1. As a result the phase shift acquires an imaginary (absorptive) part, equal to the sum over the positions of the poles of π times the residue of the integrand. A straightforward calculation gives (Bachas 1996)

$$\text{Im } \delta = \sum_{\text{multiplets}} \frac{\dim(s)}{2} \sum_{k \text{ odd}} \exp\left[-\frac{2\pi\alpha' k}{\epsilon}\left(\frac{b^2}{(2\pi\alpha')^2} + M(s)^2\right)\right] , \quad (5.22)$$

where the sum runs over all supermultiplets of dimension $\dim(s)$ and oscillator-mass $M(s)$ in the open-string spectrum, and we have restricted our attention to D-particles, i.e. we have set $p = 0$. This result has a simple interpretation: as the two D-particles move away from each other, they transfer continuously their energy to any open strings that happen to stretch between them (see figure 5). A virtual pair of open strings can thus materialize from the vacuum and stop completely, or slow down the motion.[8] The phenomenon is T-dual to the more familiar pair production in a background electric field, whose rate in open string theory has been calculated earlier by Bachas and Porrati (1992). It is worth stressing that this imaginary part cannot arise from the exchange of any finite number of closed-string states.

---

[8]Since the D-branes are extremely heavy at weak string coupling, the back reaction is a higher-order effect. For a discussion of D-brane recoil see Berenstein *et al.* (1996), and Kogan *et al.* (1996).

The onset of this dissipation puts a lower limit on the distance scales probed by the scattering,

$$b \succ \sqrt{\epsilon/T_F} \,, \tag{5.23}$$

where $T_F = (2\pi\alpha')^{-1}$ is the fundamental string tension, and the symbol $\succ$ stands for 'sufficiently larger than'. In the ultrarelativistic regime $v \to 1$, so that $\epsilon \simeq -\frac{1}{2\pi}\log(1 - v^2) \gg 1$. The D-particle behaves in this limit as a black absorptive disk, of area much bigger than string scale and growing logarithmically with energy. This typical Regge behaviour characterizes also the high-energy scattering of fundamental strings (see for example Amati *et al.* 1987, 1989). To probe substringy distances we must consider the opposite regime of low velocities, $\epsilon \simeq v/\pi \ll 1$. The stringy halo is not excited, in this regime, all the way down to impact parameters $b \succ \sqrt{v/\pi T_F}$. Quantum mechanical uncertainty ($\Delta x \Delta p \succ 1$) puts, on the other hand, an independent lower limit

$$b \, T_{(0)} v \succ 1 \,. \tag{5.24}$$

Saturating both bounds simultaneously gives

$$b_{\min}^3 \sim \frac{1}{T_F T_{(0)}} \sim \frac{1}{T_{(2)}} \,, \tag{5.25}$$

where we have used here the tension formula (4.24). We thus conclude that the dynamical size of D-particles is comparable to the (inverse cubic root of the) membrane tension, i.e. to the eleven-dimensional Planck scale of $\mathcal{M}$-theory! Since this is smaller than string length at weak string-coupling, perturbative string theory does not capture all the degrees of freedom at short scales.[9]

The fact that D-branes are much smaller than the fundamental strings at weak string coupling was conjectured early on by Shenker (1995). The appearance of the eleven-dimensional Planck scale in the matrix quantum mechanics of D-particles was first noticed by Kabat and Pouliot (1996) and by Danielsson and Ferretti (1996). A systematic analysis of the above kinematic regime, and of the validity of the approximations, was carried out by Douglas *et al.* (1996). Needless to say that this small dynamical scale of D-particles cannot be seen by using fundamental-string probes – one cannot probe a needle with a jelly pudding, only with a second needle! This is confirmed by explicit calculations of closed-string scattering off target D-branes (see Hashimoto and Klebanov 1997, Thorlacius 1998 and references therein).

---

[9]One should contrast this with the example of magnetic monopoles in $N = 2$, $d = 4$ Yang–Mills theory, whose size is comparable to the Compton wavelength of the fundamental quanta. Thus, even though monopoles are very heavy at weak coupling, the high-energy behaviour of the theory is still correctly captured by the (super)gauge bosons.

One other striking feature of the low-velocity dynamics of D-particles follows from an analysis of the real part of the phase shift. Expanding expressions (5.14, 5.17) for $\epsilon \simeq v \to 0$, and for any impact parameter, we find

$$\delta(b, v) \simeq -\epsilon^3 \left(\frac{2\pi^2 \alpha'}{b^2}\right)^3 + o(\epsilon^7) . \qquad (5.26)$$

Notice that δ must flip sign under time reversal[10] and is hence an odd function of velocity, and that the interaction time blows up as $1/|v|$ because two slow particles stay longer in the vicinity of each other. The generic form of the low-velocity expansion therefore is: $\delta(v, b) = \delta_0(b)/v + \delta_1(b)v + \delta_2(b)v^3 + \cdots$. Comparing with eq. (5.26) we conclude that, not only the static, but also the $o(v^2)$ force between two D-particles is zero. Since the $o(v^2)$ scattering of heavy solitons can be described by geodesic motion in the moduli space of zero modes (Manton 1982), what we learn is that the moduli space of D-particles is (at least to this order of the genus expansion) completely flat. Furthermore, as first recognized clearly by Douglas *et al.* (1997), the leading $o(v^4)$ interaction has the same power-law dependence on the impact parameter in the supergravity regime ($b \gg \sqrt{\alpha'}$) as at substringy scales ($b \ll \sqrt{\alpha'}$). Both of these facts are a result of space-time supersymmetry. As will become, indeed, clear in the following section, our phase-shift calculation could be rephrased as a one-loop calculation of the effective quantum action for a vector multiplet, in a theory with sixteen unbroken supercharges. Velocity is related by supersymmetry to the field strength, so that the $o(v^{2k})$ force between D-branes can be read off the $2k$-derivative terms in the quantum action. Using helicity-supertrace formulae it can be shown that, at one-loop order, the two-derivative terms are not corrected, while the only contributions to the four-derivative terms come from short (BPS) supersymmetry multiplets (Bachas and Kiritsis 1997). Since all excited states of an open string are non-BPS, this explains why the leading $o(v^4)$ interaction has a trivial dependence on the string scale.

This result implies that the (matrix) quantum mechanics of D-particles, obtained by truncating the open-string theory to its lightest modes, captures correctly the leading $o(v^4)$ supergravity interactions. It was, furthermore, shown recently that these interactions are not modified by higher-loop and non-perturbative corrections (Paban *et al.* 1998; see also Becker and Becker 1997, Dine and Seiberg 1997, Dine *et al.* 1998). This is an important ingredient of the conjecture by Banks *et al.* (1997), which will not be pursued here any further. We will instead go back now and discuss the classical worldvolume actions of D-branes.

---

[10]This would not be true if the scattering branes carried electric and magnetic charge.

6 Worldvolume actions

Although the full dynamics of a soliton cannot be separated from the field theory in which it belongs, its low-energy dynamics can be approximated by quantum mechanics in the moduli-space of zero modes. For an extended p-brane defect, the zero modes give rise to massless worldvolume fields, and the quantum mechanics becomes a $(p + 1)$-dimensional field theory. Similar considerations apply to a Dp-brane, whose long-wavelength dynamics we will analyze in this section. Two striking features of this analysis are (i) the natural emergence of a noncommutative space-time, and (ii) the power of the combined constraints of T-duality and Lorentz invariance.

6.1 Noncommutative geometry

The perturbative excitations of a static planar Dp-brane are described by an open string theory, interacting with the closed strings in the bulk. The Dirichlet boundary conditions do not modify the usual spectrum of the open string, but force its center-of-mass momentum to lie along the D-brane. The low-energy excitations make up, therefore, a vector supermultiplet dimensionnally reduced from ten down to $(p + 1)$ dimensions,

$$A^\mu(\zeta^\beta) \rightarrow A^{\alpha \, = \, 0,\dots,p}(\zeta^\beta) \, , \; Y^{m \, = \, p+1,\dots,9}(\zeta^\beta) \; . \tag{6.1}$$

The worldvolume scalars $Y^m(\zeta^\beta)$ are the transverse space-time coordinates of the Dp-brane, i.e. the Goldstone modes of broken translation invariance. There are no physical degrees of freedom in the longitudinal coordinates, which in the natural 'static' gauge are used to parametrize the worldvolume, $Y^\alpha = \zeta^\alpha$. The extra physical bosonic excitations of the open string correspond instead to a worldvolume vector gauge field – a feature that characterizes all D-branes, and which was overlooked in the earlier supersymmetric 'brane scans' (see Duff 1997).

That much can be in fact deduced from an analysis of the low-energy supergravity solutions. String theory becomes, however, important when one considers multiple, closely-spaced D-branes. In addition to the massless vector multiplets describing the dynamics of each defect, there are now extra potentially-light fields corresponding to the (ground states of) open strings stretching between different D-branes. In the simplest case of n parallel identical Dp-branes, the ensuing low-energy field theory is a dimensionally-reduced maximally-supersymmetric Yang–Mills theory as above, but with a non-abelian gauge group

$$U(n) \simeq U(1)_{\text{CM}} \times SU(n)_{\text{relative}} \; . \tag{6.2}$$

This is, indeed, the low-energy limit of an oriented open-string theory with a Chan–Paton index $i = 1, \cdots n$, labelling the n possible string endpoints.

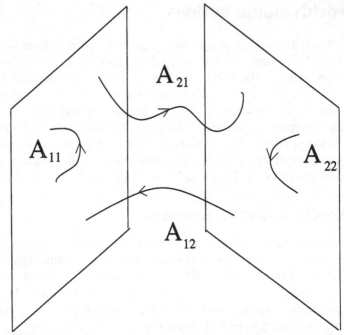

Figure 6: A D-brane sandwich, and the four types of oriented open strings giving rise to massless states in the coincidence limit. Each multiplet A_{ij} contains a worldvolume vector and $9-p$ worldvolume scalars. These latter are the non-commuting transverse coordinates of the D-branes.

The special case $n = 2$ is illustrated in figure 6. The scalar fields of the $U(1)$ vector multiplet, $Y_{\text{CM}}^m = (Y_{11}^m + Y_{22}^m)/2$, are the transverse center-of-mass coordinates of the 'sandwich', while the relative motion is described by (matrix-valued) coordinates in the Lie algebra of $SU(2)$. The scalar potential, $V \sim \text{tr}[Y^m, Y^r]^2$, is flat for any mutually commuting expectation values $\langle Y^m \rangle$. These correspond precisely to the arbitrary positions of the two D-branes, consistently with the fact that the net static force vanishes. At non-zero separation, the complex vector multiplet $A_{12} + iA_{21}$ acquires a mass, and the $SU(2)$ gauge theory is in a spontaneously-broken, Coulomb phase. It is intriguing that the permutation of the brane positions is an element of the Weyl subgroup of $SU(2)$ – quantum indistinguishability of the excitations is thus part of the local, gauge symmetry in this picture. The non-abelian nature of the D-brane coordinates, first recognized clearly by Witten (1996a), puts in a precise context earlier more general ideas about the possible role of noncommutative geometry in physics (see for instance Connes 1994, Madore 1995).

In the type-I theory, the above $U(n)$ vector multiplet is truncated by the orientifold projection. The projection must antisymmetrize the Chan–Paton indices of the worldvolume vector and symmetrize those of the worldvolume scalars, or vice versa. The reason is that the corresponding vertex operators,

$$V^\alpha = \int d\xi^a\, \partial_a X^\alpha\, e^{ip\cdot X}\,, \quad V^m = \frac{1}{2\pi\alpha'} \int d\xi_a\, \epsilon^{ab}\partial_b X^m\, e^{ip\cdot X}\,, \qquad (6.3)$$

have opposite parity under worldsheet orientation reversal. Tadpole cancellation forces an antisymmetric projection for the 9-branes, giving the standard $SO(32)$ gauge group. Consistency of the operator-product expansions then requires (Gimon and Polchinski 1996) an $SO(n)$ gauge group for the D-strings and a $USp(n)$ group for the D5-branes, with the worldvolume scalars in the symmetric, respectively antisymmetric, $n \otimes n$ representations. A single D-string, in particular, has no worldvolume gauge fields, consistently with the fact that it is dual to the heterotic string (Polchinski and Witten 1996). Likewise a single D5-brane has no transverse coordinates – the minimal dynamical excitation, dual to the heterotic five-brane, is a pair of D5-branes with a worldvolume $USp(2) \simeq SU(2)$ gauge field (Witten 1996b).

Figure 6 summarizes in itself many of the new insights brought by D-branes. The light states of stretched open strings, generically invisible in the effective supergravity, are the important degrees of freedom in various settings. They are responsible, in particular, for the thermodynamic properties of near-extremal black holes (Strominger and Vafa 1996, Callan and Maldacena 1996), and for the richness of the D-particle spectrum which lies at the heart of the (M)atrix-model conjecture (Banks *et al.* 1997). Furthermore, the realization of supersymmetric gauge field theories as worldvolume theories has led to an improved understanding of the former through brane constructions (Hanany and Witten 1997, Banks *et al.* 1996, Elitzur *et al.* 1997a,1997b), while more recently the connection with supergravity has raised new hopes of solving certain large-n superconformal gauge theories in the planar, 't Hooft limit (Maldacena 1997, Gubser *et al.* 1998, Witten 1998).

6.2 Dirac–Born–Infeld and Wess–Zumino terms

The effective action of a D-brane, used in the force calculation of section 3, can be generalized to take into acount the dynamics of the worldvolume gauge field, and the coupling to arbitrary supergravity backgrounds. The action for a single D-brane can be written as

$$I_{Dp} = \int d^{p+1}\zeta\, (\mathcal{L}_{\text{DBI}} + \mathcal{L}_{\text{WZ}} + \cdots)\,, \qquad (6.4)$$

where the Dirac–Born–Infeld and Wess–Zumino (or Chern–Simons) Lagrangians are given by

$$\mathcal{L}_{\text{DBI}} = T_{(p)}\, e^{-\Phi}\, \sqrt{-\det\left(\widehat{G}_{\alpha\beta} + \widehat{B}_{\alpha\beta} + 2\pi\alpha' F_{\alpha\beta}\right)}, \tag{6.5}$$

and

$$\mathcal{L}_{\text{WZ}} = T_{(p)}\, \widehat{C} \wedge e^{2\pi\alpha' F} \wedge \mathcal{G}\,\Big|_{(p+1)-\text{form}}. \tag{6.6}$$

Here $\widehat{B}_{\alpha\beta}$ is the pull-back of the Neveu–Schwarz two-form,

$$\widehat{C} \equiv \sum_n \frac{1}{n!}\, \widehat{C}_{\alpha_1\cdots\alpha_n}\, d\zeta^{\alpha_1} \wedge \cdots \wedge d\zeta^{\alpha_n} \tag{6.7}$$

is the sum over all electric and magnetic RR-form potentials, pulled back to the worldvolume of the D-brane, and $F = dA$ is the worldvolume field-strength two-form, normalized so that the coupling on a boundary of the fundamental-string worldsheet is $\oint A_\alpha dX^\alpha$. The geometric part of the Wess–Zumino action reads

$$\mathcal{G} = \sqrt{\mathcal{A}(\mathcal{T})/\mathcal{A}(\mathcal{N})} = 1 - \frac{(4\pi^2\alpha')^2}{48}\left[p_1(\mathcal{T}) - p_1(\mathcal{N})\right] + \cdots \tag{6.8}$$

where $\mathcal{T}$ and $\mathcal{N}$ are the tangent and normal bundles of the brane, $\mathcal{A}$ is the appropriately-normalized 'roof genus',[11] and p_1 is the first Pontryagin class (see for instance Milnor and Stasheff 1974, Eguchi *et al.* 1980, or Nakahara 1990 for definitions). The next term in the expansion of (6.8) is an eight-form, whose presence in the D-brane action has not been explicitly checked. All multiplications in $\mathcal{L}_{\text{WZ}}$, including those in the Taylor expansions of the square root and of the exponential, must be understood in the sense of forms – what one integrates in the end is the coefficient of the $d\zeta^0 \wedge \cdots \wedge d\zeta^p$ term in the expansion. Strictly-speaking, since the RR potentials cannot be globally defined in the presence of D-branes, one must use the fact that $(e^{2\pi\alpha' F} \wedge \mathcal{G} - 1)$ is an exact form, and integrate by parts to express all but Polchinski's coupling in terms of the RR field-strengths. Note, finally, that since $\mathcal{T} \oplus \mathcal{N} = \widehat{\mathcal{S}}$, the pullback of the space-time tangent bundle, we can use the multiplicative property of the roof genus,

$$\mathcal{A}(\mathcal{T}) \wedge \mathcal{A}(\mathcal{N}) = \mathcal{A}(\widehat{\mathcal{S}}) = 1 + \frac{(4\pi^2\alpha')^2}{192\pi^2}\, \text{tr}(\widehat{\mathcal{R}} \wedge \widehat{\mathcal{R}}) + \cdots, \tag{6.9}$$

to trade the dependence on either $\mathcal{T}$ or $\mathcal{N}$ for dependence on the (pulled back) target-space curvature two-form, $\widehat{\mathcal{R}}_{\alpha\beta}\, d\zeta^\alpha \wedge d\zeta^\beta$.

---

[11]The conventional normalization amounts to choosing units $4\pi^2\alpha' = 1$, in which all type-II D-branes have the same tension $\sqrt{\pi}/\kappa_{(10)}$. The roof genus is also frequently denoted $\widehat{\mathcal{A}}$, but I here reserve the use of hats to denote pullbacks on the worldvolume.

The Dirac–Born–Infeld Lagrangian is a generalization of the geometric volume of the brane trajectory, in the presence of Neveu–Schwarz antisymmetric tensor and worldvolume gauge fields (Leigh 1989). It was first derived in the context of type-I string theory in ten dimensions (Fradkin and Tseytlin 1985). The Wess–Zumino Lagrangian, on the other hand, generalizes Polchinski's coupling of Dp-branes to Ramond–Ramond $(p + 1)$-forms. The gauge-field dependence was derived by Li (1996a) and Douglas (1996), and the gravitational terms for trivial normal bundle by Bershadsky *et al.* (1996) and Green *et al.* (1997). The extension to non-trivial normal bundles was given in special cases by Witten (1997b) and Mourad (1997), and more generally by Cheung and Yin (1997) and Minasian and Moore (1997). Unlike $\mathcal{L}_{\text{WZ}}$, which is related as we will see to anomalies and is thus believed to be exact, the 'kinetic' action is known to receive corrections involving acceleration terms and derivatives of the field-strength background (Andreev and Tseytlin 1988, Kitazawa 1987). These reflect the non-local nature of the underlying open-string theory. The fermionic completion of the action (6.5-6.6), compatible with space-time supersymmetry and with worldvolume κ-symmetry, has been derived by several authors (Bergshoeff and Townsend 1997, Cederwall *et al.* 1997a, 1997b, Cederwall 1997, Aganagic *et al.* 1997a, 1997b, Abou Zeid and Hull 1997), but will not be discussed in this lecture.

The generalization of this action to multiple D-branes is non-trivial. The transverse fluctuations Y^m, the 'tangent frame' $\partial_\alpha Y^\mu$ used to pull back tensors to the worldvolume, and the field strength $F_{\alpha\beta}$, all now take their values in the Lie algebra of $U(n)$. The tree-level action, obtained from the disk diagram, must be given by a single trace in the fundamental representation of the gauge group, but the ordering of the various terms is a priori ambiguous. Things simplify considerably if the supergravity backgrounds do not depend on the coordinates x^m that are transverse to the unperturbed D-branes.[12] T-duality reduces in this case the problem to that of finding the non-abelian extension of the gauge-field action only. This is straightforward for $\mathcal{L}_{\text{WZ}}$, in which we need only make the replacement

$$e^{2\pi\alpha' F} \rightarrow \text{tr}_n\, e^{2\pi\alpha' F} . \tag{6.10}$$

The proper non-abelian generalization of the Born–Infeld action, on the other hand, is not known. The leading quadratic term in the α'-expansion of this action is unambiguous, thanks to the cyclic property of the trace. The ordering ambiguities in the next-to-leading, quartic, term are resolved by the well-known fact that the 4-point disk-amplitude has total symmetry under permutations of the external states (Green, Schwarz and Witten 1987). One natural generalization (Tseytlin 1997) is to evaluate all higher-order terms

---

[12]The more general case has been considered by Douglas (1997) and Douglas, Kato and Ooguri (1997).

with the same symmetrized-trace prescription, but this fails to reproduce some known facts about the open-string effective action (Hashimoto and Taylor 1997, see also Brecher and Perry 1998, Brecher 1998). Since commutators of field strengths cannot be separated in invariant fashion from higher-derivative ('acceleration') terms, there is in fact no reason to expect that a non-abelian D-brane action in a simple, closed form will be found.

Having summarized the known facts about effective D-brane actions, we will spend the remainder of this review justifying them, and exhibiting some of their salient features.

6.3   Type-I theory

We first consider the special case $p = 9$, which allows contact with the familiar action of the type-I string theory. The type-I background has 32 D9-branes plus an orientifold (Sagnotti 1988, Hořava 1989a), which truncates $U(32)$ to $SO(32)$ and projects out of the spectrum all antisymmetric-form fields other than $C^{(2)}$ and its dual $C^{(6)}$. Since the gauge fields live on the D9-branes, their action should be given entirely by I_{D9}, after appropriate truncation of the unnecessary fields. Note, in contrast, that the purely-gravitational part of the type-I, tree-level Lagrangian has contributions from three distinct diagrams: sphere, disk and real projective plane. The gauge-field independent pieces in I_{D9} – representing the contributions of the disk – cannot, therefore, be directly matched to the effective supergravity action.

Expanding out the Born–Infeld action in powers of the field strength, neglecting the (leading) cosmological term which is removed by the orientifold projection, and using the total symmetry of the 4-point function, we find

$$I_{\text{BI}} = T^I_{(9)}(\pi\alpha')^2 \int d^{10}x \, e^{-\Phi} \left[\operatorname{tr}(F_{\mu\nu}F^{\mu\nu}) - \frac{(\pi\alpha')^2}{12} \operatorname{tr}(t_8 F^4) + \cdots \right] \quad (6.11)$$

where the dots stand for higher orders in α', the $F_{\mu\nu}$ are Hermitean, and t_8 is the well-known eight-index tensor of string theory (without its ϵ piece),

$$\begin{aligned} t_8 F^4 = {}& 16 \, F_{\mu\nu}F^{\nu\rho}F^{\lambda\mu}F_{\rho\lambda} + 8 \, F_{\mu\nu}F^{\nu\rho}F_{\rho\lambda}F^{\lambda\mu} \\ & - 4 \, F_{\mu\nu}F^{\mu\nu}F_{\rho\lambda}F^{\rho\lambda} - 2 \, F_{\mu\nu}F_{\rho\lambda}F^{\mu\nu}F^{\rho\lambda} \, . \end{aligned} \quad (6.12)$$

The quartic piece can be written alternatively as a symmetrized trace,

$$\operatorname{tr}(t_8 F^4) = 24 \operatorname{Str}\left(F^4 - \frac{1}{4}(F^2)^2 \right) , \quad (6.13)$$

with matrix multiplication of the Lorentz indices implied. Expanding out similarly the Wess–Zumino action of the D9-branes, which have of course a trivial normal bundle, we find

$$I_{\text{GS}} = T^I_{(9)} \, (\pi\alpha')^2 \int d^{10}x \left[C^{(6)} \wedge X_4 + (\pi\alpha')^2 \, C^{(2)} \wedge X_8 \right] , \quad (6.14)$$

where

$$X_4 = 2 \operatorname{tr} F^2 + \cdots$$

$$X_8 = \frac{2}{3} \operatorname{tr} F^4 + \frac{1}{12} \operatorname{tr} F^2 \operatorname{tr} \mathcal{R}^2 + \cdots . \qquad (6.15)$$

Here the dots stand for purely gravitational corrections to X_4 and X_8, multiplication is in the sense of forms, and we recall that $\mathcal{L}_{WZ}$ was defined only up to total derivatives.

One can recognize in the above expressions many of the standard features of $SO(32)$ superstring theory. The terms in I_{GS} are the Green–Schwarz couplings that cancel the hexagon anomaly, as shown in figure 7 (Green and Schwarz 1984, 1985a,b) They have been calculated directly from the disk diagram by Callan *et al.* (1988). They are often expressed in terms of traces ('Tr') in the adjoint representation of $SO(32)$, via the relations

$$\operatorname{Tr} F^2 = 30 \operatorname{tr} F^2 , \quad \operatorname{Tr} F^4 = 24 \operatorname{tr} F^4 + 3 \left(\operatorname{tr} F^2 \right)^2 . \qquad (6.16)$$

This is less economical but more natural from the point of view of the effective supergravity. The anomalous Green–Schwarz couplings are, furthermore, related, through space-time supersymmetry, to the two leading terms which we have exhibited in the expansion of the Born–Infeld action (de Roo *et al.* 1993, Tseytlin 1996a,b, see also Lerche 1988). Comparing the coefficients of the various terms is a non-trivial check of our normalizations.[13] For instance, the tensor structure mutliplying $\operatorname{tr} F^4$ has the correct supersymmetric form, $t_8 - \frac{1}{4}\epsilon_{10} C^{(2)}$ (Tseytlin 1996a) . The two terms in the X_8 polynomial also have the right relative weights (Green *et al.* 1987). Finally, we can put the quadratic Yang–Mills Lagrangian in the standard form, $\operatorname{tr}(F_{\mu\nu} F^{\mu\nu})/2g^2_{(10)}$, with

$$\frac{g^4_{(10)}}{\kappa^2_{(10)}} = 2^{11} \pi^7 \alpha'^2 . \qquad (6.17)$$

This is, indeed, the relation between the type-I gauge and gravitational coupling constants (Sakai and Abe 1988, see also Shapiro and Thorn 1987, Dai and Polchinski 1989 for the bosonic case).

6.4   The power of T-duality

A stringent requirement on the effective D-brane actions is that they be compatible with the T-duality trasformations discussed in subsection 4.2. Indeed, being a discrete gauge invariance, T-duality must leave the entire string theory, and its low-energy limit in particular, unchanged. We have seen that the simplest T-duality – inversion of the radius of a circle – transforms Dp-branes winding around this circle to D$(p-1)$-branes transverse to it, and vice versa.

---

[13] I thank J. Conrad for discussions on this point.

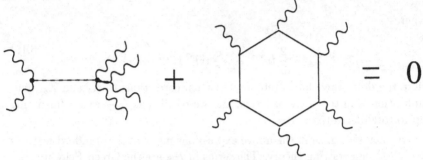

Figure 7: The Green–Schwarz anomaly-cancellation mechanism: the classical interference of the two vertices in I_{GS}, through the exchange of a RR two-form, cancels the one-loop hexagon anomaly.

It also transforms the corresponding component of the worldvolume gauge field to the extra transverse coordinate of the lower-dimensional brane,

$$Y_p{}' = 2\pi\alpha' A_p , \qquad (6.18)$$

Here x^p is the dualized dimension, and $\zeta^p = Y^p$ is the last coordinate of the wrapped brane in static gauge. The above transformation is consistent with the form (6.3) of the vertex operators. Strictly speaking, the fields on the original Dp-brane may depend on the coordinate ζ^p, or equivalently on its conjugate momentum, which transforms to winding under radius inversion. This dependence drops out in the limit $R_p \to 0$, or equivalently $R'_p \to \infty$, since momentum (or winding prime) modes become infinitely massive in this case. Thus a non-compact transverse D-brane coordinate can be thought of as a gauge field in some vanishingly small internal dimension.

Let us consider for instance the case $p = 1$. The action of a type-IIB D-string, in flat space-time and vanishing Φ and $B_{\mu\nu}$ backgrounds, reads

$$I = T_{(1)} \int d^2\zeta \left[\sqrt{1 - (\partial_0 Y^m)^2 - (2\pi\alpha'\, F_{01})^2} + \widehat{C}_{01} + 2\pi\alpha' F_{01}\, \widehat{C} \right] . \quad (6.19)$$

We have assumed that the transverse string coordinates $Y^{m=2,\cdots 9}$, as well as the gauge field and (pullback) RR backgrounds, are independent of $\zeta^1 = Y^1$. The T-duality $R'_1 = \alpha'/R_1$ transforms the D-string to a D-particle, the worldvolume electric field $F_{01} = \partial_0 A_1$ to a velocity, and the RR backgrounds as follows:

$$C'_1 = C , \quad C'_\mu = C_{\mu 1} \quad (\mu \neq 1). \qquad (6.20)$$

A straightforward calculation gives thus the following transformed action,

$$I' = T'_{(0)} \int d\zeta^0 \left(\sqrt{1 - (\partial_0 \vec{Y}{}')^2} + \widehat{C}'_0 \right) . \qquad (6.21)$$

This is indeed the effective action of a relativistic D-particle, coupling in Lorentz-invariant fashion to a background RR one-form.

We could have in fact run the argument backwards. Starting with the point-particle Lagrangian (6.21), fixed uniquely by relativistic invariance, we can use T-duality and locality to derive the (abelianized) gauge dynamics on the worldvolume of a D-string or of higher Dp-branes (Bachas 1996, Douglas 1995, Bergshoeff *et al.* 1996, Bergshoeff and de Roo 1996, Green *et al.* 1996). Thi is rather striking, since it provides a simple kinematic explanation of some of the most intriguing features of open-string gauge dynamics. The anomaly-cancelling Green–Schwarz couplings on a D9-brane, for example, can be traced to the covariantizion of Polchinski's coupling. Likewise the limiting electric field of the Born–Infeld action can be mapped under T-duality to the speed of light. This latter appears as a dynamical limit, for which there exists a natural dissipation mechanism – the pair production of subsection 5.3 – even in the limit of vanishingly weak string coupling in which gravity decouples.

7 Topological aspects of brane dynamics

The effective gauge theories on the worldvolume of D-branes have a rich spectrum of both perturbative and non-perturbative excitations. These are worldvolume projections of the various branes from the bulk which, like fundamental strings, can terminate on the D-branes of interest, or form with them stable bound states. Much can be learned about these dynamics by simple topological considerations of the worldvolume fluxes and charges, and of their spacetime counterparts. We conclude this guided tour of D-branes with a brief discussion of such issues.

7.1 Branes inside branes

One immediate consequence of the Wess–Zumino action (6.6) is that the worldvolume gauge fields and the geometry can induce different RR charges on D-branes. We will illustrate this phenomenon with some concrete examples of Dp-branes wrapped around a compact cycle Σ_k, such as a k-torus or a supersymmetric k-cycle of a Calabi–Yau space. To simplify the discussion we assume that the target space is a direct product of d-dimensional Minkowski space ($\mathbf{R}^d$) times a compactification manifold (Σ_{10-d}), and that the brane worldvolume can be factorized accordingly, $\mathcal{W}_{p+1} = \Sigma_k \times \mathcal{W}_{p-k+1}$ with $\Sigma_k \subset \Sigma_{10-d}$ and $\mathcal{W}_{p-k+1} \subset \mathbf{R}^d$. We also assume antisymmetric-tensor backgrounds that are covariantly constant on the compactification manifold, as well as a vanishing dilaton field.

Our first example is a D2-brane wrapped around a two-cycle Σ_2, on which
we turn on a (quantized) magnetic flux,

$$\frac{1}{2\pi} \int_{\Sigma_2} F = k \ . \tag{7.1}$$

This gives rise to a Wess–Zumino coupling

$$I_{\mathrm{WZ}} = T_{(2)} \int_{\mathcal{W}_3} \widehat{C}^{(3)} + k\, T_{(0)} \int_{\mathcal{W}_1} \widehat{C}^{(1)} \ , \tag{7.2}$$

showing that the D2-brane has acquired k units of RR one-form charge. We
are therefore describing a configuration of a D2-brane bound to k type-IIA
D-particles, or equivalently of a (transverse) membrane boosted along the
hidden eleventh dimension of $\mathcal{M}$-theory.

To confirm this latter interpretation, notice that the Dirac–Born–Infeld
action in 2+1 dimensions reads

$$I_{\mathrm{DBI}} = T_{(2)} \int d^3\zeta \, \sqrt{-\widehat{G}} \, \sqrt{1 + 2\pi^2 \alpha'^2 F_{\alpha\beta} F^{\alpha\beta}} \ , \tag{7.3}$$

with indices raised by the induced metric. Extremizing this action for a static
D2-brane yields a magnetic field proportional to the volume form (ω_{Σ_2}) of the
2-cycle,

$$F = 2\pi k \, \frac{\omega_{\Sigma_2}}{\int \omega_{\Sigma_2}} \ . \tag{7.4}$$

Using the relations between type-IIA and $\mathcal{M}$-theory scales we can write the
energy of this configuration (for zero RR backgrounds) as

$$\mathcal{E} = \sqrt{M^2 + \left(\frac{k}{R_{11}}\right)^2} \quad \text{with} \quad M = T_{(2)} \int \omega_{\Sigma_2} \ . \tag{7.5}$$

This is the expected energy of an excitation of mass M, carrying k units of
momentum in the eleventh dimension. The scalar dual to the vector field on
the worldvolume can be in fact interpreted as the eleventh coordinate of the
membrane (Townsend 1996a, Schmidhuber 1996),

$$\frac{2\pi\alpha' F_{\alpha\beta}}{(1 + 2\pi^2 \alpha'^2 F^2)^{1/2}} = \sqrt{-\widehat{G}} \, \epsilon_{\alpha\beta\gamma} \, \partial^\gamma Y^{11} \ . \tag{7.6}$$

This duality transformation indeed maps the magnetic field (7.4) to a uniform
motion along x^{11}, as should be the case.

Our second example consists of n coincident D4-branes wrapping around
a four-cycle Σ_4. A non-abelian k-instanton configuration,

$$-\frac{1}{8\pi^2} \int_{\Sigma_4} \mathrm{tr} F \wedge F = k \ , \tag{7.7}$$

induces RR one-form charge equal to that carried by

$$k + \frac{n}{48} \left(p_1(\mathcal{T}) - p_1(\mathcal{N}) \right) \tag{7.8}$$

(anti)D-particles. This is a very fruitful interpretation, which allows an identification of certain multi-instanton moduli spaces with vacuum manifolds of appropriate supersymmetric gauge models (Witten 1996b, Bershadsky *et al.* 1996b, Vafa 1996, Douglas 1995, Douglas and Moore 1996). The dimension of such a moduli space enters, in particular, in the simplest microscopic derivation of the Bekenstein–Hawking entropy from D-branes (Strominger and Vafa 1996, Callan and Maldacena 1996).

Consider, for example, the case of Σ_4 a four-torus, so that our configuration consists of n flat D4-branes and k (anti)D-particles. As explained in subsection 5.1, such a configuration leaves eight unbroken supersymmetries, and has the following content of low-lying open-string excitations: (i) a ten-dimensional $U(n)$ vector multiplet reduced down to the worldvolume of the four-branes, (ii) a ten-dimensional $U(k)$ vector multiplet reduced likewise to the worldline of the particles, and (iii) a six-dimensional hypermultiplet, in the (n, k) representation of the gauge group, and living also on the particle worldline. In terms of the unbroken supersymmetries, the adjoint fields decompose into vector plus hypermultiplets. The vacuum manifold of this effective theory has a Coulomb branch along which the gauge group breaks generically to $U(1)^n \times U(1)^k$, and a Higgs branch along which only a single $U(1)$ remains unbroken. These correspond, respectively, to D-particles separated from the D4-branes in the transverse space, or bound to them as finite-size instantons. The dimension around a generic point on the Higgs branch is given by the total number of scalar fields in the hypermultiplets, minus the number of gauge transformations and of D-flatness conditions (see for example Hassan and Wadia 1997)

$$\dim \mathcal{M}_k^n = 4(nk + n^2 + k^2) - (n^2 + k^2 - 1) - 3(n^2 + k^2 - 1) = 4(nk + 1). \tag{7.9}$$

This is indeed the dimension of the moduli space of k $U(n)$ instantons on a flat torus. Similar constructions work for instantons on a $K3$ surface, for which the first Pontryagin class $p_1(K3) = -48$ (Bershadsky *et al.* 1996b, Vafa 1996), as well as for instantons on a ALE space (Douglas and Moore 1996).

Our last example is a type-IIB D-string winding around a stable cycle Σ_1 of unit radius. Turning on a worldsheet electric field gives a coupling linear in the Neveu–Schwarz antisymmetric tensor,

$$I_{D1} = \frac{1}{2\pi\alpha'} \int d^2\zeta \, \Pi_1 \, \hat{B}_{01} + \cdots \tag{7.10}$$

where $\Pi_1 = \delta\mathcal{L}/\delta\partial_0 A_1$ is the momentum conjugate to A_1. We have here gone to the $A_0 = 0$ gauge, used ζ^1 to parametrize the stable cycle, and set for simplicity the RR backgrounds to zero. Since the Wilson line $\oint d\zeta^1 A_1$ is a periodic variable with period 2π, its conjugate momentum in the quantum theory is an integer, $\oint d\zeta^1 \Pi_1 = 2\pi q$. The coupling (7.10) then describes precisely the gauge charge carried by q fundamental winding strings, bound to the D-string under consideration (Witten 1996a, Callan and Klebanov 1996, Schmidhuber 1996). This is in accordance with the prediction of $SL(2,Z)$ duality, which requires the existence of subthreshold bound states of p D-strings and q fundamental strings, for all pairs (p,q) of relatively prime integers (Schwarz 1995).

7.2   Branes ending on branes

The coupling of D-branes to $B_{\mu\nu}$ can be understood from a simple worldsheet argument. Under a gauge transformation $\delta B = d\Xi$, with Ξ an arbitrary one-form, the action of a fundamental string changes by a boundary term

$$\delta_\Xi I_F = \frac{1}{2\pi\alpha'} \oint_{\partial\Sigma} d\xi^a \, \hat{\Xi}_a \ . \tag{7.11}$$

To cancel this variation, we must assume that the gauge fields living on the worldvolumes of D-branes have also a universal transformation $\delta A_\alpha = -\hat{\Xi}_\alpha/2\pi\alpha'$. This explains the appearance of the gauge-invariant combination $\hat{B} + 2\pi\alpha' F$ in the Dirac–Born–Infeld action. Invariance of the Wess–Zumino action, on the other hand, requires that the (sum over all) RR potentials transform as

$$\delta_\Xi C = C \wedge e^{d\Xi} \ . \tag{7.12}$$

The RR antisymmetric tensors have of course their own independent gauge transformations,

$$\delta_\Lambda C = d\Lambda \ , \tag{7.13}$$

with Λ (a sum of) arbitrary forms. Redefining the RR potentials, $\tilde{C} \equiv C \wedge e^{-B}$, so as to make them invariant under the Ξ-transformations, modifies the Λ-transformations which would in this case mix the RR forms and the Neveu–Schwarz tensor.[14]

This argument confirms what we have used from the very beginning, i.e. that a fundamental string may end on any D-brane, on whose worldvolume it couples as an elementary electric charge. The dynamics of such open strings can in fact be analyzed from the viewpoint of the worldvolume Born–Infeld action (Callan and Maldacena 1998, Gibbons 1998). We can, however, also generalize the argument to see what other branes can end on D-branes (Strominger 1996). Consider indeed the variation of the Wess–Zumino action of a

---

[14]I thank E. Kiritsis for this argument.

Dp-brane under a gauge transformation of the RR $(p+1)$-form,

$$\delta_\Lambda I_{Dp} = T_{(p)} \int_{\partial W_{p+1}} \hat\Lambda^{(p)} , \qquad (7.14)$$

where ∂W_{p+1} is the boundary of the brane worldvolume. We may cancel this variation if the boundary happens to lie on the worldvolume of a D$(p+2)$-brane, on which it appears as the trajectory of a $(p-1)$–dimensional magnetic charge! Indeed, the anomalous Bianchi identity on the worldvolume of the D$(p+2)$-brane reads

$$d \wedge F = 2\pi\delta^{(3)}(\partial W_{p+1}) , \qquad (7.15)$$

where we have used the normalization of electric charge to one. It can be checked that the variation of the Wess–Zumino action of the higher brane will then cancel precisely (7.14).

We thus conclude that D-strings can terminate on D3-branes, on whose worldvolume they appear as magnetic monopoles, that D2-branes can terminate on D4-branes, and so on for higher p. Notice that if the branes are orthogonal, the number of dimensions along one or other of the branes but not both is exactly four. These configurations leave therefore one quarter of unbroken supersymmetries. Various duality transformations map the above examples to other configurations of branes. Note in particular that lifted to $\mathcal{M}$-theory, the D2–D4 brane configuration teaches us that the $\mathcal{M}$-theory membrane can terminate on a $\mathcal{M}$-theory five-brane.

7.3 Branes created by crossing branes

The final point I want to discuss has to do with the role of Wess–Zumino terms in cancelling chiral anomalies. We already saw this in the context of type I theory, but the phenomenon is more general and can in fact be used to fix completely the form of the Wess–Zumino couplings (Green *et al.* 1997). Consider for example two stacks of n and n' D5-branes, spanning worldvolumes W_6 and W_6', which have generically a two-dimensional intersection, $\mathcal{I} = W_6 \cap W_6'$. Let us concentrate on the gauge part of the Wess–Zumino action of the first stack. Under a worldvolume gauge transformation, this has an anomalous variation given by

$$\delta_\xi I_{D5} = T_{(5)}(2\pi\alpha')^2 \int_{W_6} d\hat{H}^{(3)} \wedge \mathrm{tr}(\xi F) = \frac{n'}{2\pi} \int_{\mathcal{I}} \mathrm{tr}(\xi F) , \qquad (7.16)$$

with ξ in the Lie algebra of $U(n)$. We have here used the standard descent formulae

$$\mathrm{tr}(F \wedge F) = d\omega_3(A) \quad \text{and} \quad \delta_\xi\omega_3(A) = d\,\mathrm{tr}(\xi F) , \qquad (7.17)$$

with $\omega_3(A)$ the Chern–Simons three-form, as well as the (anomalous) Bianchi identity

$$d\hat{H}^{(3)} = 2\kappa_{(10)}^2 T_{(5)}\, n'\delta^{(2)}(\mathcal{I}) . \qquad (7.18)$$

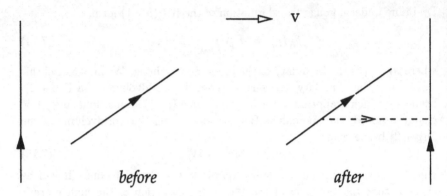

Figure 8: Anomalous creation of a stretched string when two orthogonal D4-branes cross.

This is the projection on $\mathcal{W}_6$ of the bulk Bianchi identity showing that the prime branes are magnetic sources for the RR two-form.

Thus gauge invariance seems to be violated on the intersection, but the anomaly can be precisely cancelled by n' chiral fermions in the fundamental representation of $U(n)$. But as we have already explained in subsection 5.1, string theory provides us precisely with the required fermions – the massless states of the fundamental strings stretching from one to the other stack of D5-branes, and transforming in the $(n, \bar{n}')$ representation of the $U(n) \times U(n')$ gauge group. Reversing the argument, since the embedding theory is non-anomalous, the presence of the Wess–Zumino couplings is necessary to cancel the apparent violation of charge conservation on the intersection, by inflow from the bulk of the D5-branes. The gravitational anomaly of the intersection fermions cancels similarly the anomalous variation of the gravitational Wess–Zumino action. To see how, one must use the (Whitney sum) decompositions of the tangent bundles (since the branes fill all dimensions)

$$\mathcal{T}_{\mathcal{W}_6} = \mathcal{T}_{\mathcal{I}} \oplus \mathcal{N}_{\mathcal{W}_6'}, \tag{7.19}$$

and similarly for the prime brane, together with the multiplicative property of the roof genus. It follows that the anomalous variations of the pullback normal bundles cancel between the two stacks of D5-branes, while those of the bundle tangent to the intersection add up and cancel against the anomalous fermion loop.

The anomalous inflow of charge, required to cancel the chiral anomaly on the intersection, has an interesting T-dual interpretation (Bachas *et al.* 1997, Danielsson *et al.* 1997, Bergman *et al.* 1998). Consider indeed two (stacks of) D4-branes oriented so as to have a unique common transverse dimension, say x_1. The lowest-lying state of an open string stretching between the two stacks is a chiral fermion, but since it is completely localized in space the role of momentum is played by its (oriented) stretching. It thus satisfies the T-dual Weyl equation,

$$p_0 = T_F \delta x_1 , \qquad (7.20)$$

with δx_1 the transverse displacement of the two D4-branes. Now as the D4-branes move continuously past each other, the above energy level crosses continuously the zero axis. Thus in the second-quantized theory a string must be anomalously created or destroyed, as illustrated in figure 8.

Since this is a topological phenomenon, we should expect it to commute with any (sequence of) duality transformations. Consider in particular the following chain,

(IIA) $\qquad D(2345) \otimes D(6789) \hookrightarrow F(1)$

$$\Big\downarrow T(6)$$

(IIB) $\qquad D(23456) \otimes D(789) \hookrightarrow F(1)$

$$\Big\downarrow S$$

(IIB) $\qquad NS(23456) \otimes D(789) \hookrightarrow D(1)$

$$\Big\downarrow T(56)$$

(IIB) $\qquad NS(23456) \otimes D(56789) \hookrightarrow D(156)$

Here F, D and NS denote a fundamental string, a D-brane and a Neveu–Schwarz five-brane, the dimensions which these branes span are indicated inside parentheses, and $X \otimes Y \hookrightarrow Z$ means 'Z is being created when X crosses Y'. The sequence of T- and S-duality transformations tells us that a D3-brane must be created when a NS-brane and a D5-brane, sharing two common dimensions, cross each other. From the fermionic character of the original stretched fundamental string, we also know that only a single streched brane in a supersymmetric state is allowed (Bachas *et al.* 1998). These two basic rules of brane engineering have indeed been postulated by Hanany and Witten (1997), in order to avoid immediate contradictions with the known behaviour of three-dimensional supersymmetric gauge models.

I stop here because time is up – not because the subject has been exhausted. The reader has hopefully acquired the tools, as well as the motivation, to move on to some of the exciting applications of D-branes to gauge theories and black-hole physics.

Acknowledgments

I thank David Olive, Pierre van Baal and Peter West for organizing a wonderful Newton Institute workshop. These lectures were also presented at the 1997 Trieste Spring School on 'String Theory, Gauge Theory and Quantum Gravity', and in a shorter format at the '31st International Symposium Ahrenshoop' in Buckow. I thank the organizers for the invitations to speak, and present to them my sincere apologies for failing to meet their (generously elastic) proceedings deadlines. I am finally indebted to P. Bain, J. Conrad, E. Cremmer, G. Gibbons, M.B. Green, A. Greenspoon, M. Henneaux, A. Kehagias, E. Kiritsis, J.X. Lu, S. Mukhi, H. Partouche, B. Pioline, A. Polychronakos and P. Vanhove for comments and discussions that helped improve this manuscript.

Note on conventions

Throughout the text I have used $X^\mu(\xi^a)$ for the coordinates of a fundamental string, $Y^\mu(\zeta^\alpha)$ for those of a D-brane, and x^μ for the space-time coordinates. I reserve the capital N to count supersymmetries, and the little n for the number of D-branes. Hats denote pullbacks of supergravity fields from the bulk onto the worldvolumes of branes. $T_F = 1/2\pi\alpha'$ is the tension of the fundamental string, not to be confused with the worldsheet supercurrent which I have denoted J_F. I use 'et al.' when referring to papers with three or more coauthors, and indicate in parentheses the publication year when available, or the year of submission to the archives otherwise. All authors and all archive numbers can be found in the bibliography at the end.

References

Abou Zeid, M., Hull, C.M. (1997) 'Intrinsic Geometry of D-Branes', *Phys. Lett.* **B404**, 264–270, hep-th/9704021.

Aganagic, M., Popescu, C., Schwarz, J. (1997a) 'D-Brane Actions with Local Kappa Symmetry', *Phys. Lett.* **B393**, 311–315, hep-th/9610249.

Aganagic, M., Popescu, C., Schwarz, J. (1997b) 'Gauge-Invariant and Gauge-Fixed D-Brane Actions', *Nucl. Phys.* **B495**, 99–126, hep-th/9612080.

Alvarez, E., Barbon, J.L.F., Borlaf, J. (1996) 'T-duality for open strings' *Nucl. Phys.* **B479**, 218–242, hep-th/9603089.

Alvarez, O. (1981) 'Static Potential in String Models', *Phys. Rev.* **D24**, 440–449.

Alvarez-Gaumé, L., Zamora, F. (1997) 'Duality in Quantum Field Theory (and String Theory)', lectures at The Workshop on Fundamental Particles and Interactions, Vanderbilt University, and CERN-La Plata-Santiago de Compostela School of Physics, hep-th/9709180.

Amati, D., Ciafaloni, M., Veneziano, G. (1987) 'Superstring Collisions at Planckian Energies', *Phys. Lett.* **B197**, 81–87.

Amati, D., Ciafaloni, M., Veneziano, G. (1989) 'Can Spacetime be Probed below the String Size?', *Phys. Lett.* **B216**, 41–47.

Andreev, O.D., Tseytlin, A.A. (1988) 'Partition Function Representation for the Open Superstring Effective Action', *Nucl. Phys.* **B311**, 205–252.

Antoniadis, I., Partouche, H., Taylor, T.R. (1998) 'Lectures on Heterotic-Type I Duality', *Nucl. Phys. Proc. Suppl.* **61A**, 58–71, hep-th/9706211.

Arvis, J.F. (1983) 'Deux Aspects des Théories de Cordes Duales: la Théorie de Liouville Supersymétrique et le Potentiel $q\bar{q}$ dans le Modèle de Nambu', PhD thesis, University of Paris.

Aspinwall, P. (1996) '$K3$ Surfaces and String Duality', in *Fields, Strings and Duality, TAS 1996*, World Scientific , 421–540, hep-th/9611137.

Bachas, C., Porrati, M. (1992) 'Pair Creation of Open Strings in an Electric Field', *Phys. Lett.* **B296**, 77–84, hep-th/9209032.

Bachas, C. (1996) 'D-brane Dynamics', *Phys. Lett.* **B374**, 37–42, hep-th/9511043.

Bachas, C., Kiritsis, E. (1997) 'F^4 Terms in $N = 4$ String Vacua', *Nucl. Phys. Proc. Suppl.* **55B**, 194–199, hep-th/9611205.

Bachas, C. (1997a) '(Half) a Lecture on D-branes', *Proceedings of the Workshop on Gauge Theories, Applied Supersymmetry and Quantum Gravity*, Sevrin A. et al. eds., Imperial College Press, pp. 3–22, hep-th/9701019.

Bachas, C. (1997b) 'Heterotic versus Type I', talk at 'Strings 97', *Nucl. Phys. B*, (Proc. Suppl.) **68** 348–354, hep-th/9710102.

Bachas, C., Douglas, M.R., Green, M.B. (1997) 'Anomalous Creation of Branes', *JHEP* **7**, 2, hep-th/9705074.

Bachas, C., Green, M.B., Schwimmer, A. (1998) '(8,0) Quantum Mechanics and Symmetry Enhancement in Type I' Superstrings', *JHEP* **1**, 6, hep-th/9712086.

Balasubramanian, V., Larsen, F. (1997) 'Relativistic Brane Scattering', *Nucl. Phys.* **B506**, 61–83, hep-th/9703039.

Banks, T, Susskind, L. (1995) 'Brane–Anti-Brane Forces', hep-th/9511194.

Banks, T., Douglas, M.R., Seiberg, N. (1996) 'Probing F-theory with Branes', *Phys. Lett.* **B387**, 278–281, hep-th/9605199.

Banks, T., Fischler, W., Shenker, S.H., Susskind, L. (1997) 'M Theory As A Matrix Model: A Conjecture', *Phys. Rev.* **D55**, 5112–5128, hep-th/9610043.

Banks, T., Seiberg, N., Silverstein, E. (1997) 'Zero and One-dimensional Probes with $N = 8$ Supersymmetry', *Nucl. Phys.* **B401**, 30–37, hep-th/9703052.

Banks, T. (1998) 'Matrix Theory', *Nucl. Phys. Proc. Suppl.* **B67**, 180–224, hep-th/9710231.

Bautier, K., Deser, S., Henneaux, M., Seminara, D. (1997) 'No Cosmological $D = 11$ Supergravity', *Phys. Lett.* **B406**, 49–53, hep-th/9704131.

Becker, K., Becker, M. (1997) 'A Two Loop Test of Matrix Theory', *Nucl. Phys.* **B506**, 48–60, hep-th/9705091.

Berenstein, D., Corrado, D.R., Fischler, W., Paban, S., Rozali, M. (1996) 'Virtual D-Branes', *Phys. Lett.* **B384**, 93–97, hep-th/9605168.

Bergman, O., Gaberdiel, M.R., Lifschytz, G. (1998) 'Branes, Orientifolds and the Creation of Elementary Strings', *Nucl. Phys.* **B509** 194–215, hep-th/9705130.

Bergshoeff, E., Sezgin, E., Townsend, P.K. (1987) 'Supermembranes and Eleven-dimensional Supergravity', *Phys. Lett.* **B189** , 75–78.

Bergshoeff, E., de Roo, M., Green, M.B., Papadopoulos, G., Townsend, P.K. (1996) 'Duality of Type II 7-branes and 8-branes', *Nucl. Phys.* **B470**, 113–135, hep-th/9601150.

Bergshoeff, E., de Roo, M. (1996) 'D-branes and T-duality', *Phys. Lett.* **B380**, 265–272, hep-th/9603123.

Bergshoeff, E., Townsend, P.K. (1997) 'Super D-branes', *Nucl. Phys.* **B490**, 145–162, hep-th/9611173.

Bergshoeff, E., Lozano, Y., Ortín, T. (1997) 'Massive Branes', *Nucl. Phys. B* **518**, 363–423, hep-th/9712115.

Berkooz, M., Douglas, M.R., Leigh, R.G. (1996) 'Branes Intersecting at Angles', *Nucl. Phys.* **B480**, 265–278, hep-th/9606139.

Bershadsky, M., Sadov, V., Vafa, C. (1996a) 'D-Strings on D-Manifolds', *Nucl. Phys.* **B463**, 398–414, hep-th/9510225.

Bershadsky, M., Sadov, V., Vafa, C. (1996b) 'D-Branes and Topological Field Theories', *Nucl. Phys.* **B463**, 420–434, hep-th/9511222.

Bianchi, M. (1997) 'Open Strings and Dualities', Talk at the V Korean-Italian Meeting on Relativistic Astrophysics, hep-th/9712020.

Bigatti, D., Susskind, L. (1997) 'Review of Matrix Theory', lectures at Cargese 97, hep-th/9712072.

Bilal, A. (1996) 'Duality in $N = 2$ SUSY SU(2) Yang–Mills Theory: A pedagogical introduction to the work of Seiberg and Witten', hep-th/9601007.

Bilal, A. (1997) 'M(atrix) Theory : a Pedagogical Introduction', hep-th/9710136.

Billo, M., Cangemi, D., Di Vecchia, P. (1997) 'Boundary states for moving D-branes', *Phys. Lett.* **B400**, 63–70, hep-th/9701190.

Brax, P., Mourad, J. (1997) 'Open supermembranes in eleven dimensions', *Phys, Let.* **B408**, 142–150, hep-th/9704165.

Brax, P., Mourad, J. (1998) 'Open Supermembranes Coupled to M-Theory Five-branes', *Phys. Lett.* **B416**, 295–302, hep-th/9707246.

Brecher, D., Perry, M.J. (1998) 'Bound States of D-Branes and the Non-Abelian Born–Infeld Action', hep-th/9801127.

Brecher, D. (1998) 'BPS States of the Non-Abelian Born–Infeld Action', hep-th/9804180.

Callan, C.G., Lovelace, C., Nappi, C.R., Yost, S.A. (1988) 'Loop Corrections to Superstring Equations of Motion', *Nucl. Phys.* **B308**, 221–284.

Callan, C.G., Harvey, J.H., Strominger, A. (1991a) 'Worldsheet Approach to Heterotic Instantons and Solitons', *Nucl. Phys.* **B359**, 611–634.

Callan, C.G., Harvey, J.H., Strominger, A. (1991b) 'Worldbrane Actions for String Solitons', *Nucl. Phys.* **B367**, 60–82.

Callan, C.G., Klebanov, I. (1996) 'D-Brane Boundary State Dynamics', *Nucl. Phys.* **B465**, 473–486, hep-th/9511173.

Callan, C.G., Maldacena, J. (1996) 'D-brane Approach to Black Hole Quantum Mechanics', *Nucl. Phys.* **B472**, 591–610, hep-th/9602043.

Callan, C.G., Maldacena, J. (1998) 'Brane Dynamics from the Born–Infeld Action', *Nucl. Phys.* **B513**, 198–212, hep-th/9708147.

Cederwall, M., von Gussich, A., Nilsson, B., Sundell, P., Westerberg, A. (1997a) 'The Dirichlet Super–Three-brane in Ten–Dimensional Type IIB Supergravity', *Nucl. Phys.* **B490**, 163–178, hep-th/9610148.

Cederwall, M., von Gussich, A., Nilsson, B., Sundell, P., Westerberg, A. (1997b) 'The Dirichlet Super–p-branes in Ten–Dimensional Type IIA and IIB Supergravity', *Nucl. Phys.* **B490**, 179–201, hep-th/9611159.

Cederwall, M. (1997) 'Aspects of D-brane actions', *Nucl. Phys. Proc. Suppl.* **B56**, 61–69, hep-th/9612153.

Cheung, Y-K.E., Yin, Z. (1997) 'Anomalies, Branes, and Currents', *Nucl. Phys. B* **517** 69–91, hep-th/9710206.

Coleman, S. (1981) 'The Magnetic Monopole Fifty Years Later', in 'Gauge Theories in High Energy Physics', Les Houches proceedings.

Connes, A. (1994) *Noncommutative Geometry*, Academic Press.

Conrad, J.O. (1997) 'Brane Tensions and Coupling Constants from within M-Theory', hep-th/9708031.

Cremmer, E., Julia, B., Scherk, J. (1978) 'Supergravity Theory in Eleven Dimensions', *Phys. Lett.* **B76**, 409–412.

Dai, J., Leigh, R.G., Polchinski, J. (1989) 'New Connections Between String Theories', *Mod. Phys. Lett.* **A4**, 2073–2083.

Danielsson, U.H., Ferretti, G., Sundborg, B. (1996) 'D-particle Dynamics and Bound States', *Int. Jour. Mod. Phys.* **A11**, 5463-5478, hep-th/9603081.

Danielsson, U.H., Ferretti, G. (1997) 'The Heterotic Life of the D-particle', *Int. J. Mod. Phys.* **A12**, 4581–4596, hep-th/9610082.

Danielsson, U.H., Ferretti, G., Klebanov, I. (1997) 'Creation of Fundamental Strings by Crossing D-branes', *Phys. Rev. Lett.* **79**, 1984–1987, hep-th/9705084.

de Alwis, S.P. (1996) 'A note on brane tension and M-theory', *Phys. Lett.* **B388**, 291–295, hep-th/9607011.

de Alwis, S.P. (1997) 'Coupling of branes and normalization of effective actions in string/M-theory', *Phys. Rev.* **D56**, 7963-7977, hep-th/9705139.

de Azcarraga, J.A., Gauntlett, J.P., Izquierdo, J.M., Townsend, P.K. (1989) 'Topological Extensions of the Supersymmetry Algebra for Extended Objects', *Phys. Rev. Lett.* **63**, 2443–2446.

de Roo, M., Suelmann, H., Wiedemann, A. (1993) 'The Supersymmetric Effective Action of the Heterotic String in Ten Dimensions', *Nucl. Phys.* **B405**, 326–366, hep-th/9210099.

Deser, S., Gomberoff, A., Henneaux, M., Teitelboim, C. (1997) 'p-brane Dyons and Electric-magnetic Duality' hep-th/9712189.

Deser, S., Henneaux, M., Schwimmer, A. (1998) 'p-brane Dyons, theta-terms and Dimensional Reduction', hep-th/9803106.

de Wit, B. (1998) 'Supermembranes and Super Matrix Theory', Lecture at the 31st International Symposium Ahrenshoop on the 'Theory of Elementary Particles', Buckow, to appear in *Fortschritte der Physik*, hep-th/9802073.

de Wit, B., Louis, J. (1998) 'Supersymmetry and Dualities in Various Dimensions', lectures at Cargese 97, *Nucl. Phys. B* (Proc. Suppl.) **67** 117–157, hep-th/9801132.

Dienes, K. (1997) 'String Theory and the Path to Unification: A Review of Recent Developments', *Phys. Rep.* **287**, 447–525, hep-th/9602045.

Dijkgraaf, R. (1997) 'Les Houches Lectures on Fields, Strings and Duality', Summer School Session 64, 3–147, hep-th/9703136.

Dijkgraaf, R., Verlinde E., Verlinde H. (1998) 'Notes on Matrix and Micro Strings', *Nucl. Phys. Proc. Suppl.* **62**, 348–362, hep-th/9709107.

Dine, M., Seiberg, N. (1997) 'Comments on Higher Derivative Operators in Some SUSY Field Theories', *Phys. Lett.* **B409**, 239–244, hep-th/9705057.

Dine, M., Echols, R., Gray, J. (1998) 'Renormalization of Higher Derivative Operators in the Matrix Model', hep-th/9805007.

Dirac, P.A.M. (1931) 'Quantized Singularities in the Electromagnetic Field', *Proc. Roy. Soc. London* **A133**, 60–72.

Di Vecchia, P. (1998) 'Duality in supersymmetric $N = 2, 4$ gauge theories', hep-th/9803026.

Dorn, H., Otto, H-J. (1996) 'On T-duality for open strings in general abelian and nonabelian gauge field backgrounds', *Phys. Lett.* **B381**, 81–88, hep-th/9603186.

Douglas, M.R. (1995) 'Branes within Branes', hep-th/9512077.

Douglas, M.R., Moore, G. (1996) 'D-branes, Quivers, and ALE Instantons', hep-th/9603167.

Douglas, M.R. (1996) 'Superstring Dualities, Dirichlet Branes and the Small Scale Structure of Space', in the proceedings of the LXIV Les Houches session on 'Quantum Symmetries', hep-th/9610041.

Douglas, M.R., Kabat, D., Pouliot, P., Shenker, S.H. (1997) 'D-branes and Short Distances in String Theory', *Nucl. Phys.* **B485**, 85–127, hep-th/9608024.

Douglas, M.R. (1997) 'D-branes and Matrix Theory in Curved Space', talk at 'Strings 97', hep-th/9707228.

Douglas, M.R., Kato, A., Ooguri, H. (1997) 'D-brane Actions on Kähler Manifolds', hep-th/9708012.

Duff, M.J., Khuri, R.R., Lu, J.X. (1995) 'String Solitons', *Phys. Rep.* **259**, 213–326, hep-th/9412184.

Duff, M.J., Minasian, R. (1995) 'Putting String/String Duality to the Test', *Nucl. Phys.* **B436**, 507–528, hep-th/9406198.

Duff, M.J. (1995) 'Strong/Weak Coupling Duality from the Dual String', *Nucl. Phys.* **B442**, 47–63, hep-th/9501030.

Duff, M.J., Liu, J.T., Minasian, R. (1995) '11-Dimensional Origin of String/String Duality: a One Loop Test', *Nucl. Phys.* **B452**, 261–282, hep-th/9506126.

Duff, M. (1997) 'Supermembranes', in *Proceedings of the Workshop on Gauge Theories, Applied Supersymmetry and Quantum Gravity*, Sevrin A. *et al.* eds., Imperial College Press, hep-th/9611203.

Eguchi, T., Gilkey, P.B., Hanson, A.J. (1980) 'Gravitation, Gauge Theories and Differential Geometry', *Phys. Rep.* **66**, 213–393.

Elitzur, S., Giveon, A., Kutasov, D. (1997a) 'Branes and $N = 1$ Duality in String Theory', *Phys. Lett.* **B400**, 269–274, hep-th/9702014.

Elitzur, S., Giveon, A., Kutasov, D., Rabinovici, E., Schwimmer, A. (1997b) 'Brane Dynamics and $N = 1$ Supersymmetric Gauge Theory', *Nucl. Phys.* **B505**, 202–250, hep-th/9704104.

Ferrara, S., Harvey, J.A., Strominger, A., Vafa C. (1995) 'Second-Quantized Mirror Symmetry', *Phys. Lett.* **B361**, 59–65, hep-th/9505162.

Förste, S., Louis, J. (1997) 'Duality in String Theory', in *Proceedings of the Workshop on Gauge Theories, Applied Supersymmetry and Quantum Gravity*, Sevrin A. *et al.* eds., Imperial College Press, hep-th/9612192.

Fradkin, E.S., Tseytlin, A. (1985) 'Non-linear Electrodynamics from Quantized Strings', *Phys. Lett.* **B163**, 123–130.

Friedan, D., Martinec, E., Shenker, S. (1986) 'Conformal Invariance, Supersymmetry and String Theory', *Nucl. Phys.* **B271**, 93–165.

Gauntlett, J.P. (1997) 'Intersecting Branes', Lectures at APCTP Winter School on 'Dualities of Gauge and String Theories', Korea, hep-th/9705011.

Gauntlett, J.P., Gibbons, G.W., Papadopoulos, G., Townsend, P.K. (1997) 'Hyper-Kahler Manifolds and Multiply-intersecting Branes', *Nucl. Phys.* **B500**, 133–162, hep-th/9702202.

Gibbons, G.W., Green, M.B., Perry, M.J. (1996) 'Instantons and Seven-branes in Type IIB Superstring Theory', *Phys. Lett.* **B370**, 37–44, hep-th/9511080.

Gibbons, G.W., Horowitz, T., Townsend, P.K. (1995) 'Higher-dimensional resolution of dilatonic black hole singularities', *Class. Quant. Grav.* **12**, 297-318, hep-th/9410073.

Gibbons, G.W. (1998) 'Born–Infeld particles and Dirichlet p-branes', *Nucl. Phys.* **B514**, 603–639, hep-th/9709027.

Gimon, E., Polchinski, J. (1996) 'Consistency Conditions for Orientifolds and D-Manifolds', *Phys. Rev.* **D54**, 1667–1676, hep-th/9601038.

Giveon, A., Porrati, M., Rabinovici, E. (1994) 'Target Space Duality in String Theory', *Phys. Rep.* **244** , 77–202, hep-th/9401139.

Giveon, A., Kutasov, D. (1998) 'Brane Dynamics & Gauge Theory', hep-th/9802067.

Green, M.B., Schwarz, J.H. (1984) 'Anomaly Cancellations in Supersymmetric D=10 Gauge Theories and Superstring Theory', *Phys. Lett.* **B149**, 117–122.

Green, M.B., Schwarz, J.H. (1985a) 'Infinity Cancellations in SO(32) Superstring Theories', *Phys. Lett.* **B151**, 21–25.

Green, M.B., Schwarz, J.H. (1985b) 'The Hexagon Gauge Anomaly in Type 1 Superstring Theory', *Nucl. Phys.* **B255**, 93–114.

Green, M.B., Schwarz, J.H., Witten, E. (1987) *Superstring Theory*, Cambridge University Press, two volumes.

Green, M.B. (1991) 'Duality, Strings and Point-like Structure', talk at 25th Rencontres de Moriond, Trân Thanh Vân, J. (ed.), editions Frontières.

Green, M.B., Gutperle, M. (1996) 'Light-cone supersymmetry and D-branes', *Nucl. Phys.* **B476**, 484–514, hep-th/9604091.

Green, M.B., Hull, C.M., Townsend, P.K. (1996) 'D-brane Wess–Zumino actions, T-duality and the cosmological constant', *Phys. Lett.* **B382**, 65–72, hep-th/9604119.

Green, M.B., Harvey, J.A., Moore, G. (1997) 'I-Brane Inflow and Anomalous Couplings on D-Branes', *Class. Quant. Grav.* **14**, 47–52, hep-th/9605033.

Green, M.B. (1997) 'Connections between M-theory and superstrings', talk at Cargese 97, hep-th/9712195.

Gross, D., Perry, M. (1983) 'Magnetic Monopoles in Kaluza-Klein Theories', *Nucl. Phys.* **B226**, 29–48.

Gubser, S., Klebanov, I., Polyakov, A. (1998) 'Gauge Theory Correlators from Noncritical String Theory', hep-th/9802109.

Gutperle, M. (1997) 'Aspects of D-Instantons', talk at Cargese 97, hep-th/9712156.

Güven, R. (1992) 'Black p-brane Solutions of $D = 11$ Supergravity Theory', *Phys. Lett.* **B276**, 49–55.

Hanany, A., Witten, E. (1997) 'Type IIB Superstrings, BPS Monopoles and Three-Dimensional Gauge Dynamics', *Nucl. Phys.* **B492**, 152–190, hep-th/9611230.

Harvey, J.A. (1996) 'Magnetic Monopoles, Duality, and Supersymmetry', hep-th/9603086.

Hashimoto, A., Klebanov, I. (1997) 'Scattering of Strings from D-branes', *Nucl. Phys. Proc. Suppl.* **55B**, 118–133, hep-th/9611214.

Hashimoto, A., Taylor, W. (1997) 'Fluctuation Spectra of Tilted and Intersecting D-branes from the Born–Infeld Action', *Nucl. Phys.* **B503**, 193–219, hep-th/9703217.

Hassan, S.F., Wadia, S.R. (1997) 'Gauge Theory Description of D-brane Black Holes: Emergence of the Effective SCFT & Hawking Radiation', hep-th/9712213.

Henneaux, M., Teitelboim, C. (1986) 'Quantization of Topological Mass in the Presence of a Magnetic Field', *Phys. Rev. Lett.* **56**, 689–692.

Hořava, P. (1989a) 'Strings on Worldsheet orbifolds', *Nucl. Phys.* **B327**, 461–484.

Hořava, P. (1989b) 'Background Duality of Open-string Models', *Phys. Lett.* **B231**, 251–257.

Hořava, P., Witten, E. (1996a) 'Heterotic and Type I String Dynamics from Eleven Dimensions', *Nucl. Phys.* **B460**, 506–524, hep-th/9510209.

Hořava, P., Witten, E. (1996b) 'Eleven-Dimensional Supergravity on a Manifold with Boundary', *Nucl. Phys.* **B475**, 94–114, hep-th/9603142.

Howe, P.S., Lambert, N.D., West, P.C. (1998) 'A New Massive Type IIA Supergravity From Compactification', *Phys. Lett.* **B416**, 303–308, hep-th/9707139.

Hull, C.M., Townsend, P.K. (1995) 'Unity of Superstring Dualities', *Nucl. Phys.* **B438**, 109–137, hep-th/9410167.

Hussain, F., Iengo, R., Núñez, C., Scrucca, C. (1997) 'Interaction of Moving D-branes on Orbifolds' *Phys. Lett.* **B409**, 101–108, hep-th/9706186.

Intriligator, K., Seiberg, N. (1996) 'Lectures on Supersymmetric Gauge Theories and Electric-magnetic Duality', *Nucl. Phys. Proc. Suppl.* **45B**, 1–28, hep-th/9509066.

Julia, B.L. (1998) 'Dualities in the Classical Supergravity Limits', talk at Cargese 97, hep-th/9805083.

Kabat, D., Pouliot, P. (1996) 'A Comment on Zero-brane Quantum Mechanics', *Phys. Rev. Lett.* **77**, 1004-1007, hep-th/9603127.

Kachru, S., Vafa, C. (1995) 'Exact Results for $N = 2$ Compactifications of Heterotic Strings', *Nucl. Phys.* **B450**, 69–89, hep-th/9505105.

Kiritsis, E. (1998) *Introduction to Superstring Theory*, Leuven University Press, hep-th/9709062.

Kitazawa, Y. (1987) 'Effective Lagrangian for the Open Superstring from a 5-point Function', *Nucl. Phys.* **B289**, 599–608.

Kogan, I.I., Mavromatos, N.E., Wheater, J.F. (1996) 'D-Brane Recoil and Logarithmic Operators', *Phys. Lett.* **B387**, 483–491, hep-th/9606102.

Leigh, R.G. (1989) 'Dirac–Born–Infeld Action for Dirichlet σ–Models', *Mod. Phys. Lett.* **A4**, 2767–2772.

Lerche, W. (1988)'Elliptic Index and the Superstring Effective Action', *Nucl. Phys.* **B308**, 102–126.

Lerche, W. (1997) 'Introduction to Seiberg–Witten Theory and its Stringy Origin', *Nucl. Phys. Proc. Suppl.* **55B**, 83–117, *Fortsch. Phys.* **45**, 293–340, hep-th/9611190.

Li, M. (1996a) 'Boundary States of D-Branes and Dy-Strings', *Nucl. Phys.* **B460**, 351–361, hep-th/9510161.

Li, M. (1996b) 'Dirichlet Boundary State in Linear Dilaton Background', *Phys. Rev.* **D54**, 1644–1646 hep-th/9512042.

Lifschytz, G. (1996) 'Comparing D-branes to Black-branes', *Phys. Lett.* **B388**, 720–726, hep-th/9604156.

Lifschytz, G. (1997) 'Probing Bound States of D-branes', *Nucl. Phys.* **B499**, 283–297, hep-th/9610125.

Lifschytz, G., Mathur, S.D. (1997) 'Supersymmetry and Membrane Interactions in M(atrix) Theory', *Nucl. Phys.* **B507**, 621–644, hep-th/9612087.

Lu, J.X. (1997) 'Remarks on M-theory Coupling Constants and M-brane Tension Quantizations', hep-th/9711014.

Lüscher, M., Symanzik, K., Weisz, P. (1980) 'Anomalies of the Free Loop Equation in the WKB Approximation', *Nucl. Phys.* **B173**, 365–396.

Lüst, D. (1998) 'String Vacua with $N = 2$ Supersymmetry in Four Dimensions', hep-th/9803072.

Madore, J. (1995) *An Introduction to Noncommutative Differential Geometry and its Physical Applications*, Cambridge University Press.

Maldacena, J.M. (1996) 'Black Holes in String Theory', Princeton University PhD thesis, hep-th/9607235.

Maldacena, J.M. (1997) 'Branes Probing Black Holes', talk at STRINGS'97, hep-th/9709099.

Maldacena, J.M. (1997) 'The Large N Limit of Superconformal Field Theories and Supergravity', *Advances in Theor. and Math. Physics* **2** 231–252, hep-th/9711200.

Manton, N. (1982) 'A Remark on the Scattering of BPS Monopoles', *Phys. Lett.* **B110**, 54–56.

Matusis, A. (1997) 'Interaction of non-parallel D1-branes', hep-th/9707135.

Milnor, J.W., Stasheff, J.D. (1974) *Characteristic Classes*, Princeton U. Press.

Morales, J.F., Scrucca, C.A., Serone, M. (1997) 'A Note on Supersymmetric D-brane Dynamics', hep-th/9709063.

Morales, J.F., Scrucca, C.A., Serone, M. (1998) 'Scale Independent Spin effects in D-brane Dynamics', hep-th/9801183.

Minasian, R., Moore, G. (1997) 'K-theory and Ramond–Ramond Charge', *JHEP* **11**, 2, hep-th/9710230.

Mourad, J. (1998) 'Anomalies of the SO(32) Five-brane and their Cancellation', *Nucl. Phys.* **B512**, 199–208, hep-th/9709012.

Nakahara, M. (1990) *Geometry, Topology and Physics*, Graduate Student Series in Physics, Inst. of Phys. Publ.

Nepomechie, R.I. (1985) 'Magnetic Monopoles from Antisymmetric-Tensor Gauge Fields', *Phys. Rev.* **D31**, 1921–1924.

Olive, D. (1996) 'Exact Electromagnetic Duality', *Nucl. Phys. Proc. Suppl.* **45A**, 88–102, hep-th/9508089.

Ooguri, H., Yin, Z. (1996) 'TASI Lectures on Perturbative String Theories', in *Fields, Strings and Duality, TAS 1996* World Scientific, 5–81, hep-th/9612254.

Paban, S., Sethi, S., Stern, M. (1998) 'Constraints from Extended Supersymmetry in Quantum Mechanics', hep-th/9805018.

Periwal, V., Tafjord, Ø. (1996) 'D-brane Recoil', *Phys. Rev.* **D54**, 3690–3692, hep-th/9603156.

Peskin, M. (1997) 'Duality in Supersymmetric Yang–Mills Theory', in *Fields, Strings and Duality, TAS 1996* World Scientific, 729–809, hep-th/9702094.

Polchinski, J. (1995) 'Dirichlet-Branes and Ramond–Ramond Charges', *Phys. Rev. Lett.* **75**, 4724–4727, hep-th/9510017.

Polchinski, J., Chaudhuri, S., Johnson, C. (1996) 'Notes on D-Branes', lectures at ITP–Santa Barbara, hep-th/9602052.

Polchinski, J. (1996) 'TASI Lectures on D-Branes', in *Fields, Strings and Duality, TAS 1996* World Scientific, 193–356, hep-th/9611050.

Polchinski, J., Witten, E. (1996) 'Evidence for Heterotic-Type I Duality', *Nucl. Phys.* **B460**, 525–540, hep-th/9510169.

Polchinski, J. (1998) *String Theory*, Cambridge University Press., two volumes.

Polychronakos, A. (1987) 'Topological Mass Quantization and Parity Violation in $2 + 1$-Dimensional QED', *Nucl. Phys.* **B281**, 241–252.

Porrati, M., Rozenberg, A. (1998) 'Bound States at Threshold in Supersymmetric Quantum Mechanics', *Nucl. Phys.* **B515**, 184–202, hep-th/9708119.

Pradisi, G., Sagnotti, A. (1989) 'Open String Orbifolds', *Phys. Lett.* **B216**, 59–67.

Polyakov, D. (1996) 'RR-Dilaton Interaction in a Type IIB Superstring' *Nucl. Phys.* **B468**, 155–162, hep-th/9512028.

Rey, S-J. (1997) 'Heterotic M(atrix) Strings and their Interactions', *Nucl. Phys.* **B502**, 170–190, hep-th/9704158.

Romans, L. (1986) 'Massive $N = 2a$ Supergravity in Ten Dimensions', *Phys. Lett.* **B169**, 374–380.

Sagnotti, A. (1988) in *Non-Perturbative Quantum Field Theory*, eds. Mack *et al.*, Pergamon Press, p.521.

Sagnotti, A. (1997) 'Surprises in Open-String Perturbation Theory', *Nucl. Phys. Proc. Suppl.* **B56**, 332–343, hep-th/9702093.

Schmidhuber, C. (1996) 'D-brane Actions', *Nucl. Phys.* **B467**, 146–158, hep-th/9601003.

Schwarz, J.H. (1995) 'An SL(2,Z) Multiplet of Type IIB Superstrings', *Phys. Lett* **B360**, 13–18, erratum *ibid.* **B364**, 252, hep-th/9508143

Schwarz, J.H. (1997a) 'Lectures on Superstring and M-theory Dualities', *Nucl. Phys. Proc. Suppl.* **55B**, 1–32; see also *Fields, Strings and Duality, TAS 1996* World Scientific, 359–418, hep-th/9607201.

Schwarz, J.H. (1997b) 'The Status of String Theory', hep-th/9711029.

Schwinger, J. (1968) 'Sources and Magnetic Charge', *Phys. Rev.* **173**, 1536–1544.

Sen, A. (1996) 'U-duality and Intersecting D-branes', *Phys. Rev.* **D53**, 2874–2894, hep-th/9511026.

Sen, A. (1998) 'An Introduction to Non-perturbative String Theory', in this volume, hep-th/9802051.

Sethi, S., Stern, M. (1998) 'D-Brane Bound States Redux', *Commun. Math. Phys.* **194**, 675–705, hep-th/9705046.

Shenker, S.H. (1995) 'Another Length Scale in String Theory?', hep-th/9509132.

Sorkin, R. (1983) 'Kaluza–Klein Monopole', *Phys. Rev. Lett.* **51**, 87–90.

Stelle, K. (1997) 'Lectures on Supergravity p-branes', lectures at the 1996 ICTP Summer School, Trieste, hep-th/9701088.

Stelle, K. (1998) 'BPS Branes in Supergravity', lectures at the 1997 ICTP Summer School, Trieste, hep-th/9803116.

Strominger, A. (1995) 'Massless Black Holes and Conifolds in String Theory', *Nucl. Phys.* **B451**, 96–108, hep-th/9504090.

Strominger, A. (1996) 'Open p-Branes', *Phys. Lett.* **B383**, 44–47, hep-th/9512059.

Strominger, A., Vafa, C. (1996) 'Microscopic Origin of the Bekenstein–Hawking Entropy', *Phys. Lett.* **B379**, 99–104, hep-th/9601029

Taylor, W. (1998) 'Lectures on D-branes, Gauge Theory and M(atrices)', lectures at the 1997 ICTP Summer School, Trieste, hep-th/9801182.

Teitelboim, C. (1986a) 'Gauge Invariance for Extended Objects', *Phys. Lett.* **B167**, 63–68.

Teitelboim, C. (1986b) 'Monopoles of Higher Rank', *Phys. Lett.* **B167**, 69–72.

Thorlacius, L. (1998) 'Introduction to D-branes', *Nucl. Phys. Proc. Suppl.* **61A**, 86–98, hep-th/9708078.

Townsend, P.K. (1995) 'The Eleven–dimensional Supermembrane Revisited', *Phys. Lett* **B350**, 184–187, hep-th/9501068.

Townsend, P.K. (1996a) 'D-branes from M-branes', *Phys. Lett* **B373**, 68–75, hep-th/9512062.

Townsend, P.K. (1996b) 'Four Lectures on M-theory', lectures at the 1996 ICTP Summer School, Trieste, hep-th/9612121.

Townsend, P.K. (1997) 'M-theory from its Superalgebra', lectures at Cargese 97, hep-th/9712004.

Tseytlin, A. (1996a) 'On SO(32) heterotic-type I superstring duality in ten dimensions', *Phys. Lett.* **B367**, 84–90, hep-th/9510173.

Tseytlin, A. (1996b) 'Heterotic-type I superstring duality and low-energy effective actions', *Nucl. Phys.* **B467**, 383–398, hep-th/9512081.

Tseytlin, A. (1997) 'On Non-abelian Generalisation of Born–Infeld Action in String Theory', *Nucl. Phys.* **B501**, 41–52 hep-th/9701125.

Vanhove, P. (1997) 'BPS Saturated Amplitudes and Non-perturbative String Theory', talk at Cargese 97, hep-th/9712079.

Vafa, C. (1996) 'Instantons on D-branes', *Nucl. Phys.* **B463**, 435–442, hep-th/9512078.

Vafa, C. (1997) 'Lectures on Strings and Dualities', hep-th/9702201.

Veltman, M. (1975) 'Quantum Theory of Gravitation', in *Methods in Field Theory*, Les Houches XXVIII, Balian R. *et al.* eds., North Holland.

West, P. (1998) 'Introduction to Rigid Supersymmetric Theories', lectures at TASI97, hep-th/9805055.

West, P. (1998) 'Supergravity, String Duality and Brane Dynamics', in this volume.

Witten, E. (1979) 'Dyons of charge $e\theta/2\pi$', *Phys. Lett.* **B86**, 283–287.

Witten, E. (1995) 'String Theory Dynamics In Various Dimensions', *Nucl. Phys.* **B443**, 85–126, hep-th/9503124.

Witten, E. (1996a) 'Bound States of Strings and *p*-branes', *Nucl. Phys.* **B460**, 335–350, hep-th/9510135.

Witten, E. (1996b) 'Small Instantons in String Theory', *Nucl. Phys.* **B460**, 541–559, hep-th/9511030.

Witten, E. (1997a) 'On Flux Quantization in M-Theory and the Effective Action', *J. Geom. Phys.* **22**, 1–13, hep-th/9609122.

Witten, E. (1997b) 'Five-brane Effective Action In M-Theory', *J. Geom. Phys.* **22**, 103–133, hep-th/9610234.

Witten, E. (1998) 'Anti De Sitter Space and Holography', *Advances in Theor. and Math. Physics* **2** 253–291, hep-th/9802150.

Yi, P. (1997) 'Witten Index and Threshold Bound States of D-Branes', *Nucl. Phys.* **B505**, 307–318, hep-th/9704098.

Youm, D. (1997) 'Black Holes and Solitons in String Theory', hep-th/9710046.

Zwanzinger, D. (1968) 'Quantum Field Theory of Particles with both Electric and Magnetic Charges', *Phys. Rev.* **176**, 1489–1495.